D0909050

Biology of
Serotonergic Transmission

MEDICAL COLLEGE OF PENNSYLVANIA
AND HAHNEMANN UNIVERSITY
UNIVERSITY LIBRARY, CENTER CITY

Biology of Serotonergic Transmission

Edited by
NEVILLE N. OSBORNE
University Lecturer
Nuffield Department of Ophthalmology
Oxford University

A Wiley–Interscience Publication

1807 1982

JOHN WILEY & SONS
Chichester · New York · Brisbane · Toronto · Singapore

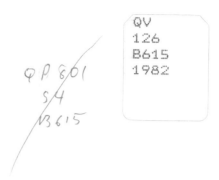

QV
126
B615
1982

Copyright © 1982 by John Wiley & Sons Ltd.

All rights reserved.

No part of this book may be reproduced by any means, nor transmitted, nor translated into a machine language without the written permission of the publisher

Library of Congress Cataloging in Publication Data:
Main entry under title:

Biology of serotonergic transmission.
   'A Wiley–Interscience Publication.'
   Includes index.
   1. Serotonin–Physiological effect. 2. Neural transmission. I. Osborne, Neville N.
QP801.S4B56 591.1′88 81-14671
                AACR2

ISBN 0 471 10032 3

British Library Cataloguing in Publication Data:

Biology of serotonergic transmission.
   1. Serotonin
   2. Neurochemistry
   I. Osborne, Neville N.
   612′.814   QP951.S4

   ISBN 0 471 10032 3

Typeset by Activity, Salisbury, Wilts.
Printed by Page Bros. (Norwich) Limited.

To my parents and Jane

Contents

List of Contributors

H. G. BAUMGARTEN   *Department of Neuroanatomy, and Electron Microscopy, Free University of Berlin, West-Berlin, (GFR).*

M. S. BERRY   *Department of Zoology, University of Swansea, Swansea SA2 8PP, UK.*

H. BLASCHKO   *Department of Pharmacology, South Parks Road, University of Oxford, Oxford OX1 3QT, UK.*

MARGARET C. BOADLE-BIBER   *Department of Physiology, Medical College of Virginia, Virginia Commonwealth University, Richmond, Virginia, USA.*

A. BJÖRKLUND   *Department of Histology, University of Lund, Lund, Sweden.*

S. BOURGOIN   *Groupe NB, INSERM U.114, Collège de France, 11, Place Marcelin Berthelot, F-75231 Paris cedex 05, France.*

ADRIANA CONSOLAZIONE   *Department of Pharmacology, South Parks Road, University of Oxford, Oxford OX1 3QT, UK.*

A. C. CUELLO   *Department of Pharmacology, South Parks Road, University of Oxford, Oxford, OX1 3QT, UK.*

A. DRAY   *MRC Neurochemical Pharmacology Unit, Medical Research Council Centre, Medical School, Hills Road, Cambridge, CB2 2QH, UK.*

R. W. FULLER   *The Lilly Research Laboratories, Eli Lilly and Company, Indianapolis, Indiana 46285, USA.*

M. D. GERSHON   *Department of Anatomy and Cell Biology, Columbia University, College of Physicians and Surgeons, New York, NY 10032, USA.*

D. J. GOLDBERG   *Departments of Pharmacology, Physiology and Neurology and Center for Neurobiology and Behavior, College of Physicians and Surgeons, Columbia University, New York, NY 10032, USA.*

M. HAMON   *Groupe NB, INSERM U.114, Collège de France, 11, Place Marcelin Berthelot, F-75231 Paris cedex 05, France.*

S. JENNER   *Department of Neuroanatomy and Electron Microscopy, Free University of Berlin, West-Berlin (GFR)*

H. P. KLEMM   *Department of Neurology, University of Hamburg, Universitätskrankenhaus Eppendorf, Hamburg, GFR.*

C. M. LENT   *Center for Neuroscience, Brown University, Providence, Rhode Island 02912, USA.*

L. L. MARTIN   *Department of Pharmacology, Vanderbilt University School of Medicine and Tennessee Neuropsychiatric Institute, Nashville, Tennessee, USA.*

L. M. NECKERS   *National Cancer Institute, Laboratory of Pathology, National Institute of Health, Building 10, Room 2N-115, Bethesda, Maryland 20205, USA.*

S. O. ÖGREN   *Research Laboratories, Astra Läkemedel AB, S-151 85 Södertälje, Sweden.*

N. N. OSBORNE   *Nuffield Laboratory of Ophthalmology, University of Oxford, Walton Street, Oxford OX2 6AW, UK.*

V. W. PENTREATH   *Department of Biology, University of Salford, Salford M5 4WT, UK.*

S. J. PEROUTKA

Johns Hopkins Medical School, Department of Pharmacology, 725 North Wolfe Street, Baltimore, Maryland 21205, USA.

S. B. ROSS

Research and Development Laboratories, Astra Läkemedel AB, S-151 85 Södertälje, Sweden.

ELAINE SANDERS-BUSH

Department of Pharmacology, Vanderbilt University School of Medicine and Tennessee Neuropsychiatric Institute, Nashville, Tennessee, USA.

H. G. SCHLOSSBERGER

Max-Planck-Institute for Biochemistry, Martinsried, GFR.

J. H. SCHWARTZ

Departments of Pharmacology, Physiology and Neurology and Center for Neurobiology and Behavior, College of Physicians and Surgeons, Columbia University, New York, NY 10032, USA.

S. H. SNYDER

Johns Hopkins Medical School, Department of Pharmacology, 725 North Wolfe Street, Baltimore, Maryland 21205, USA.

MARTHE VOGT

A.R.C. Institute of Animal Physiology, Babraham, Cambridge CB2 4AT, UK.

Preface

Synthetic serotonin first became available for scientific use in 1951. However, historically it can be traced back considerably earlier to studies published in 1901 by T. G. Brodie on blood clotting. In the last 15 years, serotonin research has advanced significantly, due in part to the discovery of *p*-chlorophenylalanine, which specifically inhibits the enzyme tryptophan hydroxylase. There is now unambiguous proof that serotonin has certain functions in the nervous system and this is documented by the contents of this book. The purpose of this work is to provide an up-to-date report on the state of serotonin research in the 1980s and to focus on the trends of future research. Since the length of this publication is of necessity limited, it cannot be claimed that all important topics are adequately covered. Nor has the editor attempted to reconcile or censor any differences in opinion held by the contributors of the various chapters, each of whom is an expert in his or her own field. The subject which has recently aroused most controversy relates to the finding that serotonin and the pharmacologically active peptide, substance P, co-exist in certain neurones, where it is not known whether both of these substances function as transmitter substances. Another problem which has not been adequately solved is the lack of proof of the existence of a specific MAO species for the decarboxylation of 5-hydroxytryptophan to form serotonin.

As editor I wish to thank the individual authors who worked so diligently and effectively to produce their chapters. I also acknowledge with gratitude the support given to me by Dr S. D. Thornton, the publisher, in every phase of the preparation of this book. My thanks are due to the publishers and authors, who have given permission for the reproduction of various figures, and finally also to my wife, for her steady encouragement and co-operation.

NEVILLE OSBORNE

Biology of Serotonergic Transmission
Edited by N. N. Osborne
© 1982, John Wiley & Sons Ltd.

Introduction

H. Blaschko

Department of Pharmacology, South Parks Road,
University of Oxford, Oxford OX1 3QT, UK

The great leap forward that began with the discovery and characterization of Otto Loewi's Vagus and Accelerans substances, was followed by a period of intense activity during which the concept of chemical transmission of nerve impulses was consolidated and extended by the mapping of peripheral cholinergic and adrenergic neurones, particularly in mammals. Evidence in favour of cholinergic transmission at central synapses remained scanty (see the full discussion in Feldberg's review of 1945), and the existence of adrenergic transmission had to wait until the discovery of 'brain sympathin' (Vogt, 1954).

The beginnings of work on serotonin also go far back, into an early period in which there was at first no connection with the nervous system. There were two streams of research that preceded the isolation of serotonin. The first was a study of the toxic substances that appeared in blood plasma upon clotting. Gaddum (1936), in his monograph on the vasodilator substances of the tissues, refers to the work by Freund (1920), who confirmed earlier suggestions by O'Connor (1912), that the blood platelets were the source of the toxic material; Freund found that the latter was released during the breakdown of platelets.

It was this line of research that was taken up and continued in the laboratory of I. H. Page, and that eventually led to the isolation of pure serotonin and subsequently to its chemical characterization as 5-hydroxytryptamine by M. M. Rapport (see Page, 1954, 1968).

The second line of enquiry stems from histological studies of cells present in the lining of the mammalian gastro-intestinal tract and variously described as Kulschitzky cells or as argentaffin or enterochromaffin cells. This latter name was used by M. Vialli, (1966) whose student, V. Erspamer (1966), succeeded in isolating a substance that he called enteramine and that gave the characteristic colour reactions of the intact cells. After the identification of serotonin with 5-hydroxytryptamine, Erspamer soon showed that the latter was also identical with enteramine (see Erspamer, 1966).

There were early indications that material similar to serotonin occurred in invertebrates, and particularly in molluscs. Erspamer found large amounts of the amine in the posterior salivary glands of *Octopus vulgaris*.

Soon afterwards J. H. Welsh (1954) described the extremely high sensitivity of some molluscan preparations to serotonin; in particular, he used the heart of *Venus mercenaria* as an assay preparation for serotonin. By the help of this method Twarog and Page (1953) found serotonin in the mammalian brain. Soon the possibility of the existence of 'serotoninergic' (referred to as 'serotonergic') neurones was discussed (see Welsh, 1957), and this work received an enormous impetus with the development of the fluorescence-microscopic study of nerve tissue.

The possibility of an existence of serotonergic neurones was reinforced by a number of pharmacological studies:

(1) B. B. Brodie and his colleagues (Pletscher *et al.*, 1955) discovered that the serotonin content of neuronal tissue was depleted by reserpine, a centrally acting drug that had been introduced as an anti-hypertensive agent.

(2) J. H. Gaddum (1953) found that lysergic acid diethylamide (LSD), an extremely potent psychotomimetic substance, acted as a specific antagonist of serotonin in a number of test preparations, e.g. the rat's stomach. He suggested that LST might exert its central effects by a similar blocking action at the sites where serotonin was normally effective.

(3) In 1960 A. Todrick and his colleagues (Marshall *et al.*, 1960) found that the tranquillizing drug imipramine depleted the blood platelets of serotonin. Again the question arose if the central actions of imimpramine were due to similar causes.

These early discoveries lead on to findings that represent the material discussed in this volume. For the biochemists, it is a satisfying experience to see that their early work has made substantial contributions to this field of knowledge. Historically, work on biosynthesis and on degradation shows certain parallels. For instance, at first the central nervous system was not particularly implicated. The decarboxylase that forms serotonin was first found by Holtz and his colleagues in the guineapig kidney, and the enzyme monoamine oxidase (MAO) was first studied in the mammalian liver, although J. H. Quastel appreciated the importance of MAO in the central nervous system at an early stage.

One problem that still exercises the minds of students of these enzymes is that of their specificity. Both the decarboxylase and MAO play an important part in the metabolism of both catecholamines and serotonin. There is no satisfactory evidence at present of a specific decarboxylase for 5-hydroxytryptophan; specificity seems to reside in the hydroxylation of tryptophan. As to MAO, the discovery of the different isoenzymes might have raised hopes of finding some degree of specificity, but there seems to be one isoenzyme that preferentially acts on all neurotransmitters.

The catecholamines were the first mediators for which a localization in characteristic intracellular particles was definitely established. For serotonin, it was first shown that the intestinal localization was in particulate elements (Baker, 1959), and soon afterwards (Prusoff, 1960) it was made likely that these intestinal particles were rich in adenosine triphosphate (ATP). However, the main work on particulate elements containing serotonin was carried out on the blood platelets. The discovery of large amounts of ATP in the platelets (Born, 1956) was soon followed by attempts to isolate the particulate serotonin-carrying elements (Baker, Blaschko, and Born, 1959), but it was mainly through the work of Pletscher (1968) that the platelet granules carrying both serotonin and ATP were fully characterized. Quite recently I saw a report according to which these particles are rich also in inorganic pyrophosphate (Fukami, Dangelmaier, Bauer, and Holmsen, 1980).

Masson (1914) was probably the first to compare the system of argentaffin cells to that of the chromaffin cells. The main difference between these two endocrine systems was that the former was scattered over almost the whole of the gastro-intestinal tract whereas the latter had in the adrenal medulla one principal site. Later on, he qualified his views somewhat (Masson and Berger, 1923), because he had discovered a close connection between the argentaffin cell and the submucous plexus of nerves. He called the argentaffin system 'neurocrine' because he believed that the argentaffin cells delivered that secretion product directly into the neurones of the submucous plexus. This statement was probably the first that postulated a connection between serotonin and nerve tissue.

The catecholamines were the prototypes of substances that serve the dual function of both hormones and transmitters. By now this is a familiar situation, and serotonin has been joined by many of the neuronal peptides who follow the same pattern. These peptides also share with serotonin frequently an occurrence in the gastrointestinal tract.

At a symposium held in London 10 years ago both A. Dahlström (1971) and A. D. Smith (1971) quoted early work by Scott, who had emphasized the similarity between neuronal and secretory function. In modern times it was the concept of neurosecretion, as established by Bargmann and Scharrer, that underlined the same relationship. With all the new information that has accumulated in the past few years one wonders if it is still possible to single out a small number of observations and reserve to them the term 'neurosecretory'.

A more recent nomenclature was introduced by Pearse (1968); he has used the term 'APUD' system for cells that have two abilities: that to release a pharmacologically active peptide and that to form an amine (e.g. serotonin) by enzymic decarboxylation of the amino acid precursor. Pearse stresses the common origin of many cells of this system in the same area, the neural crest. It will be interesting to learn how many 'serotonergic' systems conform with such a classification. It seems worth while to remember that amine-producing cells are not confined to the vertebrates.

One of the aspects clearly recognized by Pearse was the fact that amine formation (and amine release) are not the only functions of cells that contain these substances. Long ago we reported on the release of soluble proteins characteristic of chromaffin granules from the adrenal medulla when the gland secreted amine in response to stimulation of the splanchnic nerve (Blaschko, Comline, Schneider, Silver, and Smith, 1967).

In the intervening years there have been many reports indicating that more than one substance can be released from an excitable cell (Burnstock, 1976). It has already been pointed out that these reports are expressions of the fact that all cells of one body carry the same genes and that what we should really enquire into, is what causes some biochemical abilities in one cell from becoming manifest whereas in another cell this does not happen (Blaschko, 1977).

The interaction of serotonin release with the release of other biologically active agents is surely one of the topics that will be discussed in the near future.

REFERENCES

Baker, R. V. (1959). Mitochondria and storage granules for 5-hydroxytryptamine. *J. Physiol. (Lond.),* **145,** 473–481.

Baker, R. V., Blaschko, H., and Born, G. V. R. (1959). The isolation from blood platelets of particles containing 5-hydroxytryptamine and adenoside triphosphate. *J. Physiol. (Lond.),* **149,** 55P–56P.

Blaschko, H. (1977). Biochemical aspects of transmitter formation and storage. Comments on relationship in *The Synapse* (Eds. G. A. Cottrell and P. N. R. Usherwood), pp. 102–116, Blackie, Glasgow and London.

Blaschko, H., Comline, R. S., Schneider, F. H., Silver, M. S., and Smith, A. D. (1967). Secretion of a chromaffin granule protein, chromogranin, from the adrenal gland after splanchnic stimulation. *Nature (Lond.),* **215,** 58–59.

Born, G. V. R. (1956) Adenosine triphosophate (ATP) in blood platelets. *Biochem. J.,* **62,** 33P.

Burnstock, G. (1976). Do some nerve cells release more than one transmitter?, *Neuroscience,* **1,** 239–248.

Dahlström, A. (1971). Axoplasmic transport (with particular respect to adrenergic neurons), *Phil. Trans. Roy. Soc. (Lond.) B.* **261,** 325–358.

Erspamer, V. (1966). *5-Hydroxytryptamine and Related Indolealkylamines.* Handbook of experimental Pharmacology, Vol. XIX, pp. 1–928. Springer, Berlin, Heidelberg, New York.

Feldberg, W. (1945). Present views on the mode of action of acetylcholine in the central nervous system, *Physiol. Rev.,* **25,** 596–642.

Freund, H. (1920). Ueber die pharmakologischen Wirkungen des defibrinierten Blutes, *Arch. f. exp. Path. Pharmak.* **86,** 266–280.

Fukami, M. H., Dangelmaier, C. A., Bauer, J. S., and Holmsen, H. (1980). Secretion, subcellar localization and metabolic status of inorganic pyrophosphate in human platelets, *Biochem. J.* **192,** 99–105.

Gaddum, J. H. (1936). *Gefässerweiternde Stoffe der Gewebe,* Thieme-Verlag, Leipzig.

Gaddum, J. H. (1953) Antagonism between lysergic acid diethylamide and 5-hydroxytryptamine, *J. Physiol. (Lond.),* **121,** 15P.

Marshall, E. F., Stirling, G. S., Tait, A. C., and Todrick, A. (1960). The effect of iproniazid and imipramine on the blood platelet 5-hydroxytryptamine level in man, *Brit. J. Pharmacol.* **15,** 35–41.

Masson, P. (1914). La glande endocrine de l'intestin chez l'homme, *C. R. Acad. Sci (Paris)*, **158**, 59–61.

Masson, P., and Berger, L. (1923). Sur un nouveau mode de sécrétion interne: La neurocrinie, *C. R. Acad. Sci. (Paris)*, **176**, 1748–1751.

O'Connor, J. M. (1912). Ueber den Adrenalingehalt des Blutes, *Arch. f. exp. Path. Pharmak.* **67**, 195–232.

Page, I. H. (1954). Serotonin (5-hydroxytryptamine), *Physiol. Rev.* **34**, 563–588.

Page, I. H. (1968). *Serotonin,* Year Book Medical Publishers Inc., Chicago.

Pearse, A. G. E. (1968). Common cytochemical and ultrastructural characteristics of cells producing polypeptide hormones (the APUD series) and their relevance to thyroid and ultimobranchial C-cells and calcitonin, *Proc. Roy. Soc. (Lond.) B*, **170**, 71–80.

Pletscher, A. (1968). Metabolism, transfer and storage of 5-hydroxytryptamine in blood platelets, *Brit. J. Pharmacol.* **32**, 1–16.

Pletscher, A., Shore, P. A., and Brodie, B. B. (1955). Serotonin release as a possible mechanism of reserpine action, *Science,* **122**, 374–375.

Prusoff, W. H. (1960). The distribution of 5-hydroxytryptamine and adenosine triphosphate in cytoplasmic particles of the dog's small intestine, *Brit. J. Pharmacol.* **15**, 520–524.

Smith, A. D. (1971). Summing up: some implications of the neuron as a secreting cell, *Phil. Trans. Roy. Soc. (Lond.). B,* **261**, 423–437.

Twarog, B. M. and Page, I. H. (1953). Serotonin content of some mammalian tissues and urine and a method for its determination, *Am. J. Physiol.* **175**, 157–161.

Vialli, M. (1966) Histology of the enterochromaffin cell system, in *5-Hydroxytryptamine and Related Indolealkylamines* (Ed. V. Erspamer). Handbook of Experimental Pharmacology, Vol. XIX, pp. 1–65, Springer, Berlin, Heidelberg, New York.

Vogt, M. (1954). The concentration of sympathin in different parts of the central nervous system under normal conditions and after the administation of drugs, *J. Physiol. (Lond.),* **123**, 451–481.

Welsh, J. H. (1954) Marine invertebrate preparation useful in the bioassay of acetylcholine and 5-hydroxytryptamine, *Nature (Lond.),* **173**, 955–956.

Welsh, J. H. (1957). Serotonin as a possible neurohumoral agent. Evidence obtained in lower animals, *Annals of the N.Y. Acad. Sci.* **66**, 618–630.

[handwritten marginal notes: "Cauthle", "Culture of APUD concept"]

Biology of Serotonergic Transmission
Edited by N. N. Osborne
© 1982, John Wiley & Sons Ltd.

Chapter 1

Assay, Distribution and Functions of Serotonin in Nervous Tissues

NEVILLE N. OSBORNE
Nuffield Laboratory of Ophthalmology,
University of Oxford, Walton Street, Oxford OX2 6AW, UK

INTRODUCTION

Although serotonin (5-hydroxytryptamine) was identified in animal tissues in 1949 (Rapport, 1949), it was not until Twarog and Page (1953) pointed out its existence in the mammalian brain that neurobiologists became interested in the substance. Later observations, in particular the observation that the action of serotonin on the rat uterus could be antagonized by minute quantities of *d*-lysergic acid diethylamide (LSD) (Fingl and Gaddum, 1953), which was already known to elicit a mental

state resembling schizophrenia (Stoll, 1947), resulted in the suggestion that serotonin might have a specific role in cerebral functions (Gaddum, 1954). Wooley and
Shaw (1945a, b, 1957) were responsible for popularizing the idea and even postulated that human schizophrenia might be due to a serotonin deficiency. We now
know that serotonin may indeed be a transmitter, as originally suggested by Brodie
and Shore (1957), but whether it is involved in causing schizophrenia seems debatable (Stoll, 1947).

The most convincing evidence for serotonin being a transmitter substance came
in the mid-1950s, from studies on a variety of invertebrate preparations, in particular the anterior byssus retractor muscle of *Mytilus* (Twarog, 1954) and the heart of
the clam (Welsh, 1953, 1957). In the past 20 years, the invertebrate nervous system
has continued to provide the best material for the unequivocal demonstration that
serotonin is a neurotransmitter substance (see Osborne, 1978; Osborne and Neuhoff,
1980; Cottrell, 1977).

ASSAY OF SEROTONIN

The serotonin content of a tissue sample may be assayed by any of a variety of
methods. All of the methods available have advantages and disadvantages, ranging
from the two major problems of specificity and sensitivity to the questions of cost
and the labour involved. The methods available can be subdivided into the following categories: bioassay procedures, spectrophotometric procedures, fluorometric
procedures, chromatographic procedures and enzymatic procedures.

Bioassay procedures

Several bioassay systems have been reported (Page and Green, 1948; Dalgliesh *et
al.*, 1953; Erspamer, 1954; Twargo and Page, 1953; Cottrell and Osborne, 1969),
and while they are very sensitive to serotonin, they lack specificity. For example,
the isolated perfused snail-heart preparation is sensitive to as little as 0.5 ng of serotonin (see Fig. 1), and has been used to determine the serotonin content of a single
neurone (Cottrell and Osborne, 1970). Although bioassay procedures for assaying
serotonin have become obsolete in recent years, not only because of their lack of
specificity but also because of the variation in sensitivity of different preparations,
they have the advantage of being sensitive and inexpensive to perform. If used
correctly, bioassay techniques still have a place in modern research.

Spectrophotometric analysis

The ultraviolet absorption spectra of serotonin and 5-HTP in alkaline solution are
different from their non-hydroxyindole analogues (see Lovenberg and Engelman,
1971). This property may therefore be exploited to assay serotonin in the absence
of any 5-HTP.

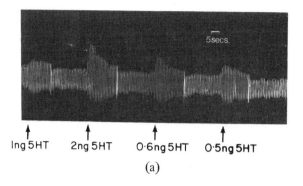

(a)

Substance	Relative stimulatory effect
Serotonin	4.78±1.00
5-Methoxytryptamine	3.02±.067
Tryptamine	2.40±0.80
5, 6-Dihydroxytryptamine	2.26±0.28
6-Hydroxytryptamine	1.64±0.18
5, 7-Dihydroxytryptamine	1.58±0.18
N, N'-Dimethyltryptamine	1.16±0.15
5-Hydroxyindole	1.14±0.17

All of the substances were made up at a concentration of 10^{-7} M, and the effect of 100 μl of each substance was tested on the isolated snail heart. The relative stimulatory effect is the ratio between the maximal stimulation produced by the substance compared with the normal contraction of the heart. A relative effect of 1 would therefore mean that the substance does not stimulate the heart. Results are the mean ±SEM (n = 5)

(b)

Fig. 1 (a) The dose-response effect of different amounts of serotonin (5-HT) on the isolated snail (*Helix*), heart. (b) A comparison of the relative effects of serotonin and the same amounts of other tryptamine analogues on the isolated snail heart

It is also possible to make use of the colour reaction of the 5-hydroxyindole nucleus to assay serotonin. For example, 1-nitroso-2-napthol reacts with 5-hydroxy-indoles in the presence of sodium nitrate and dilute sulphuric acid (Udenfriend *et al.*, 1955a) to yield a violet chromophore. The nature of the chromophore is not known but it is reproducible. Unfortunately, as all hydroxyindole compounds produce the same chromophore, it is necessary to isolate the serotonin before reaction. This is a major disadvantage.

Fluorometric methods

Fluorometric techniques for measuring serotonin have been used extensively, particularly because they are much more sensitive than spectrophotometric methods. These procedures are based on the extraction of serotonin and the analysis of the natural fluorescence of the substance (Bogdanski *et al.*, 1956), the fluorescence of the substance at different pH values (Udenfriend *et al.*, 1955b), the fluorescence of the product formed between ninhydrin and serotonin (Venable, 1963) or the fluorescence of the product formed between O-ophthaldialdehyde and serotonin (Maickel *et al.*, 1968, Curzon and Green, 1970). The main disadvantage of all fluorometric assays is that they require a purification procedure prior to fluorometric assay. This diminishes the overall sensitivity of the methods. Nevertheless, fluorometric procedures still remain very popular and the reader is referred to the excellent article by Atack (1977) for further details.

Chromatographic procedures

Paper and thin-layer chromatography are still used extensively for the qualitative identification of serotonin. The major disadvantage in this procedure lies essentially in the identification of serotonin. Good detecting compounds for serotonin on chromatograms are O-phthalaldehyde (see Aures *et al.*, 1968) or paraformaldehyde vapour (see Osborne, 1971). Generally it has been found better to use thin layer chromatography because more 'extract' can be applied to the chromatograms. For details of various chromatographic solvent systems used for separating serotonin from related compounds, the reader is directed to the papers by Aures *et al.*, (1968) and Lovenberg and Engelman (1971).

A number of more sophisticated chromatographic procedures have been developed within the last 10 years for assaying serotonin. These combine sensitivity and specificity but have the disadvantage of demanding expensive equipment and a great deal of expertise on the part of the experimenter. The most popular of these procedures is high pressure liquid chromatography (HPLC) in conjunction with electrochemical detection (Chilcote, 1974; McMurtrey *et al.*, 1976; Mefford and Barchas, 1980; Reinhard *et al.*, 1980). With this method, picogram quantities of serotonin can be detected and a previous purification of the extract is not required (see Fig. 2). Gas-liquid chromatography (GLC) coupled with either ionization detection, mass spectrometry, ion detectors or electron chemical detection has also been used for determining serotonin levels (see Costa *et al.*, 1968; Cattabeni *et al.*, 1972; Atack, 1977), but is less popular than HPLC because of the complexity of each procedure.

A simple, very sensitive and inexpensive way of assaying serotonin is first to react the amine with dansyl chloride and then isolate chromatographically the dansyl derivatives of serotonin. The reason for using dansyl chloride is that it reacts readily with aliphatic NH_2 and OH groups at alkaline pH to form very stable fluorophores (Gray, 1967), and the dansylated derivatives can be easily separated from each other chromatographically. For example, Fig. 3 shows the separation of a number of compounds usually present in brain tissue and which react with dansyl chloride. Separation is achieved on a single chromatogram measuring only 3 X 3 cm using two solvent systems. A flow diagram of the methodology of the procedure is shown in Fig. 4. Details of the procedure, together with its application, have been

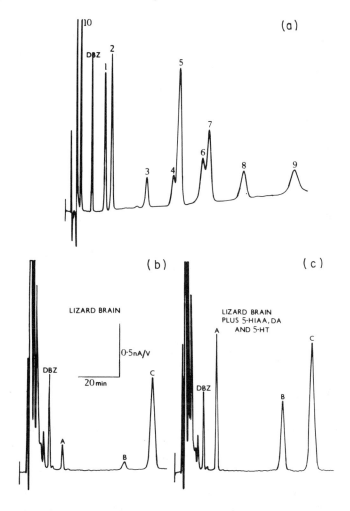

Fig. 2   Detection and estimation of serotonin in the brain of the lizard using high performance liquid chromatography (HPLC). The apparatus used was from DuPont equipped with a C-18 reverse-phase column and an LC-4 amperometric detector. The electrochemical detector was set at 2 nA/V with a potential of 0.8 V. The mobile phase was 0.02 mM sodium citrate buffer pH 3.5 containing 0.2 mM octane-sulfonic acid and 6.5 per cent methanol. With this system (Fig. 2a) a complete separation of serotonin (7) from the following substances was achieved: dopamine (1), dihydroxyphenylacetic acid (2), 5,6-dihydroxytryptamine (3), 3-methoxy-tyramine (4) 5-hydroxyindoleacetic acid (5), homovanillic acid (6), 6-hydroxy-tryptamine (8) and bufotenine (9). Dihydroxybenzylamine (DBZ) is an internal standard which was included in all analyses. Fig. 2b shows a chromatographic separation of a perchloric acid extract of lizard brain to which was added an internal standard of DBZ. Three peaks identified in the extract were for dopamine (A), 5-hydroxy-indoleacetic acid (B) and serotonin (C). When standard amounts of these three substances were added to the brain extract (Fig. 2c) this could be confirmed

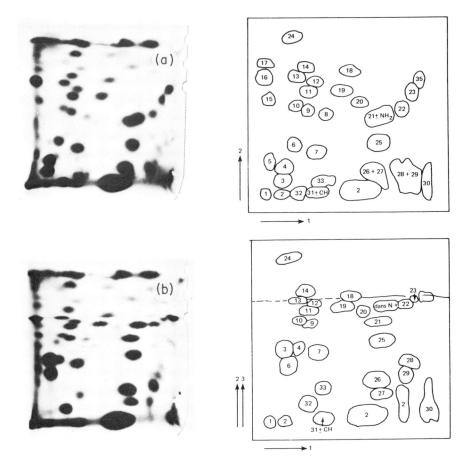

Fig. 3   Autoradiograms from extracts of Retzius cells of *Hirudo medicinalis* after
having been reacted with [^{14}C]dansyl chloride and chromatographed in either two
(a) or three (b) solvent systems. The direction of chromatography is indicated by
the arrows. First direction, water/formic acid (100:3 v/v); second direction, ben-
zene/acetic acid (9:1 v/v); third direction, ethylacetate/methanol/acetic acid (20:1:1
v/v). Identification keys are presented to assist in the interpretation of the dansyl
derivatives on the microchromatograms: 1, starting point; 2, dansyl-OH; 3, dansyl-
N-serotonin; 4, dansyl-tryptophan; 5, dansyl-bis-5-HTP; 6, dansyl-bis-lysine; 7,
dansyl-ornithine; 8, dansyl-methionine; 9, dansyl-phenylalanine; 10, dansyl-bis-
histidine; 11, dansyl-leucine; 12, dansyl-isoleucine; dansyl-bis-tyrosine; 16 dansyl-
bis-serotonin; 17, dansyl-5-hydroxyindole; 18, dansyl-proline; 19, dansyl-valine; 20,
dansyl-GABA; 21, dansyl-alanine; 22, dansyl-ethanolamine; 25, dansyl-glycine; 26,
dansyl-glutamic acid; 27, dansyl-aspartic acid; 28, dansyl-threonine + dansyl-
glutamine + dansyl-asparagine; 29, dansyl-serine; 30, dansyl-arginine + dansyl-*α*-
amino-histidine + dansyl-*ε*-lysine; 31, dansyl-unknown 3 + dansyl-taurine, 32,
dansyl-*N*-5-HTP; 33, dansyl-*N*-tyrosine, 13, 14, 23, 24, 34, 35, 36, 37 and 38 are all
unknown substances

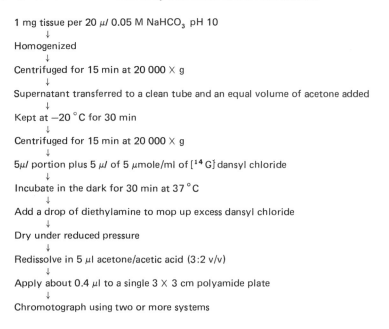

1 mg tissue per 20 μl 0.05 M NaHCO$_3$ pH 10
↓
Homogenized
↓
Centrifuged for 15 min at 20 000 × g
↓
Supernatant transferred to a clean tube and an equal volume of acetone added
↓
Kept at −20 °C for 30 min
↓
Centrifuged for 15 min at 20 000 × g
↓
5μl portion plus 5 μl of 5 μmole/ml of [^{14}C] dansyl chloride
↓
Incubate in the dark for 30 min at 37 °C
↓
Add a drop of diethylamine to mop up excess dansyl chloride
↓
Dry under reduced pressure
↓
Redissolve in 5 μl acetone/acetic acid (3:2 v/v)
↓
Apply about 0.4 μl to a single 3 × 3 cm polyamide plate
↓
Chromotograph using two or more systems

Fig. 4 A summary of the method used to obtain microchromatograms for the analysis of nervous tissue

described elsewhere (Osborne, 1973, 1974). From the flow diagram (Fig. 4) it can be seen that [^{14}C-] dansyl chloride is employed. This enables one simply to elute the dansyl derivatives from the chromatogram (dansyl chloride reacts with serotonin to form two derivatives) and determine the radioactivity for some quantitative results. Radioactivity need not be used, and the amount of substance can be determined by fluorometry. It is possible to determine the fluorescence associated with each derivative either by scanning the chromatogram directly (Kronberg *et al.*, 1978) or by eluting individual derivatives from the chromatogram and analysing fluorometrically (Seiler and Bruder, 1975). The major disadvantage of using dansyl chloride is that the substance does not react maximally with serotonin (see Osborne, 1973, 1974), thus making quantitative analysis difficult. This can therefore only be achieved by using some form of internal standard (see Snodgrass and Iversen, 1973; Osborne, 1974, Joseph and Halliday, 1975; Kronberg *et al.*, 1978), which lengthens the whole assay method.

Enzymatic procedures

A number of sensitive enzymatic-isotopic methods have been developed for the determination of amines (see Saavedra, 1978). These methods are based on the incubation of amines, e.g. serotonin with methyl transferase enzymes together with the donor or radioactive methyl groups, S-adenosyl-methionine. The radioactive methyl derivatives formed are then separated by means of solvent extraction and the radioactivity counted. Specificity of the assays is achieved by, (a) use of methylating enzymes with marked specificity for a given amine, (b) use of solvents of different degrees of polarity to separate radioactive products formed in the reaction, (c) use of selective evaporating techniques to eliminate radioactive but volatile contaminating substances.

A flow diagram and the procedure for determining serotonin are shown in Fig. 5 from the work of Saavedra *et al.,* (1973). The method is very sensitive—the detection limit being about 50 pg—relatively inexpensive and simple to perform. The main disadvantages are that it is time-consuming, involves the use of carcinogenic organic solvents and the enzyme hydroxyindole-O-methyl transferase is very unstable.

DISTRIBUTION OF SEROTONIN

No attempt will be made to review all the studies that have been carried out on the quantitative estimation of serotonin in nervous systems of animals. The work on mammals, in particular, is too extensive to be covered fully in this article. Instead, an attempt will be made merely to summarize the amine's distribution. For further information the reader is referred to the articles by Bogdanski *et al.* (1956), Welsh and Moorhead (1960), Brodie *et al.* (1964), Gerattini and Valzelli (1965), Erspamer (1966), Welsh (1968), Quay (1969), Green and Grahame-Smith (1975).

Invertebrates

In their survey on the distribution of serotonin in invertebrates, Welsh and Moorhead (1960) showed that it exists in all invertebrate animals. However, they analysed a variety of tissues, and not specifically the nervous systems of different animals. Today there is still a lack of conclusive evidence for the association of serotonin with nervous elements of Porifera (sponges), Coelenterata (Jelly-fish), Platyhelminthes (planarians) or Echinodermata (star-fish). Welsh and Moorhead did show that the nervous systems of a variety of worms belonging to the phylum Annelida contain between 3 and 10 μg serotonin/g fresh tissue. This has been confirmed in a number of other studies on animals from this phylum (Rude *et al.*, 1969; Osborne *et al.*, 1972).

The nervous systems of Mollusca are particularly rich in serotonin, which is most concentrated (40 μg/g fresh tissue) in the ganglia of *Venus mercenaria* (Welsh and Moorhead, 1960). In general, the serotonin content of molluscan ganglia is less, averaging 7 μg/g fresh tissue (see Welsh and Moorhead, 1960; Kerkut and Cottrell, 1963; Welsh, 1968; Sweeney, 1968).

Nervous tissues of Arthropoda also contain serotonin although the level is much lower than in the Annelida or Mollusca. A number of authors have detected levels

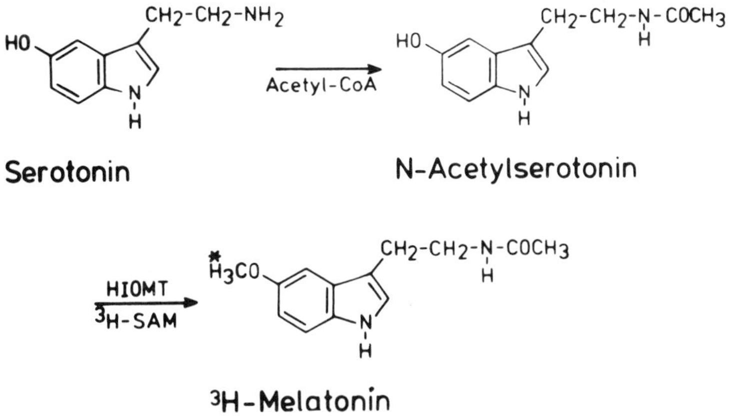

$$\text{Serotonin} \xrightarrow{\text{Acetyl-CoA}} \text{N-Acetylserotonin}$$

Serotonin (HO-indole with CH$_2$-CH$_2$-NH$_2$)

N-Acetylserotonin (HO-indole with CH$_2$-CH$_2$-N-COCH$_3$, H)

$$\xrightarrow[^3\text{H-SAM}]{\text{HIOMT}}$$

^{3}H-Melatonin (H$_3$CO-indole with CH$_2$-CH$_2$-N-COCH$_3$, H)

(a)

Homogenize tissue (0.1 N HCl)
↓
Separate aliquot for protein determination
↓
Centrifuge
↓
Separate duplicate aliquots of supernatant
↓
Add serotonin internal standard
↓
Incubate with *N*-acetyl transferase and acetyl coenzyme A
↓
Incubate with hydroxyindole-O-methyl transferase and [^{3}H]-SAM
↓
Stop with borate buffer and melatonin
↓
Extract into toluene
↓
Evaporate toluene at 90 °C
↓
Add phosphor
↓
Count the amount of [^{3}H]-melatonin formed

(b)

Fig. 5 Theoretical basis for the enzymatic-isotopic assay for serotonin. (b) Flow sheet of the procedure used. (Method of Saavedra *et al.*, 1973)

ranging between 0.01 and 0.03 μg serotonin/g fresh nervous tissue (see Welsh and Moorhead, 1960; Gerschenfeld, 1973; Klemm, 1976).

Vertebrates

Serotonin levels in the vertebrate CNS vary greatly in different parts, but there is a relatively uniform pattern of distribution in all classes. Generally, the mid-brain and hypothalamus have the highest levels (excluding the pineal gland), whereas the cerebellum contains less serotonin than other parts. The regional distribution of serotonin in a variety of vertebrate nervous tissues has been reviewed by Brodie *et al.* (1964). Serotonin is present in all vertebrate classes. Its concentration in the CNS of Reptilia and Amphibia is generally high in relation to the mammals. In the brain of the toad *Bufo americanus*, the level is extraordinarily high, averaging 9 µg/g fresh tissue. However, in the brain of two fish species, the concentration of serotonin is even less than in the mammals (see Brodie *et al.*, 1964; Welsh, 1968). The birds, in contrast, have about twice as much serotonin in their CNS (ca 1 µg serotonin/g fresh tissue) as do fish, which have approximately the same as the mammalian brain (see Welsh, 1968).

With the development of sensitive enzymatic-isotopic procedures (see above), it has become possible to analyse the serotonin content in minute areas of the CNS (see Saavedra *et al.*, 1973). This is illustrated by an analysis of the serotonin concentration in various parts of the adult cat brain by Gaudin-Chazal *et al.* (1979). The results of the study are shown in Table 1 together with the tryptophan and 5-hydroxyindole acetic acid concentrations of the tissue. It can be seen from Table 1 that high concentrations of serotonin were measured in all parts of the brain stem. The greatest amount was found in the superior colliculi and the parts of the mesencephalon, which includes the dorsalis and centralis superior raphe nuclei (see Table 1). The hypothalamus is also very rich in serotonin.

Saavedra *et al.* (1974a, b), using the sensitive enzyme-isotopic technique (which the same authors developed) to its maximum limit, analysed in detail the serotonin content of individual nuclei of the rat limbic system (Saavedra *et al.*, 1974a) and rat hypothalamus and preoptic region (Saavedra *et al.*, 1974b). A summary of these data is given in Tables 2 and 3. For additional information on the use of microdissection and microanalytical procedures for measuring serotonin levels in minute brain areas, (especially the raphe nuclei), the article by Saavedra (1979) on 'Microquantitation of neurotransmitters in specific areas of the central nervous system' is recommended.

FUNCTIONS OF SEROTONIN

In this section, discussion of the possible functional roles of serotonin will be restricted to the mammalian CNS. No attempt will be made to review the evidence available but I shall briefly summarize our present knowledge and select a few examples which implicate the participation of serotonin in certain mechanisms. Detailed information may be found in the articles by van Praag (1970), Coppen (1973), Schildkraut (1973), Chase and Murphy (1973), Jouvet and Pujol (1974), Gillin *et al.* (1978), Mandell and Knapp (1978) and Murphy *et al.* (1978).

Sleep/wake relationship

The raphe nuclei in the brain stem, which is very rich in serotonin, is regarded as an important sleep-inducing system (see Jouvet, 1967). This idea was instigated

Table 1. Concentrations, in μg/g of wet weight, of tryptophan (TRY), serotonin (5-HT) and 5-hydroxyindole acetic acid (5-HIAA), in various structures in the brain of the adult cat. Each value is the mean ± S.E.M. of 24 determinations, from 12 amimals. (From Gaudin-Chazal *et al.*, 1979.)

Tissue	TRY (μg/g)	5-HT (μg/g)	5-HIAA (μg/g)
Spinal cord	2.141±0.538	0.538±0.061	0.396±0.095
Median medulla		0.641±0.071	0.518±0.109
Right lateral medulla	2.820±0.376	0.557±0.049	0.400±0.102
Left lateral medulla		0.574±0.050	0.554±0.122
Inf. colliculi + mesen.	3.090±0.317	0.836±0.064	1.059±0.129
Sup. colliculi + mesen.		1.026±0.071	0.978±0.142
Cerebellum	3.609±0.653	0.143±0.022	0.135±0.038
Hypothalamus	2.858±0.704	1.255±0.097	0.698±0.071
Caudate nucleus	3.683±0.657	0.830±0.055	0.592±0.089
Hippocampus	3.594±0.495	0.481±0.026	0.586±0.052
Thalamus	2.976±0.333	0.574±0.026	0.675±0.083
Sigmoid gyrus	3.116±0.457	0.262±0.022	0.276±0.045
Proreus gyrus	3.248±0.522	0.291±0.017	0.292±0.043
Olfactory gyrus	2.866±0.396	1.051±0.067	0.770±0.095
Piriform gyrus	2.909±0.349	0.806±0.036	0.504±0.044
Coronal gyrus	3.437±0.492	0.236±0.016	0.220±0.018
Ant. lateral gyrus	3.743±0.538	0.144±0.011	0.185±0.051
Post. lateral med. gyrus	3.502±0.517	0.119±0.011	0.128±0.032
Post. lateral post. gyrus	3.525±0.514	0.159±0.015	0.202±0.041
Post. suprasylvian gyrus	3.343±0.448	0.129±0.012	0.161±0.031
Mid. suprasylvian gyrus	3.149±0.420	0.096±0.008	0.129±0.024
Ant. suprasylvian gyrus	3.639±0.437	0.136±0.010	0.237±0.037
Ant. ectosylvian gyrus	3.551±0.511	0.188±0.010	0.273±0.053
Mid. ectosylvian gyrus	2.806±0.216	0.177±0.018	0.194±0.027
Post. ectosylvian gyrus	3.605±0.548	0.191±0.013	0.222±0.037
Post. sylvian gyrus	3.318±0.475	0.219±0.020	0.306±0.040
Ant. sylvian gyrus	3.081±0.482	0.313±0.018	0.315±0.036

by the finding that, (a) when injected into patients, drugs with a selective effect on central serotonin metabolism influence the patients' sleep/wake rhythm, and (b) the raphe nuclei have an abundance of serotonin-containing neurones. This theory was corroborated by studies where it was shown, for example, that selective destruction of the raphe system leads to a selective decrease in the serotonin content and persistent wakefulness. The loss of the serotonin content caused by lesions was proportional to the hours of sleep lost (Jouvet and Delorme, 1965).

Temperature regulation

A number of experiments have been interpreted as suggesting that serotonin has a role in regulating temperature. This is exemplified in studies which have shown that the body temperature is increased in cats, dogs and monkeys following intracranial

Table 2. Distribution of serotonin in the limbic system of the rat. Serotonin was measured as described in the text. Results are expressed as means ±S.E.M. N represents the number of different samples analysed in each group. (From Saavedra *et al.*, 1974a.)

Area of nucleus	Serotonin concentration (ng/mg protein)	N
Limbic cortex		
Cingulate cortex	5.4±0.9	(17)
Piriform cortex	12.5±1.3	(10)
Gyrus dentatus	11.2±0.9	(6)
Hippocampus	3.8±0.5	(11)
Entorhinal cortex	8.7±0.5	(10)
Rostral limbic nuclei		
Tuberculum olfactorium	26.1±2.6	(18)
Nucleus tractus diagonalis	12.8±1.2	(12)
Nucleus accumbens	14.7±1.2	(16)
Nucleus interstitialis striae terminalis dorsal	18.9±3.4	(16)
Nucleus interstitialis striae terminalis ventral	16.4±1.7	(16)
Nucleus interstitialis striae medullaris	19.0±2.3	(12)
Nucleus interstitialis commissurae hippocampi	9.8±1.3	(12)
Medial forebrain bundle		
Medial forebrain bundle (preoptic)	19.0±2.1	(12)
Medial forebrain bundle (anterior)	21.6±5.8	(9)
Medial forebrain bundle (posterior)	30.7±8.0	(9)
Septal nuclei		
Nucleus septalis medialis	13.7±1.9	(16)
Nucleus septalis dorsalis	9.6±1.2	(16)
Nucleus septalis intermediate	9.5±1.2	(16)
Nucleus septalis lateralis	14.5±2.6	(13)
Nucleus septalis fimbrialis	7.2±1.1	(15)
Nucleus septalis triangularis	18.6±2.8	(15)
Amygdaloid nuclei		
Nucleus tractus olfactorii lateralis	17.8±2.4	(12)
Area amygdaloid anterior	19.0±1.9	(12)
Nucleus amygdaloideus medialis	19.9±4.0	(12)
Nucleus amygdaloideus corticalis	17.3±1.3	(12)
Nucleus amygdaloideus lateralis	20.7±2.7	(12)
Nucleus amygdaloideus centralis	16.8±2.2	(12)
Nucleus amygdaloideus basalis	17.5±1.6	(12)
Nucleus amygdaloideus medialis posterior	17.6±2.7	(11)
Nucleus amygdaloideus basalis posterior	23.0±2.1	(12)
Nucleus amygdaloideus posterior	15.9±1.6	(11)
Habenula		
Nucleus habenularis medialis	14.1±1.7	(10)

Table 2. *(Cont.)*

Area of nucleus	Serotonin concentration (ng/mg protein)	*N*
Nucleus habenularis lateralis	9.7±1.2	(12)
Mamillary body		
Nucleus mamillaris medialis	15.2±1.9	(6)
Nucleus mamillaris lateralis	15.7±1.1	(6)
Mesencephalic limbic area		
Nucleus interpeduncularis	17.3±2.2	(6)
Area tegmentalis ventralis (Tsai)	22.2±1.8	(6)

Table 3. Serotonin content of preoptic and hypothalamic nuclei. (From Saavedra *et al.,* 1974b.)

	Serotonin content (ng/mg protein: X ± S.E.M.)	Number of samples
Nucleus preopticus medialis	11.6±1.2	(16)
Nucleus preopticus lateralis	14.4±1.4	(15)
Nucleus preopticus periventricularis	7.4±0.8	(16)
Nucleus preopticus suprachiasmatis	24.6±3.1	(14)
Nucleus suprachiasmatis (total)	25.4±8.0	(9)
(internal)	37.2±4.7	(12)
(external)	17.0±1.3	(15)
Area retrochiasmatica	15.9±2.3	(15)
Nucleus periventricularis	10.9±4.1	(8)
Nucleus supraopticus	9.5±2.8	(9)
Nucleus paraventricularis	13.5±3.1	(9)
Nucleus hypothalamicus anterior	10.2±2.3	(7)
Nucleus arcuatus	36.4±9.8	(7)
Nucleus ventromedialis	8.5±3.9	(9)
Nucleus dorsomedialis	13.6±5.3	(9)
Nucleus perifornicalis	30.0±10.9	(8)
Nucleus hypothalamicus posterior	24.5±7.6	(9)
Nucleus premammillaris dorsalis	22.9±8.8	(7)
Nucleus premammilaris ventralis	18.3±8.2	(7)
Medial forebrain bundle (anterior)	21.6±5.8	(9)
Medial forebrain bundle (posterior)	30.7±8.0	(9)
Median eminence	15.3±3.2	(9)
Cortex	5.7±0.6	(13)

installation of serotonin, but decreased in rats, mice, rabbits, oxen and sheep after the same treatment (see Chase and Murphy, 1973). It is generally thought that the hypothalamus is involved in thermoregulation (Bligh, 1966), and this area of the brain contains high levels of serotonin (see Table 1). While the experimental data show that serotonin is involved in thermoregulation, it is also clear from literature that noradrenaline plays an important part in this function (Feldberg and Myers, 1965; Giarman *et al.*, 1968).

Aggression, hypersexuality

A number of relationships have been observed between serotonin levels in the brain and patterns of behaviour. For example, rats given *p*-chlorophenylalanine exhibit aggressive and mouse-killing activity (Sheard, 1969; Di Chiari *et al.*, 1971), which can be reduced by treatment with either 5-hydroxytryptophan or pargyline. However, although monkeys treated with *p*-chlorophenylalanine do not exhibit behavior changes (Weizman *et al.*, 1968; Gralla and Rubin, 1970), the same treatment in cats produces an increase in the mounting behaviour of adult males (Hoyland *et al.*, 1970). This important area of research clearly requires expanding.

Neurological influences

Phenylketonuric patients show disorders of serotonin metabolism and an overall reduction in circulating (platelet) serotonin (Pare, 1968), leading to severe mental deficiency. It is thought that inhibition of tryptophan-hydroxylase caused by excess phenylalanine in the tissue reduces the circulating serotonin levels in sufferers from phenylketonuria (Jequier, 1968). In Down's syndrome, circulating serotonin levels are also diminished, although this is ascribed to a defect in the uptake and binding of the amine to platelets, caused by the ATP of the cell (Boullin and O'Brien, 1971; Lott *et al.*, 1972).

Biochemical and pharmacological observations indicate that a defect in serotonin-containing neuronal systems plays a part in the pathogenesis of Parkinsonism. The serotonin levels of the basal ganglia are reduced and the 5-hydroxyindole acetic-acid content in the cerebrospinal fluid of patients with this disorder is diminished (Hornykiewicz, 1966; Johansson and Roose, 1967). The work by Poirer *et al.* (1966) may be taken as support for this idea, as they found that destruction of a small pathway in the cerebral peduncle led to a tumour in the contralateral limbs of monkeys (a Parkinson like syndrome) and loss of serotonin in the striatum.

Several authors have reported an increase in renal 5-hydroxyindole acetic acid excretion during migraine (see for example, Curzon *et al.*, 1966). A diminution of plasma serotonin levels at the onset of migraine attacks in patients has also been found (Anthony, 1968). These observations, together with the finding that methysergide is an excellent anti-migrainous agent, support the impression that serotonin is involved in producing migraine.

Reduced levels of 5-hydroxyindole acetic acid in cerebral fluid have also been observed in some studies on depressed and manic patients (Ashcrot and Sharman, 1960). It has been found that serotonin levels in the brain are lower in suicide than in control subjects (Shaw *et al.*, 1967; Pare *et al.*, 1969). While evidence that serotonin is involved in depression is persuasive, the mechanism of its action is still unknown.

The opinion is held that serotonin is also involved in the following neurological afflictions: Mongolism, Hartnups disease, schizophrenia and Huntington's disease. Again the mechanisms of action of serotonin in these diseases is uncertain.

Pain perception

It has been reported that *p*-chlorophenylalanine treatment of rats increases sensitivity to painful stimuli (Tenen, 1967), though this is not the case for mice or rabbits (Saarnivaara, 1969; Major and Pleuvry, 1971). The systemic administration of 5-hydroxytryptophan to rats abolishes the *p*-chlorophenylalanine effect. It is also known that lesions placed in the mid-brain raphe nuclei of rats diminish fore-brain serotonin levels and this increases sensitivity to shock and reduces the analgesic effect of morphine (Samanin *et al.*, 1970). Experiments along these lines have lent weight to the belief that serotonin is involved in pain perception although the precise mechanisms are not yet understood.

Neuroendocrine functions

There are many instances quoted which suggest that serotonin participates in the control of the releasing factor cells, possibly by acting on synapses in these cells or on the multisynaptic pathways leading to them. One interesting example is the in-involvement of serotonin with the suckling-induced prolactin release. In this particular case, restoration of hypothalamic serotonin levels after inhibiting biosynthesis of the amine does not, by itself, induce the hormonal response unless the adequate physiological stimulus, viz. the suckling reflex, is concomitantly applied (Kordon *et al.*, 1973). Serotonin also plays a part in gonadotropin regulation. Injection of the amine into ventricular cerebro-spinal fluid decreases serum levels of luteinizing hormone (Schneider and McCann, 1970). It is also known that serotonin administration produces an inhibitory effect on ovulation in rodents (Labhsetwar, 1971). Futhermore, it has been demonstrated that serotonin inhibits the release of a luteinizing hormone-releasing hormone from the mediobasal hypothalamic region but does not affect the secretion of this same hormone situated in terminals in the anterior hypothalamus (Charli *et al.*, 1978).

There are many data suggesting that the inhibitory feedback action of glucocorticoids on adrenocorticotropic hormone (ACTH) secretion is mediated by central serotonin-containing neurones. This could be due to the influence of the adrenal steroids on tryptophan-hydroxylase. Serotonergic neurones in the CNS may therefore be influenced by the activity of the pituitary-adrenal axis and the same neurones may, in turn, participate in the modulation of ACTH regulation (see Krieger and Krieger, 1970; Azmitia *et al.*, 1970).

REFERENCES

Anthony, M. (1968). Plasma serotonin levels in migraine, *Advance Pharmacol.* **6B**, 203.

Ashcroft, G. W. and Sharman, D. F. (1960). 5-Hydroxyindoles in human cerebrospinal fluid, *Nature* (Lond.), **186**, 1050–1051.

Atack, C. (1977). Measurement of biogenic amines, *Acta Physiol. Scand.* Suppl. 451.

Aures, D., Fleming R., and Hakanson, R. (1968). Separation and detection of biogenic amines by thin-layer chromatography, *. Chromatog.* **33**, 480–493.

Azmitia, E. C., Algeri, S., and Costa, E. (1970). In vivo conversion of ^{3}H-L-tryptophan into ^{3}H-serotonin in brain areas of adrenalectomized rats, *Science*, **169**, 201–203.

Bligh, J. (1966). The thermosensitivity of the hypothalamus and thermoregulation in mammals, *Biol. Rev.* **41**, 317–367.

Bogdanski, D. F., Pletscher, A., Brodie, B. B., and Udenfriend, S. (1956). Identification and assay of serotonin in brain, *J. Pharmacol.* **117**, 82–88.

Boullin, D. J. and O'Brien, R. A. (1971). Abnormalities of 5-hydroxytryptamine uptake and binding by blood platelets from children with Down's syndrome, *J. Physiol.* **212**, 287–297.

Brodie, B. B. and Shore, P. A. (1957). A concept for a role of serotonin and norepinephrine as chemical mediators in the brain, *Ann. N.Y. Acad. Sci.* **66**, 631–642.

Brodie, B. B., Bogdanski, D. F., and Bonomi, L. (1964). Formation, storage and metabolism of serotonin (5-hydroxytryptamine) and catecholamines in lower vertebrates, in *Comparative Neurochemistry* (Ed.: D. Richter,)' pp. 367–378, Pergamon Press, Oxford.

Cattabeni, F., Koslow, S. H., and Costa, E. (1972). Gas chromatographic-mass spectrometric assay of four indole alkylamines in rat pineal, *Science*, **178**, 166–168.

Charli, J. L., Rotsztejin, W. H., Patton, E., and Kordon, C. (1978). Effect of neurotransmitters on in vitro release of luteinizing-hormone-releasing hormone from the mediobasal hypothalamus of male rats, *Neuroscience Letters*, **10**, 159–163.

Chase, T. N. and Murphy, D. L. (1973). Serotonin and central nervous system function, *Ann. Rev. Pharmacol.* **13**, 181–197.

Chilcote, D. D. (1974). Column-chromatographic analysis of naturally fluorescing compounds. III Rapid analysis of serotonin and tryptamines, *Clin. Chem.* **20**, 421–423.

Coppen, A. J. (1973). Role of serotonin in affective disorders in *Serotonin and Behaviour* (Eds. J. Barchas and R. Usdin), pp. 523–527, Academic Press, New York.

Costa, E., Spario, P. F., Groppetti, A., Algeri, S., and Neff, N. H. (1968). Simultaneous determination of tryptophan, tyrosine, catecholamines and serotonin specific activity in rat brain, *Attidella Academica Medica Lombarda*, **23**, 1100–1104.

Cottrell, G. A. (1977). Identified amino-containing neurones and their synaptic connections, *Neuroscience*, **2**, 1–18.

Cottrell, G. A. and Osborne, N. N. (1969). Localisation and mode of action of cardioexcitatory agents in molluscan hearts, in *Comparative Physiology of the Heart: Current Trends.* (Ed. F. V. McCann, pp. 220–231, Experientia Suppl. 15.

Cottrell, G. A. and Osborne, N. N. (1970). Subcellular localisation of serotonin in an identified serotonin-containing neuron, *Nature* (Lond.), **225**, 470–472.

Curzon, G. and Green, A. R. (1970). Rapid method for the determination of 5-hydroxytryptamine and 5-hydroxyindoleacetic acid in small regions of rat brain, *Brit. J. Pharmacol.* **39**, 653–655.

Curzon, G., Theaker, P., and Phillips, B. (1966). Excretion of 5-hydroxyindole-acetic acid (5-HIAA) in migraine, *J. Neurol. Neurosurg. Pyschiat.* **29**, 85–90.

Dalgliesh, C. E., Toh, C. C., and Work, T. S. (1953). Fractionation of the smooth muscle stimulants present in extracts of gastrointestinal tract. Identification of 5-hydroxytryptamine and its distinction from substance P, *J. Physiol.* **120**, 298–310.

Di Chiara, G., Camba, G., and Spano, P. F. (1971). Evidence for inhibition by brain serotonin of mouse killing behaviour in rats, *Nature* (Lond.), **233**, 272–273.

Erspamer, V. (1954). *Pharmacology of indolealkylamines, Pharmacol. Rev.* **6**, 425–487.

Erspamer, V. (1966). 5-Hydroxytryptamine and related indolealkylamines, *Handbook of Experimental Pharmacology*, XIX, pp. 1–928, Springer Verlag, Berlin.

Fedlberg, W. and Myers, R. D. (1965). Changes in temperature produced by micro-injections of amines into the anterior hypothalamus of cats, *J. Physiol.* **177**, 239–245.

Fingl, E. and Gaddum, J. H. (1953). Hydroxytryptamine blockade by dihydroer-gotamine *in vitro, Fed. Proc.* **12**, 320–321.

Gaddum, J. H. (1954). Drugs antagonistic to 5-hydroxytryptamine, in *Ciba Foundation Symposium on Hypertension, Humoral and Neurogenic Factors* (Eds. G. E. W. Wolstenholme and M. P. Cameron), pp. 75–78, Little Brown, Boston.

Gaudin-Chazal, G., Daszuta, A., Faudon, M., and Ternaux, J. P. (1979). 5-HT concentrations in cat brain, *Brain Research,* **160**, 281–293.

Gerattini, S. and Valzelli, L. (1965). *Serotonin,* Elsevier, Amsterdam.

Gerschenfeld, H. M. (1973). Chemical transmission in invertebrate central nervous system and neuromuscular junction, *Phys. Rev.* **53**, 1–119.

Giarman, N. J., Tanaka, C., Mooney, J., and Atkins, E. (1968). Serotonin, nore-pinephrine, and fever, *Advanc. Pharmacol.* **6A**, 307–317.

Gillin, J. C., Meldelson, W. B., Sitaram, N., and Wyatt, R. J. (1978). The neuro-pharmacology of sleep and wakefulness, *Ann. Rev. Pharmacol. Toxicol.* **18**, 563–579.

Green, A. R. and Grahame-Smith, D. G. (1975). 5-Hydroxytryptamine and other indoles in the central nervous system, in *Handbook of Psychopharmacology* (Eds. L. L. Iversen, S. D. Iversen and S. H. Snyder,) Vol. 3 pp. 145–169, Plenum Press, New York.

Gralla, E. J. and Rubin L. (1970). Ocular studies with parachlorophenylalanine in rats and monkeys, *Arch. Opthal.* **83**, 734–740.

Gray, W. R. (1967). Dansyl-chloride procedure, in *Methods in Enzymology* (Ed. C. H. W. Hirs), pp. 139–151, Academic Press, New York.

Hornykiewicz, O. (1966). Dopamine (3-hydroxytryptamine) and brain function, *Pharmacol. Rev.* **18**, 925–964.

Hoyland, V. J., Shilliton, E. E., and Vogt, M. (1970). The effect of parachloro-phenylalanine on the behaviour of cats, *Brit. J. Pharmacol.* **40**, 659–667.

Jequier, E. (1968). *Tryptophan Hydroxylation in Phenylketonuria, Advance Pharmacol.* **6B**, 169–170.

Johansson, B. and Roose, B. E. (1967). 5-Hydroxyindoleacetic and homovanillic acid levels in the cerebrospinal fluid of healthy volunteers and patients with Parkinson's syndrome, *Life Science,* **6**, 1449-1454.

Joseph, M. H. and Halliday, J. (1975). A dansylation microasssay of some amino acids in brain, *Analyt. Biochem.* **64**, 389-402.

Jouvet, M. (1967). Neurophysiology of the states of sleep, *Physiol. Rev.* **47**, 117-177.

Jouvet, M. and Delorme, F. (1965). Locus coeruleus et Sommeil paradoxal, *C. R. Soc. Biol.* **159**, 895-899.

Jouvet, M. and Pujol, J. F. (1974). Effects of central alterations of serotonergic neurons upon the sleep-waking cycle, in *Advances in Biochemical Psychopharmacology,* Vol. II, *Serotonin – New Vistas* (Eds. E. Costa, G. L. Gessa, and M. Sandler), pp. 199-209. Raven Press, New York.

Kerkut, G. A. and Cottrell, G. A. (1963). Acetylcholine and 5-hydroxytryptamine in the snail brain, *Comp. Biochem. Physiol.* **8**, 53-63.

Klemm, N. (1976). Histochemistry of putative neurotrasmitter substances in the insect brain, *Prog. in Neurobiology* **7**, 99-169.

Kordon, C., Blake, C. A., Terkel, J., and Sawyer, C. H. (1973). Participation of serotonin-containing neurons in the suckling-induced rise in plasma prolactin levels in lactating rats, *Neuroendocrinology,* **13**, 213-223.

Krieger, H. P. and Krieger, D. T. (1970). Chemical stimulation of the brain: Effect on adrenal cortical release, *American J. Physiol.* **218**, 1632-1641.

Kronberg, H., Zimmer, H.-G., and Neuhoff, V. (1978). Automatische Fluorimetrie von Mikro-Dünnschict-Chromatogrammen, *Fresenius Z. Anat. Chem.* **290**, 133-134.

Labhsetwar, A. P. (1971). Effects of serotonin on spontaneous ovulation in rats, *Nature (Lond.),* **229**, 203-204.

Lott, I., Murphy, D. L., and Chase, T. N. (1972). Down's syndrome. Central monoamine turnover in patients with diminished platelet serotonin, *Neurology,* **22**, 967-972.

Lovenberg, W. and Engelman, K. (1971). Assay of serotonin, related metabolites and enzymes. In *Methods of Biochemical Analysis.* Suppl. volume on biogenic amines (Ed. D. Glick), pp. 1-34, Interscience, New York.

Maickel, R. P., Cox, R. H., Saillant, J., and Miller, F; P. (1968). A method for the determination of serotonin and norepinephrine in discrete areas of rat brain, *Int. J. Neuropharmacol.* **7**, 275-281.

Major, C. T. and Pleuvry, B. J. (1971). Effects of α-methyl-*p*-tryosine, *p*-chlorophenylalanine, L-c-(3,4 dihydroxyphenyl) alanine, 5, hydroxytryptophan and diethyldithiocarbonate on the analgesic activity of morphine and methylamphetamine in the mouse, *Brit. J. Pharmacol.* **42**, 512-521.

Mandell, A. J. and Knapp, S. (1978). Current research in the indoleamine hypothesis of affective disorders, in *Biochemistry of Mental Disorder, New Vistas,* Vol. 13, *Modern Pharmacology-Toxicology* (Eds. S. Usdin and A. J. Mandell), pp. 61-81, Dekker, Basel.

McMurtrey, K. D., Meyerson, L. R., Cashaw, J. L., and Davis, V. E. (1976). High pressure cation exchange chromatography of biogenic amines, *Anal. Biochem.* **72**, 566-572.

Mefford, I. N. and Barchas, J. D. (1980). Determination of tryptophan and metabolites in rat brain and pineal tissue by reverse phase HPLC and electrochemical detection. *J. Chromatog. Biomed. Applications,* **181**, 187-193.

Murphy, K., Campbell, I. C., and Costa, J. L. (1978). The brain serotonergic system in the affective disorders. *Prog. Neuro-Psychopharmacol.* **2**, 1-31.

Osborne, N. N. (1971). A microchromatographic method for the detection of biologically active monamines from isolated neurones, *Experientia*, 27, 1502-1513.
Osborne, N. N. (1973). The analysis of amines and amine acids in microquantities of tissue, *Progress in Neurobiology*, 1, 299-329.
Osborne, N. N. (1974). *Microchemical Analysis of Nervous Tissue*, Pergamon Press, Oxford.
Osborne, N. N. (1978). The neurobiology of a serotonergic neuron, in *Biochemistry of Characterised Neurons* (Ed. N. N. Osborne), pp. 47-80, Pergamon Press, Oxford.
Osborne, N. N. and Neuhoff, V. (1980). Identified serotonin neurons, in *International Review of Cytology*, Vol. 67, (Eds. G. H. Bourne and J. F. Danielli), pp. 259-290, Academic Press, New York.
Osborne, N. N., Briel, G., and Neuhoff, V. (1972). The amine and amino acid composition in the Retzius cells of the leech *Hirudo medicinalis*, *Experientia*, 28, 1015-1018.
Page, I. H. and Green, A. A. (1948). Perfusion of rabbits' ear for study of vasonstrictor substances, in *Methods in Medical Research*, Vol. 1, (Ed. V. R. Potter), pp. 123-128, Year Book Publishers, Chicago.
Pare, C. M. B. (1968). 5-Hydroxyindoles in phenylketonuric and nonphenylketonuric mental defectives, *Advance Pharmacol.* 6B, 159-165.
Pare, C. M. B., Young, D. P. U., Price, K. and Stacey, R. S. (1969). 5-Hydroxytryptamine, noradrenaline and dopamine in brainstem, hypothalamus and caudate nucleus of controls and of patients committing suicide by coal-gas poisoning, *Lancet*, 2, 133-135.
Poirier, L. J., Sourkes, T. L., Bonvier, G., Butcher, R., and Carabin, S. (1966). Striatal amines, experimental tremor and the effect of harmaline in the monkey, *Brain*, 89, 37-52.
Quay, W. B. (1969). Catecholamines and tryptamines, *J. Neuro-Visceral Relations*, Suppl. 9, 212-235.
Rapport, M. M. (1949). Serum vasoconstrictor (serotonin). V. Presence of creatine in the complex. A proposed structure of the vasoconstrictor principle, *J. Biol. Chem.* 180, 961-969.
Reinhard, J. F., Moskowitz, Sved, A. F., and Fernstrom, J. D. (1980). A simple, sensitive and reliable assay for serotonin and 5-HIAA in brain tissue using liquid chromatography and electrochemical detection, *Life Sciences*, 27, 905-911.
Rude, S., Coggeshall, R. E., and van Ordern, L. S. (1969). Chemical and ultrastructural identification of 5-hydroxytryptamine in an identified neuron. *J. Cell Biol.* 41, 832-854.
Saarnivaara, L. (1969). Effect of 5-hydroxytryptamine on morphine analgesia in rabbits, *Ann. Med. Exp. Fenn.* 47, 113-123.
Saavedra. J. M. (1978). Microassay of biogenic amines in neurons of *Aplysia*, the coexistence of more than one transmitter molecule in a neuron, in *Biochemistry of Characterised Neurons*, (Ed. N. N. Osborne), pp. 217-238, Pergamon Press, Oxford.
Saavedra, J. M. (1979). Microquantitation of neurotransmitters in specific areas of the central nervous system, *Int. Rev. of Neurobiol.* 21, 259-274.
Saavedra, J. M., Brownstein, M., and Axelrod, J. (1973). A specific and sensitive enzymatic-isotopic microassay for serotonin in tissues, *J. Pharmacol. exp. Therap.* 186, 508-515.
Saavedra, J. M., Palkovits, M., and Brownstein, M. J. (1974a). Serotonin distribution in the limbic system of the rat, *Brain Res.* 79, 437-441.
Saavedra, J. M., Palkovits, M., Brownstein, M. J., and Axelrod, J. (1974b). Sero-

tonin distribution in the nuclei of the rat hypothalamus and preoptic region, *Brain Res.* **77**, 157–165.

Samanin, R., Gumulka, W., and Valzelli, L. (1970). Reduced effect of morphine in midbrain raphe lesioned rats, *Europ. J. Pharmacol.* **10**, 339–343.

Schildkraut, J. J. (1973). Neuropharmacology of the affective disorders, *Ann. Rev. Pharmacol. Toxicol.* **13**, 427–454.

Schneider, H. P. and McCann, S. M. (1970). Mono- and indolamines and control of LH secretion, *Endocrinology*, **86**, 1127–1133.

Seiler, N. and Bruder, K. (1975). Determinations of serotonin and bufotenin as their dansyl derivates, *J. Chromatog.* **106**, 159–173.

Shaw, D. M., Camps, F. E., and Eccleston, E. C. (1967). 5-Hydroxytryptamine in the hind-brain of depressed suicides, *Brit. J. Psychiat.* **113**, 1407–1411.

Sheard, M. H. (1969). The effect of *p*-chlorophenylalanine on behaviour in rats: relation to brain serotonin and 5-hydroxyindoleacetic acid, *Brain Res.* **15**, 524–528.

Snodgrass, S. R. and Iversen, L. L. (1973). A sensitive double isotope derivative assay to measure release of amino acids from brain in vitro. *Nature (New Biology)*, **241**, 154–156.

Stoll, W. A. (1947). Lysergsäure-diäthylamide, ein Phantastikum aus der Mutter-komgruppe, *Schweiz. Arch. Neurol. Neurochi. Psychiat.* **68**, 279–323.

Sweeney, D. (1968). The anatomical distribution of monoamines in the fresh-water bivalve mollusc *Sphaerium sulcatum*, *Comp. Biochem. Physiol.* **25**, 601–613.

Tenen, S. S. (1967). The effects of *p*-chlorophenylalanine, a serotonin depletor, on avoidance acquisition, pain sensitivity and related behaviour in the rat, *Psychopharmacologia*, **10**, 204–219.

Twarog, B. M. (1954). Responses of a molluscan smooth muscle to acetylcholine and 5-hydroxytryptamine, *J. Cellulua Comp. Physiol.* **44**, 141–163.

Twarog, B. M. and Page, I. H. (1953). Serotonin content of some mammalian tissues and urine and a method for its determination, *Am. J. Physiol.* **174**, 157–161.

Udenfriend, S., Weissbach, H., and Clark, C. T. (1955a). The estimation of 5-hydroxytryptamine (serotonin) in biological tissues, *J. Biol. Chem.*, **215**, 337–344.

Udenfriend, S., Bogdanski, D. F., and Weissbach, H. (1955b). Fluorescence characteristics of 5-hydroxytryptamine (serotonin), *Science*, **122**, 972–973.

van Praag, H. M. (1970). Indoleamines in the central nervous system, *Psychiat. Neurol. Neurochir.* **73**, 9–36.

Venable, J. W. (1963). A ninhydrin reaction giving a sensitive quantitative fluorescence assay for 5-hydroxytryptamine, *Anal. Biochem.* **6**, 393–403.

Weitzman, E. D., Rapport, M. M., McGregor, P., and Jacoby, J. (1968). Sleep patterns of the monkey and brain serotonin concentration: Effect of *p*-chlorophenylalanine. *Science*, **160**, 1361–1363.

Welsh, J. H. (1953). The occurrence of an excitor amine in the nervous system of *Buccinum* and its action on the heart, *Anat. Record*, **117**, 637–638.

Welsh, J. H. (1957). Serotonin as a possible neurohumoral agent, evidence obtained in lower animals, *Ann. N.Y. Acad. Sci.* **66**, 618–630.

Welsh, J. H. (1968). Distribution of serotonin in the nervous systems of various animal species, *Advance Pharmacol.* **6A**, 171–188.

Welsh, J. H. and Moorhead, M. (1960). The quantitative distribution of 5-hydroxytryptamine in the invertebrates, especially in the nervous system, *J. Neurochem.* **6**, 146–169.

Wooley, D. W. and Shaw, E. (1954a). A biochemical and pharmacological suggestion about certain mental disorders, *Proc. Nat. Acad. Sci.* **40**, 228–231.

Wooley, D. W. and Shaw, E. (1954b). A biochemical and pharmacological suggestion about certain mental disorders, *Science,* **119**, 587–588.
Wooley, D. W. and Shaw, E. (1957). Evidence for the participation of serotonin
in mental processes, *Ann. N.Y. Acad. Sci.* **66**, 649–667.

Biology of Serotonergic Transmission
Edited by N. N. Osborne
© 1982, John Wiley & Sons Ltd.

Chapter 2

CNS Serotonin Pathways

ADRIANA CONSOLAZIONE and A. C. CUELLO
Department of Pharmacology, South Parks Road,
University of Oxford, Oxford OX1 3QT, UK

INTRODUCTION

Serotonin was found to be unevenly distributed in the brain nearly thirty years ago (Amin, 1954; Bogdanski *et al.,* 1957). The advent of the 'Falck and Hillarp' technique (Falck *et al.,* 1962; Corrodi and Jonsson, 1967) allowed Dahlstrom and Fuxe (1964) to demonstrate the neuronal localization of this amine. Nevertheless, detailed mapping of the serotonin-containing cell bodies and terminals in the mammalian central nervous system (CNS) progressed slowly. With the Falck and Hillarp method only small pieces of brain can be investigated and the fluorophore obtained fades very rapidly under ultraviolet light. Other techniques have also assisted in the neuroanatomical tracing of the central serotonergic system. In recent years besides the histochemical fluorescence techniques the following procedures were applied: quantitative biochemical analysis in combination with mechanical or chemical lesions; autoradiographic demonstration of [^{3}H] serotonin and radiolabelled amino acids orthogradely transported; immunohistochemistry using antibodies to tryptophan hydroxylase or dopa decarboxylase, as markers for serotonin; and finally, the immunocytochemical detection of serotonin itself.

The aim of this chapter will be to review all these morphological techniques and to summarize the data obtained with these methods in an attempt to give a comprehensive anatomical picture of the serotonergic system in the mammalian CNS.

METHODOLOGICAL CONSIDERATIONS

Formaldehyde induced fluorescence technique (Falck–Hillarp).

The introduction of the formaldehyde induced fluorescence methodology by Falck *et al.* (1962) made possible the direct localization of serotonin-containing neurones. For this, freeze-dried tissue is treated with slightly humid formaldehyde vapours at 80 °C for 1–2 h, to convert monoamines into a condensation product, a 1,2,3,4-tetrahydrobetacarboline. This product undergoes dehydrogenation to form a very fluorescent 6-hydroxy-3,4-dihydro-beta-carboline, with an activation peak for serotonin at 420 mμ and an emission peak around 520–525 mμ. The serotonin fluorophore has a lower fluorescent yield than those formed from catecholamines and a very rapid rate of photodecomposition under ultraviolet light. In addition, catecholamine fluorescence can mask the weaker fluorescence of the indoleamine. Because of these difficulties, several approaches have been used in order to improve serotonin histochemical detection. Fuxe and Jonsson (1967) modified histochemical reaction conditions, incubating the dried blocks of tissue twice, first with formaldehyde vapour at low humidity (70 per cent) for 1 h, and then at high humidity (97.5 per cent) for 1–2 h. Alternatively, catecholamines can be selectively depleted with 6-hydroxydopamine, prior to the fluorescence histochemistry (Fuxe and Jonsson, 1974; Jonsson *et al.*, 1969). Several pharmacological manipulations have been attempted for the preferential demonstration of serotonin, such as monoamine oxydase inhibition coupled with pretreatment with reserpine and 5-hydroxytryptophan (Dahlstrom and Fuxe, 1964; Dalhstrom and Fuxe, 1965; Jonsson *et al.*, 1974) or, more specifically, with tryptophan (Aghajanian and Asher, 1971; Kuhar *et al.*, 1972). The lesioning of presumptive serotonin pathways results in a 'build-up' of the transmitter, proximal to the lesion and a decrease distally (Ungerstedt, 1971; Dahlstrom *et al.*, 1973). A number of dihydroxytryptamines cause a preferential degeneration of serotonergic cells resulting in a diminution or disappearance of induced fluorescence in the CNS neurones, (Fuxe and Johnsson, 1974; Daly *et al.*, 1973; Baumgarten and Lachenmayer, 1972; Baumgarten *et al.*, 1976).

Autoradiographic methods

Radioautography has been applied successfully for the localization of indoleamine-containing neurones following the intraventricular or topical application of [^{3}H] serotonin ($[^3H]$ 5-HT). This technique is based on the property of monoaminergic neurones to take up selectively its own transmitter by a high-affinity transport mechanism. The radioactivity retained in the tissue is then demonstrated auto-radiographically as a deposit of silver grains. This requires a substantial retention of the labelled substance in the tissue. Usually this is obtained by the intravascular perfusion of the brain with glutaraldehyde or glutaraldehyde and paraformaldehyde. Following post-fixation and embedding, tissue sections are coated with a photo-graphic emulsion and exposed for 2–4 weeks. The characteristics of the emulsions, conditions of development and fixation can be very critical. Different types of emulsions have been used. Kodak NTB 2 and Ilford L4 have been successfully applied by various authors. Exposure time has to be established experimentally. Generally for electron microscopy longer intervals are required. For further comments on the use of this technique to trace transmitter specific pathways see Iversen (1978).

The possibility of unspecific uptake of tritiated serotonin by non-serotonin containing elements must be taken into account. Serotonin in high concentrations can also be incorporated into catecholaminergic neurones in significant proportions (Shaskan and Snyder, 1970). In order to overcome this problem, Bloom *et al.* (1972) produced neurotoxic lesion of catecholaminergic terminals with 6-hydroxy-dopamine, while Chan-Palay, (1977) opted to use competitive monoamine uptake by infusing 10^{-5} M $[^3H]$ 5-HT together with 10^{-4} cold *d*-1-norepinephrine. In order to prevent the inactivation of the exogenous $[^3H]$ 5-HT a monoamine oxidase inhibitor is generally used. The exogeneous labelled transmitter can be delivered in various ways. Stereotaxic micro-injections into specific brain areas seem to result in an adequate labelling. The volume of the fluid injected and the rate of infusion should be strictly controlled in order to minimize local mechanical damage to the tissue.

Autoradiography has also been a valuable technique in following efferent fibres from the raphe nuclei through a long distance using $[^3H]$ proline (Azmitia and Segal, 1974, 1978; Halaris *et al.,* 1976) or $[^3H]$ leucine (Bobillier *et al.,* 1976). This fibre tracing technique is based on the orthograde axoplasmic transport of macro-molecular material, labelled by the neuronal incorporation of the radioactive precursor, that can be demonstrated by autoradiogaphy (Droz and Leblond, 1963; Lasek *et al.,* 1968). Nevertheless, the interpretation of results thus obtained have to be taken with caution as these isotopes will label serotonergic as well as non-serotonergic neurones.

Immunohistochemical methods

Immunohistochemistry is based on the ability of antibodies to bind specific antigen compounds. Once the antibody binds its specific antigen in the fixed tissue, immunostaining is necessary to detect the immunoreaction, and hence the antigen under investigation. For immunostaining, direct or indirect approaches have been attempted. (For review see Cuello, 1978). In the direct technique the antibody molecule is conjugated with a detectable marker. This approach has been abandoned since inactivation of the antibody can result following conjugation procedures with the marker. Since the work of Coons and Kaplan (1950) indirect methods have been more successfully applied. This indirect technique is sometimes referred to as the 'sandwich' technique, and it implies the use of labelled anti-immunoglobulin (second antibody) against the species of the primary antibody. The specific primary (or first) antibody thus binds the tissue antigen, but also acts as an antigen to the labelled anti-antibody immunoglobulin. The markers, attached to the anti-immunoglobulin, can be heavy metals, enzymes or fluorescent substances. Fluorescein isothiocyanate (FITC) and tetramethyl rhodamine isothiocyanate (TRITC) are the most used fluorescent substances. They display a very clear and relatively stable fluorescence and cause limited inactivation of antibodies following conjugation procedures (Coons, 1958). Horseradish peroxidase is the most commonly used enzyme marker. The enzyme labelled antibody techniques were introduced by Nakane and Pierce (1966, 1967) and by Avrameas (1970). The presence of the enzyme is demonstrated by diaminobenzidine in the presence of hydrogen peroxide, basically as originally described by Graham and Karnovsky (1966). The oxidation product formed after this reaction can be visualized both at light and electron microscopy level. A further development of this technique, the peroxidase-antiperoxidase (PAP) procedure, developed by Sternberger et al. (1970) has been shown to be more sensitive than the enzyme-labelled techniques. In this version of the procedure the second antibody is unlabelled and is added in excess in order to bind the first antibody with only one of the combining sites while the second site should remain free to react with a third antibody, an antiperoxidase antibody, produced in the same species as the first one. Detection of this final complex is carried out as described above, either for light microscopy or for electron microscopy. (See Sternberger 1974 for full details.)

For immunohistochemistry, tissue must be previously fixed. The choice of fixative depends on the nature of the antigen. In most cases freshly prepared 4 per cent paraformaldehyde renders satisfactory results. For electron microscopy the addition of glutaraldehyde to the fixative improves the cell ultrastructure (Pickel, 1980; Vaughn et al., 1981; Priestley and Cuello, 1981). The procedure can be briefly summarized (a) fixation by intracardiac perfusion, (b) dissection of the desired structure, (c) post-fixation of this structure in the same fixative for 1-2 h at 4 °C, (d) rinsing in phosphate buffered saline (PBS) for 18-24 h at 4 °C, (e) sectioning by cryostat at −25 °C for immunofluorescence or by rotary microtome in the case of tissue previously embedded in paraffin for light microscopy, or by vibrating microtome for electron microscopy, (f) rinsing of the sections in PBS, sometimes supplemented with Triton-X-100 as Hartman (1973) suggested in order to facilitate penetration of the antibodies, (g) incubation with the specific antibody (first antibody); time and temperature conditions for this step must be found experimentally, (h) incubation with the developing antibodies and washes in between. For immunofluorescence the sections are mounted in glycerin /PBS and they are then ready for observation. In the case of applying immunoenzyme techniques, after a 10-30 min incubation with diaminobenzidine sections are mounted in Permount of DPX.

Antibody production: Geffen *et al*., (1969) introduced immunocytochemical techniques for the study of neurotransmitter compounds, when they obtained an antibody against the catecholamine biosynthetic enzyme dopamine-beta-hydroxylase. Since then many neurotransmitter biosynthetic enzymes have been purified and used for the production of antisera. For the serotonin immunocytochemical localization, antibodies against dopa-decarboxylase and tryptophan hydroxylase have been successfully utilized. Dopa-decarboxylase has been purified from bovine adrenal glands (Goldstein *et al.*, 1972) and from hog kidneys (Christenson *et al.*, 1970) by DEAE Sephadex column chromatography and polyacrylamide disc gel electrophoresis. This enzyme catalyses not only the decarboxylation of L-dopa to dopamine but also the decarboxylation of all naturally occurring amino acids, including L-5-hydroxytryptophan, according to Lovenberg *et al.* (1962). It is therefore better referred to as aromatic L-amino acid decarboxylase. Purified preparation of this enzyme was used to raise antibodies in rabbits and the specificity tested by immunoelectrophoresis (Hökfelt *et al.*, 1973a). Immunocytochemical application of this antiserum showed a positive immunoreactivity in catecholamine as well as serotonin-containing structures of the central and peripheral nervous system (Goldstein *et al.*, 1971, 1972). This further supports the idea that L-amino decarboxylase is probably the same enzyme for these two types of neurones. The enzyme tryptophan hydroxylase catalyses the first step of the biosynthesis of serotonin and is presumed to be present only within neurones which contain serotonin. This enzyme is therefore a more specific marker of serotonin-containing cells. Tryptophan hydroxylase has been isolated from the region of the raphe nuclei of the rat midbrain by sequential column chromatography and disc gel electrophoresis (Joh *et al.*, 1975). Extracts of the three single bands were injected with Freund's complete adjuvant in rabbits. The antisera were tested by immunoelectrophoresis or double immunodiffusion. Antisera able to inhibit tryptophan hydroxylase activity of the purified enzyme or from rat midbrain homogenates were used for the immunohistochemical localization of serotonin neurones in the midbrain by light and electron microscopy (Pickel *et al.*, 1976).

Recently, antibodies against serotonin have been applied successfully in immunohistochemistry (Steinbusch *et al.*, 1978). Serotonin is a low molecular weight molecule and not immunogenic 'per se'. Antibodies to serotonin have been developed in rabbits by different groups (Ranadive and Sehon, 1967; Grota and Brown, 1974, Peskar and Spector, 1973; Spector *et al.*, 1973) conjugating serotonin to albumin either by formaldehyde or carbodiimide. Nevertheless, not one of these authors utilized these antibodies for immunocytochemical localization of serotonin. More recently, the earlier procedure by Ranadive and Sehon (1967) was followed to lead a better characterized antibody, suitable for immunohistochemical studies (Hökfelt *et al.*, 1978; Steinbusch *et al.*, 1978; Consolazione *et al.*, 1981). Steinbusch and collaborators have characterized their antisera by the immunodiffusion method, immunoelectrophoresis and immunofluorescence inhibition tests. By these techniques it was shown that the antibody against serotonin might have a 2 per cent cross reactivity to 5-methoxytryptamine and dopamine and less than 1 per cent to noradrenaline and adrenaline. By applying the hybridoma-myeloma strategy we have recently developed a rat X rat monoclonal antibody which detects serotonin-immunoreactive sites in fixed tissue sections (Köhler and Milstein, 1975; Galfre *et al.*, 1979). This technique allows the continuous production of a single species of immunoglobulin. This monoclonal antibody against serotonin sites was characterized by immunohistofluorescence and haemagglutination and coded YC5/45 HLK (Consolazione *et al.*, 1981; Cuello and Milstein, 1981).

SEROTONIN LOCALIZATION

Serotonin-containing cell bodies

The very first mapping of serotonin-containing cell bodies was presented by Dahlstrom and Fuxe (1964) using the formaldehyde induced fluorescence technique. These authors showed yellow fluorescent cell bodies in the medulla oblongata, pons and mesencephalon of rats treated with nialamide, a potent MAO inhibitor. They described nine groups of cell bodies, B1-9, most of them within the so-called 'raphe nuclei'. This classification is still largely accepted. Further confirmation of this distribution came later from other groups using induced fluorescence, immunohistochemistry, radioautography or biochemical techniques (Pin *et al.*, 1968; Felton *et al.*, 1974; Hubbard and DiCarlo, 1974; Aghajanian *et al.*, 1973; Daly *et al.*, 1974; Fuxe and Jonsson, 1974; Hökfelt *et al.*, 1973a,b; Joh *et al.*, 1975; Sladek and Walker, 1977; Chan-Palay, 1977; Levitt and Moore, 1978; Jacobowitz and McLean, 1978; Beaudet and Descarries, 1979; Steinbusch and Nieuwenhuys, 1980; Steinbusch, 1981) and current studies applying the monoclonal antibody YC5/45HL as illustrated in Figs. 1-4. From all these studies it has emerged that there is good agreement in the overall distribution of the serotonin-containing cell bodies in different mammalian species. It is worth noting that while the large majority of the serotonin-containing neurones lie within the limit of the so-called 'raphe nuclei' system (i.e. midline neuronal groups of the brain stem) a good number of these occur beyond those limits.

A summary of the distribution of the serotonergic cell groups follows below:

B1. Cell bodies of this group are localized mainly but not only within the nucleus raphe pallidus in the medulla oblongata. Other cell bodies are found in the formatio reticularis dorsally to the nucleus olivaris accessorius dorsalis. Caudally this group extends to the medial part of the pyramidal decussation, outlining the medial surface of the pyramidal tract. Rostrally, cell bodies belonging to this group are not clearly separated from those of the B3 group (Fig. 1).

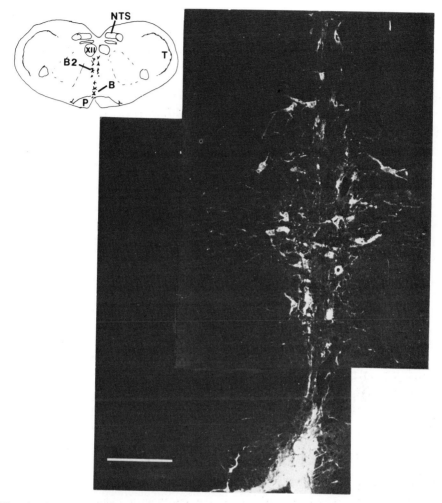

Fig. 1 Serotonin-immunofluorescent cell bodies corresponding to groups B1 and B2 in the lower medulla oblongata of the rat (nuclei raphe pallidus and obscurus). P, tractus cortico-spinalis (pyramidal tract); NTS, nucleus of the tractus solitarius; XII, nucleus originis nervi hypoglossi. Scale bar = 100 μm

B2. Serotonin immunoreactive neurones of this group are largely included within the limits of the nucleus raphe obscurus. This group is characterized by two symmetrical paramedian rows of cell bodies, dorsal to the nucleus olivaris inferior (Fig. 1).

B3. Cell bodies of this group are located both in the midline and laterally. They extend from a level just caudal to the raphe magnus to the anterior pole of the nucleus olivaris superior, rostrally. Most cells of this group are present within the nucleus raphe magnus in between the fibres of the lemniscus medialis (Fig. 2).

B4. A very small group of cells not clearly associated to any raphe nucleus. They are situated in the midline, under the fourth ventricle, at the lower levels of the nucleus vestibularis medialis.

B5. Most cells of this group are in the nucleus raphe pontis and more dorsally, medially to the fasciculus longitudinalis medialis (Fig. 3).

B6. This group is composed of a few cells lying medially on the floor of the fourth ventricle at levels where the locus coeruleus is present. These cells do not correspond to the raphe system (Fig. 3).

B7. This is a large group of serotonergic neurones located within the nucleus raphe dorsalis. Most cells of this group are located in the substantia grisea centralis (Fig. 4). The B.7 group extends rostrally to the nucleus Edinger–Westphal.

B8. This group can be identified with the nucleus centralis superior of the lower mesencephalon (Fig. 4). Some additional cells can be observed dorsally to this nucleus, scattered in the formatio reticularis and as rostally as the beginning of nucleus interpeduncularis.

B9. The cell bodies belonging to this group spread out dorsally and within the lemniscus medialis and in the zone between the lemniscus medialis and the cortico-spinal tract. This group extends as a horizontal band and does not correspond with any raphe nucleus (Fig. 4).

Other serotonergic neurones

Using various methodologies, some serotonergic cell bodies have also been identified in other brain regions. They are found in the area postrema (Fuxe and Owman, 1965) locus coeruleus and subcoeruleus (Sladek and Walker, 1977; Steinbusch *et al.*, 1979; Steinbusch, 1981), the interpeduncular nucleus (Chan-Palay, 1977) and some hypothalamic nuclei such as the nucleus arcuatus, periventricular (Kent and Sladek, 1978) and dorsomedial (Descarries and Beaudet, 1978).

Local circuit serotonergic neurones have also been described in the retina. These are a population of axonless amacrine neurones in which cell bodies lie in the internal nuclear layer and with profuse dendritic arborization in the inner plexiform layer (Osborne *et al.*, 1981) (see Fig. 13).

Serotonin-containing nerve fibres

An attempt to map the CNS distribution of serotonergic fibres was made by Fuxe as early as 1965. More recent studies, applying histochemical as well as biochemical and radioautographic tracing techniques (Fuxe, 1965; Aghajanian and Bloom, 1967; Fuxe and Ungerstedt, 1968; Fuxe *et al.*, 1968; Calas *et al.*, 1976; Beaudet and Descarries, 1976; 1979; Chan-Palay, 1975; 1977; Palkovits *et al.*, 1977; Terneaux *et al.*, 1977; Saavedra, 1977; Oliveras *et al.*, 1977; Descarries and Beaudet,

Fig. 2 5-HT-immunoreactive cell bodies corresponding to the group B3 in the upper medulla oblongata of the rat. P, tractus cortico-spinalis (pyramidal tract); NTS, nucleus tractus solitarius; T, tractus spinalis nervi trigemini; Ch, nucleus cochlearis dorsalis. Scale bar = 100 μm

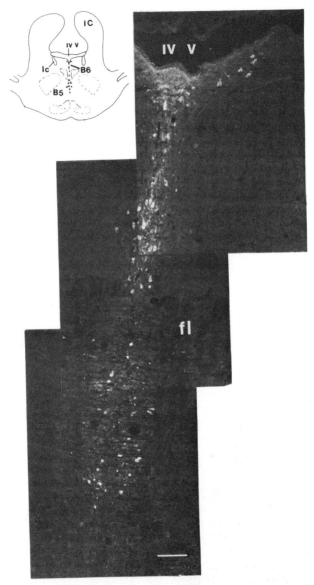

Fig. 3   Groups B5 in the nucleus raphe pontis and B6 in the lower mesencephalon
of the rat as revealed by YC5/45. IC, inferior colliculus; IV V, fourth ventricle; lc,
locus coeruleus; fl, fasciculus longitudinalis medialis. Scale bar = 100 μm

1978; Halasz *et al.,* 1978; Léger and Descarries, 1978; Segu and Calas, 1978; Ruda
and Gobel, 1980) have expanded this map and provided new data, including the
presence of fibres in areas such as the median eminence (Calas *et al.,* 1974; Baum-

Fig. 4 Composite micrograph showing the distribution of serotonergic cell groups B7 (in the nucleus raphe dorsalis), B8 and B9 (within the lemniscus medialis). IC, inferior colliculus, po, nuclei pontis; f1, fasciculus longitudinalis medialis; sgc, substantia grisea centralis. Scale bar = 100 μm

garten and Lachenmayer, 1972), the subcommisural organ (Møllgård and Wiklund, 1979), and a system of supraependymal serotonergic fibres (Lorez and Richards, 1973; Richards *et al.,* 1973; Richards, 1977; Chan-Palay, 1976). A precise account of the serotonin-containing fibres in the CNS as revealed by immunofluorescence has been produced by Steinbusch and Nieuwenhuys (1980), Steinbusch, (1981) and for cortical structures by Lidov *et al.,* (1980). Readers are referred to those publications for more detailed information.

In Table 1 we summarize the main CNS territories which receive 5-HT afferents as described by the above authors and illustrate these in Figs. 5-12 with our preliminary observations applying YC5/45 HL.

Serotonergic pathways

The prevailing ideas on the organization of the main pathways originated from the 'B' serotonergic groups are outlined below, based on results obtained by Carlsson *et al.,* 1964; Dahlstrom and Fuxe, 1964; Ungerstedt, 1971; Kuhar *et al.,* 1971, 1972; Morgane and Stern, 1974; Lorens and Guldberg, 1974; Anden *et al.,* 1966, 1967; Fuxe and Jonsson, 1974; Moore and Halaris, 1975; Conrad *et al.,* 1974; Azmitia and Seagal, 1974, 1978; Bobillier *et al.,* 1976, 1979; Taber-Pierce *et al.,* 1976; Kellar *et al.,* 1977; Palkovits *et al.,* 1977; Jacobs *et al.,* 1978; Moore *et al.,* 1978; Basbaum *et al.,* 1978; Nikesley *et al.,* 1978; Leichnetz *et al.,* 1978; Van de Kar and Lorens, 1979; Azmitia 1978; Parent *et al.,* 1981; Steinbusch *et al.,* 1980; Steinbusch 1981; Priestley *et al.,* 1981; Osborne *et al.,* 1981; Del Fiacco and Cuello (in preparation). A schematic representation of the main CNS serotonergic pathways is illustrated in Fig. 17.

Projections from B1, B2 and B3 groups

Cells of these groups send their fibres down to the spinal cord and end with terminals in the grey matter. These axons descend in the dorsal part of the lateral funiculus to innervate the dorsal horn and probably the substantia gelatinosa of the spinal nucleus of the trigeminal nerve. Other axons descend in the medial part of the anterior funiculus and in the anterior part of the lateral funiculus to innervate the ventral horn. This bulbospinal serotonin pathway has been confirmed by the orthograde axoplasmic transport of labelled substances and the retrograde transport of enzyme horseradish peroxidase. Nevertheless the differential contribution of each nuclei is still uncertain. More detailed information is available about the projections from B3 group, namely the raphe magnus nucleus. By autoradiography of previously incorporated [^{3}H] 5-HT some ascending and descending projections from this cell group have been traced in the cat. The ascending projections are organized into two main systems: a dorsal bundle and a main vertical ascending bundle. The dorsal bundle brings fibres to the substantia grisea centralis, to the locus coeruleus and the dorsolateral and ventromedial part of the brachium con-

Table 1

I. *TELENCEPHALON*

(a) Olfactory areas:
nuclei olfactorii anteriores
tuberculum olfactorium, inner layers
olfactory bulb (delicate network in lamina glomerulosa, see Fig. 9)
(b) Cortex
neocortex (uniformly throughout the layers)
gyrus cinguli (in posterior region restricted to layers I and III)
claustrum (moderate)
cortex entorhinalis and cortex piriformis
(c) Hippocampus
(d) Amygdala—stria terminalis
nucleus amygdaloideus anterior
nucleus amygdaloideus medialis (moderate to dense)
nucleus amygdaloideus corticalis
nucleus amygdaloideus basalis (dense network)
nucleus amygdaloideus lateral
nucleus interstitialis striae terminalis
(e) Basal ganglia
globus pallidus (very dense network)
caudate putamen (moderate density)
nucleus accumbens (medial aspects)
nucleus entopeduncularis
(f) Septum
nucleus lateralis septi (abundant fibres)
nucleus medialis septi

II. *DIENCEPHALON*

(a) Thalamus, metathalamus and epithalamus
nuclei habenulae
nucleus reuniens
mamillary body
nucleus dorsalis corporis geniculati lateralis (moderate)
nucleus ventralis corporis geniculati lateralis (moderate)
nucleus corporis geniculati medialis
nucleus anterior ventralis thalami (moderate)
nucleus periventricularis (moderate to dense)
nucleus lateralis thalami
(b) Nucleus subthalamicus
(c) Hypothalamus
nucleus mamillaris (dense network)
nucleus preopticus suprachiasmaticus (dense network)
nucleus suprachiasmaticus (dense network) (Fig. 6)
nucleus ventromedialis hypothalami (moderate to dense network)
nucleus preopticus lateralis
nucleus preopticus medialis
medial forebrain bundle

(Cont. overleaf)

Table 1 *(Cont.)*

III. *MESENCEPHALON*

 pretectal region
 colliculus inferior (superficial layers)
 colliculus superior
 substantia nigra (very dense homogeneous network in pars reticulata)
 nucleus interpeduncularis
 substantia grisea centralis

IV. *RHOMBOENCEPHALON*

 nucleus tegmenti dorsalis
 nucleus coeruleus (dense network)
 nucleus olivaris inferior
 nucleus cuneatus
 nucleus gracilis
 nucleus linearis caudalis
 nucleus intercalatus
 nucleus commissuralis
 nucleus tractus solitarius (Fig. 11)
 nucleus dorsalis nervi vagi
 nucleus salivatorius
 nucleus ambigus
 nucleus nervi facialis
 nucleus motorius nervi trigemini
 nucleus nervi hypoglossi
 nucleus nervi abducenti
 reticular formation (Fig. 10)
 nuclei raphe (all neuclei of this system display an almost continuous network of fibres)
 substantia gelatinosa trigemini

V. *SPINAL CORD*

 dorsal horn (restricted mostly to substantia gelatinosa)
 lateral horn
 ventral horn (loose fibre network) (Fig. 12)

junctivum. The main ascending ventral bundle goes to the interpeduncular nucleus and to the central area of the reticular formation. Branches of this bundle seem also to innervate the superior colliculus, the substantia grisea centralis, the pretectal area, the thalamic area, the hypothalamic region and the diagonal band of Broca. The inferior cerebellar peduncle seems to have a contribution of serotonergic fibres coming from the B3 group which terminate in the cerebellar cortex.

 No data are available about the serotonergic projections of the B4 group.

Fig. 5 Serotonin immunoreactive fibres in the rat neocortex. Scale bar = 50 μm

Fig. 6 Intense immunofluorescence in the preoptic region
of the rat hypothalamus restricted to the nucleus suprachiasmaticus (NSCh)

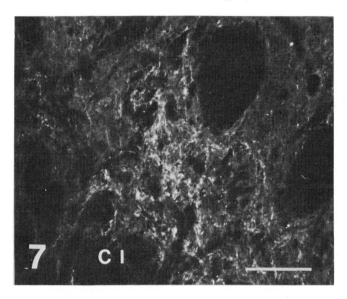

Fig. 7 Serotonin immunofluorescence in the rat globus pallidus. CI, capsula interna. Scale bar = 100 μm

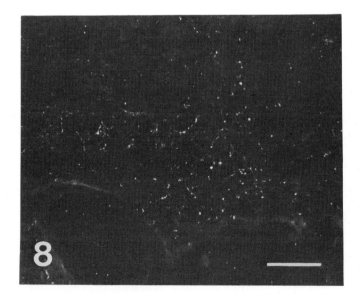

Fig. 8 5-HT-immunoreactive fibres in the olfactory tuberculum of the rat. Scale bar = 50 μm

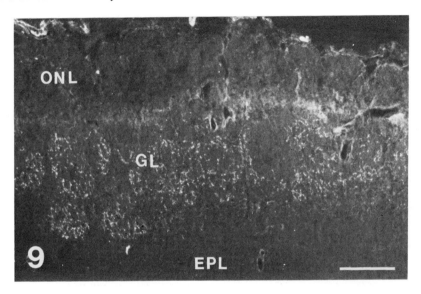

Fig. 9   Serotonin-immunofluorescent fibres restricted to the glomerular layer (GL) of the rat olfactory bulb. ONL, olfactory nerve layer; EPL, external plexiform layer. Scale bar = 100 μm

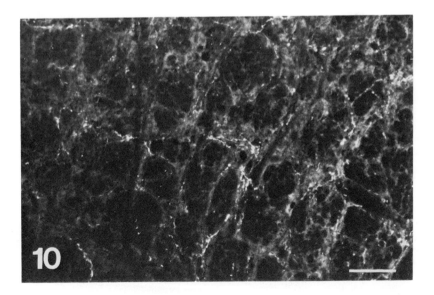

Fig. 10   Network of serotonergic fibres in the reticular formation (nucleus reticularis gigantocellularis) of the rat brain stem. Scale bar = 50 μm

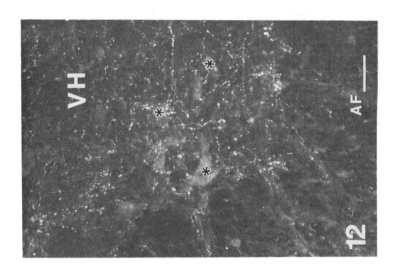

Fig. 11 Serotonergic innervation of the lower medulla oblongata around the ependimal canal (E). Abundant fluorescent fibres in the nucleus of the tractus solitarius (NTS), FG, fasciculus gracilis. Scale bar = 100 μm

Fig. 12 Network of 5-HT-immunoreactive fibres in the ventral horn (VH) of the rat spinal cord. Asterisks denote location of motoneurones. AF, anterior fasciculus. Scale bar = 50 μm

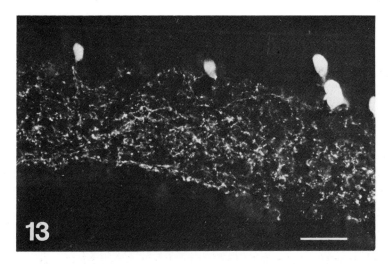

Fig. 13   Amacrine cell bodies immunoreactive to YC5/45 in the internal nuclear layer of the frog retina. Dendrites of these cells have profuse arborizations in the internal plexiform layer. Scale bar = 20 µm

Projections from B5 group

The serotonin containing axons from B5 group, localized in the nucleus raphe pontis, have been considered in more detail. The ascending projections are spread through various systems. A dorsal ascending bundle, running in the raphe and in the fasciculus longitudinalis medialis leaves fibres to the substantia grisea centralis, the locus coeruleus, the brachium conjunctivum, the nucleus periventricularis anterior and the preoculomotor complex. A cerebellar bundle, going to the cerebellar cortex and cerebellar nuclei through the brachium pontis. Fibres passing through the mesencephalic reticular formation ramify in the substantia grisea centralis, in the nucleus raphe dorsalis and in the superior colliculus. Other fibres of this main bundle reach the pretectal region and the thalamic nuclei, the nucleus paraventricularis hypothalami and the nuclei supraopticus and suprachiasmaticus, the septum and the diagonal band of Broca. Few fibres project to the neocortex and few others to the nucleus amygdaloideus centralis medialis.

The descending projections from raphe pontis seem to be also organized in three bundles. The dorsal fibres reach along the fasciculus longitudinalis medialis the periventricular grey matter. The lateral fibres disseminate in most of the nuclei of the pontine and bulbar reticular formation. The paramedian fibres descend in the cervical medulla to supply, via the funiculus ventralis, the ventral horn along the tractus cerebrospinalis ventralis, and the tractus tectospinalis. As for the B4 group, B6 group projections have not been clearly established.

Projections from the B7 group

We can schematically distinguish three ascending systems of fibres:

(1) The dorsal system follows the lateral and ventral part of the aqueductus Silvii and terminates in the substantia grisea centralis of the mesencephalon and diencephalon.

(2) The medial system is characterized by fibres which cross the midline through the decussatio brachii conjunctivi to ramify either in the reticular formation of mesancephalon or in the medial lemniscus, along the dorsal border of the substantia nigra.

(3) The ventrolateral system includes the most important bundle leaving the B7 group. This bundle ascends through the fasciculus longitudinalis medialis to run vertically along each side of the midline, down to the dorsolateral border of the nucleus interpeduncularis and to join then the ventral tegmental area of Tsai and the substantia nigra (see Fig. 14).

At the level of the mesencephalic-diencephalic junction this bundle, going always ventrally and laterally, passes the zona incerta where it splits into two components in the fields of Forel to distribute in the hypothalamic regions following the medial forebrain bundle. In the hypothalamic region some fibres radiate ventrally to the nucleus supraopticus and suprachiasmaticus. The dorsal and lateral parts of this bundle contribute to innervate the habenula lateralis and the medial thalamic nuclei. Other lateral fibres distribute either through the zona incerta in the nucleus geniculatus lateralis or through the ansa lenticularis in the area amygdaloidea anterior and the cortex pyriformis, with some collaterals diffusing through the nucleus entopeduncularis in the globus pallidus. The ventrolateral system is also responsible for the innervation, with its dorsal part of the nuclei septi, nucleus accumbens, the nucleus caudate-putamen, the gyrus cinguli, the olfactory tuberculum, the olfactory bulb and the neocortex. The localization of cells from this group projecting unilaterally to the corpus striatum is represented in Fig. 15.

The descending projections from B7 group (caudal part) distribute to the locus coeruleus, the nucleus tegmenti dorsalis and the brachium conjuntivum. A small contigent, following the fasciculus longitudinalis medialis, reach the nucleus olivaris inferior.

Projections from B8 group

The ascending systems of the B8 group are similar to those of B7 group.

(1) The dorsal system ascends vertically to reach the substantia grisea centralis of the mesencephalon.

(2) The medial system is mainly responsible for the innervation of the reticular formation of mesencephalon.

(3) The ventrolateral system is the most important and complex.

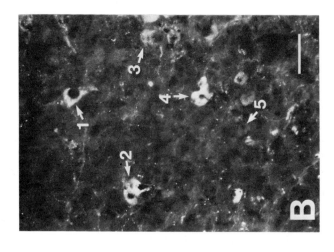

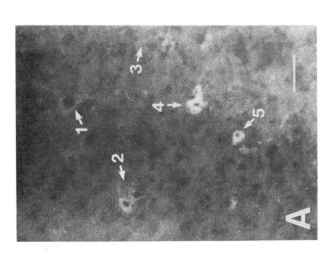

Fig. 14 (A) Neuronal cell bodies in the nucleus raphe dorsalis of the rat, filled with retrogradely transported True Blue injected in the ipsilateral substantia nigra. (B) Same section and field as (A). 5-HT-immunofluorescence is observed in some but not all neurones projecting to the substantia nigra. Arrows and numbers indicate position of the same neurones in (A) and (B). Scale bar = 50 μm. (From M. del Fiacco and ACC, unpublished observations)

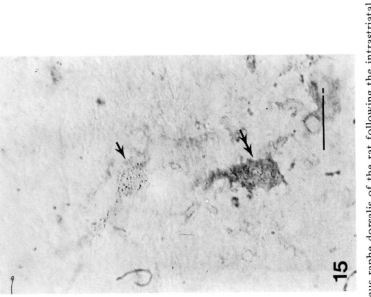

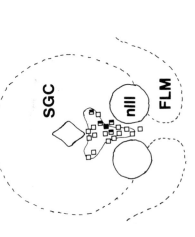

Fig. 15 Nucleus raphe dorsalis of the rat following the intrastriatal injection of horseradish peroxidase (HRP). Identification of HRP retrogradely labelled and 5-HT immunostained neurones with the peroxidase antiperoxidase method (PAP) in a single section. Interference contrast illumination. Single arrow head: HRP retrogradely labelled neurone without immunostaining. Double arrow head: 5-HT immunostained cell with retrograde HRP labelling. Scale bar = 50 μm. Diagram represents approximate localization of HRP retrogradely labelled and 5-HT immunostained neurones within the nucleus raphe dorsalis. Open square: 5-HT immunostained cells without retrograde labelling. Filled squares: HRP retrogradely labelled cells without immunostaining. Half-filled squares: 5-HT-immunostained cells with retrograde HRP labelling (dually labelled neurones). SGC, substantia grisea centralis; FLM, fasciculus longitudinalis medialis; nIII, nucleus originis nervi oculomotorii. Scale bar = 50 μm. (From Priestly et al., 1981)

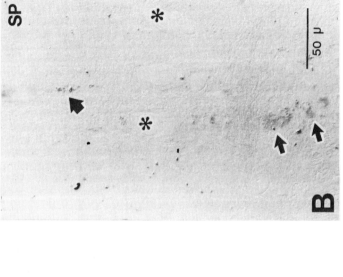

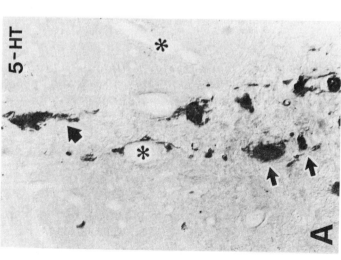

Fig. 16   (A) Neurones from the group B3 immunostained (PAP) for serotonin. (B) Sequential 5 μm-thick section immunostained for substance P, using monoclonal antibodies YC5/45 and NC1/34, respectively. Arrow indicates position of neurones containing both serotonin (5-HT) and substance P (SP). Asterisks denote positions of same blood vessels in (A) and (B). (From Priestley *et al*, 1980)

Several fibres ascend through the decussatio brachii conjunctivi, deviating laterally towards the ventral area of Tsai. The nucleus interpeduncularis, the substantia nigra and the mamillary body are richly innervated. This system supplied even a dense innervation of various thalamic structures, the nucleus entopeduncularis, the nucleus habenulae lateralis and, more rostrally, the gyrus cinguli, hippocampus and cortex entorhinalis. The descending projections from B8 group can be grouped into three bundles.

(i) The medial cerebellar bundles, originating from the ventral cells of this group, moves laterally in a rostro-caudal direction, to distribute into the nuclei of the cerebellum and the cortex cerebelli.

(ii) The dorsal bundle ascends from the dorsal cells of this group inside the fasciculus longitudinalis medialis and distributes in the substantia grisea centralis. This system sends several collaterals which reach the nucleus tegmenti dorsalis, the locus coeruleus, the dorsolateral and ventromedial nuclei of the brachium conjunctivum, and more caudally, the genu nervi facialis and the nucleus intercalatus.

(iii) The paramedian bundle includes an important contingent of fibres coming from the caudal part of the B.8 group. These fibres cross the reticular formation, the nucleus raphe pontis and raphe magnus. They run dorsally along the fasciculus longitudinalis medialis. More caudally they divide into two distinct contingents, the first one going to the nucleus raphe pallidus and obscurus, and the second one diffusing into the reticular region. At the level of the pons and medulla many collaterals reach some cranial nerve nuclei and the nucleus olivaris inferior.

Projections from B9 group

Projections from this group are not certain. It seems that fibres from the cells of this group ascend laterally and through the internal capsule to reach higher structures, including the basal ganglia.

CONCLUSIONS

While most of the 5-HT-containing cell bodies are restricted to the brain stem, the nerve terminal networks extend from the olfactory bulb to the spinal cord. Besides some dorsally ascending projections there is at least one well-defined ventral ascending tract which runs in the proximity of the medial forebrain bundle. This tract is supplied by various nuclei of the 'B' classification (see scheme of Fig. 17) and distributes fibres to a large variety of structures in the diencephalon and telencephalon: the cortex, hippocampus, the basal ganglia, olfactory bulb and tubercle, the amygdala and thalamic hypothalamic nuclei. Descending fibres are innervating various structures of the medulla oblongata and spinal cord, very prominently the substantia gelatinosa of the spinal nucleus of the trigeminal nerve and spinal cord. Fibres

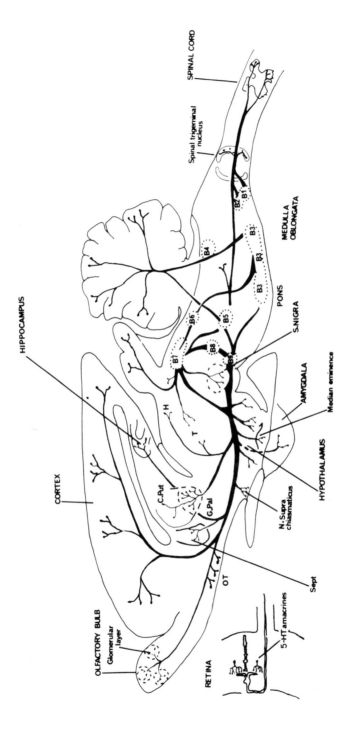

Fig. 17 Schematic drawing representing the location of the 'B-groups' in a sagittal section of the rat central nervous system and their major projections. OT, olfactory tuberculum; sept, septum; C.Put, nucleus caudate-putamen; G.Pal globus pallidus; T, thalamus; H, habenula

from B3 and B5 arrive in the cerebellum via the inferior and middle cerebellar peduncles. The complicated system, and various mesencephalic and positive nuclei are less well defined. The ventral ascending tract provides fibres to the substantia nigra originating in B7 and probably B5 and B8. Methodological limitations have, up until now, prevented the production of a detailed and uncontroversial account for every serotonergic projection. Recent developments such as dual tracing of retrogradely transported substances and immunocytochemistry (Steinbusch *et al.*, 1980; Priestley *et al.*, 1981) could contribute towards a more definitive account of these CNS pathways.

Finally, the most relevant task will be to ascribe, whenever possible, a functional role to each of the projections. Serotonin seems to be involved in a variety of physiological functions such as the control of sleep rhythms, sexual behaviour, mood, thermoregulation and pain. There are hints that some of these projections could have a defined role in the above functions. For instance it is increasingly clear that the descending serotonergic projections exert an inhibitory role on pain processing mechanisms (Akil and Liebeskind, 1975; Yaksh *et al.*, 1976), but the fine synaptology of this is still unsettled.

The fact that neurones containing serotonin also contain peptides with transmitter-like characteristics (Chan-Palay *et al.*, 1978; Hökfelt *et al.*, 1978; Priestley *et al.*, 1980) adds further interest to the anatomical functional organization of the CNS serotonergic pathways (see Fig. 16). It is therefore very likely that future analysis of the CNS serotonergic pathways will have to be considered taking into account which subsets of 5-HT-containing neurones contain which peptide and the pattern of distribution throughout the mammalian CNS.

ACKNOWLEDGEMENTS

We would like to thank Drs Del Fiacco and Priestley for providing unpublished material. The technical assistance of S. Bramwell, T. Barclay, T. Richards, and B. Archer, and the secretarial help of Mrs Ella Iles is also gratefully acknowledged. ACC wishes to acknowledge grants from the Medical Research Council (UK). The Wellcome Trust and the Royal Society. AC was a British Council Scholar and permanent address is 'Mario Negri', Institute of Pharmacological Research, Milan, Italy.

REFERENCES

Aghajanian, G. K. and Asher, I. M. (1971). Histochemical fluorescence of raphe neurons: selective enhancement by tryptophan, *Science,* **172,** 1159-1161.

Aghajanian, G. K. and Bloom, F. E. (1967). Localisation of tritiated serotonin in rat brain by electron-microscopic autoradiography, *J. Pharmacol. Exp. Ther.* **156,** 23-30.

Aghajanian, G. K., Kuhar, M. J., and Roth, R. H. (1973). Serotonin-containing neuronal perikarya and terminals: differential effects of *p*-chlorophenylalanine, *Brain Res.* **54,** 85-101.

Akil, H. and Liebeskind, J. C. (1975) Monoaminergic mechanisms of stimulation-produced analgesia, *Brain Res.* **94,** 279-296.

Amin, A. H., Crawford, T. B. B., and Gaddum, J. H. (1954). The distribution of substance P and 5-hydroxytryptamine in the central nervous system of the dog, *J. Physiol.* **126**, 596–618.

Anden, N. E., Dahlstrom, A., Fuxe, K., Larsson, K., Olsen, L., and Ungerstedt, U. Y. (1966). Ascending monoamine neurones to the telencephalon and diencephalon, *Acta Physiol. Scand.* **67**, 313–326.

Anden, N. E., Fuxe, K., and Ungerstedt, U. (1967). Monoamine pathways to the cerebellar and cerebral cortex. *Experientia*, **23**, 838–839.

Avrameas, S. (1970). Immunoenzyme techniques: enzymes as markers for the localisation of antigens and antibodies. *Int. Rev. Cytol.* **27**, 349–385.

Azmitia, E. C. (1978). The serotonin-producing neurones in the midbrain median and dorsal raphe nuclei, in *Handbook of Psychopharmacology* Vol. 9. (Eds. L. L. Iversen, S. D. Iversen, and S. H. Snyder) pp. 233–314, Plenum Press, New York.

Azmitia, E. C. and Segal, M. (1974). Biochemical and autoradiographic study of median and dorsal raphe projection and transport rate using radioactive proline, *Neuroscience 4th Annual Meeting*, Abs. 32. pp. 124.

Azmitia, E. C. and Segal, M. (1978). An autoradiographic analysis of the differential ascending projections of the dorsal and median raphe nuclei in the rat,*J. Comp. Neurol.* **179**, 641–668.

Basbaum, A. I., Clanton, C. H., and Field, H. L. (1978). Three bulbospinal pathways from the rostral medulla of the cat: an autoradiographic study on pain modulating systems,*J. Comp. Neurol.* **178**, 209–224.

Baumgarten, H. G. and Lachenmayer, L. (1972). Chemically induced degeneration of indoleamine-containing nerve terminals in rat brain, *Brain Res.* **38**, 228–232.

Baumgarten, H. G., Bjorklund, A., Nobin, A., Rosengren, E., and Schlossberger, H. G. (1976). Neurotoxicity of hydroxylated tryptamines: structure-activity relationships. 1. Long-term effects on monamine content and fluorescence morphology of central monoamine neurons, *Acta Phys. Scand.* **96**, Suppl. 429, 5–27.

Beaudet, A. and Descarries, L. (1976). Quantitative data on serotonin nerve terminals in adult rat neocortex, *Brain Res.* **111**, 301–309.

Beaudet, A. and Descarries, L. (1979). Radioautographic characterisation of a serotonin-accumulating nerve cell group in adult rat hypothalamus, *Brain Res.* **160**, 231–243.

Bloom, F. E., Hoffer, B. J., Siggins, G. R., Barker, J. L., and Nicoll, R. A. (1972). Effects of serotonin on central neurons: microintophoretic administration. *Fed. Proc.* **31**, 97–106.

Bobillier, P., Seguin, S., Degueurce, A., Lewis, B. D., and Pujol, J. F. (1979). The efferent connections of the nucleus raphe centralis superior in the rat as revealed by radioautography. *Brain. Res.* **166**, 1–8.

Bobillier, P., Seguin, S., Petithjean, F., Salvert, D., Touret, M., and Jouvet, M. (1976). The raphe nuclei of the cat brainstem: a topographical atlas of their efferent projections as revealed by autoradiography. *Brain Res.* **113**, 449–486.

Bogdanski, D. F., Weissbach, H., and Udenfriend, S. (1957). The distribution of serotonin, 5-hydroxytryptophan decarboxylase and monoamine oxidase in brain,*J. Neurochem.* **1**, 272–278.

Calas, A., Alonso, G., Arnauld, E., and Vincent, J. D. (1974). Demonstration of indolaminergic fibres in the median eminence of the duck, rat and monkey, *Nature*, **250**, 242–243.

Calas, A., Besson, M. J., Gauchy, C., Alonso, G., Glowinski, J., and Cheramy, A. (1976). Radioautographic study of in vivo incorporation of [^{3}H] monoamines in the cat caudate nucleus: identification of serotonergic fibers, *Brain Res.* **118**, 1–14.

Carlsson, A., Falck, B., Fuxe, K., and Hillarp, N. A. (1964). Cellular localisation of monoamines in the spinal cord, *Acta Physiol. Scand.* **60**, 112-119.

Chan-Palay, V. (1975). Fine structure of labelled axons in the cerebellar cortex and nuclei of rodents and primates after intraventricular infusions with tritiated serotonin. *Anat. Embryol.* **148**, 235-265.

Chan-Palay, V. (1976). Serotonin axons in the supra- and subependymal plexuses and in the leptomeninges; their roles in local alterations of cerebrospinal fluid and vasomotor activity. *Brain Res.* **102**, 103-130.

Chan-Palay, V. (1977). Indoleamine neurons and their processes in the normal rat brain and chronic diet-induced thiamine deficiency demonstrated by uptake of [^{3}H]serotonin, *J. Comp. Neurol.* **176**, 467-494.

Chan-Palay, V., Jonsson, G., and Palay, S. L. (1978). Serotonin and substance P coexist in neurons of the rat's central nervous system, *Proc. Natl. Acad. Sci. (USA)*, **75**, 1582-1586.

Christenson, J. G., Dairman, W., and Udenfriend, S. (1970). Preparation and properties of a homogenous aromatic L-amino acid decarboxylase from hog kidney, *Arch. Biochem. Biophys.* **141**, 356-367.

Conrad, L. C. A., Leonard, C. M., and Pfaff, D. W. (1974). Connections of the median and dorsal raphe nuclei of the rat: an autoradiographic and degeneration study, *J. Comp. Neurol.* **156**, 179-206.

Consolazione, A., Milstein, C., Wright, B., and Cuello, A. C. (1981). Immunocytochemical detection of serotonin with monoclonal antibodies, *J. Histochem. Cytochem.* (in press).

Coons, A. H. (1958) Fluorescent antibody methods, in *General Cytochemical Methods* (Ed. J. F. Danielli) pp. 399-422, Academic Press, New York.

Coons, A. H. and Kaplan, M. H. (1950). Localisation of antigens in tissue cells. II. Improvements in a method for the detection of antigen by means of fluorescent antibody, *J. Exp. Med.* **91**, 1-9.

Corrodi, H. and Jonsson, G. (1967). The formaldehyde fluorescence method for the histochemical demonstration of biogenic monoamines: a review on the methodology. *J. Histochem. Cytochem.* **15**, 65-72.

Cuello, A. C. (1978). Immunocytochemical studies of the distribution of neurotransmitters and related substances in the CNS, in *Handbook of Psychopharmacology* Vol. 9, (Eds. L. L. Iversen, S. D. Iversen and S. H. Snyder), pp. 69-137, Plenum Press, New York.

Cuello, A. C. and Milstein, C. (1981). Monoclonal antibodies against neurotransmitter substances, in *Monoclonal Antibodies Against Neural Antigens*, Cold Spring Harbor, New York (in press).

Daly, J., Fuxe, K. and Jonsson, G. (1973). Effects of intracerebral injections of 5,6-dihydroxytryptamine on central monoamine neurons: evidence for selective degeneration of central 5-hydroxytryptamine neurons, *Brain Res.* **49**, 476-482.

Daly, J., Fuxe, K., and Jonsson, G. (1974). 5,7-dihydroxytryptamine as a tool for the morphological and functional analysis of central 5-hydroxytryptamine neurons, *Res. Comm. Chem. Pathol. Pharmacol.* **1**, 175-187.

Dahlstrom, A. and Fuxe, K. (1964). Evidence for the existence of monoamine-containing neurons in the central nervous system. I. Demonstration of monoamines in the cell bodies of brainstem neurons, *Acta Physiol. Scand.* **62**, suppl. 232, 1-55.

Dahlstrom, A. and Fuxe, K. (1965). Evidence for the existence of monoamine neurons in the central nervous system. II. Experimentally induced changes in the intraneuronal amine levels of bulbospinal neuron system, *Acta Physiol. Scand.* **64**, Suppl. 247, 7-36.

Dahlstrom, A., Haggendal, J., and Atack, C. (1973). Localisation and transport of serotonin, in *Serotonin and Behaviour*. (Eds. J. Barchas and E. Usdin) pp. 87-96, Academic Press, New York.

Descarries, L. and Beaudet, A. (1978). The serotonin innervation of adult rat hypothalamus, in *Cell Biology of Hypothalamic Neurosecretion* (Eds J. D. Vincent and C. Kordon) Vol. 80 pp. 135-153, Editions du CNRL, Paris.

Droz, B. and Leblond C. P. (1963). Axonal migration of proteins in the central nervous system and peripheral nerves as shown by radioautography, *J. Comp. Neurol.* 121, 325-346.

Falck, B., Hillarp, N. A., Thieme, G., and Thorp, A. (1962). Fluorescence of catecholamines and related compounds with formaldehyde, *J. Histochem. Cytochem.* 10, 348-354.

Felten, D. L., Laties, A. M., and Carpenter, M. B. (1974). Monoamine-containing cell bodies in the Squirrel monkey brain, *Am. J. Anat.* 139, 153-166.

Fuxe, K. (1965). Evidence for the existence of monoamine neurons in the central nervous system. IV. Distribution of monoamine nerve terminals in the central nervous system, *Acta Physiol. Scand.* 64, Suppl. 247, 39-85.

Fuxe, K., Hökfelt, T. and Ungersted, U. (1968). Localisation of indolealkylamines in CNS, in *Advances in Pharmacology* (Eds S. Garattini and P. A. Shore) Vol. 6, Part A, pp. 235-251, Academic Press, New York.

Fuxe, K. and Jonsson, G. (1967) A modification of the histochemical fluorescence method for the improved localisation of 5-hydroxytryptamine, *Histochemie,* 11, 161-166.

Fuxe, K. and Jonsson, G. (1974). Further mapping of central 5-hydroxytryptamine neurons: studies with the neurotoxic dihydroxytryptamines, in *Serotonin—New Vistas. Histochemistry and Pharmacology* Vol. 10 pp. 1-12 (Eds E. Costa, G. L. Gessa and M. Sandler) Raven Press, New York.

Fuxe, K. and Owman, C. (1965). Cellular localisation of monoamines in the area postrema of certain mammals, *J. Comp. Neurol.* 125, 337-354.

Fuxe, K. and Ungerstedt, U. (1968). Histochemical studies on the distribution of catecholamines and 5-hydroxytryptamine after intraventricular injections, *Histochemie,* 13, 16-28.

Galfré, G., Milstein, C., and Wright, B. (1979). Rat x rat hybrid myeloma and a monoclonal anti-Fd portion of mouse IgG, *Nature* (London), 277, 131-133.

Geffen, L. B., Livett, B. G., and Rush, R. A. (1969). Immunohistochemical localisation of protein components of catecholamine storage vesicles, *J. Physiol.* 204, 593-604.

Goldstein, M., Fuxe, K., and Hökfelt, T. (1972). Characterisation and tissue localisation of catecholamine synthesising enzymes, *Pharmacol. Rev.* 24, 293-309.

Goldstein, M., Fuxe, K., Hökfelt, T., and Joh, T. H. (1971). Immunohistochemical studies on phenylethanolamine N-methyltransferase, dopa-decarboxylase and dopamine-beta-hydroxylase, *Experientia,* 27, 951-952.

Graham, R. C. and Karnovsky, M. J. (1966). The early stages of absorption of injected horseradish peroxidase in the proximal tubules of mouse kidney: ultrastructural cytochemistry by a new technique, *J. Histochem. Cytochem.* 14, 291-304.

Grota, L. J. and Brown, G. M. (1974). Antibodies to indolealkylamines: serotonin and melatonin, *Can. J. Biochem.* 52, 196-202.

Halaris, A. E., Jones, B. E., and Moore, R. Y. (1976). Axonal transport in serotonin neurons of the midbrain raphe, *Brain Res.* 107, 555-574.

Halasz, N., Ljungdahl, A., and Hökfelt, T. (1978). Transmitter histochemistry of the rat olfactory bulb. II. Fluorescence histochemical, autoradiographic and electron microscopic localisation of monoamines, *Brain Res.* 154, 253-271.

Hartman, B. K. (1973). Immunofluorescence of dopamine-beta-hydroxylase. Application of improved methodology to the localisation of the peripheral and central noradrenergic nervous system. *J. Histochem. Cytochem.* **21**, 312-332.

Hökfelt, T., Fuxe, K., and Goldstein, M. (1973b). Immunohistochemical localisation of aromatic L-aminoacid decarboxylase (DOPA-decarboxylase) in central dopamine and 5-hydroxytryptamine nerve cell bodies of the rat brain, *Brain Res.* **53**, 175-180.

Hökfelt, T., Fuxe, K., Goldstein, M., and Joh, T. H. (1973a). Immunohistochemical localisation of three catecholamine synthesising enzymes: aspects on methodology, *Histochemie*, **33**, 231-254.

Hökfelt, T., Ljungdahl, A., Steinbusch, H., Verhofstad, A., Nilsson, G., Brodin, E., Pernow, B., and Goldstein, M. (1978). Immunohistochemical evidence of substance P-like immunoreactivity in some 5-hydroxytryptamine-containing neurons in the rat central nervous system, *Neuroscience,* **3**, 517-538.

Hubbard, J. E. and DiCarlo, V. (1974). Fluorescence histochemistry of monoamine-containing cell bodies in the brainstem of the Squirrel monkey (Saimiri siureus). III. Serotonin containing groups, *J. Comp. Neurol.* **153**, 385-398.

Iversen, L. L. (1978). Identification of transmitter-specific neurons in CNS by autoradiography, in *Handbook of Psychopharmacology* Vol. 9, (Eds L. L. Iversen, S. D. Iversen and S. H. Snyder) pp. 41-68, Plenum Press, New York.

Jacobs, B. L., Foote, S. L., and Bloom, F. E. (1978). Differential projections of neurones within the dorsal raphe nucleus of the rat: a horseradish peroxidase (HRP) study, *Brain Res.* **147**, 149-153.

Jacobowitz, D. M. and MacLean, P. D. (1978). A brain stem atlas of catecholaminergic perikarya in a pigmy primate (Cebuella pygmaea), *J. Comp. Neurol.* **177**, 397-416.

Joh, T. H., Shikimi, T., Pickel, V. M., and Reis, D. J. (1975). Brain tryptophan hydroxylase: purification of, production of antibodies to and cellular and ultrastructural localisation in serotonergic neurons of rat midbrain, *Proc. Natl. Acad. Sci. USA,* **72**, 3575-3579.

Jonsson, G., Einarsson, P., Fuxe, K., and Hallman, H. (1974). Microspectrofluorimetric studies on central 5-hydroxytryptamine neurons, in *Serotonin—New Vistas.* Vol. 10. *Histochemistry and Pharmacology.* (Eds E. Costa, G. L. Gessa, and M. Sandler) Raven Press, New York.

Jonsson, G., Fuxe, K., Hamberger, B., and Hökfelt, T. (1969). 6-Hydroxytryptamine—a new tool in monamine fluorescence histochemistry, *Brain Res.* **13**, 190-195.

Kellar, K. J., Brown, P. A., Madrid, J., Bernstein, M., Vernikos-Dannelis, J., and Mehler, W. R. (1977). Origins of serotonin innervation of forebrain structures, *Exp. Neurol.* **56**, 52-62.

Kent, D. L. and Sladek, J. R. (1978). Histochemical, pharmacological and microspectrofluorimetric analysis of new sites of serotonin localisation in the rat hypothalamus, *J. Comp. Neurol.* **180**, 221-236.

Kneisley, L. W., Biber, M. P., and La Vail, J. H. (1978). A study of the origin of brainstem projections to monkey spinal cord using the retrograde transport method, *Exp. Neurol.* **60**, 116-139.

Köhler, G. and Milstein, C. (1975). Continuous cultures of fused cells secreting antibody of predefined specificity, *Nature,* (London), **256**, 495-497.

Kuhar, M. J., Aghajanian, G. K., and Roth, R. H. (1972). Tryptophan hydroxylase activity and synaptosomal uptake of serotonin in discrete brain regions after midbrain raphe lesions: correlations with serotonin level and histochemical fluorescence, *Brain Res.* **44**, 165-176.

Kuhar, M. J., Roth, R. H., and Aghajanian, G. K. (1971). Selective reduction of tryptophan hydroxylase activity in rat forebrain after midbrain raphe lesions, *Brain Res.* **35**, 167–176.

Lasek, R. J., Joseph, B. S., and Whitlock, D. G. (1968). Evaluation of a radioautographic neuroanatomical tracing method, *Brain Res.* **8**, 319–336.

Leichnetz, G. R., Watkins, L., Griffin, G., Murfin, R., and Mayer, D. J. (1978). The projections from nucleus raphe magnus and other brainstem nuclei to the spinal cord in the rat: a study using the HRP blue reaction, *Neurosci. Lett.* **8**, 119–124.

Léger, L. and Descarries, L. (1978). Serotonin nerve terminals in the locus coeruleus of adult rat: a radioautographic study, *Brain Res.* **145**, 1–13.

Levitt, P. and Moore, R. Y. (1978). Developmental organisation of raphe serotonin neuron groups in the rat, *Anat. Embryol.* **154**, 241–251.

Lidov, H. G. W., Grzanna, R., and Molliver, M. E. (1980). The serotonin innervation of cerebral cortex in the rat—an immunohistochemical analysis, *Neuroscience* **5**, 207–227.

Lorez, H. P. and Richards, J. G. (1973). Distribution of indolealkylamine nerve terminals in the ventricles of the rat brain. *Z. Zellforsch.* **144**, 511–522.

Lorens, S. A. and Guldberg, H. C. (1974). Regional 5-hydroxytryptamine following selective midbrain raphe lesions in the rat, *Brain Res.* **78**, 45–56.

Lovenberg, W., Weisback, H., and Udenfriend, S. (1962). Aromatic-L-amino acid decarboxylase, *J. Biol. Chem.* **237**, 89–93.

MølIgård, K. and Wiklund, L. (1979). Serotonergic synpases on ependymal and hypendymal cells of the rat subcommissural organ, *J. Neurocytol.* **8**, 445–467.

Moore, R. Y. and Halaris, A. (1975). Hippocampal innervation by serotonin neurons of the midbrain raphe in the cat, *J. Comp. Neurol.* **164**, 171–184.

Moore, R. Y., Halaris, A. E., and Jones, B. E. (1978). Serotonin neurons of the midbrain raphe: ascending projections, *J. Comp. Neurol.* **180**, 417–438.

Morgane, P. J. and Stern, W. C. (1974). Chemical anatomy of brain circuits in relation to sleep and wakefulness, in *Advances in Sleep Research.* Vol. 1 (Ed E. Weitzman) pp. 1–131, Spectrum, New York.

Nakane, P. K. and Pierce, G. B. (1966). Enzyme-labelled antibodies: preparations and application for the localisation of antigens, *J. Histochem. Cytochem.* **14**, 929–931.

Nakane, P. K. and Pierce, G. B. (1967). Enzyme-labelled antibodies for the light and electron microscopic localisation of tissue antigens,*J. Cell. Biol.* **33**, 307–318.

Oliveras, J. L., Bourgoin, S., Hery, F., Besson, J. M., and Hamon, M. (1977). The topographical distribution of serotonergic terminals in the spinal cord of the cat: biochemical mapping by combined use of microdissection and microassay procedures, *Brain Res.* **138**, 393–406.

Osborne, N. N., Nesselhut, T., Nicholas, D. A., and Cuello, A. C. (1981). Serotonin: a transmitter candidate in the vertebrate retina,*Neurochem. Internat.* **3**, 171–176.

Palkovits, M., Saavedra, J. M., Jacobowitz, D. M., Kizer, J. S., Zaborzsky, L., and Brownstein, M. J. (1977). Serotonergic innervation of the forebrain: effect of lesions on serotonin and tryptophan hydroxylase levels, *Brain Res.* **130**, 121–134.

Parent, A., Descarries, L., and Beaudet, A. ((1981). Organisation of ascending serotonin systems in the adult rat brain: A radioautographic study after intraventricular administration of [3]H 5HT, *Neuroscience* **6**, 115–138.

Peskar, B. and Spector, S. (1973). Serotonin: radioimmunoassay, *Science,* **179**, 1340–1341.

Pickel, V. M. (1980). Immunocytochemical methods, in *Neuroanatomical Tract Tracing Methods* (Eds L. Heimer and M. J. Robards). Plenum Press, New York.

Pickel, V. M., Joh, T. H., and Reis, D. J. (1976). Monoamine synthesising enzymes in central dopaminergic noradrenergic and serotonergic neurons. Immunocytochemical localisation by light and electron microscopy, *J. Histochem. Cytochem.* **24,** 792–806.

Pin, C., Jones, B., and Jouvet, M. (1968). Topographie des neurons monoaminergie ques du tronc cerebral du chatetude par histofluorescence, *Compt. Rend. Soc. Biol.* **162,** 2136–2141.

Priestley, J. V., Consolazione, A., and Cuello, A. C. (1980). Identification of substance P and serotonin containing neurons in the CNS with monoclonal antibodies. *VIth International Histochemistry Cytochemistry Congress.* Abstract.

Priestley, J. V. and Cuello, A. C. (1981). Electronmicroscopic immunocytochemistry. CNS transmitters and transmitter markers, in *IBRO Series of Methods in Neurosciences, Immunohistochemistry.* (Ed. A. C. Cuello), John Wiley, Chichester (in press).

Priestley, J. V., Somogyi, P., and Cuello, A. C. (1981). Neurotransmitter specific projection neurons revealed by combining PAP immunocytochemistry with retrograde transport of HRP, *Brain Res.* (In press.)

Ranadive, N. S. and Sehon, A. H. (1967). Antibodies to serotonin, *Can. J. Biochem.* **45,** 1701–1710.

Richards, J. G. (1977). Autoradiographic evidence for the selective accumulation of ³H 5-HT by supra-ependymal nerve terminals, *Brain Res.* **134,** 151–157.

Richards, J. G., Lorez, H. P., and Transzer, J P. (1973). Indolealkylamine nerve terminals in cerebral ventricles: identification by electron microscopy and fluorescence histochemistry, *Brain Res.*, **57,** 277–288.

Ruda, M. A. and Gobel, S. (1980). Ultrastructural characterisation of aconal endings in the substantia gelatinosa which take up ³H-serotonin, *Brain Res.* **184,** 57–83.

Saavedra, J. M. (1977). Distribution of serotonin and synthesising enzymes in discrete areas of the brain, *Fed. Proc.* **36,** 2134–2141.

Segu, L. and Calas, A. (1978). The topographical distribution of serotonergic terminals in the spinal cord: quantitative radio-autographic studies, *Brain Res.* **153,** 449–464.

Shaskan, E. A. and Snyder, S. H. (1970). Kinetics of serotonin uptake into slices from different regions of rat brain, *J. Pharmacol. Exp. Ther.* **175,** 404–418.

Sladek, J. R. and Walker, P. (1977). Serotonin-containing neuronal perikarya in the primate locus coeruleus and subcoeruleus, *Brain Res.* **134,** 354–366.

Spector, S. Berkowitz, B., Flynn, E. J., and Peskar, B. (1973). Antibodies to morphine barbiturates and serotonin, *Pharmacol. Rev.* **25,** 281–291.

Steinbusch, H. W. M. (1981). Distribution of serotonin-immunoreactivity in the central nervous system of the rat. Cell bodies and terminals, *Neuroscience,* **6,** 557–618.

Steinbusch, H. W. M. and Nieuwenhuys, R. (1980). Localisation of serotonin-like immunoreactivity in the central nervous system and pituitary of the rat, with special references to the innervation of the hypothalamus, in *Serotonin: Current Aspects of Neurochemistry and Functions.* (Eds B. Haber, S. Gabay, M. Issidorides, and S. G. A. Alivisatos) Plenum, New York (in press).

Steinbusch, H. W. M., Van der Kooy, D., Verhofstad, A. A. J., and Pellegrino, A. (1980). Serotonergic and non-serotonergic projections from the nucleus raphe dorsalis to the caudate-putamen complex in the rat, studied by a combined immunofluorescence and fluorescent retrograde axonal labelling technique, *Neurosci. Lett.* **19,** 137–142.

Steinbusch, H. W. M., Verhofstad, A. A. J., and Joosten, H. W. J. (1978). Localisation of serotonin in the central nervous system by immunohistochemistry: description of a specific and sensitive technique and some applications, *Neuroscience,* **3,** 811–819.

Steinbusch, H. W. M., Verhofstad, A. A. J., Hoosten, H. W. J., Penke, B., and Varga, J. (1979). Immunohistochemical characterisation of monoamine-containing neuron populations in the central nervous system using antibodies to serotonin and noradrenaline. A comparative study in the Lamprey (Lampetra fluivatilis) and the rat. *XXI Symposion der Gesellschaft für Histochemie. (Osterraich). 19–22 Sept. 1979.* Abs. v/2.

Sternberger, L. A. (1974). *Immunocytochemistry.* Prentice Hall, Englewood Cliffs, New Jersey.

Sternberger, L. A., Hardy, P. H., Cuculis, J. J., and Meyer, H. G. (1970). The unlabelled antibody-enzyme method of immunohistochemistry-preparation and properties of soluble antigen–antibody complex (horseradish antihorseradish peroxidase) and its use in identification of spirochetes, *J. Histochem. Cytochem.* **18,** 315–333.

Taber-Pierce, E., Foote, W., and Hobson, A. (1976). The efferent connections of the nucleus raphe dorsalis, *Brain Res.* **107,** 137–144.

Terneaux, J. P., Héry, G., Bourgoin, S., Adrien, J., Glowinski, J., and Hamon, M. (1977). The topographical distribution of serotonergic terminals in the neostriatum of the rat and the caudate nucleus of the cat, *Brain Res.* **121,** 311–326.

Ungerstedt, U. (1971). Stereotaxic mapping of the monoamine pathways in the rat brain, *Acta Phys. Scand.* **82,** Suppl. 367, 1–48.

Van der Kar, L. D. and Lorens, S. A. (1979). Differential serotonergic innervation of individual hypothalamic nuclei and other forebrain regions by the dorsal and median midbrain raphe nuclei, *Brain Res.* **162,** 45–54.

Vaughn, J. E. Barber, R. P., Ribak, C. E., and Houser, C. R. (1981). Methods for the immunocytochemical localisation of proteins and peptides involved in neurotransmission in *Current Trends in Morphological Techniques* Vol. III (Ed. J. E. Johnson). C.R.C. Press, Florida (in press).

Yaksh, T. L., Du Chateau, J. C., and Rudy, T. A. (1976). Antagonism by methysergide and cinanserin of the antinociceptive action of morphine administered into the periaqueduct gray, *Brain Res.* **104,** 367–372.

Biology of Serotonergic Transmission
Edited by N. N. Osborne
© 1982, John Wiley & Sons Ltd.

Chapter 3

Biosynthesis of Serotonin

MARGARET C. BOADLE-BIBER

Department of Physiology, Medical College of Virginia,
Virginia Commonwealth University, Richmond, Virginia, USA

INTRODUCTION

In recent years it has become apparent that the rate at which a transmitter is formed in the terminal projections of a particular chemical class of neurones can be modulated over the short or long term by the level of activity (i.e. firing rate) of those neurones. In other words it appears as if the synthesis of transmitters is in some way coupled to transmitter release, possible through a common mediator or signal, and can be adjusted to match the extent of transmitter release and loss. Thus the study of the formation of neurotransmitters aims not only to understand the biosynthetic pathway and chemistry of the individual enzyme reactions involved but also to identify the processes that determine the actual rate of transmitter formation in intact neurones in the living organism under different physiological conditions, and to establish the molecular mechanisms whereby these processes exert their regulatory effects. In the discussion that follows it will be clear that while progress has been made in characterizing the two enzymes involved in the synthesis of serotonin in nervous tissue, much remains to be learnt about the molecular events which modulate its formation *in vivo*. In this account of 5-HT synthesis, emphasis will be placed on areas which have provided new insights into regulatory processes or which seem likely to do so in the near future. A brief account will also be given of serotonin metabolism in the pineal gland. Most of the experimental evidence that will be discussed has been obtained with the rat brain.

PATHWAY AND ENZYMES

Serotonin is formed from L-tryptophan in a two step reaction in which L-tryptophan is first converted to L-5-hydroxytryptophan (5-HTP) by the aromatic amino acid hydroxylase. L-Tryptophan-5-monooxygenase, or tryptophan hydroxylase (EC 1.14.16.4) and the 5-hydroxytryptophan is then decarboxylated to 5-hydroxytryptamine (serotonin) by aromatic L-amino acid decarboxylase (EC 4.1.1.28, Fig. 1). The initial hydroxylation reaction involves two substrates, L-tryptophan and molecular oxygen, as well as a reduced pterin cofactor, now considered to be L-erythro-tetrahydrobiopterin (2-amino-4-hydroxy-6(L-erythro-1'2'-dihydroxypropyl)-5,6,7,8-tetrahydropteridine (Fig. 2). In this hydroxylation reaction one atom of molecular oxygen is used in the 5-hydroxylation of tryptophan while the other is reduced to water. Electrons are donated by the tetrahydrobiopterin cofactor, and the unstable quinonoid dihydropterin that results is immediately regenerated to the tetrahydropterin form by NADH (NADPH)-linked quinonoid dihydropteridine reductase (Fig. 3) (Kaufman, 1974; Kaufman and Fisher, 1974; Kaufman, 1977a). For every mole of tetrahydrobiopterin that is oxidized, one mole of 5-hydroxtryptophan is formed so that the reaction can be represented thus:

$$\text{L-tryptophan} + BH_4 + O_2 \rightarrow \text{L-5-hydroxytryptophan} + \text{quinonoid } BH_2 + H_2O.$$

Since the cofactor constantly shuttles between oxidized and reduced forms it acts as a true catalyst in this reaction (Friedman *et al.* 1972). Tryptophan hydroxylase can utilize a number of different unconjugated reduced pterin cofactors including 6-methyl and 6,7-dimethyl-5,6,7,8-tetrahydropterin (Fig. 2; Friedman *et al.*, 1972; Kato *et al.*, 1980). However, of all the congeners and their isomers tested so far, the enzyme from rat brain stem has the highest affinity for L-erythro-tetrahydrobiopterin. Although the affinity of the enzyme for 6-MPH$_4$ is only about one fourth of that for tetrahydrobiopterin, the V_{max} observed in the presence of this synthetic

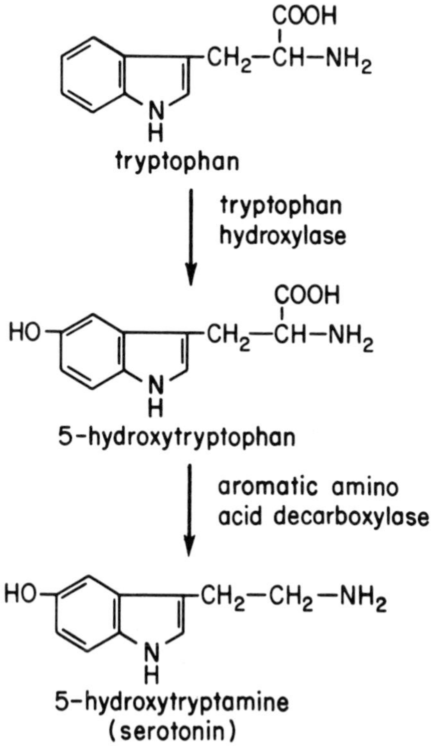

Fig. 1 Biosynthesis of serotonin

cofactor is nearly as high as that with tetrahydrobiopterin (Kato *et al.*, 1980). Thus 6-MHP$_4$ is a useful cofactor for *in vitro* assays.

One of the most important characteristics of tryptophan hydroxylase is that its affinity for its two substrates, oxygen and tryptophan, depends on the structure of the reduced pterin cofactor used in the assay (Kaufman, 1974; Table 1). The lowest K_m values for the two substrates are observed in the presence of L-erythro-tetra-hydrobiopterin. This discovery has proved important for understanding the role substrate concentration plays in regulating 5-HT formation *in vivo* (Kaufman, 1974).

Aromatic L-α-amino acid decarboxylase is a pyridoxal-5'-phosphate dependent enzyme that is found in serotonergic neurones where it converts 5-HTP to 5-HT and in catecholaminergic neurones where it converts 3,4-dihydroxyphenylalanine (DOPA) to dopamine (Sourkes, 1977). Very different conditions of pH and concentrations of substrate and cofactor are, however, required for optimum activity of brain homogenates with these two substrates (Sims *et al.*, 1973). Since the affinity of the enzyme for 5-HTP is at least two orders of magnitude greater than that for tryptophan, 5-HTP is preferentially decarboxylated in serotonergic neurones (Ichiyama *et al.*, 1970). Aromatic amino acid decarboxylase activity is present in brain extracts in far greater amounts than tryptophan hydroxylase (Ichiyama *et al.*,

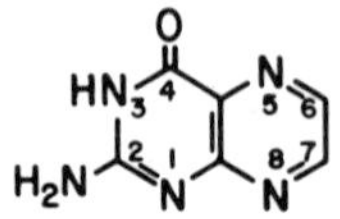

2-amino-4-hydroxypteridine (pterin)

The structure given is that of the
favoured 4-oxo tautomer (Blakley,1969)

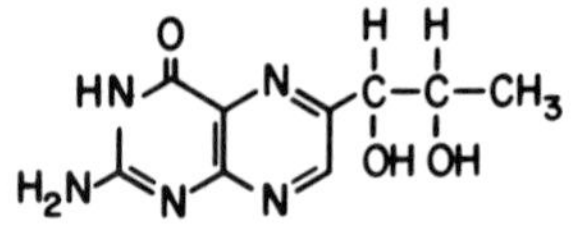

6-(L-<u>erythro</u>-1', 2',-dihydroxypropyl) pterin
(L-<u>erythro</u>biopterin)

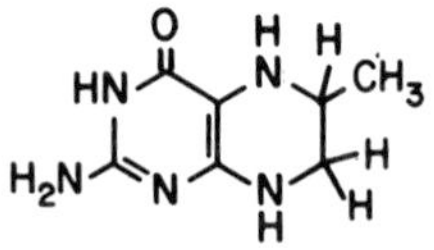

6-methyl-5,6,7,8-tetrahydropterin
($6MPH_4$)

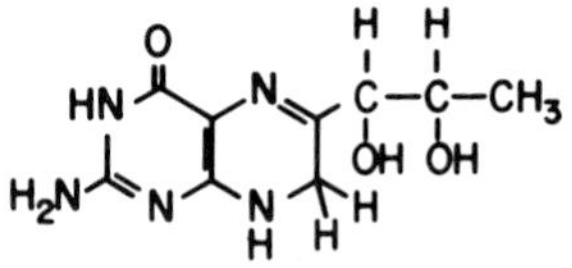

L-<u>erythro</u>-7,8-dihydrobiopterin
(BH_2)

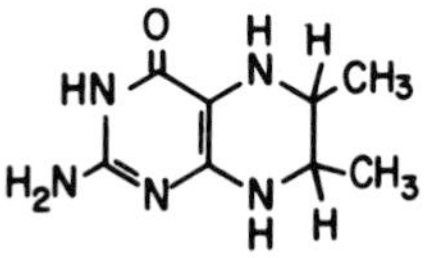

6,7-dimethyl-5,6,7,8-tetrahydropterin
($DMPH_4$)

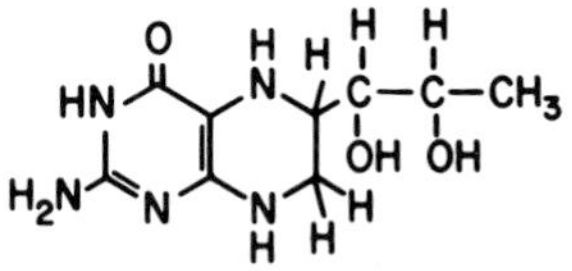

L-<u>erythro</u>-5,6,7,8-tetrahydrobiopterin
(BH_4)

Fig. 2   Structures of pterin and various derivatives

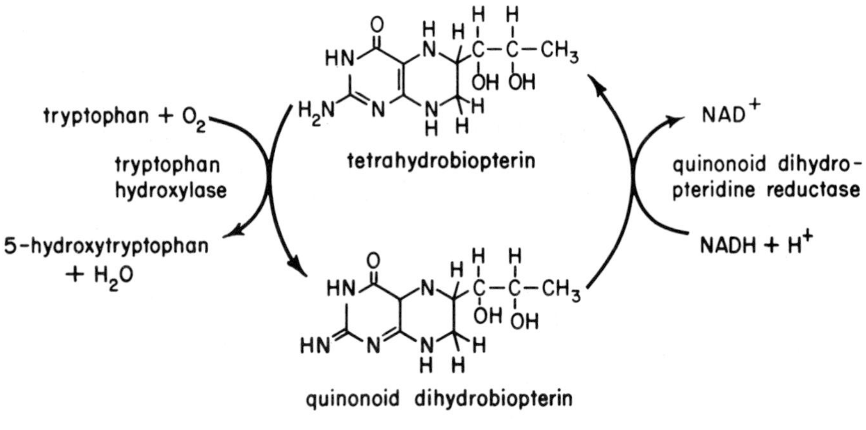

tryptophan + O_2

tryptophan
hydroxylase

5-hydroxytryptophan
+ H_2O

tetrahydrobiopterin

quinonoid dihydrobiopterin

NAD^+

quinonoid dihydro-
pteridine reductase

$NADH + H^+$

Fig. 3   Regeneration of tetrahydrobiopterin

Table 1. Kinetic properties of tryptophan hydroxylase

| Substrate | Cofactor | | | Enzyme source |
	BH_4 K_m (μM)	6-MPH$_4$ K_m (μM)	DMPH$_4$ K_m (μM)	
Pterin	31	67	130	Rabbit hindbrain[1]
	57	191	–	Rat brainstem[2]
L-tryptophan	50	78	290	Rabbit hindbrain[1]
	48	152	–	Rat brainstem[2]
Oxygen	2.5%	–	20%	Rabbit hindbrain[1]

[1] Friedman *et al.*, 1972
[2] Kato, *et al.*, 1980

1968) and for this reason the hydroxylation of tryptophan rather than the decarboxylation of 5-HTP is presumed to be the rate limiting step in 5-HT formation. The fact that 5-HTP is found in brain tissue in only trace amounts (Tappaz and Pujol, 1980) indicates that it is decarboxylated almost as rapidly as it is formed and also suggests that the hydroxylation reaction is rate limiting in 5-HT formation. This view prompted detailed studies of factors that are important for determining the overall rate of this reaction including concentrations of cofactor and substrates, as well as the kinetic behaviour of the enzyme molecule itself. One outcome of these studies has been the finding that the kinetic properties of tryptophan hydroxylase in low speed supernatant preparations can be altered by a variety of agents including phospholipids, detergents, a calcium activated protease, trypsin, and phosphorylating conditions (Table 2). Although the kinetic changes are not all identical after these treatments, an increase in affinity for the synthetic cofactor 6-MPH$_4$ is always observed. Changes in the affinity of an enzyme for its substrates or cofactor will of course only influence the rate of product formation *in vivo* if these concentrations are normally subsaturating. Since this appears to be the situation in the serotonergic neurones these observations on crude enzyme extracts may very well be relevant to 5-HT synthesis in the *in vivo* situation. The activation observed under phosphorylating conditions is of special interest for the regulation of 5-HT synthesis in the intact serotonergic neurone, as is the finding that the effect of phosphorylating conditions is enhanced by micromolar concentrations of calcium, but not by cyclic AMP or any other cyclic nucleotides (Hamon *et al.*, 1978a; Kuhn *et al.*, 1978). Recent experiments indicate that calmodulin may be involved in mediating the effect of calcium (Yamauchi and Fujisawa, 1979a) and that the calcium dependent activation of the enzyme observed under phosphorylating conditions is reversible (Yamauchi and Fujisawa, 1979b).

Table 2. Alterations to the kinetic properties of low speed supernatant preparations of rat brain tryptophan hydroxylase subjected to various treatments

Treatment	L-tryptophan K_m	6-MPH$_4$ K_m	V_{max}
Phosphorylating conditions: 0.5 mM ATP, 10 mM Mg^{2+}, 0.2 mM cyclic AMP*	No change	Decrease	No change[1]
0.5 mM ATP, 5 mM Mg^{2+}, 50 μM Ca^{2+}	Decrease	Decrease	No change[2]
Phosphatidylserine	No change	Decrease	No change[3]
Sodium dodecylsulfate	Decrease	Decrease	Doubled when 6-MPH$_4$ varied[4]
5 mM Ca^{2+} (activation of protease)	Decrease	Decrease	No change[5]
Trypsin (30-min pre-incubation)	Decrease	Decrease	Increased when 6-MPH$_4$ varied[4]

*Rat midbrain used as enzyme source; omission of cyclic AMP did not alter enzyme actitivity.

[1] Kuhn *et al.*, 1978
[2] Hamon *et al.*, 1978a
[3] Hamon *et al.*, 1978b
[4] Hamon *et al.*, 1976
[5] Hamon *et al.*, 1977

METHODS

Assay of tryptophan hydroxylase

In the most direct method used for the assay of tryptophan hydroxylase, the formation of 5-HTP from tryptophan is measured in the presence of a decarboxylase inhibitor which prevents the conversion of 5-HTP to 5-HT. The sensitivity of this assay is largely determined by the method used to detect the 5-HTP formed. The usual source of enzyme is a low-speed supernatant preparation (20,000–30,000 $\times$ g) made by homogenizing brain stem or midbrain in about 3 volumes of a hypotonic Tris HCl or Tris acetate buffer pH 7.4, containing a thiol reagent such as dithiothreitol (see Hamon *et al.*, 1979a). Tryptophan hydroxylase is an extremely unstable enzyme and crude extracts lose all their activity within a few hours at 37 °C and more than 70 per cent if kept overnight at 0 °C. (Dithiothreitol slows up this inactivation process and is therefore routinely added to enzyme preparations, Friedman *et al.*, 1972; Kuhn *et al.*, 1980). One reason for this rapid loss of catalytic activity is the susceptibility of the enzyme to inactivation by molecular oxygen. The extent to which activity is lost depends on the partial pressure of the oxygen to which the enzyme is exposed and the temperature at which exposure occurs. The inactivation is fully reversible if the enzyme is incubated for a prolonged period under anaerobic conditions in the presence of ferrous ion and dithiothreitol, and it can be avoided during storage by freezing the enzyme in a nitrogen atmosphere in the presence of dithiothreitol (Kuhn *et al.*, 1980). The standard tryptophan hydroxylase assay mixture contains L-tryptophan (D-tryptophan in blanks) a reduced pterin cofactor, usually 6-MPH$_4$, catalase to destroy any hydrogen peroxide that may be formed, a decarboxylase inhibitor (e.g. 3-hydroxybenzylhydrazine (NSD 1015) or N'-(3-hydroxybenzyl)-N'-methylhydrazine (NSD 1034) (Pletscher *et al.*, 1966) to prevent conversion of 5-HTP to 5-HT and a system for regenerating the fully reduced form of the cofactor, either sheep liver NADH (NADPH) dependent quinonoid dihydropteridine reductase (Friedman *et al.*, 1972), or a high concentration of mercaptoethanol or dithiothreitol (see Hamon *et al.*, 1979a). The 5-HTP formed can be detected by its native fluorescence in 3N HCl (emission 595 nm; excitation 295 nm), or by the fluorescence of the condensation product with *o*-phthalaldehyde (Gal and Patterson, 1973). The sensitivity of the assay is greatly improved (0.5 ng detection limit) if the 5-HTP is separated from tryptophan with high-pressure liquid chromatography (HPLC) to reduce the blank fluorescence (Meek and Neckers, 1975). This allows measurement of enzyme activity in single midbrain raphe nuclei.

Some investigators have used L-tryptophan radiolabelled with ^{14}C on the side chain in the enzyme assay and have then separated the radiolabelled 5-HTP formed by thin layer chromatography. This allows measurement of enzyme activity in single midbrain raphe nuclei (Kan *et al.*, 1975; Kan *et al.*, 1977).

The other assays for tryptophan hydroxylase activity are indirect and not as suitable for kinetic studies of the enzyme as the procedure developed by Friedman *et al.* (1972) because they employ additional enzyme reactions (decarboxylation, *N*-acetylation 5-*O*-methylation) to convert 5-HTP quantitatively to other reaction products, which can be more readily detected. Nevertheless they have proved very useful for measuring low levels of enzyme activity.

These assays are carried out in the absence of a decarboxylase inhibitor. One depends on the release of [14]CO_2 in the decarboxylation by added aromatic amino acid decarboxylase of 5-HTP formed from ^{14}C carboxyl-labelled L-tryptophan. The assay is based on the preferential decarboxylation of 5-HTP that occurs in a mixture of 5-HTP and tryptophan because of the much higher affinity of the aromatic amino acid decarboxylase for 5-HTP than for tryptophan. In practice, if one is to avoid the decarboxylation of significant amounts of tryptophan, and thus maintain the validity of the assay, the concentration of tryptophan must not exceed 20 μM. For this reason the assay cannot be carried out under conditions where tryptophan hydroxylase is saturated with tryptophan (Ichiyama *et al.*, 1968, 1970). This procedure is not only used to assay soluble preparations of tryptophan hydroxylase but is widely applied to measure the capacity of synaptosomal preparations to hydroxylate tryptophan in the presence of endogenous cofactor and endogenous aromatic amino acid decarboxylase (see Knapp *et al.*, 1974).

Another assay for tryptophan hydroxylase involves the conversion of 5-HTP formed from non-radioactive tryptophan to 5-HT in the presence of excess added

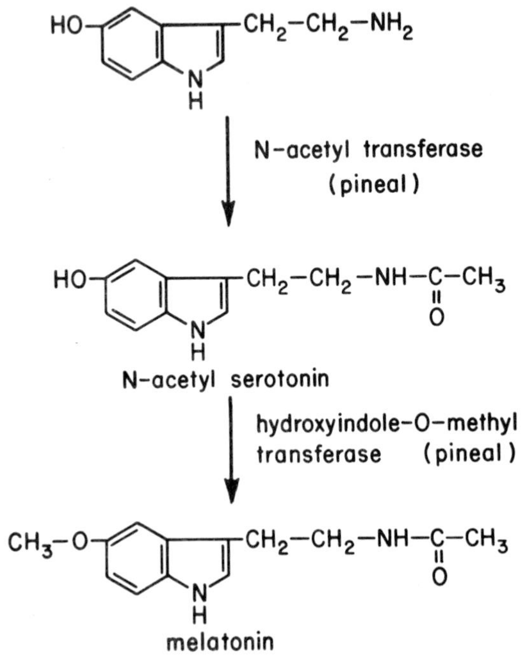

Fig. 4   Synthesis of melatonin

aromatic amino acid decarboxylase and a monoamine oxidase inhibitor. The 5-HT is separated from tryptophan by procedures similar to those outlined for the assay of aromatic amino acid decarboxylase in the next section (Hamon *et al.*, 1979a).

Finally, the most sensitive assay for tryptophan hydroxylase depends on the conversion of the 5-HTP formed in the hydroxylation reaction to [^{3}H]melatonin. The tryptophan hydroxylase reaction is carried out in the presence of saturating concentrations of L-tryptophan and 6-MPH$_4$, a monoamine oxidase inhibitor, albumin, to stabilize the extremely dilute enzyme extracts, dithiothreitol and HEPES buffer, pH 7.4. The hydroxylation reaction is stopped with HCl and an aliquot of the reaction is then incubated with a mixture of rat liver aromatic amino acid decarboxylase and *N*-acetyl transferase in the presence of a phosphate buffer at pH 8.4 and acetyl coenzyme A. Under these conditions 5-HTP is converted to 5-HT, and this is in turn acetylated on the side chain amino group (Fig. 4). The *N*-acetyl serotonin is then converted to [^{3}H]melatonin by hydroxy indole-*O*-methyl transferase and [^{3}H]S-adenosyl methionine, and the [^{3}H]melatonin is extracted and quantitated by liquid scintillation counting (Kizer *et al.*, 1975). With this procedure tryptophan hydroxylase activity can be detected in as little as 5 µg of a crude homogenate of rat brainstem, and in individual brain nuclei.

Assay of aromatic amino acid decarboxylase

There are two assays for this enzyme commonly in use. One is based on the release of $^{14}CO_2$ from [^{14}C] carboxyl labelled amino acid (Sims *et al.*, 1973) and the other on the detection of the amine product after its separation from the amino acid precursor by ion exchange chromatography (Bouchard and Roberge, 1979). The assay for 5-HTP decarboxylase activity in brain homogenates is routinely carried out between pH 8 and pH 8.3 in the presence of a phosphate buffer, added pyridoxal-5'-phosphate cofactor, non-radioactive L-5-HTP (D-5-HTP in the blanks) or the [^{14}C] carboxyl labelled L-5-HTP. (Heat inactivated enzyme blanks are used in this case). A monoamine oxidase inhibitor is included in the non-radioactive assay to prevent breakdown of 5-HT. At the end of the reaction, trapped $^{14}CO_2$ is counted, or 5-HT is separated from 5-HTP by chromatography on Amberlite G-50 Type I resin in the sodium form, buffered with sodium acetate at pH 6.5. The 5-HT which is eluted from the column with HCl can be quantitated by native fluorescence of the *o*-phthalaldehyde derivative of 5-HT (see Gal and Patterson, 1973). The newer procedures of high performance liquid chromatography and electrochemical detection, after addition of an internal standard such as *N*-acetyl serotonin (Mefford and Barchas, 1980; Reinhard *et al.*, 1980). Allow separation and quantitation of picogram amounts of 5-HT and have been successfully applied to the assay of DOPA decarboxylase activity (Nagatsu *et al.*, 1979).

Methods for measuring synthesis of 5-HT *in vivo*

A number of methods are used to estimate the synthesis of 5-HT in brain tissue *in vivo* (see Morot-Gaudry *et al.*, 1974). They include

(1) the accumulation of endogenous 5-HT with time after blocking its metabolism by inhibition of monamine oxidase
(2) the accumulation of endogenous 5-HTP with time after inhibition of aromatic amino acid decarboxylase, a measure of *in vivo* tryptophan hydroxylase activity
(3) the conversion of radiolabelled tryptophan to radiolabelled 5-HT after intravenous infusion or a pulse injection of radiolabelled amino acid.

The first two methods involve the administration of high concentrations of drugs which may have effects besides those intended. For example some monoamine oxidase inhibitors increase brain levels of tryptophan and therefore perturb the very process they are being used to study. Estimates of 5-HT synthesis with this method are much higher during the initial 10 min following drug administration than at later times (see Morot-Gaudry *et al.*, 1974). This may be partly due to a fall in amino acid levels at longer time intervals but could also result from the increased levels of endogenous 5-HT which have been found to reduce 5-HT synthesis *in vivo* at least in some brain regions (Carlsson and Lindquist, 1973; Tappaz and Pujol, 1980).

The decarboxylase inhibitors used for *in vivo* studies of 5-HT synthesis, NSD 1015, NSD 1034 and Ro4-4602, (N'[DL-seryl]-N^2-[2,3,4 trihydroxybenzyl] hydrazine) all derivatives of benzylhydrazine, produce a small increase in the level of endogenous tryptophan (Carlsson and Lindquist, 1973) and may also inhibit other enzymes that are involved in amino acid metabolism such as transaminases, because they bind to pyridoxine (Wurtman, 1974). Furthermore, since 5-HTP can inhibit tryptophan hydroxylase *in vitro* (Kaufman, 1974), the 5-HTP that accumulates as a result of decarboxylase inhibition may inhibit its own synthesis if it reaches high enough local concentrations. In spite of these potential problems this method is now widely used for estimating *in vivo* hydroxylation of tryptophan and hence, indirectly, 5-HT formation. Carlsson (1974) has shown that the accumulation of 5-HTP is linear for 45 min after administration of the decarboxylase inhibitor, and furthermore that 5-HTP is not lost by diffusion out of the brain in significant amounts during this time. By elimination of endogenous 5-HT and conversion of the 5-HTP formed to [^{3}H]melatonin, Tappaz and Pujol (1980) have now succeeded in applying this method to the measurement of 5-HTP synthesis in single brain nuclei. It should be possible to adapt recently published methods for measurement of 5-HT and 5-HIAA by the HPLC and electrochemical detection for the determination of 5-HTP as well (Reinhard *et al.*, 1980).

Radiolabelled L-tryptophan has been widely used for the study of 5-HT synthesis *in vitro* in brain slices (e.g. Elks *et al.*, 1979a, b). However, its use *in vivo* is expensive because of the high specific activities required. It also necessitates precise determinations of the specific activity in all brain regions under study since there are regional variations in the endogenous brain tryptophan concentration (Neckers *et al.*, 1977). On he other hand the isotopic method has the advantage that it avoids the use of drugs.

REGULATION OF 5-HT SYNTHESIS

One striking feature of the synthesis of 5-HT in intact neurones *in vivo* or in synaptosomal or slice preparations *in vitro* is the low rate of conversion of tryptophan to 5-hydroxytryptophan compared with the level of tryptophan hydroxylase activity that can be measured *in vitro* in supernatant preparations of the enzyme made from the same brain tissue and assayed under saturating conditions of substrate and cofactor. For dorsal raphe the difference exceeds two orders of magnitude (Meek and Neckers, 1975; Meek and Lofstrandh, 1976), and indicates clearly that the activity of tryptophan hydroxylase within serotonergic neurones is severely restricted. The factors responsible for regulating tryptophan hydroxylase activity *in vivo* are not fully understood, though some have been identified. For example tryptophan levels in brain are known to be subsaturating for tryptophan hydroxylase and thus presumably have an important role in determining the ongoing rate of 5-HT synthesis *in vivo*. On the other hand the extent to which the concentrations of oxygen or cofactor also control the ongoing rate of 5-HT synthesis *in vivo* is far less well understood, but may likewise be of considerable significance. Finally the possibility exists that the kinetic properties of tryptophan hydroxylase can themselves be modulated *in vivo*, and in this way provide a very rapid and sensitive mechanism for changing the rate of 5-HT synthesis in the presence of limiting concentrations of substrates and cofactor under a variety of conditions. These and other factors which may influence 5-HT synthesis will be examined. The discussion will be limited to studies in mammalian brain.

Regulation of tryptophan availability and its effects on 5-HT synthesis

The K_m of partially purified tryptophan hydroxylase for tryptophan in the presence of tetrahydrobiopterin is 50 μM (Friedman *et al.,* 1972) or somewhat higher than the overall concentration of tryptophan in brain of 30 μM (Young and Sourkes, 1977). Thus if the concentration of tryptophan in serotonergic neurones is assumed to be the same as that calculated for whole brain, tryptophan hydroxylase would be less than half saturated with its substrate and the formation of 5-HT would be expected to rise as the brain concentration of tryptophan increased, until saturating levels were reached. This is in fact what is observed when doses of tryptophan, low enough to keep circulating tryptophan levels within the physiological range, are injected into rats and the brain levels of tryptophan and serotonin measured 1h later. The brain concentration of 5-HT increases with increasing tryptophan levels and reaches a maximum at a concentration between 15 and 20 μg/g (Fernstrom *et al.,* 1974).

Since tryptophan is an essential amino acid it must be derived from food sources and delivered to the tissues via the circulation. The concentration of tryptophan in the brain is thus determined by the combined effects of its plasma concentration and its transport into the brain. This involves passage across the blood brain barrier into the extracellular fluid of the brain and uptake at the neuronal or glial cell membrane, processes which are brought about by carrier mediated as well as non-saturable transport (diffusion) (Pratt, 1979; Young and Sourkes, 1977). A number of important factors can significantly alter the concentration of tryptophan in the plasma or the efficiency of the carrier-mediated transport systems for handling tryptophan and thereby influence brain tryptophan and serotonin levels. One of these factors is the amount of tryptophan in the diet. For example, weanling rats fed for about six weeks on a diet of corn, a source of protein which is highly deficient in tryptophan, have lower brain levels of tryptophan and serotonin and synthesize 5-HT (measured by 5-HTP accumulation) at less than half the rate of controls. The depressed rate of 5-HT formation in this case is due to low substrate levels and not to some other change resulting from malnutrition, since 5-HT synthesis in control (casein fed) and experimental animals is not significantly different after a tryptophan load (Fernstrom and Hirsch, 1976). Abnormalities in plasma tryptophan levels can also result from metabolic disorders rather than improper tryptophan intake. One striking example occurs in hepatic encephalopathy in which loss of liver function leads to extraordinarily high plasma and brain tryptophan levels and a large rise in brain serotonin content (Sourkes, 1978; Jellinger *et al.*, 1978).

Another important factor which influences brain tryptophan levels is the relative proportions of other neutral amino acids in the diet, besides tryptophan. Although administration of *pure* L-tryptophan to animals increases serotonin synthesis, the ingestion of tryptophan in protein does not have this effect since protein contains a variety of amino acids some of which use the same transport system at the blood brain barrier and at the cell membrane as tryptophan. The K_m of the blood brain barrier amino acid transport system for each of the neutral amino acids that share it (e.g. leucine, tyrosine, phenylalanine, isoleucine, alanine, tryptophan) is about the same as their individual concentrations in the plasma (0.1-0.6 mM). These amino acids must therefore compete with each other for their carrier (Pardridge, 1979a). Competition not only occurs between amino acids for uptake at the blood brain barrier, but also for uptake at the membrane of brain cells (Young and Sourkes, 1977). However, it is the blood brain barrier transport system that is rate limiting (Pardridge, 1977) and therefore the interactions between amino acids at this site will presumably affect brain concentrations most profoundly.

The other regulatory site for tryptophan entry into serotonergic neurones is the uptake at the neuronal membrane itself. In addition to the saturable uptake process for tryptophan, which is found throughout the brain and is shared with other neutral amino acids, (Young and Sourkes, 1977) serotonergic neurones may possess a carrier which is more selective for tryptophan. The evidence for this comes from experiments in which lesions of 5-HT neurones produced a reduction in the uptake of tryptophan by synaptosomal preparations, without any change in the uptake of alanine or isoleucine, amino acids that normally compete for the tryptophan carrier (Denizeau and Sourkes, 1978).

Carlsson and Lindquist (1978a) have exploited the competition between tryptophan and other neutral amino acids to reduce brain tryptophan to very low levels with injections of various neutral amino acids. By combining this technique with that of tryptophan loading they were able to study regional 5-HT synthesis by 5-HTP accumulation in the presence of a decarboxylase inhibitor over a wide range of tryptophan concentrations. In this way they determined a K_m for tryptophan *in vivo* of 25 μM which is reasonably close to that of 50 μM obtained *in vitro*.

Another factor that influences the passage of tryptophan into the brain is binding of tryptophan to serum albumin. Unlike other amino acids tryptophan circulates in the blood largely in bound form and this has lead to the view that only unbound or free tryptophan can cross the blood brain barrier and enter brain tissue (Young and Sourkes, 1977). However, the amino acid carrier of the blood brain barrier has both a greater capacity and a higher affinity for tryptophan than does circulating albumin and recent evidence indicates that it can effectively compete with albumin for tryptophan as the blood passes through the cerebral circulation (Pardridge, 1979a,b). The extent of albumin binding is influenced by plasma pH, by non-esterified fatty acids which will displace tryptophan and by a variety of drugs that have similar effects (Young and Sourkes, 1977; Yuwiler *et al.*, 1977). Any significant fall in the binding of tryptophan to albumin would be expected to enhance uptake by the blood brain barrier transport system as well as to increase the contribution made by diffusion, since albumin binding normally limits the amount of freely diffusible tryptophan (Pardridge, 1979a). Quantitatively, however, the proportion of competing neutral amino acids present in the plasma is of far more importance in determining the transport of tryptophan into the brain, than the extent of serum albumin binding (Yuwiler *et al.*, 1977). The effect of a fall in the plasma concentration of competing neutral amino acids on brain tryptophan levels and 5-HT synthesis is strikingly illustrated when a carbohydrate-fat meal is given to fasting rats. The insulin secreted in response to the rise in plasma glucose increases plasma tryptophan levels, but reduces the plasma concentrations of the other neutral amino acids that compete with tryptophan for uptake into the brain. The result is that brain concentrations of tryptophan, 5-HT and 5-HIAA rise and the conversion of tryptophan to 5-HTP measured in the presence of a decarboxylase inhibitor is significantly increased in most brain regions (Colmenares *et al.*, 1975).

There is another aspect to the role of tryptophan in regulating 5-HT synthesis. This is the ability of the substrate to alter the behaviour of the tryptophan hydroxylase molecule itself. Substrate inhibition is observed *in vitro* with purified enzyme preparations at tryptophan concentrations above 200 μM in the presence of tetrahydrobiopterin (Kaufman, 1974) and could serve to reduce 5-HT synthesis *in vivo* after administration of massive amounts of tryptophan. So far, however, no evidence has been obtained which suggests that the conversion of tryptophan to 5-HT is reduced *in vivo* even with tissue tryptophan concentrations as high as 1.5 mM (Eccleston *et al.*, 1965). On the other hand there is evidence that the rate of 5-HT synthesis may be enhanced when tryptophan levels are severely depleted as for example in experimentally induced diabetes or following chlorimipramine treatment. In both instances tryptophan hydroxylase isolated from the brains of treated animals shows a marked increase in V_{max} compared with enzyme from non-diabetic or drug-free animals (Neckers *et al.*, 1977; Trulson and Mackenzie, 1980). This alteration in the kinetic properties of the enzyme disappears when brain tryptophan levels are raised by tryptophan administration, or by insulin injection in the case of diabetic rats. Such a regulatory mechanism indicates that 5-HT synthesis may in fact be buffered against the effects of severe substrate depletion and that transmitter availability therefore may not be solely dependent on the delivery of tryptophan to the brain. Indeed the 5-HT and 5-HIAA levels in the brains of diabetic rats do not differ significantly from those of control animals in spite of marked decreases (~40 per cent) in brain tryptophan levels (Trulson and Mackenzie, 1980).

The effect of oxygen availability on 5-HT synthesis

The K_m for oxygen of partially purified rabbit hind brain tryptophan hydroxylase in the presence of tetrahydrobiopterin is 2.5 per cent (19 Torr) (Friedman *et al.*, 1972), a value which lies within the range of tissue oxygen tensions measured for cerebral cortex in several species (Lübbers, 1968). Values range from 1–90 Torr in the anaesthetized cat with 70 per cent of values less than 30 Torr and 50 per cent between 5 and 20 Torr (Smith *et al.*, 1977). Thus the oxygen concentration within serotonergic neurones may be limiting for tryptophan hydroxylase under normal conditions *in vivo*. Furthermore the oxygen concentration may actually vary over the length of the diffusely projecting 5-HT neurones themselves, as judged by the striking non-uniformity of PO_2 values measured over very small distances of cerebral cortex (Smith *et al.*, 1977).

Cerebral PO_2 tensions drop to low levels extremely rapidly under anoxic or hypoxic conditions (Leniger-Follert *et al.*, 1975). Although tensions of 1–2 Torr will sustain mitochondrial respiration (Lübbers, 1968; Clark *et al.*, 1976) they are clearly suboptimal for tryptophan hydroxylase with a K_m for oxygen of 19 Torr. Thus 5-HT synthesis may be vulnerable to even mild degrees of hypoxia. The effect of reduced arterial PO_2 on 5-HT synthesis was examined in experiments in which anaesthetized adult rats, treated with an aromatic amino acid decarboxylase inhibitor, were exposed to different air nitrogen mixtures for 30 min and the accumulation of 5-HTP used as an index of 5-HT synthesis (Davis *et al.*, 1973). There was a decrease in 5-HT synthesis as arterial PO_2 fell below 60 Torr and a linear correlation was observed between cerebral venous PO_2 values (calculated from arterial PO_2) and the accumulation of 5-HTP in whole brain. Over the time period of these experiments no alteration was observed in brain tryptophan levels indicating that the reduction in 5-HTP accumulation could not be ascribed to a decrease in tryptophan availability (Davis and Carlsson, 1973a). A similar reduction in 5-HTP accumulation was also observed in unanaesthetized adult rats (Davis and Carlsson, 1973b) and more recently has also been reported for whole brains and brain regions in young unanaesthetized rats (Hedner *et al.*, 1978; Hedner and Lundborg, 1979). Tryptophan hydroxylase thus appears to be equally affected by hypoxia at all ages and in all brain regions examined.

One problem in the interpretation of the effects of hypoxia is the acidosis that accompanies it at arterial PO_2 values below 50 Torr and the effect that the resulting decrease in intracellular pH may have on enzyme activity. An attempt was therefore made to dissociate effects secondary to a fall in intracellular pH and those due to hypoxia by altering the arterial PO_2 and PCO_2 values independently in artificially respired, anaesthetized animals. Under these conditions 5-HT accumulation was closely correlated with venous PO_2 but was independent of the prevailing arterial PCO_2 and hence tissue pH (Carlsson *et al.*, 1977).

If tissue oxygen levels are normally at or below the K_m of tryptophan hydroxylase for oxygen, then raising tissue PO_2 would be expected to increase 5-HT synthesis. However, since the haemoglobin in arterial blood is nearly saturated with oxygen when an animal breathes air, raising the inhaled oxygen to 100 per cent produces only a slight increase in the amount of oxygen delivered to the tissues (Lambertson, 1980). Presumably this is the reason only small increases in 5-HT synthesis are ever observed *in vivo* when 100 per cent oxygen is breathed (Diaz *et al.*, 1968; Davis and Carlsson, 1973a) whereas 5-HT formation in brain homogenates can be increased about five fold in the presence of 100 per cent oxygen (Green and Sawyer, 1966).

There have been several studies made of the effect of hyperbaric oxygen on 5-HT levels in brain and these have shown either no change or a decrease in 5-HT levels (Sourkes, 1979). In view of the known sensitivity of isolated tryptophan hydroxylase to oxygen (Kuhn *et al.*, 1980) it is possible that hyperbaric oxygen may inactivate tryptophan hydroxylase and thus reduce 5-HT synthesis, though it may have other effects which could affect 5-HT synthesis indirectly. Brain oxygen tensions increase to 800 Torr at 4 atmospheres oxygen (Jamieson and van den Brenk, 1965) so one might have expected to see an increase in 5-HT formation similar to that observed *in vitro*.

The role of cofactor

With the use of HPLC and flow fluorimetry biopterin has now been identified as the only unconjugated pterin present in rat brain in the reduced state, and more than 90 per cent of it occurs in the quinonoid dihydro or tetrahydro form (Fukushima and Nixon, 1980). Since the quinonoid dihydro form is extremely unstable and rearranges spontaneously to 7,8-dihydrobropterin, most biopterin is presumably in the fully reduced state.

The concentration of the hydroxylase cofactor ranges between 0.5 and 1.0 μM if one assumes a uniform distribution in brain tissue. However, the regional levels of tetrahydrobiopterin-like activity in rat brain (determined by the phenylalanine hydroxyglase bioassay) show a high correlation with the combined activities of tryptophan and tyrosine hydroxylases within the same brain region (excluding the neuroendocrine tissues, the pineal and pituitary, which have extraordinarily high levels of cofactor) (Levine *et al.*, 1979). Thus there is a real possibility that the cofactor may be associated with monoaminergic neurones in which case its concentration could be considerably higher. However, even if the neuronal concentration of tetrahydrobiopterin were ten fold higher it would still lie around the K_m of tryptophan hydroxylase for tetrahydrobiopterin (30 μM) and would therefore remain rate limiting for the conversion of tryptophan to 5-HTP.

One way to test whether the concentration of tetrahydrobiopterin is limiting in serotonergic neurones is to see whether 5-HT synthesis can be increased by supplying exogenous cofactor to the neurones. All attempts to enhance 5-HT synthesis in synaptosomal preparations either by supplying exogenous cofactor in the incubation medium or by attempting to increase cofactor levels in brain tissue via intraventricular injections of tetrahydrobiopterin prior to isolation of synaptosomes have failed (Bullard *et al.*, 1978; Hamon *et al.*, 1979a). One reason may be that the cofactor does not enter the synaptosomes in sufficient amounts *in vitro* or cannot be accumulated in sufficient quantities *in vivo*. Another possibility is that cofactor levels are actually saturating in serotonergic neurones. However, as Hamon *et al.* (1979a) point out, this seems unlikely in view of the fact that the rate of 5-HT synthesis *in vivo* still falls below *in vitro* tryptophan hydroxylase activity when brain tryptophan levels are made saturating.

The identification of tetrahydrobiopterin in rat brain leaves no doubt that this pterin serves as the natural cofactor for both tryptophan hydroxylase and tyrosine hydroxylase in this species. In man circumstantial evidence for this comes from a rare form of hyperphenylalaninaemia in which the formation of tyrosine from phenylalanine in liver is blocked not by the usual deficiency of phenylalanine hydroxylase (Kaufman, 1977b) but by the absence of tetrahydrobiopterin, presumably due to a defect in one or more of the enzymes involved in its synthesis. Patients with this disorder have greatly reduced levels of monoamine metabolites in urine (Kaufman *et al.*, 1978).

Tetrahydrobiopterin can be synthesized in brain from guanosine triphosphate via D-erythro-7,8-dihydroneopterin triphosphate and L-erythro-7,8-dihydrobiopterin as well as other as yet unidentified intermediates (see Nixon *et al.*, 1980a; Nixon *et al.*, 1980b) and this endogenously produced tetrahydrobiopterin presumably serves as the source of cofactor for both aromatic amino acid hydroxylases in brain. There is no evidence that dietary sources of biopterin can be utilized since animals placed on a biopterin-free diet continue to excrete biopterin in their urine (Kraut *et al.*, 1963). It is also unlikely that tetrahydrobiopterin in brain is synthesized elsewhere (e.g. liver) since it penetrates the blood brain barrier very poorly indeed (Kettler *et al.*, 1974). If concentrations of tetrahydrobiopterin in serotonergic neurones are, in fact well below the K_m of tryptophan hydroxylase for this cofactor then one way of regulating the activity of tryptophan hydroxylase would be to induce changes in the concentration of the reduced cofactor in response to specific physiological stimuli by varying the rate of *de novo* synthesis of tetrahydrobiopterin or its breakdown. There is at present no evidence that a control mechanism of this type exists in nervous tissue. However, changes in the levels of tetrahydrobiopterin which are clearly related to *de novo* synthesis of cofactor have been reported in adrenal gland of the rat after drug or hormone treatments *in vivo*. In the adrenal medulla depletion of the gland by insulin-induced hypoglycemia results in induction of the enzymes of catecholamine biosynthesis (tyrosine hydroxylase and dopamine-β-hydroxylase) and a marked increase in the formation of catecholamines (Viveros *et al.*, 1969). These same treatments have now been found to increase the tetrahydrobiopterin content of the adrenal medulla by 40 per cent (Abou-Donia *et al.*, 1980), and to more than double the activity of the initial enzyme involved in the conversion of GTP to tetrahydrobiopterin, GTP cyclohydrolase, apparently by formation of new enzyme protein (Nixon *et al.*, 1980). As cofactor levels are also rate limiting for catecholamine synthesis in the medulla, the increases in GTP cyclohydrolase activity and tetrahydrobiopterin content presumably ensure the availability of adequate amounts of cofactor to permit the enhanced levels of tyrosine hydroxylase to be expressed as an increase in 3,4-dihydroxyphenylalanine formation. The increases in GTP cyclohydrolase activity and tetrahydrobiopterin content induced by reserpine or insulin treatment can be prevented by cycloheximide or denervation respectively and are thus clearly related to functional activity of the medullary cells. These findings are very provocative and may well serve as a model system for monamine neurones. They raise the possibility that increases in the functional activity of such neurones may also trigger an increase in the formation of the cofactor and by so doing enhance transmitter synthesis and availability. Studies on the regulation of tetrahydrobiopterin synthesis and how this may influence the ongoing rate of monoamine formation will require a detailed knowledge of the pathways, enzymes and rate limiting steps in its formation, something which has not yet been achieved. The manipulation of cofactor concentration would be an extremely sensitive mechanism for regulating 5-HT synthesis. Future work in this area promises important new insights into strategies for modulating transmitter formation by serotonergic and other monoamine-containing neurones.

Tetrahydrobiopterin is regenerated by NADH dependent quinonoid dihydropteridine reductase (Fig. 3). The obligatory role of this enzyme in the tryptophan hydroxylating system, at least in human brain was recently demonstrated with the discovery of yet another form of hyperphenylalaninemia in which the conversion of phenylalanine to tyrosine in liver is impaired not by a deficiency of phenylalanine hydroxylase or by the inability to synthesize biopterin, but by the absence of quinonoid dihydropteridine reductase and hence of the reduced form of the cofactor, tetrahydrobiopterin. In one patient the enzyme was also shown to be absent from brain tissue and 5-HT levels there were extremely low (Butler *et al.,* 1978). Such individuals excrete biopterin in the 7,8-dihydro and fully oxidized forms whereas normal individuals eliminate 80–90 per cent in the tetrahydro-form (Milstein *et al.,* 1980). This observation, taken together with the inability of these individuals to hydroxylate phenylalanine in liver, or tryptophan in brain (Kaufman *et al.,* 1975; Butler *et al.,* 1978), indicates that no effective route besides quinonoid dihydropteridine reductase exists for cofactor regeneration *in vivo* (Kaufman, 1979).

Quinonoid dihydropteridine reductase is not confined to monoamine containing neurones, but is rather uniformly distributed throughout the brain. The fact that its specific activity exceeds that of the combined activities of tryptophan and tyrosine hydroxylase in different brain regions by more than a thousand-fold (Bullard *et al.,* 1978) makes it unlikely that regeneration of reduced cofactor could be rate limiting for the hydroxylation of tryptophan or tyrosine unless for some reason NADH (or NADPH) levels were to become rate limiting. Then a situation might arise where the proportion of biopterin present in the fully reduced form would decrease through the failure to reduce quinonoid dihydrobiopterin to tetrahydrobiopterin sufficiently rapidly. The unstable quinonoid dihydrobiopterin would then rearrange spontaneously to 7,8-dihydrobiopterin which is not a substrate for quinonoid dihydropteridine reductase (Kaufman and Fisher, 1974). The fact that most biopterin in rat brain is present in the fully reduced form (Fukushima and Nixon, 1980) indicates that this does not normally occur. On the other hand a rapid albeit modest decrease in tetrahydrobiopterin levels without any change in total biopterin has been reported following *d*-amphetamine treatment in rats (Mandell *et al.*, 1980) indicating that alterations in the proportion of biopterin present in the reduced form can occur under abnormal conditions and may provide yet another mechanism for altering the *in vivo* activity of the aromatic amino acid hydroxylases.

The effect of electrical stimulation of 5-HT neurones

The synthesis of 5-HT from tryptophan in 5-HT nerve endings is increased in a frequency dependent manner in response to electrical stimulation of the midbrain raphe nuclei in which clusters of 5-HT cell bodies are found. A maximal increase in the conversion of radiolabelled tryptophan to 5-HT of 100 per cent was observed at a stimulation frequency of 10 Hz. At higher frequencies of stimulation synthesis rates fell off though they still remained significantly above unstimulated controls (Shields and Eccleston, 1972). This increase in amine synthesis results from the enhanced conversion of tryptophan to 5-HTP (Herr *et al.*, 1975; Bourgoin *et al.*, 1980) which can be demonstrated in the presence of a decarboxylase inhibitor. However, the mechanism for this increase in tryptophan hydroxylase activity remains unknown. Electrical stimulation does not increase brain tryptophan levels over control values so it is unlikely that the increased conversion of tryptophan to 5-HTP is due to a rise in tryptophan concentration within serotonergic neurones though this possibility cannot be tested directly. Studies on brain slices have revealed some other important characteristics of the stimulation-induced increase in 5-HT synthesis. For example it persists when tissue tryptophan concentrations are increased from 230 to 2000 pmol/mg tissue protein by incubation of the slices in media containing increasing concentrations of L-tryptophan. In addition the stimulation-induced increase in 5-HT synthesis has an absolute dependance on extracellular calcium. It can be completely abolished by omission of calcium ions from the incubation medium or by addition of magnesium ions (Elks *et al.*, 1979a, b), which block the voltage-dependent calcium channels present in excitable tissues (Baker, 1975). Since such manipulations also block transmitter release both processes may be closely coupled by a common triggering mechanism, the depolarization induced entry of calcium ions into nerves. Although electrical stimulation of the slices does not alter overall tissue tryptophan levels it does increase the uptake of radiolabelled tryptophan. However, the enhanced uptake persists during depolarization in a calcium free medium and is therefore not associated with the increase in 5-HT synthesis which is abolished under these conditions (Elks *et al.*, 1979b). Whether an increase in the concentration of reduced pterin cofactor is produced under these conditions and contributes to the stimulation induced increase in 5-HT synthesis has not been examined. Several investigators have now looked to see whether the properties of tryptophan hydroxylase are changed in response to depolarization. Exposure of brain stem slices to a potassium-enriched incubation medium increases the activity of tryptophan hydroxylase in low-speed supernatant preparations assayed in the presence of subsaturating or barely saturating concentrations of tryptophan and 6-MPH$_4$. The increase in activity is not observed when calcium is omitted from the incubation medium. A kinetic analysis shows an increase in V_{max} and in one study a small decrease in the K_m of the enzyme for both substrate and and 6-MPH$_4$ cofactor. (Boadle-Biber, 1978; Hamon *et al.*, 1979). Similar changes in the kinetic properties of the enzyme are also produced by incubating the slices with agents which increase the free intracellular calcium levels in nerve, such as sodium free medium, ouabain and metabolic inhibitors (Boadle-Biber, 1979) an observation that lends support to the hypothesis that calcium ions trigger the alterations in the kinetic properties of the enzyme. One interesting observation is that the activity of enzyme prepared from depolarized slices cannot be further increased by incubation of the enzyme supernatant under phosphorylating conditions. The view has been expressed that the depolarization-induced increase in enzyme activity may be triggered by a calcium dependent phosphorylation of the enzyme itself or of an activator protein (Hamon *et al.*, 1979). However, the changes in the kinetic properties of the enzyme following potassium depolarization are not identical with those observed under phosphorylating conditions. Whether the alterations in the properties of the enzyme resulting from potassium depolarization are relevant to the increase in 5-HT synthesis observed with electrical stimulation is not known. In a recent study involving electrical stimulation of the raphe magnus no change was observed in the activity of tryptophan hydroxylase prepared from extracts of spinal cord, which receives the projections of 5-HT neurones of the raphe magnus, even though 5-HT synthesis was increased by 50 per cent (Bourgoin *et al.*, 1980). This failure to detect any change in enzyme activity may be due to the instability of the changes induced by electrical stimulation. However, the enhanced synthesis of 5-HT persists for half an hour after stimulation stops (Herr *et al.*, 1975). Moreover the altered kinetic properties of the enzyme from depolarized brain slices persist after freezing and thawing (Boadle-Biber, unpublished). Thus other factors besides activation of tryptophan hydroxylase may be responsible for the stimulation induced increase in 5-HT synthesis.

Effect of inhibition of 5-HT neuronal firing

In contrast to the effects of electrical stimulation, inhibition of the ongoing firing of 5-HT neuonres by acute lesions of the midbrain raphe nuclei or acute transections of tracts which carry ascending or descending 5-HT fibers reduces the synthesis of 5-HT (Carlsson, *et al.*, 1973; Herr and Roth, 1976; Kehr and Speckenbach, 1978). As in the case of electrical stimulation no changes are observed in the tissue levels of endogenous tryptophan following acute lesions. Furthermore, the decrease in synthesis persists when brain tryptophan levels are made saturating with a tryptophan load (Carlsson *et al.*, 1973). So far no studies have been carried out to determine whether the kinetic properties of tryptophan hydroxylase from 5-HT neurones in which firing has been blocked are altered in any way.

A variety of drugs that are known to inhibit the firing of the 5-HT neurones of the midbrain raphe such as lysergic acid diethylamide, and chlorimipramine (Aghajanian and Wang, 1978) have also been found to reduce 5-HT synthesis in rat brain (Carlsson and Lindquist, 1978b; Schubert *et al.*, 1970; Kehr and Speckenbach, 1978). However, other drugs, such as monoamine oxidase inhibitors, which also shut off raphe cell firing (Aghanjanian and Wang, 1978) do not reduce 5-HT synthesis in all brain regions (Tappaz and Pujol, 1980). Furthermore it is by no means clear that the observed decrease in 5-HT synthesis is necessarily related to the suppression of 5-HT neuronal firing. Chlorimipramine for example also reduces brain tryptophan levels (Neckers *et al.*, 1977) and this in turn can alter the kinetic properties of tryptophan hydroxylase. Thus the association between lowered 5-HT neuronal firing rate and a reduction in 5-HT synthesis may in this instance be purely fortuitous.

Diurnal variations in 5-HT synthesis

One characteristic feature of 5-HT neurones is the existence of a circadian rhythm of 5-HT levels. This is observed with whole brain or in discrete regions and is most marked in the cortex, hypothalamus and lower brain stem of the rat (Quay, 1968), where 5-HT levels decrease markedly during the dark period and increase during the light period of the 24-h cycle. One explanation for such cycles of transmitter content is that utilization outstrips synthesis during the period when amine content falls, a time which corresponds to the awake and active state of the nocturnal rodent, and also to the highest rates of 5-HT neuronal firing in the dorsal raphe of the cat (Trulson and Jacobs, 1979). Others on the other hand, have suggested that brain 5-HT levels simply follow the cyclic changes in plasma and brain tryptophan levels (Fernstrom *et al.*, 1974), though recent studies indicate that the correlation between regional brain levels of tryptophan and 5-HT is often rather poor (Morgan *et al.*, 1975; Hery *et al.*, 1977). Although brain tryptophan levels do determine the overall level of 5-HT synthesis as already discussed, there are other important factors which influence 5-HT synthesis such as the presence or absence of neuronal activity. A number of investigators have therefore tried to ascertain whether these rhythms of 5-HT content are paralleled by changes in 5-HT synthesis and tryptophan hydroxylase activity during the 24-h cycle. In one study, Hery *et al.* (1972) could find no difference between 5-HT synthesis at night or during the day when they corrected for the increased uptake of radiolabelled tryptophan into brain observed at night. On the other hand marked cyclic variations were found in the activity of tryptophan hydroxylase measured *in vitro* in homogenates of discrete brain regions under saturating concentrations of substrate and artificial cofactor (6-MPH$_4$). The levels of enzyme activity were, however, out of phase in different brain regions, and even in the cell bodies and terminal projections of the same 5-HT neurones. Since the changes in enzyme activity in the terminals occurred too rapidly to be explained by axonal transport of new enzyme molecules from the cell bodies, the possibility was raised that the terminal region may regulate its enzyme activity independently of the cell body (Kan *et al.*, 1977); Natali *et al.*, 1980). Since these measurements of enzyme activity were made under zero order kinetics they suggest that a change in apparent V_{max} may have occurred. However, a detailed kinetic study of a supernatant preparation of enzyme from rat midbrain which is derived largely from 5-HT cell bodies revealed a night-time activation of tryptophan hydroxylase expressed as a decrease in the K_m for tetrahydrobiopterin, but no change in V_{max} (McLennan and Lees, 1978). At present it remains unclear which of these factors influence the rate of 5-HT synthesis in 5-HT neurones during the night/day cycle, or indeed whether alterations in enzyme activity measured *in vitro* are actually reflected in altered rates of 5-HT synthesis *in vivo* at all.

Inhibition of 5-HT synthesis by 5-HT

Although preparations of tryptophan hydroxylase are inhibited by 5-HT and 5-HTP, extremely high concentrations of these substances (millimolar range) must be added to the enzyme reaction medium to see an effect and 5-HTP is much more potent than 5-HT (Kaufman, 1974). On the other hand studies *in vivo* indicate that synthesis of 5-HT may be depressed in some brain regions after intraneuronal levels of 5-HT have been increased by inhibition of monoamine oxidase. For example Tappaz and Pujol (1980) found that 5-HT synthesis 3 h after pargyline treatment was reduced to 16 per cent of control values in cortex, which contains terminal branches of 5-HT neurones, but was virtually unchanged in the dorsal and median raphe nuclei, which contain 5-HT cell bodies. A decrease in 5-HT synthesis also occurred in slices of rat brain stem when 2 μM 5-HT was included in the incubation medium, a manipulation that raises neuronal 5-HT levels significantly (Hamon *et al.*, 1979b). However monoamine oxidase inhibition left the rates of 5-HT synthesis in control or field stimulated slices unchanged (Elks *et al.*, 1979a). Thus the question as to whether 5-HT reduced its own synthesis remains confused. Nevertheless the fact that inhibition was observed only in serotongeric terminals *in vivo* may explain some of these discrepancies and also provide a clue as to the way 5-HT exerts its inhibitory effect. Recent studies indicate that 5-HT reduces its own calcium-dependent release from 5-HT terminals in depolarized cortical slices (Göthert and Weinheimer, 1979) and also decreases the voltage dependant calcium current of action potentials in dorsal root ganglion cells (Dunlap and Fischbach, 1978). If, as seems likely, 5-HT interferes with its own release, through an interaction with presynaptic autoreceptors, that reduces the inward calcium current evoked by action potentials, then it could also reduce the calcium-dependent increase in 5-HT synthesis observed during depolarization through the same mechanism. Such an effect would be unmasked following treatment with a monoamine oxidase inhibitor since the 5-HT release into the environment of tonically firing neurones would remain around longer and in higher concentrations than normal. It is conceivable that such a mechanism could also have a physiological role in the control of 5-HT synthesis.

Corticosteroids and stress

Tryptophan hydroxylase appears to be an inducible enzyme since its levels increase by a cycloheximide sensitive mechanism 1-2 days after treatment with reserpine, a drug that markedly increases plasma levels of corticosteroids (Zivcovic *et al.*, 1974). This increase in enzyme activity results in enhanced 5-HT formation which can be demonstrated *in vitro* in slice preparations (Hamon *et al.*, 1979b). Although these findings suggest that corticosteroids may regulate enzyme activity directly, with the exception of the work of one group (Azmitia and McEwen, 1974) neither adrenalectomy nor corticosteroid administration has been found to alter tryptophan hydroxylase activity in adult rats (Kizer, *et al.*, 1976; Sze *et al.*, 1976). In neonatal rats, on the other hand, the adrenal gland is required for the normal developmental rise in tryptophan hydroxylase activity (Sze *et al.*, 1976). It should be noted that corticosteroid administration does increase 5-HT synthesis, but that this occurs as a result of enhanced tryptophan uptake and not from changes in the level of tryptophan hydroxylase (Neckers and Sze, 1975).

Stressful stimuli such as footshock, ether anaesthesia and cold exposure which raise circulating levels of corticosteroids, were found to increase tryptophan hydroxylase activity, an effect that was blocked by bilateral adrenalectomy (Azmitia and McEwen, 1974). However, immobilization stress was without effect on enzyme activity in another study (Palkovits *et al.*, 1976). Although these results are not at all in agreement, they do suggest that corticosteroids may mediate stress-induced changes in enzyme activity.

Effects of drugs on 5-HT synthesis

A great number of drugs, particularly psychoactive ones such as hallucinogens and antidepressants, have been examined for their effects on 5-HT synthesis. From a consideration of what has already been said about 5-HT synthesis it is clear that a variety of mechanisms exist by which a given drug can potentially alter the synthesis of 5-HT, excluding direct inhibition of either of the enzymes involved. These include changing the brain levels of tryptophan or reduced cofactor, altering the firing rate of 5-HT neurones through direct actions on the 5-HT perikarya or effects mediated over other synapses (e.g. Aghajanian and Wang, 1978), interactions with presynaptic 5-HT autoreceptors in the terminal region, and modifications to the kinetic behaviour of the rate limiting enzyme tryptophan hydroxlyase. A given drug may obviously exert one or more of these actions. In addition other types of interaction probably exist, but remain to be identified. Examples of the effects on 5-HT synthesis of drugs that alter 5-HT neuronal firing rate or brain tryptophan levels are well known and have already been discussed. Numerous other instances of substances that increase brain tryptophan levels have also been documented (Gessa and Tagliamonte, 1974). Whether, in fact, drug induced alterations in the level of reduced cofactor (Mandell *et al.,* 1980) or interactions with presynaptic autoreceptors on the 5-HT nerve terminals that regulate 5-HT release (e.g. Göthert and Weinheimer, 1979), actually influence 5-HT synthesis has not been determined. There are, however, several examples of drug induced alterations in the kinetic properties of tryptophan hydroxylase. One case involves methiothepin, which is believed to block 5-HT receptors in brain; *in vivo* treatment with this drug enhances 5-HT synthesis measured in slices of brain stem and produces an increased affinity of low-speed supernatant preparations of tryptophan hydroxylase for tryptophan (Hamon *et al.,* 1976). Dibutyryl cyclic AMP (N^6, O^2-dibutyryladenosine-3′,5′-cyclic monophosphate) also increases 5-HT synthesis when infused into the cerebral ventricles (Debus and Kehr, 1979). Part of this effect may arise from enhanced tryptophan uptake (Gessa and Tagliamonte, 1974) and part from the activation of tryptophan hydroxylase observed when enzyme is prepared from brain slices pretreated with dibutyryl cyclic AMP (Boadle-Biber, 1980).

Inhibitors of 5-HT synthesis

Serotonin synthesis can be reduced by inhibition of either tryptophan hydroxylase or aromatic amino acid decarboxylase. In practice little reduction in 5-HT levels occurs after *in vivo* decarboxylase inhibition with the widely used inhibitors of the hydrazine class (e.g. NSD 1015; Ro4-4602). A single dose of Ro4-4602 results in a decrease in 5-HT levels of 30 per cent 1 h later, but levels return to normal within 4–8 h. Furthermore the effect of the drug is not enhanced by repeated administration (Pletscher *et al.,* 1966). On the other hand an irreversible inhibitor (suicide inactivator) of aromatic amino acid decarboxylase, α-monofluoromethyldopa, was recently developed (Maycock *et al.,* 1980) and this drug produces a profound fall in monoamine levels after repeated dosage *in vivo* (Jung *et al.,* 1979). It may be a useful tool for studying both tryptophan hydroxylation and 5-HT depletion in the same brain areas, particularly if its use can be combined with the new analytical tools of HPLC and electrochemical detection.

Inhibitors of tryptophan hydroxylase are widely used as experimental tools to estimate transmitter turnover and have the advantage that they produce a much more selective depletion of 5-HT, thereby permitting the study of 5-HT turnover in the absence of major alterations in other monoamine systems. The inhibitors of tryptophan hydroxylase fall into three classes, ring-substituted phenylalanine, 6-substituted tryptophan derivatives, and catechols. (McGeer and Peters, 1969). Inhibitors in widespread use include parachlorophenylalanine (Koe and Weissman, 1966) which appears to cause an irreversible inhibition of the enzyme as well as a profound and long-lasting reduction in 5-HT levels; 6-fluoro and 6-chloro tryptophan which induce a somewhat more selective but briefer inhibition of tryptophan hydroxylase (Peters, 1971) and finally a catechol, α-propyldopacetamide, (α-propyl-3,4-dihydroxyphenylacetamide) which is the least specific of these inhibitors (see Hamon *et al.*, 1979a).

THE PINEAL GLAND

The pineal gland contains both tryptophan hydroxylase and aromatic amino acid decarboxylase and synthesizes and stores large amounts of serotonin. The serotonin content of the gland exhibits a very pronounced diurnal rhythm with a dramatic fall in levels taking place soon after the onset of darkness and a return to high levels occurring during the daytime (see Axelrod, 1978). The serotonin in the pineal gland has a special function as the precursor for 5-methoxy-*N*-acetylserotonin or melatonin, the factor isolated by Lerner which blanches amphibian skin and suppresses the function of the gonads in mammals. Melatonin formation is at its peak during darkness when serotonin levels decline rapidly. The synthesis of melatonin involves a two-step reaction shown in Fig. 4 in which 5-HT is first converted to *N*-acetylserotonin by *N*-acetyltransferase and acetyl coenzyme A and the *N*-acetylserotonin is then *O*-methylated by hydroxyindole-*O*-methyltransferase which transfers a methyl group from *S*-adenosylmethionine. At the start of the dark phase of the day/night cycle there is an increase in the activity of both of these enzymes, and particularly *N*-acetyltransferase, in response to norepinephrine released from the terminals of sympathetic postganglionic neurones from the superior cervical ganglion which innervate the pinealocytes. The 50- to 100-fold increase in pineal *N*-acetyltransferase activity that occurs is mediated by an increase in cyclic AMP levels in response to beta adrenergic receptor activation, and involves extremely rapid synthesis of new enzyme protein (see Zatz, 1978).

Other metabolites of serotonin besides melatonin are also formed during the dark period and include 5-methoxytryptophol which presumably arises from oxidative deamination of serotonin by monoamine oxidase followed by reduction and *O*-methylation (Wurtman and Ozaki, 1978).

The diurnal rhythm of pineal melatonin content follows an endogenous clock. The rhythm persists in blinded animals, but is abolished if the sympathetic innervation is destroyed or decentralized, if a beta blocking agent is administed or if the animals are kept in light 24 h a day. Blood borne catecholamines secreted from the adrenal medulla in response to stress will also increase melatonin formation even in a denervated gland. There is also evidence that melatonin formation may be affected by ovarian steroid hormones (Wurtman and Ozaki, 1978).

SUMMARY

Synthesis of serotonin in serotonergic neurones *in vivo* proceeds at a rate which is much lower than would be predicted from the *in vitro* activity of the rate-limiting enzyme, tryptophan hydroxylase, measured in air and in the presence of saturating concentrations of substrate and cofactor. This observation leads naturally to the conclusion that the activity of this enzyme must be severely restricted *in vivo*. In fact it is likely that the enzyme operates at or below the K_m for tryptophan and possibly oxygen as well and the cofactor concentration may be even lower. Other factors, such as the kinetic properties of the enzyme itself, may also contribute to the regulation of tryptophan hydroxylase activity, and hence 5-HT synthesis *in vivo*. For example, a variety of treatments *in vitro* and *in vivo* have been observed to alter the kinetic behaviour of this enzyme, thus suggesting that it may exist in different states of activity *in vivo*, or possibly even in an inactive form. (The fact that the enzyme can be reversibly inactivated by oxygen provides an experimental basis for this second possibility.) Although 5-HT synthesis can be altered in a great many different experimental situations the mechanisms involved are only partially understood. It is unlikely that the regulatory processes involved in serotonin synthesis in intact tissue will be fully appreciated until tryptophan hydroxylase itself has been obtained in a homogeneous form and its kinetic and regulatory properties worked out. In spite of much excellent earlier work this remains a formidable task.

FUTURE DEVELOPMENTS

The recent development of specific methods for isolating and quantitating tetrahydrobiopterin has provided a tool for studying the formation and regulation of this cofactor. It should now be possible to identify the factors that may influence its formation and ultimately characterize the pathway and enzymes involved. Such studies should provide important new insights into the regulation of serotonin synthesis as well as that of other monoamine neurotransmitters. In the field of serotonin metabolism new analytical techniques (HPLC with electrochemical detection) also promise more refined approaches to long standing areas of interest. These methods allow the rapid and simultaneous analysis of serotonin and its metabolites in a single sample of tissue without the necessity of preparing derivatives or purification. Deproteination is all that is required. Their use should permit more complete studies of regional 5-HT metabolism under different physiological conditions.

ACKNOWLEDGEMENT

This chapter was written while the author was receiving partial support from grant NS14090 from NINCDS.

REFERENCES

Abou Donia, M., Wilson, S. P., and Viveros, O. H. (1980). Changes in adrenal tetrahydropbiopterin (BH$_4$) induced by insulin and reserpine and its role in BH$_4$-dependent hydroxylation reactions, *Soc. Neurosci., Abstr.* **10**, 643.

Aghajanian, G. K. and Wang, R. Y. (1978). Physiology and pharmacology of central serotonergic neurones, in *Psychopharmacology: A Generation of Progress* (Eds M. A. Lipton, A. DiMascio and K. F. Killam), pp. 171–183, Raven Press, New York.

Axelrod, J. (1978) Introductory remarks on regulation of pineal indoleamine synthesis, *J. Neural Transmission*, Suppl., 13, 73–79.

Azmitia, E. C. and McEwen, B. S. (1974). Adrenal cortical influence on rat brain tryptophan hydroxylase activity, *Brain Res.* 78, 291–302.

Baker, P. F. (1975) Transport and metabolism of calcium ions in nerve; in *Calcium Movements in Excitable Cells*, (Eds P. F. Baker and H. Reuter) pp. 7–53, Pergamon Press, Oxford.

Blakley, R. L. (1969). *The Biochemistry of Folic Acid and Related Pteridines*, pp. 58–76, John Wiley and Sons, Inc., New York.

Boadle-Biber, M. C. (1978). Activation of tryptophan hydroxylase from central serotonergic neurones by calcium and depolarization, *Biochem. Pharmacol.* 27, 1069–1079.

Boadle-Biber, M. C. (1979). Activation of tryptophan hydroxylase from slices of rat brain stem incubated with agents which promote calcium uptake or intraneuronal release, *Biochem. Pharmacol.* 28, 2129–2138.

Boadle-Biber, M. C. (1980). Activation of tryptophan hydroxylase from slices of rat brain stem incubated with $N^6,O^2{}'$-dibutyryl adenosine-3,5-cyclic monophosphate, *Biochem. Pharmacol.* 29, 669–672.

Bouchard, S. and Roberge, A. G. (1979). Biochemical properties and kinetic parameters of dihydroxyphenylalanine-5-hydroxytryptophan decarboxylase in brain, liver and adrenals of cat, *Can. J. Biochem.* 57, 1014–1018.

Bourgoin, S., Oliveras, J. L., Bruxelle, J., Hamon, M., and Besson, J. M. (1980). Electrical stimulation of the nucleus raphe magnus in the rat. Effects on 5-HT metabolism in the spinal cord, *Brain Res.*, 194, 377–389.

Bullard, W. P., Guthrie, P. B., Russo, P. V., and Mandell, A. J. (1978). Regional and subcellular distribution and some factors in the regulation of reduced pterins in rat brain, *J. Pharmacol. Exp. Ther.* 206, 4–20.

Butler, I. J., Koslow, S. H., Krumholz, A., Holtzman, N. A., and Kaufman, S. (1978). 'A disorder of biogenic amines in dihydropteridine reductase deficiency', *Ann. Neurol.* 3, 224–230.

Carlsson, A., (1974). The *in vivo* estimation of rates of tryptophan and tyrosine hydroxylation: Effects of alteration in enzyme environment and neuronal activity, in *Aromatic Amino Acids in the Brain*, Ciba Foundation Symposium 22 (Eds G. E. W. Wolstenholme and D. W. Fitzsimmons), pp. 117–125, Elsevier, Amsterdam.

Carlsson, A., Holmin, T., Lindquist, M., and Siesjö, B. (1977). Effect of hypercapnia on tryptophan and tyrosine hydroxylation in rat brain, *Acta Physiol. Scand.* 99, 503–509.

Carlsson, A. and Lindquist, M. (1973). *In vivo* measurements of tryptophan and tyrosine hydroxylase activities in mouse brain, *J. Neural Transmission*, 34, 79–91.

Carlsson, A. and Lindquist, M. (1978a). Dependence of 5-HT and catecholamine synthesis on concentrations of precursor amino acids in rat brain, *Naunyn Schmiedeberg's Arch. Pharmacol.* 303, 157–164.

Carlsson, A and Lindquist, M. (1978b). Effects of antidepressant agents on the synthesis of brain monoamines, *J. Neural Transmission*, 43, 73–91.

Carlsson, A., Lindquist, M., Magnussen, T., and Atack, C. (1973). Effect of acute transection on the synthesis and turnover of 5-HT in the rat spinal cord, *Naunyn Schmiedeberg's Archiv. Pharmacol.* 277, 1–12.

Clark, J. B., Nicklas, W. J., and Degn, H. (1976). The apparent K_m for oxygen of rat brain mitochondrial respiration, *J. Neurochem.* **26**, 409–411.

Colmenares, J. L., Wurtman, R. J., and Fernstrom, J. D. (1975). Effects of ingestion of a carbohydrate-fat meal on the levels and synthesis of 5-hydroxyindoles in various regions of the rat central nervous system, *J. Neurochem.* **25**, 825–829.

Davis J. N. and Carlsson, A. (1973a). The effect of hypoxia on monoamine synthesis levels and metabolism in rat brain, *J. Neurochem.* **21**, 783–790.

Davis J. N. and Carlsson, A. (1973b). Effect of hypoxia on tyrosine and tryptophan hydroxylation in unanaesthetized rats brain, *J. Neurochem.* **20**, 913–915.

Davis, J. N., Carlsson, A., MacMillan, V., and Siesjo, B. K. (1973). Brain tryptophan hydroxylation: Dependance on arterial oxygen tension, *Science*, **182**, 72–74.

Debus, G. and Kehr, W. (1979). Catecholamine and 5-hydroxytryptamine synthesis and metabolism following intracerebroventricular injection of dibutyryl cyclic AMP, *J. Neural Transmission*, **45**, 195–206.

Denizeau, F. and Sourkes, T. L. (1978). Regional transport of tryptophan in rat brain, *J. Neurochem.* **28**, 951–959.

Diaz, P. M., Ngai, S. H., and Costa, E. (1968). Effect of oxygen on brain serotonin metabolism in rats, *Am. J. Physiol.* **214**, 591–594.

Dunlap, K. and Fischbach, G. D. (1978). Neurotransmitters decrease the calcium component of sensory neurone action potentials, *Nature (Lond.)* **276**, 837–839.

Eccleston, D., Ashcroft, G. W., and Crawford, T. B. B. (1965). 5-hydroxyindole metabolism in rat brain. A study of intermediary metabolism using the technique of tryptophan loading-II, *J. Neurochem.* **12**, 493–503.

Elks, M. L., Youngblood, W. W., and Kizer, J. S. (1979a). Serotonin synthesis and release in brain slices: Independance of tryptophan, *Brain Res.* **172**, 471–486.

Elks, M. L., Youngblood, W. W., and Kizer, J. S. (1979b). Synthesis and release of serotonin by brain slices: Effect of ionic manipulations and cationic ionophores, *Brain Res.,* **172**, 461–469.

Fernstrom, J. D. and Hirsch, M. J. (1976). Brain serotonin synthesis: Reduction in corn malnourished rats, *J. Neurochem.* **28**, 877–879.

Fernstrom, J. D., Madras, B. K., Munro, H. M., and Wurtman, R. J. (1974). Nutritional control of the synthesis of 5-hydroxytryptamine in the brain, in *Aromatic Amino Acids in the Brain*, Ciba Foundation Symposium 22 (Eds G. E. W. Wolstenholme and D. W. Fitzsimons) pp. 153–166, Elsevier, Amsterdam.

Friedman, P. A., Kappelman, A. H., and Kaufman, S. (1972). Partial purification and characterization of tryptophan hydroxylase from rabbit hind brain, *J. Biol. Chem.* **247**, 4165–4173.

Fukushima, T. and Nixon, J. C. (1980). Analysis of reduced forms of biopterin in biological tissues and fluids, *Anal. Biochem.* **102**, 176–188.

Gal, E. M. and Patterson, K. (1973). Rapid nonisotopic assay of tryptophan-5-hydroxylase activity in tissues, *Anal. Biochem.* **52**, 625–629.

Gessa, G. L. and Tagliamonte, A. (1974). Serum free tryptophan: Control of brain concentrations of tryptophan and synthesis of 5-hydroxytryptamine, in *Aromatic Amino Acids in the Brain*, Ciba Foundation Sympsoium 22, (Eds G. E. W. Wolstenholme and D. W. Fitzsimons), pp. 207–216, Elsevier, Amsterdam.

Göthert, M. and Weinheimer, G., (1979). Extracellular 5-hydroxytryptamine inhibits 5-hydroxytryptamine release from rat brain cortex slices, *Naunyn Schmiedeberg's Arch. Pharmacol.* **310**, 93–96.

Green, H. and Sawyer, J. L. (1966). Demonstration, characterization and assay procedure of tryptophan hydroxylase in rat brain, *Anal. Biochem.* **15**, 53–64.

Hamon, M., Bourgoin, S., Artaud, F., and Glowinski, J. (1979). The role of intraneuronal 5-HT and of tryptophan hydroxylase activation in the control of 5-HT synthesis in rat brain slices incubated in K$^+$-enriched medium, *J. Neurochem.* **33**, 1031–1042.

Hamon, M., Bourgoin, S., Artaud, F., Hery, F. (1977). Rat brain stem tryptophan hydroxylase: Mechanism of activation by calcium, *J. Neurochem.* **28**, 811–818.

Hamon, M., Bourgoin, S., Hery, F., and Simmonet, G. (1978a). Activation of tryptophan hydroxylase by adenosine triphosphate; magnesium and calcium, *Mol. Pharmacol.* **14**, 99–110.

Hamon, M., Bourgoin, S., Hery, F., and Simmonet, G. (1978b). Phospholipid-induced activation of tryptophan hydroxylase from the rat brain stem, *Biochem. Pharmacol.* **27**, 915–922.

Hamon, M., Bourgoin, S., Hery, F., Ternaux, J. P., and Glowinski, J. (1976). *In vivo* and *in vitro* activation of soluble tryptophan hydroxylase from rat brain stem, *Nature (Lond.)* **260**, 61–63.

Hamon, M., Bourgoin, S., and Youdim, M. B. H. (1979a). Tryptophan hydroxylation in the central nervous system and other tissues, in *Aromatic Amino Acid Hydroxylases and Mental Disease* (Ed. M. B. H. Youdim) pp. 233–297, John Wiley, New York.

Hedner, T. and Lundborg, P. (1979). Regional changes in monoamine synthesis in the developing rat brain during hypoxia, *Acta Physiol. Scand.* **106**, 139–143.

Hedner, T., Lundborg, P., and Engel, J. (1978). Effect of hypoxia on monoamine synthesis in brains of developing rats, *Biol. Neonate,* **34**, 55–60.

Herr, B. E., Gallager, D. W., and Roth, R. H. (1975). Tryptophan hydroxylase: Activation *in vivo* following stimulation of central serotonergic neurons, *Biochem. Pharmacol.* **24**, 2019–2023.

Herr, B. E. and Roth, R. H. (1976). The effect of acute raphe lesion on serotonin synthesis and metabolism in the rat forebrain and hippocampus, *Brain Res.* **110**, 189–193.

Hery, F., Chouvet, G., Kan, J. P., Pujol, J.-F., and Glowinski, J. (1977). Daily variations of various parameters of serotonin metabolism in the rat brain. II. Circadian variations in serum and cerebral tryptophan levels: Lack of correlation with 5-HT turnover, *Brain Res.* **123**, 137–145.

Hery, F., Rouer, E., and Glowinski, J. (1972). Daily variations of serotonin metabolism in the rat brain, *Brain Res.* **43**, 445–465.

Ichiyama, A., Nakamura, S., Nishizuka, Y., and Hayaishi, O. (1968). Tryptophan-5-hydroxylase in mammalian brain, *Adv. Pharmacol.* **6A**, 5–17.

Ichiyama, A., Nakamura, S., Nishizuka, Y., and Hayaishi, O. (1970). Enzyme studies on the biosynthesis of serotonin in mammalian brain, *J. Biol. Chem.* **245**, 1699–1709.

Jamieson, D. and van den Brenk, H. A. S. (1965). Electrode size and tissue PO_2 measurements in rats exposed to air or high pressure, *J. Appl. Physiol.* **20**, 514–518.

Jellinger, K., Riederer, P., Rausch, W. D., and Kothbauer, P. (1978). Brain monoamines in hepatic encephalopathy and other types of metabolic coma, *J. Neural Transmission,* Suppl., 14, 103–120.

Jung, M. J., Palfreyman, M. G., Wagner, J., Bey, P., Ribereau-Gayon, Zraika, M., and Koch-Weser, J. (1979). Inhibition of monoamine synthesis by irreversible blockade of aromatic amino acid decarboxylase with α-monofluoromethyldopa, *Life Sci.* **24**, 1037–1042.

Kan, J. P., Buda, M., and Pujol, J. F. (1975). Tryptophan-5-hydroxylase activity in the raphe system of the rat brain stem, *Brain res.* **93**, 353–357.

Kan, J. P., Chouvet, G., Hery, F., Debilly, G., Mermet, A., Glowinski, J., and Pujol, J. F. (1977). Daily variations of various parameters of serotonin metabolism in rat brain I. Circadian variations of tryptophan-5-hydroxylase in the raphe nuclei and the striatum, *Brain Res.* **123**, 125–136.

Kato, T., Yamaguchi, T., Nagatsu, T., Sugimoto, T., and Matsuura, S. (1980). Effect of structures of tetrahydropterin cofactors on rat brain tryptophan hydroxylase, *Biochim. Biophys. Acta,* **611**, 241–250.

Kaufman, S. (1974). Properties of the pterin-dependent aromatic amino acid hydroxylases, in *Aromatic Amino Acids in the Brain,* Ciba Foundation Symposium 22, (Eds G. E. W. Wolstenholme and D. W. Fitzsimons) pp. 85–108, Elsevier, Amsterdam.

Kaufman, S. (1977a). Mixed function oxygenases-general considerations, in *Structure and Function of Monoamine Enzymes* (Eds E. Usdin, N. Weiner and M. B. H. Youdim) pp. 3–22, Marcel Dekker, Inc., New York.

Kaufman, S. (1977b). Phenylketonuria: Biochemical Mechanisms, *Adv. Neurochem.,* **2**, 1–116.

Kaufman, S. (1979). Biopterin and metabolic disease, in *Chemistry and Biology of Pteridines* (Eds R. L. Kisliuk and G. M. Brown) pp. 117–124, Elsevier, North Holland, New York.

Kaufman, S., Berlow, S., Summer, G. K., Milstein, S., Schulman, J. D., Orloff, S., Spielberg, S., and Pueschel, S. (1978). Hyperphenylalaninemia due to a deficiency of biopterin, A variant form of phenylketonuria, *New England. J. Med.,* **299**, 673–679.

Kaufman, S. and Fisher, D. B. (1974). Pterin requiring aromatic amino acid hydroxylases, in *Molecular Mechanisms of Oxygen Activation* (Ed O. Hayaishi) pp. 285–369, Academic Press.

Kaufman, S., Holtzman, N. A., Milstein, S., Butler, I. J., and Krumholz, A. (1975). Phenylketonuria due to a deficiency of dihydropteridine reductase, *New Engl. J. Med.,* **293**, 785–790.

Kehr, W. and Speckenbach, W. (1978). Effect of lisuride and LSD on monoamine synthesis after axotomy or reserpine treatment in rat brain, *Naunyn Schmiedeberg's Arch. Pharmacol.,* **301**, 163–169.

Kettler, R., Bartholini, G., and Pletscher, A. (1974). *In vivo* enhancement of tyrosine hydroxylation in rat striatum by tetrahydrobiopterin, *Nature (Lond.)* **249**, 476-478.

Kizer, J. S., Palkovits, M., Kopin, I. J., Saavedra, J. M., and Brownstein, M. J. (1976). Lack of effect of various endocrine manipulations on tryptophan hydroxylase activity of individual nuclei of the hypothalamus, limbic system, and midbrain of the rat, *Endocrinology,* **98**, 743–747.

Kizer, J. S., Zivin, J. A., Saavedra, J. M., and Brownstein, M. J. (1975). A sensitive microassay for tryptophan hydroxylase in brain, *J. Neurochem.* **24**, 779–785.

Knapp, S., Mandell, A. J., and Geyer, M. A. (1974). Effects of amphetamines on regional tryptophan hydroxylase activity and synaptosomal conversion of tryptophan to 5-hydroxytryptamine in rat brain, *J. Pharmacol. Exp. Ther.* **189**, 676–689.

Koe, B. K. and Weissman, A. (1966). *P*-chlorophenylalanine: A specific depletor of brain serotonin, *J. Pharmacol. Exp. Ther.* **154**, 499–516.

Kraut, H., Pabst, W., Rembold, H., and Wildemann, L. (1963). Über das Verhalten des Biopterins in Säugetierorganismus. I Bilanz und Wachstumersuche an Ratten, *Z. Physiol. Chem.* **332**, 101–108.

Kuhn, D. M., Ruskin, B., and Lovenberg, W., (1980). Tryptophan hydroxylase. The role of oxygen, iron and sulfhydryl groups as determinants of stability and catalytic activity, *J. Biol. Chem.,* **255**, 4137–4143.

Kuhn, D. M., Vogel, R. L., and Lovenberg, W., (1978). Calcium-dependent activation of tryptophan hydroxylase by ATP and magnesium, *Biochem. Biophys. Res. Comm.* **82**, 759–766.

Lambertson, C. J. (1980). Hypoxia altitude and acclimatization, in *Medical Physiology* Ed V. B. Mountcastle) Vol. II, pp. 1843–1872, Mosby, St. Louis.

Leniger-Follert, E., Lübbers, D. W., and Wrabetz, W. (1975). Regulation of local tissue PO_2 of the brain cortex at different arterial O_2 pressures, *Pflügers Arch.* **359**, 81–95.

Levine, R. A., Kuhn, D. M., and Lovenberg, W. (1979). The regional distribution of hydroxylase cofactor in rat brain, *J. Neurochem.* **32**, 1575–1578.

Lübbers, D. W. (1968). Tissue hypoxia: Cellular oxygen requirements with special regard to the *in vivo* PO_2 of the brain, *Scand. Clin. Lab. Invest. Suppl.* **102**, 11A.

McGeer, E. G. and Peters, D. A. V. (1969). *In vitro* screen of inhibitors of rat brain serotonin biosynthesis, *Can. J. Biochem.* **47**, 501–506.

McLennan, I. S. and Lees, G. J. (1978). Diurnal changes in the kinetic properties of tryptophan hydroxylase from rat brain, *J. Neurochem.* **31**, 557–559.

Mandell, A. J., Bullard, W. P., Yellin, J. B., and Russo, P. V. (1980). The influence of *D*-amphetamine on rat brain striatal reduced biopterin concentration, *J. Pharmacol. Exp. Ther.* **213**, 569–574.

Maycock, A. L., Aster, S. D., and Patchett, A. A. (1980). Inactivation of 3-(3,4-dihydroxyphenyl)alanine decarboxylase by 2-(fluoromethyl)-3-(3,4-dihydroxyphenyl)alanine, *Biochemistry*, **19**, 709–718.

Meek, J. L. and Lofstrandh, S. (1976). Tryptophan hydroxylase in discrete brain nuclei: Comparison of activity *in vitro* and *in vivo*, *Eur. J. Pharmacol.* **78**, 377–380.

Meek, J. L. and Neckers, L. M., (1975). Measurement of tryptophan hydroxylase in single brain nuclei by high pressure liquid chromatography, *Brain Res.* **91**, 336–340.

Mefford, I. N. and Barchas, J. D. (1980). Determination of tryptophan and metabolites in rat brain and pineal tissue by reversed phase high performance liquid chromatography with electrochemical detection. *J. Chromatography, Biomedical Appl.* **181**, 187–193.

Milstein, S., Kaufman, S., and Summer, G. K. (1980). Hyperphenylalaninemia due to dihydropteridine reductase deficiency: diagnosis by measurement of oxidized and reduced pterins in urine, *Pediatrics,* **65**, 806–810.

Morgan, W. W., Saldana, J. J., Catherine, A. Y., and Morgan, J. F. (1975). Correlation between circadian changes in serum amino acids or brain tryptophan and the contents of serotonin and 5-HIAA in regions of the rat brain, *Brain Res.* **84**, 75–86.

Morot-Gaudry, Y., Hamon, M., Bourgoin, S., Ley, J. P., and Glowinski, J. (1974). Estimation of the rate of 5-HT synthesis in the mouse brain by various methods, *Naunyn Schmiedeberg's Arch. Pharmacol.* **282**, 223–238.

Nagatsu, T., Yamamoto, T., and Kato, T. (1979). A new highly sensitive voltammetric assay for aromatic L-amino acid decarboxylase activity by high performance liquid chromatography, *Anal. Biochem.* **100**, 160–165.

Natali, J. P., McRae-Degueurce, A., Chouvet, G., and Pujol, J. F. (1980). Genetic studies of daily variations of first step enzymes of monoamine metabolism in the brain of inbred strains of mice and hybrids. I. Daily variations of tryptophan hydroxylase activity in the nuclei raphe dorsalis, raphe centralis and in the striatum, *Brain Res.* **191**, 191–203.

Neckers, L. M., Biggio, G., Moja, E., and Meek, J. L. (1977). Modulation of brain tryptophan hydroxylase activity by brain tryptophan content, *J. Pharmacol. Exp. Ther.* **201**, 110–116.

Neckers, L. M. and Sze, P. Y. (1975). Regulation of 5-hydroxytryptamine metabolism in mouse brain by adrenal glucocorticoids, *Brain Res.* **93**, 123-132.

Nixon, J. C., Lee, C. L., Abou-Donia, M., Fukushima, T., Nichol, C. A., Diliberto, Jr. E., and Viveros, O. H. (1980a). Induction of GTP-cyclohydrolase by the stimuli that increase tetrahydrobiopterin (BH$_4$) in rat adrenal medulla and cortex. *Soc. Neurosci., Abstr.* **10**, 643.

Nixon, J. C., Lee, C. L., Milsteinm, S., Kaufman, S., and Bartholome, K. (1980b). Neopterin and biopterin levels in patients with atypical forms of phenylketonuria, *J. Neurochem.* **35**, 898-904.

Palkovits, M., Brownstein, M., Kizer, J. S., Saavedra, J. M., and Kopin, I. J. (1976). Effect of stress on serotonin concentration and tryptophan hydroxylase activity of brain nuclei, *Neuroendocrinology*, **22**, 298-304.

Pardridge, W. M. (1977). Regulation of amino acid availability to the brain, in *Nutrition in the Brain* (Eds R. J. Wurtman and J. J. Wurtman), pp. 141-200, Raven Press, New York.

Pardridge, W. M. (1979a). The role of blood brain barrier transport of tryptophan and other neutral amino acids in the regulation of substrate limited pathways of brain amino acid metabolism, *J. Neural Transmission*, **Suppl. 15**, 43-54.

Pardridge, W. M. (1979b). Tryptophan transport through the blood brain barrier: *in vivo* measurement of free and albumin bound amino acid, *Life Sci.* **25**, 1519-1528.

Peters, D. A. V. (1971). Inhibition of serotonin biosynthesis by 6-halotryptophans *in vivo*, *Biochem. Pharmacol.* **20**, 1413-1420.

Pletscher, A., Gey, K. F., and Burkard, W. P. (1966). Inhibitors of monoamine oxidase and decarboxylase of aromatic amino acids, in *5-Hydroxytryptamine and Related Indolealkylamines, Handbook of Experimental Pharmacology*, XIX (Eds O. Eichler and A. Farah), pp. 652-668, Springer Verlag, New York.

Pratt, O. E. (1979). Kinetics of tryptophan transport across the blood brain barrier, *J. Neural Transmission*, Suppl. 15, 29-42.

Quay, W. B. (1968). Differences in circadian rhythms in 5-hydroxytryptamine according to brain region, *Am. J. Physiol.* **215**, 1448-1453.

Reinhard, J. F., Moskowitz, M. A., Sved A. F., and Fernstrom, J. D. (1980). A simple sensitive and reliable assay for serotonin and 5-HIAA in brain tissue using liquid chromatography with electrochemical detection, *Life Sci.* **27**, 905-911.

Schubert, J., Nybäck, H.., and Sedvall, G. (1970). Accumulation and disappearance of [^{3}H] 5-hydroxytryptamine formed from [^{3}H] tryptophan in mouse brain, Effect of LSD-25, *Eur. J. Pharmacol.* **10**, 215-224.

Shields, P. J. and Eccleston, D. (1972). Effects of electrical stimulation of rat midbrain on 5-hydroxytryptamine synthesis as determined by a sensitive radioisotope method, *J. Neurochem.* **19**, 265-272.

Sims, K. L., Davis, G. A., and Bloom, F. E. (1973). Activities of 3,4-dihydroxy-L-phenylalanine and 5-hydroxy-L-tryptophan decarboxylases in rat brain: Assay characteristics and distribution, *J. Neurochem.* **20**, 449-464.

Smith, R. H., Guilbeau, E. J., and Reneau, D. D. (1977). The oxygen tension field within a discrete volume of cerebral cortex, *Microvasc. Res.* **13**, 233-240.

Sourkes, T. L. (1977). Enzymology of aromatic amino acid decarboxylase, in *Structure and Function of Monoamine Enzymes*, Eds E. Usdin, N. Weiner, and M. B. H. Youdim), pp. 477-495, Marcel Dekker, New York.

Sourkes, T. L. (1978). Tryptophan in hepatic coma, *J. Neural Transmission,* Suppl. 14, 79-86.

Sourkes, T. L. (1979). Nutrients and the cofactors required for monoamine synthesis in nervous tissue, in *Nutrition and the Brain* (Eds R. J. Wurtman and J. J. Wurtman) Vol. 3, 265–299. Raven Press, New York.

Sze, P. Y., Neckers, L., and Towle, A. C. (1976). Glucocorticoids as a regulatory factor for brain tryptophan hydroxylase, *J. Neurochem.* **26**, 169–173.

Tappaz, M. L. and Pujol, J. F. M. (1980). Estimation of the rate of tryptophan hydroxylation *in vivo*: A sensitive microassay in discrete brain nuclei, *J. Neurochem.* **34**, 933–940.

Trulson, M. E. and Jacobs, B. L. (1979). Raphe unit activity in freely moving cats: Correlation with level of behavioural arousal, *Brain Res.* **163**, 135–150.

Trulson, M. E. and MacKenzie, R. G. (1980). Increased tryptophan hydroxylase activity may compensate for decreased brain tryptophan levels in streptozotocin diabetic rats, *J. Pharmacol. Exp. Ther.* **212**, 269–273.

Viveros, O. H., Argueros, L., Connett, R. J. and Kirshner, N. (1969). Mechanism of secretion from the adrenal medulla. IV. The fate of the storage vesicles following insulin and reserpine administration, *Mol. Pharmacol.* **5**, 69–82.

Wurtman, R. J. (1974). Discussion on tryptophan and tyrosine hydroxylation *in vivo*, in *Aromatic Amino Acids in the Brain*, Ciba Foundation Symposium 22 (Eds G. W. W. Wolstenholme and D. W. Fitzsimons), pp. 128, Elsevier, Amsterdam.

Wurtman, R. J. and Ozaki, Y. (1978). Physiological control of melatonin synthesis and secretion: Mechanisms generating rhythms in melatonin, methoxytryptophol, and arginine vasotocin levels and effects on the pineal of endogenous catecholamines the estrous cycle and environmental lighting, *J. Neural Transmission*, Suppl. 13, 59–70.

Yamauchi, T. and Fujisawa, H., (1979a). Regulation of rat brainstem tryptophan-5-monooxygenase. Calcium dependant reversible activation by ATP and magnesium, *Arch. Biochem. Biophys.* **198**, 219–226.

Yamauchi, T. and Fujisawa, H., (1979b). Activation of tryptophan-5-monooxygenase by calcium dependant regulator protein, *Biochem. Biophys. Res. Comm.* **90**, 28–35.

Young, S. N. and Sourkes, T. L. (1977). Tryptophan in the central nervous system: Regulation and significance, *Adv. Neurochem.* **2**, 133–191.

Yuwiler, A., Oldendorf, W. H., Geller, E., and Braun, L. (1977). Effect of albumin binding and amino acid competition of tryptophan uptake into brain, *J. Neurochem.* **28**, 1015–1023.

Zatz, M. (1978). Sensitivity and cyclic neucleotides in the rat pineal gland, *J. Neural Transmission*, Suppl. 13, 97–114.

Zivcovic, B., Guidotti, A., and Costa, E. (1974). On the regulation of tryptophan hydroxylase in brain, *Adv. Biochem. Psychopharmacol.* **11**, 19–30.

Biology of Serotonergic Transmission
Edited by N. N. Osborne
© 1982, John Wiley & Sons Ltd.

Chapter 4

Storage and Release of Serotonin

Elaine Sanders-Bush and L. L. Martin

Department of Pharmacology, Vanderbilt University School of Medicine and
Tennessee Neuropsychiatric Institute, Nashville, Tennessee, USA

INTRODUCTION

The generally accepted concept that intraneuronal serotonin (5-HT) is primarily stored in synaptic vesicles is based largely on indirect evidence and analogy with other transmitter systems. The characteristics of 5-HT storage structures in platelets are considered in relation to their possible modeling of brain storage sites. Several methods for studying 5-HT release exist and each has its limitations. In our opinion, the superfusion of synaptosomes is the preferred system for most studies, although conclusions should be corroborated by *in vivo* studies. Exocytosis is believed to be the mechanism of the nerve impulse-induced release of 5-HT, and we have presented the evidence for this release mechanism in some detail. Several apparent presynaptic receptors which modify the release of 5-HT may exist. We have examined the role of these receptors, of cell firing rates, and of 5-HT synthesis in the control of release. Although many agents of varied structure have been found to release 5-HT, the most well studied are reserpine-like drugs and ring-substituted phenylethylamine and amphetamine derivatives. Methods for indirectly studying 5-HT release and utilization (such as estimation of the turnover of the amine, levels of the major metabolite, 5-hydroxyindoleacetic acid, and behavioural models of 5-HT receptor stimulation) have certainly contributed to our understanding of the basic control mechanisms and the effects of drugs; however, we have emphasized those studies in which the actual release of the amine into the extracellular space was determined directly. After consideration of the general concepts, we have concentrated on the more recent literature and have attempted to point out deficiencies and gaps in present knowledge which may be the subject for future work.

STORAGE

Intraneuronal binding sites

It is generally believed that 5-HT is stored primarily in synaptic vesicles where it is protected from deamination by monoamine oxidase. The accumulating evidence for release of 5-HT by an exocytotic mechanism requires a vesicular storage form of the amine. However, the heterogeneity of intraneuronal 5-HT pools is suggested by a number of observations. For example, it has been shown that newly synthesized 5-HT is preferentially released during depolarization (Shields and Eccleston, 1973; Elks *et al.*, 1979b) suggesting that a large intraneuronal pool is relatively stable and unavailable for release. Various terms have been used to identify these two pools, perhaps the most common being 'functional' and 'reserve' pools referring to the small readily releasable pool and the large relatively stable pool, respectively. The properties and possible functions of such intraneuronal pools have been reviewed by Glowinski (1975).

The possibility of an extravesicular or soluble storage form of 5-HT is suggested in subcellular distribution studies of brain homogenates showing low vesicular, high cytoplasmic concentrations of 5-HT (Halaris and Freedman, 1977; Maynert *et al.*, 1965; Tamir and Huang, 1974). Tamir and associates have demonstrated the presence of a specific 5-HT binding protein in 5-HT neurones (Tamir and Huang, 1974; Tamir and Kuhar, 1975) and it is tempting to speculate that this protein might correspond to the proposed soluble binding site. However, recent studies suggest that the serotonin binding protein is more likely associated with the vesicular storage form. It is highly localized in synaptic vesicles (Tamir and Gershon, 1979) and is released along with 5-HT by electrical stimulation through a Ca^{2+}-dependent and probably exocytotic mechanism (Jonakait *et al.*, 1979). Further evidence for a role of the serotonin binding protein in 5-HT vesicular storage is provided by pharmacological studies showing that 5-HT binding to this protein is prevented by reserpine, which is known to impair 5-HT storage, but not by uptake or receptor blockers (Tamir *et al.*, 1976). Thus, it is entirely possible that the high cytoplasmic concentrations of 5-HT found in subcellular distribution studies are artefacts generated by the release of the amine during the preparation of the synaptic vesicles.

Platelets as models

Platelets have long been considered as useful models for studying 5-HT neuronal storage and release. In platelets, 5-HT is stored in specific cytoplasmic organelles with dense osmiophilic cores (see for example, Da Prada and Pletscher, 1974 for a review). In addition to 5-HT, these organelles contain a high concentration of ATP and it is believed that 5-HT, nucleotides and divalent cations exist as aggregates held together by intermolecular bonding. Given the evidence that the serotonin binding protein may play a role in the intraneuronal binding of 5-HT, it is noteworthy that this protein is absent from platelets (Tamir *et al.*, 1980). This suggests that the mechanism of 5-HT storage in platelets is different from that in brain. Interestingly, electron microscopic autoradiographic studies of platelets show that 85 per cent of the recently taken up 5-HT is associated with structures other than the dense core vesicles, (Lewis and Moertel, 1978), suggesting a second intracellular storage site for 5-HT in mammalian platelets. In this respect the platelet may not be different from brain since a substantial portion of the 5-HT in synaptosomes seems not to be found in vesicles (Halaris and Freedman, 1977; Maynert *et al.*, 1964). The possibility that this is an artefact in both tissues must, however, be considered.

RELEASE

Measurement of 5-HT release

While it is not possible to study the release of 5-HT *in vitro* under the same conditions in which neurones function *in vivo*, it is also not possible to study release mechanisms at the cellular, subcellular and molecular levels *in vivo* because of the complexity of the brain. For this reason, *in vitro* techniques for studying 5-HT release have been devised and in this section, we will discuss the fundamental considerations related to the study of 5-HT release *in vitro* followed by a similar discussion of *in vivo* release studies.

Serotonin release in vitro

The type of tissue preparation (i.e. slices or synaptosomes) which one chooses to use depends, in part, on what aspect of 5-HT release will be studied. For example, if one wishes to examine the local neuronal interactions involved in the control of 5-HT release, tissue slices would be most appropriate. The primary advantage of tissue slices is that they provide a fairly intact preparation with most of the nerve terminals in their natural spatial relationships with other tissue elements. On the other hand, the rapid re-uptake of released transmitter cannot be adequately prevented in slices. Therefore, it is difficult to distinguish between agents which inhibit uptake and those which accelerate release. For this reason, a 5-HT uptake inhibitor is routinely added to the incubation medium. However, many drugs apparently must interact with the 5-HT carrier in order to stimulate release, so the effects of these agents on release are not apparent in the presence of an uptake inhibitor (see p. 119). One must also consider the possibility that pharmacologically active substances (in addition to 5-HT itself) may be released into the intercellular spaces within the tissue slices and since they would not rapidly diffuse away, they might modify 5-HT release. These problems can be partially avoided when synaptosomes rather than tissue slices are used. However, even with synaptosomes, certain procedures must be employed to prevent the reuptake of 5-HT released into the incubation medium and to prevent the effects of pharmacologically active substances released into the incubation medium. Otherwise, in standard incubation-type experiments, uptake inhibition may be interpreted as release and vice versa (Baumann and Maitre, 1976; Heikkila *et al.,* 1975).

The problem of studying release under conditions where the reuptake of released transmitter is negligible can be overcome by using a superfusion technique similar to that described by Raiteri *et al.,* (1974). Briefly, synaptosomal suspensions are pipetted onto a Millipore filter positioned in the bottom of a superfusion chamber. The suspension medium and the superfusate are then drawn across the filter under a moderate vacuum using a peristaltic pump. Superfusate fractions are collected in tubes and analysed for 5-HT. At the end of the superfusion, the filters may be removed directly for analysis or they can be superfused with 0.1N HCl (following extraction for 30 min) and the acid superfusate collected for 5-HT analysis. Release of 5-HT into each fraction is expressed as a percentage of the 5-HT present in the tissue at the time of collection. A similar procedure is used by Mulder *et al.* (1975) except that a layer of Sephadex G-15 serves as a support for the synaptosomes. Using these methods, the re-uptake of 5-HT can be demonstrated to be negligible by the lack of effect of 5-HT uptake inhibitors on both the spontaneous (Cerrito and Raiteri, 1979) and potassium-evoked (Martin and Sanders-Bush, unpublished observations) release of 5-HT.

The next major consideration is how to quantify 5-HT release. Due to the development of a sensitive enzymatic-isotopic microassay for 5-HT (Saavedra *et al.,* 1973), it is now possible to study the release of endogenous 5-HT from brain slices (Reubi and Emson, 1978; Elks *et al.,* 1979a,b). Measurement of the release of endogenous 5-HT has the advantage of allowing one to study the release of 5-HT from pre-existing transmitter pools which have not been manipulated by the investigator. Nevertheless, it is generally simpler to assay for [³H]5-HT than for endogenous

5-HT and so, the use of an isotopic method for determining the rate of 5-HT release may be preferred. When tissues are prelabelled with [³H] 5-HT, however, conditions must be controlled so that only serotonergic terminals accumulate the labelled amine. The non-specific labelling of other tissue components can be minimized by using low concentrations of [³H] 5-HT and, if necessary, selective inhibitors of catecholamine uptake such as nomifensine (Tuomisto, 1977) to prevent the uptake of [³H] 5-HT into catecholaminergic terminals. The selected brain region may also be important. Most investigators have used the hypothalamus for studies of 5-HT release since, in contrast to the striatum, [³H] 5-HT uptake into slices of this region appears to be localized exclusively in serotonergic terminals (Shaskan and Snyder, 1970). However, other regions of the brain may be just as suitable for 5-HT release studies as is the hypothalamus. Since the enzymatic machinery necessary for converting tryptophan to 5-HT is found exclusively in serotonergic cells, another approach used to label 5-HT stores selectively involves incubation of tissues with [³H] tryptophan (Reubi *et al.*, 1978). However, Elks *et al.* (1979b) recently demonstrated that [³H] 5-HT newly synthesized from [³H] tryptophan is preferentially released from brain slices by electrical depolarization. Similarly, since it is not known whether newly taken up 5-HT distributes homogeneously within the various intrasynaptosomal transmitter pools, the release of [³H] amine from tissue prelabelled with [³H] 5HT might reflect release from pools largely containing transmitter accumulated through the uptake process. However, since both [³H] 5-HT newly synthesized from [³H] tryptophan (Snodgrass and Iversen, 1974) and [³H] 5-HT newly taken up by tissues (Mulder *et al.*, 1975) are released following depolarizing stimuli by a Ca^{2+}-dependent mechanism, both methods of labelling tissues appear to be physiologically relevant. It would, nevertheless, be interesting to determine if the release of newly synthesized and newly taken up 5-HT differ and, if so, how.

Various methods have been used to elicit 5-HT release from tissues. These include electrical stimulation, exposure to buffers containing a high concentration of potassium ions, and addition of veratridine to the incubation medium. All appear to produce satisfactory results. However, it should be kept in mind that a careful comparison of the effects of these stimuli on 5-HT release has not been performed. In addition, with respect to the use of electrical depolarization, it is very important to select the proper parameters for electrical stimulation of tissues to ensure that transmitter release does not occur as the result of damage to the tissues (Snodgrass and Iversen, 1974).

As will be discussed later in more detail, 5-HT may be released from tissues either by exocytosis or by carrier-mediated transport across the plasma membrane. Since the release of 5-HT induced by depolarizing stimuli is not carrier-mediated (Martin and Sanders-Bush, unpublished observations) , it is possible to study selectively the effects of drugs on the stimulus-evoked release of 5-HT by inactivating the carrier with a 5-HT uptake inhibitor added to the incubation medium. Likewise, drug-induced 5-HT release from unstimulated tissues can be demonstrated to be carrier-mediated if it is blocked by 5-HT uptake inhibitors. When an uptake inhibitor is used, it is important to demonstrate that it does not affect either basal or depolarization-induced release. For example, we have noted that chlorimipramine (at concentrations routinely used to inhibit 5-HT uptake ($\geqslant$ 1 μM), inhibits the release of 5-HT induced by 15 mM K$^+$, while at higher concentrations ($\geqslant$ 10 μM), it enhances the spontaneous release of 5-HT (Martin and Sanders-Bush, unpublished observations). At a low concentration (0.3 μM), chlorimipramine effectively inhibits uptake, but does not alter release.

Serotonin release in vivo

Various methods have been used to measure the release of 5-HT from nerve terminals *in vivo*. In many of the early (and a few recent) studies, 5-HT released from nerve terminals surrounding the ventricles was collected using a cerebroventricular perfusion technique. Push-pull cannulae, developed for measurement of transmitter release from discrete brain regions have also been used to study 5-HT release. Other perfusion techniques have been recently developed, but they have not yet gained widespread use. Yaksh and Tyce (1979, 1980) developed a spinal superfusion system for measuring release of 5-HT from the spinal cord. Also, cups have been used to superfuse the surfaces of the cerebral cortex (Aiello-Malmberg *et al.*, 1979) and the caudate nucleus (Ternaux *et al.*, 1977) for the purpose of collecting 5-HT released from these areas. However, a major disadvantage of these latter methods is that the perfused animals must be either anaesthetized or cervically transected. The effects of these procedures on 5-HT release have not been throughly examined. Furthermore, these procedures do not allow the assessment of changes in the release of 5-HT during ongoing steady-state behaviours or during the disruption of behaviour following the administration of drugs. In contrast, push-pull cannulae can be used to measure 5-HT release in essentially unrestrained animals.

Prior to 1975, most investigators determined the 5-HT content of brain perfusates using a sensitive bioassay system employing the rat stomach fundus (Vane, 1957). Since that time, more specific and highly sensitive methods have been developed. For example, central 5-HT stores may be labelled prior to the initiation of the perfusion with [^{3}H] 5-HT or [^{3}H] tryptophan or continuously with [^{3}H] tryptophan present in the perfusing medium (Gallagher and Aghajanian, 1975; Hery *et al.*, 1979, 1980; Kantak *et al.*, 1978). [^{3}H] 5-HT released into the superfusate is then chromatographically purified and quantified by liquid scintillation spectrophotometry. It has been suggested by Hery *et al.* (1979) that the possible interference by 5-HT originating from platelets is eliminated by the use of isotopic methods. However, such interference usually does not seem to present a significant problem and can be minimized by centrifugation of the perfusate and by discarding samples that are obviously contaminated with blood (Ashkenazi *et al.*, 1973; Ternaux *et al.*, 1976; Yaksh and Tyce, 1979, 1980). Since tryptophan is preferentially metabolized to 5-HT within serotonergic terminals, 5-HT stores may be more selectively labelled isotopically using [^{3}H] tryptophan rather than [^{3}H] 5-HT which may be accumulated in both serotonergic and catecholaminergic terminals. Hery *et al.* (1979) have also suggested that since newly synthesized 5-HT may be preferentially released, continuous labelling of 5-HT stores with [^{3}-H] tryptophan present in the perfusing medium might be superior to other methods of labelling 5-HT stores. However, Kantak *et al.* (1978) obtained qualitatively similar results regarding the effects of pharmacological agents on [^{3}H] 5-HT release when 5-HT stores were labelled with [^{3}H] tryptophan or [^{3}H] 5-HT prior to initiation of the perfusion or with [^{3}H]-tryptophan applied continuously during perfusion. Thus it appears that the choice of the method for isotopically labelling 5-HT stores is not a critical issue.

New methods for the measurement of the release of endogenous 5-HT into brain perfusates have also been developed. Aiello-Malmberg *et al.* (1979) and Ternaux *et al.* (1976; 1977) used the radioenzymatic method of Saavedra *et al.* (1973) to determine the 5-HT content of brain superfusates. In addition, endogenous 5-HT in superfusates has been assayed by high pressure liquid chromatography combined with electrochemical detection which permits the measurement of less than 1.0 pmole (Yaksh and Tyce, 1979, 1980; Loullis *et al.,* 1980). This technique has the advantage of allowing one to determine rapidly the endogenous release of 5-HT and a number of other neurotransmitters and metabolites simultaneously in a single sample of perfusate.

Recently, a novel approach called voltammetry or electrochemical recording, has been used to measure 5-HT release *in vivo* (Marsden *et al.,* 1979). Basically, a potential (+0.2–+1.0 V) is applied to a micrographic electrode chronically implanted sterotaxically within a specific brain region and the minute current which results from the oxidation or reduction of low molecular weight electroactive compounds in the vicinity of the electrode tip is recorded. The potential at which the oxidation occurs serves as a qualitative indication of the substance being oxidized and the amount of current measured is proportional to the concentration of the substance oxidized. Unfortunately, norepinephrine (NE), dopamine (DA), 5-HT and ascorbic acid are all oxidized at similar potentials so that neuropharmacological manipulations must be performed in order to determine which of these substances is responsible for a change in current. To compound this problem care must be taken in the choice of drugs used to manipulate release since many of the drugs commonly employed in such studies are also electroactive (Marsden *et al.,* 1979) and non-specific. For example, *p*-chloroamphetamine releases catecholamines as well as 5-HT at early times after drug administration (see for example, Sanders-Bush and Steranka, 1978). The main approach used to determine whether enhanced 5-HT release is responsible for an increase in current has been to deplete 5-HT stores with a selective inhibitor of 5-HT synthesis such as *p*-chlorophenylalanine. Thus, pretreatment with *p*-chlorophenylalanine has been demonstrated to prevent the rise in current observed in the striatum following the administration of *p*-chloroamphetamine, and in the hippocampus following electrical stimulation of the median raphe nucleus. Other neuropharmacological tools, may also be useful. For example, the administration of fluoxetine (a selective 5-HT uptake inhibitor) which blocks the *p*-chloroamphetamine-induced release of 5-HT *in vitro* (Ross and Kelder, 1977) also prevents the rise in current measured following *p*-chloroamphetamine administration (Marsden *et al.*, 1979). The specificity for 5-HT release may also be enhanced by recording from regions containing relatively high ratios of 5-HT to catecholamine concentrations. Disadvantages of the *in vivo* voltammetry technique for measuring 5-HT release are numerous: the experiments are technically complex, additional pharmacological studies must be performed to identify the substance(s) responsible for the increase in current and, because of the high backgrounds obtained, it has not yet been established whether one can detect decreases in 5-HT release. On the other hand, 5-HT release can be monitored continuously in essentially unrestrained animals and without the need for perfusion and separate biochemical assays. Results obtained with this technique have thus far looked very promising and it is hoped that in the near future, significant improvements will be made regarding both its qualitative and quantitative aspects.

Evidence for exocytotic release

Several lines of evidence suggest that the stimulus-evoked release of 5-HT is mediated by exocytosis. First, at physiological pH, about 98 per cent of the 5-HT molecules are ionized and are not sufficiently lipophilic to cross the plasma membrane by simple diffusion. Secondly, the potassium-evoked release of 5-HT from superfused synaptosomes is not inhibited by 0.3 μM chlorimipramine present in the superfusion medium (Martin and Sanders-Bush, unpublished observations). This concentration of chlorimipramine completely inhibits 5-HT transport as is demonstrated by its ability to prevent the release of [^{3}H] 5-HT induced by the addition of 0.3 μM unlabelled 5-HT to the superfusion medium. Thus the stimulus-evoked release of 5-HT is not likely to be carrier-mediated. Thirdly, as discussed earlier in this review, most intraneuronal 5-HT appears to be stored within synaptic vesicles. Fourthly, colchicine has been shown to inhibit the release of 5-HT from rat spinal cord slices induced by electrical depolarization (Snodgrass and Iversen, 1974). Colchicine disaggregates microtubules (Shelanski and Taylor, 1967) which appear to be necessary components in the exocytotic release of neurotransmitters and other substances (Axelrod, 1972). Fifthly, the depolarization-induced release of 5-HT has been demonstrated to require Ca^{2+} in studies performed both *in vitro* (Elks *et al.*, 1979a; Jonakait *et al.*, 1979; Lane and Aprison, 1977; Mulder *et al.*, 1975; Snodgrass and Iversen, 1974) and *in vivo* (Yaksh and Tyce, 1980). The spontaneous release of 5-HT *in vivo* is also Ca^{2+}-dependent (Hery *et al.*, 1979). Depolarization of synaptosomes by the addition of high potassium concentrations or veratridine (Blaustein, 1975; Blaustein *et al.*, 1972 and Blaustein and Weismann, 1970) has been shown to enhance Ca^{2+} entry into the synaptosomes. In fact, it appears that the influx of Ca^{2+} with or without depolarization of the membrane is sufficient to elicit 5-HT release since 5-HT release from brain slices is enhanced by the Ca^{2+}-specific ionophore A23187 (Elks *et al.*, 1979a). This ionophore has been shown to increase intracellular Ca^{2+} conductance without significantly altering the resting membrane potential of cells (Pressman, 1976). In addition, the Ca^{2+} which enters the synaptosomes during depolarization appears to stimulate the fusion of vesicular membranes with the plasma membrane (Baldessarini, 1975). Lastly, both serotonin and serotonin binding protein are released by electrical stimulation from the myenteric plexus through a Ca^{2+}-dependent mechanism (Jonakait *et al.*, 1979). In addition, the release of 5-HT and serotonin binding protein from the myenteric plexus is not accompanied by significant release of cytosolic proteins as demonstrated by the absence of lactate dehydrogenase (a cytosolic marker protein) activity in the tissue perfusate measured before, during, or after electrical stimulation. Together with these findings, the simultaneous release of 5-HT and serotonin binding protein, a high MW protein apparently involved in 5-HT storage suggests that both substances are released by a similar mechanism, probably exocytosis.

Regulation of 5-HT release

Role of autoreceptors

Numerous studies have demonstrated the ability of various neurotransmitters to modulate their own release or the release of other neurotransmitters from nerve terminals. It has been proposed that these effects are mediated by neurotransmitter receptors located on presynaptic terminal membranes. Of these various presynaptic receptors, one group called autoreceptors have been found to be specific for the transmitter released from the nerve terminals on which they are located. It is believed that transmitters may inhibit their own release by binding to autoreceptors. Thus, autoreceptors have been suggested to be an important mechanism by which neurones can regulate the amount of transmitter present in the synapse (for a review see: Langer *et al.*, 1980).

Several groups of investigators have proposed the existence of autoreceptors on the terminal membranes of serotonergic cells (Bourgoin *et al.*, 1977; Cerrito and Raiteri, 1979; Farnebo and Hamberger, 1974; Gothert and Weinheimer, 1979). Serotonin (Cerrito and Raiteri, 1979; Gothert and Huth, 1980; Gothert and Weinheimer, 1979) and *d*-lysergic acid diethylamide [LSD]), a putative 5-HT receptor agonist, (Bourgoin *et al.*, 1977; Chase *et al.*, 1967, 1969; Hamon *et al.*, 1974; Farnebo and Hamberger, 1971; Katz and Kopin, 1969) both inhibit the stimulation-induced release of [^{3}H]5-HT from tissue slices and synaptosomes. Furthermore, methiothepin, a putative 5-HT receptor antagonist, prevents the inhibitory effects of exogenous 5-HT (Cerrito and Raiteri, 1979; Gothert and Weinheimer, 1979) and LSD (Bourgoin *et al.*, 1977) on [^{3}H]5-HT release. The published *in vitro* experiments then are consistent with the concept of a 5-HT autoreceptor localized on presynaptic terminals which modulates release. However, our unpublished results, using synaptosomes isolated from hypothalamus and a superfusion technique similar to that described by Raiteri *et al.* (1974), have lead us to believe that this simple explanation may not be adequate to explain the *in vitro* effects of 5-HT and its interactions with presumed antagonists. In our experiments, presumed 5-HT antagonists, such as methiothepin, cyproheptadine, metergoline and mianserin, actually mimic effects of 5-HT, that is inhibit [^{3}H] 5-HT release (Martin and Sanders-Bush, unpublished observations). At present, it is not clear whether the inhibition of 5-HT release by these various agents is a receptor-mediated event. However, it is tempting to suggest that these agents may be acting at sites with properties similar to postsynaptic inhibitory 5-HT receptors identified in electrophysiological studies (Aghajanian and Wang, 1978). These studies show that cinanserin, cyproheptadine, metergoline, methysergide and methiothepin do not block the inhibitory effects of iontophoretically applied 5-HT on the firing rates of neurones in the ventral lateral geniculate and the amygdala, instead they mimicked the effects of 5-HT. Alternatively, the effects of the blockers and of 5-HT itself on release may be nonspecific, i.e. not receptor mediated.

In vivo evidence for the local control of release by 5-HT receptors localized on or near presynaptic terminals has not been forthcoming. There is, however, clear evidence for 5-HT autoreceptors localized on serotonergic raphe cell bodies and receiving input from 5-HT axon collaterals (see Aghajanian and Wang, 1978 for a review). These receptors may serve to control cell firing rate and consequently 5-HT release at terminal sites (see p. 106 below). At least for certain negative feedback mechanisms, the adaptive changes in raphe-cell firing are mediated locally in the raphe by the cell body autoreceptors with no demonstrable role for intraneuronal feedback loops from the terminal areas (Mosko and Jacobs, 1977) although interneuronal loops apparently have some role in controlling 5-HT turnover in terminal areas (Neckers *et al.,* 1979).

Role of α-adrenergic receptors

Presynaptic inhibitory α-adrenoceptors have been demonstrated to be involved in the regulation of the stimulation-evoked release of NE (Langer, 1977). Data from several studies now suggest that 5-HT release may also be regulated by α-receptors located on the terminals of serotonergic cells (Frankhuyzen and Mulder, 1980; Gothert and Huth, 1980; Starke and Montel, 1973). NE has been demonstrated to inhibit the stimulus-evoked release of 5-HT from brain slices (Frankhuyzen and Mulder, 1980; Gothert and Huth, 1980). The inhibitory effect of NE on 5-HT release appeared to be mediated by α-receptors since it was blocked by the α-antagonist, phentolamine, but not by the β-antagonist, propranolol. Furthermore, since phentolamine did not antagonize the inhibition of [^{3}H] 5-HT release from brain slices by exogenous 5-HT, the inhibitory effect of NE did not appear to be mediated by 5-HT receptors (Gothert and Huth, 1980).

Two steps should be taken to enhance the validity of these studies. First, tissues should be labelled with [^{3}H] 5-HT in the presence of a selective NE uptake inhibitor such as desipramine to exclude the possibility that [^{3}H] 5-HT is being accumulated by and released from noradrenergic terminals. Secondly, these experiments should be performed with synaptosomes rather than tissue slices. Phentolamine has been demonstrated to enhance the release of both 5-HT (Frankhuyzen and Mulder, 1980; Gothert and Huth, 1980) and NE (Dismukes and Mulder, 1976 and Wemer *et al.*, 1979) from tissue slices. Presumably, phentolamine is antagonizing the inhibition of release of [^{3}H] 5-HT and [^{3}H] NE by endogenous NE present in the intercellular spaces of the tissue slices. This is supported by the lack of effect of phentolamine on NE release from synaptosomes (Mulder *et al.*, 1978) in spite of its ability to antagonize the inhibitory effect of exogenous NE on [^{3}H] NE release from these synaptosomes. Until a similar lack of effect of phentolamine on 5-HT release from synaptosomes can be demonstrated, its ability to antagonize the inhibition of 5-HT release by NE could alternatively be explained as a physiological rather than a pharmacological (receptor-specific) antagonism.

Effects of substance P, dopamine and γ-aminobutyric acid

Substance P has been shown to enhance the sponanteous release of 5-HT from slices of the midbrain raphe (Kerwin and Pycock, 1979), substantia nigra (Reubi *et al.*, 1978) and spinal trigeminal nucleus (Reubi, 1980). However, at least in the substantia nigra, dopamine appears to mediate the effects of substance P on 5-HT release. This conclusion is based on the observation that substance P stimulates DA release from nigral slices and the DA receptor agonist, apomorphine, stimulates the release of 5-HT from these slices. In addition, the DA receptor antagonist, α-fluphenthixol, abolishes the excitatory effect of substance P on 5-HT release (Reubi *et al.*, 1978). In contrast to these findings, Hery *et al.* (1980) demonstrated *in vivo* that the nigral application of DA inhibits 5-HT release from both the ipsilateral substantia nigra and caudate nucleus. Since the nigral application of DA alters 5-HT release in the caudate, it is likely that this effect is mediated by long neuronal circuits. A similar involvement of long neuronal loop circuitry (absent in slices) in the effects of DA on 5-HT release from the substantia nigra may explain the discrepancy between the results obtained *in vitro* and *in vivo*. Additional studies of the effects of substance P on 5-HT release and its possible interaction with DA are needed.

The local application of γ-aminobutyric acid (GABA) has been reported to reduce the release of 5-HT from the substantia nigra of the cat (Hery *et al.*, 1980). However, the potassium-evoked release of 5-HT from slices of the rat substantia nigra is unaffected by GABA (Reubi *et al.*, 1978) and the spontaneous release of 5-HT from slices of the rat midbrain raphe region is enhanced by GABA (Kerwin and Pycock, 1979). The inhibitory effect of GABA in the substantia nigra *in vivo* but not *in vitro* and the excitatory effect of GABA in another brain region would suggest that the effects of GABA on 5-HT release are probably not mediated directly through receptors located on serotonergic cells.

Role of neuronal firing rate

Numerous studies have demonstrated that 5-HT release is dependent on the firing rate of serotonergic cells. Electrical stimulation of serotonergic cell bodies in the raphe nuclei increases 5-HT release from serotonergic terminals (Aiello-Malmberg *et al.*, 1979; Ashkenazi *et al.*, 1972, 1973; Chiueh and Moore, 1976; Holman and Vogt, 1972; Marsden *et al.*, 1979). Thus it is possible for substances which modify the firing rates of serotonergic cells to similarly modify 5-HT release. LSD which, inhibits raphe cell firing directly (Aghajanian *et al.*, 1972), inhibits 5-HT release *in vivo* when administered i.p. (Gallagher and Aghajanian, 1975). L-glutamate, which increases raphe cell firing directly (Aghajanian *et al.*, 1972), and morphine, which increases raphe cell firing when administered i.v. (Anderson *et al.*, 1977), both enhance 5-HT release from serotonergic terminal regions when applied locally to the raphe nuclei (Aiello-Malberg *et al.*, 1979; Hery *et al.*, 1979; Yaksh and Tyce, 1979). On the other hand, chlorimipramine, which inhibits raphe cell firing presumably through a negative feedback mechanism (Sheard *et al.*, 1972), either has no effect or increases 5-HT release (Gallagher and Aghajanian, 1975). The latter investigators suggested that this apparent discrepancy may be related to an inhibition of 5-HT uptake at low doses of chlorimipramine which obscures a reduction in release and to a direct release of 5-HT at higher doses. Indeed, we have observed that chlorimipramine (10 μM) enhances the spontaneous release of 5-HT from synaptosomes using a superfusion technique in which the reuptake of 5-HT is not significant (Martin and Sanders-Bush, unpublished observations).

Role of synthesis rate

To what extent changes in 5-HT synthesis dictate parallel changes in 5-HT release is not clear. Although, depletion of 5-HT by synthesis inhibition with *p*-chlorophenylalanine reduces 5-HT release (Marsden *et al.*, 1979), the extent which synthesis must be inhibited before a significant effect on release is observed is not known.

Recent studies with rat brain slices demonstrate that increasing the concentration of tryptophan in the incubation medium enhances the rate of 5-HT synthesis but not the rate of 5-HT release except in tissue slices pretreated with a MAO inhibitor (Elks *et al.*, 1979b). Similarly, using voltammetry to measure 5-HT release *in vivo*, Marsden *et al.* (1979) demonstrated that the administration of L-tryptophan (100 mg/kg ip) to rats produces no significant changes in current values in the striatum nor does it produce any obvious behavioural effects. However, when L-tryptophan (75 mg/mg) was administered to rats pretreated with a MAO inhibitor an increase in current and a behavioural response was observed. In contrast, Ternaux *et al.* (1976, 1977) found increases in 5-HT release in both rats and cats following the administration of L-tryptophan (100 mg/kg ip) using two different brain perfusion techniques. Animals were either anaesthetized or cervically transected for these studies and the effects of L-tryptophan on behaviour could not be evaluated. It is not clear whether the discrepancies between the results of these studies regarding the effects of L-tryptophan on 5-HT release are due to the different methods used to measure 5-HT release or to other factors. In general, however, it is thought that increases in 5-HT synthesis do not necessarily enhance 5-HT release since excess 5-HT may either be accumulated and stored by synaptic vesicles or metabolized by MAO (Green and Grahame-Smith, 1975). Only when one of these processes is compromised (e.g. by MAO inhibition) would an increase in 5-HT release occur. Furthermore, it is clear from the studies of Ternaux *et al.* (1976) that brain 5-hydroxyindoleacetic acid (5-HIAA) levels following tryptophan administration are elevated at times when increased 5-HT release is not apparent. Thus, 5-HT can be metabolized intraneuronally before being released and so, measures of brain 5-HIAA levels and rates of 5-HT metabolism are not necessarily satisfactory methods for determining changes in 5-HT release (Reinhard and Wurtman, 1977).

Effects of pharmacological agents on 5-HT release

The oldest and most well-characterized amine releasing drug is reserpine, the discovery of which could be considered the beginning of modern neuropharmacology. Much of the work related to the mechanism of action of reserpine involves studies of catecholamines and it is generally assumed that similar conclusions apply to the indoleamines. The release and depletion of biogenic amines induced by reserpine is mediated by an interference with the intraneuronal storage mechanisms. An excellent review of the action of reserpine and related drugs was recently published (Shore and Giachetti, 1978) and we will only briefly discuss it here. Using an *in vivo* push-pull cannula perfusion technique, Sulser *et al.* (1969) found that NE released by reserpine is largely metabolized by monoamine oxidase presumably before it exits from the terminal. Using similar techniques, our laboratory (unpublished results) has demonstrated that RO4-1248, a short-acting reserpine-like drug, causes a marked release of 5-HT but a large portion of the released amine appears in the perfusate as deaminated product, suggesting that substantial intraneuronal metabolism precedes release. The amine depletion following reserpine has an extremely long duration and is correlated with the presence of persistently bound drug. Recovery apparently depends on the arrival of new vesicles to the nerve terminals via axoplasmic transport (Shore, 1972). The biochemical and pharmacological effects of tetrabenazine and other benzoquinolizines apparently duplicate those of reserpine except that the duration of action is generally much shorter (Pletscher *et al.*, 1968).

The more recently characterized 5-HT releasing agents will be discussed in more detail since the releasing actions of these drugs are apparently different. This discussion will include ring-substituted phenylethylamine and amphetamine derivatives and miscellaneous compounds.

Ring-substituted phenylethylamine and amphetamine derivatives

A large number of drugs of this general class have been studied with regard to effects on 5-HT in brain. In general, meta- and para-substituted derivatives are more potent than the ortho-substituted compounds. Interestingly, the introduction of a halogen seems to confer a particular avidity for the 5-HT system.

The duration of 5-HT depletion by individual drugs in this class varies from several hours to several weeks and is complicated by the fact that some amphetamine derivatives such as *p*-chloroamphetamine, have a dual mechanism of 5-HT depletion: an initial reversible effect with a duration of 1–2 days followed by an irreversible effect (Fuller *et al.*, 1975; Sanders-Bush *et al.*, 1975). Evidence suggests that the delayed irreversible effect is mediated by a neurotoxic action (Harvey *et al.*, 1975; Massari *et al.*, 1978a) although direct neuroanatomical evidence for nerve terminal degeneration has not been found (Lorez *et al.*, 1976; Massari *et al.*, 1978b; Powers *et al.*, 1979). The initial effects of *p*-chloroamphetamine on 5-HT neurones are complex and include release of 5-HT, inhibition of MAO, inhibition of the high affinity uptake of 5-HT, and a decrease in tryptophan hydroxylase activity (see the review of Sanders-Bush and Steranka, 1978). The first three effects can be demonstrated by the *in vitro* addition of the drug to model systems but the reduction of tryptophan hydroxylase activity only occurs *in vitro* at concentrations greater than 1 mM (Koe and Corkey, 1976; Sanders-Bush *et al.*, 1972) which are well above those concentrations found in brain. It is possible therefore that this effect on tryptophan hydroxylase is secondary to the other actions on 5-HT neurones, all of which would tend to increase 5-HT transmission. The report that the *p*-chloro-amphetamine-induced decrease in tryptophan hydroxylase is blocked by presumed 5-HT receptor antagonists (Ross and Fröden, 1977) agrees with this interpretation. Clearly, there is a need for additional studies of the role of 5-HT release and consequent receptor stimulation in the fall in tryptophan hydroxylase activity. Similar studies of other substituted amphetamines and phenylethylamines, which do not induce the long-term, neurotoxic action, would be very interesting.

The first, direct demonstration that *p*-chloroamphetamine releases 5-HT from preloaded brain synaptosomes came from the laboratory of Fuller (Wong *et al.*, 1973) in incubation-type experiments and have been confirmed more recently using similar assay conditions (Ross *et al.*, 1977; Chun-Hwang and Van Woert, 1979). Although the long-term effects of *p*-chloroamphetamine on 5-HT are selective, the drug also releases the catecholamines and has initial biochemical effects on NE and DA turnover which can be explained by this mechanism (Sanders-Bush and Steranka, 1978). A rapid, dramatic increase in release of 5-HT into the perfusate of a push–pull cannula is induced by *p*-chloroamphetamine administration (Sanders-Bush *et al.*, 1974). In unpublished studies in our laboratory, we have demonstrated that the administration of RO4-1284 also induces a several-fold increase in the release of total radioactivity in the perfusate; however, unlike *p*-chloroamphetamine which does not change the ratio of 5-HT/5-HIAA in the perfusate, RO4-1284 reduces it. Thus, a substantial portion of the 5-HT released by RO4-1284, but not *p*-chloroamphetamine, is metabolized by monoamine oxidase before entering the perfusate. These data then suggest that the mechanisms of the releasing actions of these two drugs are not the same.

The psychotomimetic methoxyamphetamine derivatives also release 5-HT as demonstrated initially using an *in vitro* incubation-type technique (Tseng *et al.,* 1976). However, as discussed earlier, interpretation of results using this technique is confounded by reuptake of the amine so that apparent release may merely reflect uptake inhibition. Indeed, the IC_{50} values and the rank order of potency for uptake inhibition and release are similar. Furthermore, these investigators found that amphetamine increases the release of NE, while others in superfusion experiments, have found that amphetamine does not modify the release of NE, while the release of DA and 5-HT is enhanced (Raiteri *et al.,* 1975; Ross *et al.,* 1977). Other substituted phenylethylamine derivatives have been found to decrease the retention of 5-HT in synaptosomes, presumably by enhancing release (Chun-Hwang and Van Woert, 1979; Ross *et al.,* 1977; Tessell and Rutledge, 1976). In recent studies, Tseng *et al.* (1978) have used superfusion-type techniques and have clearly shown that *p*-methoxyamphetamine releases 5-HT from synaptosomes and, furthermore, in elegant ventricular perfusion studies, have confirmed that the drug has this effect *in vivo*. Presumably, the other substituted amphetamines and phenylethylamines have similar effects although this has not been directly demonstrated to our knowledge. *In vivo* pharmacological and behavioural experiments do, however, demonstrate drug effects which are apparently mediated by a release of 5-HT, i.e. blocked by depletion of 5-HT (Chun-Hwang and Van Woert, 1979; 1980; Clineschmidt and McGuffin, 1978; Menon *et al.,* 1976; Ögren and Ross, 1977; Sloviter *et al.,* 1980a). Furthermore, based on this pharmacological criterion, amphetamine, in high doses, induces a release of 5-HT (Lees *et al.,* 1979; Menon *et al.,* 1976; Sloviter *et al.,* 1978; 1980b) an interpretation which is supported by *in vitro* data (Ross *et al.,* 1977; Raiteri *et al.,* 1975). Interestingly, although β-phenylethylamine has behavioral effects that suggest 5-HT stimulation, these effects are apparently mediated by a direct receptor stimulation rather than release of 5-HT (Sloviter *et al.,* 1980b). Some phenylethylamine and amphetamine derivatives may also have direct effects on 5-HT receptors (Tseng *et al.,* 1978; Sloviter *et al.,* 1980b).

The mechanism of 5-HT release by *p*-chloroamphetamine and other releasing agents has been the subject of recent investigations. The release mechanism is clearly different from exocytosis since it is Ca-independent (Arnold *et al.,* 1977; Schwarz *et al.,* 1980). Since the displaced amine is positively charged and not lipophilic, passive diffusion is not likely. Many investigators have alternatively suggested that the amine exits via the carrier-mediated uptake system acting in reverse. The ability of uptake blockers to prevent the chemical release of catecholamines and 5-HT agrees with this interpretation (Azzaro *et al.,* 1974; Fuller and Snoddy, 1979; Raiteri *et al.,* 1977, 1979; Ross and Kelder, 1977; Tseng *et al.,* 1976). Uptake blockers also prevent the depletion of 5-HT induced by the administration of *p*-chloroamphetamine and related drugs (Fuller *et al.,* 1975; Meek *et al.,* 1971; Ross, 1976). Although inhibition of a carrier-mediated efflux of the released 5-HT may explain these observations, Fuller (1980) has presented compelling arguments that the ability of uptake blockers to antagonize the *p*-chloroamphetamine-induced de-

pletion of brain 5-HT is related to a blockade of the uptake of p-chloroamphetamine into the 5-HT neurone via the membrane carrier. However, it has been found
that uptake blockers also prevent the efflux of 5-HT induced by low external
sodium and by veratridine presumably by preventing the efflux of 5-HT by the
reversed 5-HT uptake mechanism (Ross and Kelder, 1977). Furthermore, attempts
to demonstrate directly that p-chloroamphetamine is actively accumulated in synaptosomes have failed (Ross, 1976; Ross and Ask, 1980; Sanders-Bush and Steranka,
1978) and since it is extremely lipid-soluble and should readily diffuse through
membranes, an obligatory role for carrier-mediated transport of the drug to its
inteneuronal site of action seems superfluous. Alternatively, perhaps both carrier-
mediated drug influx and carrier-mediated amine efflux take place and an accelerated exchange diffusion model is appropriate to explain the data. In this model, an
interaction of the releasing agent with the carrier accelerates the efflux of the amine
out of the neurone by increasing the shuttling of the transport carrier from the outside to the inside. There is evidence to suggest that such a model may explain the
drug-induced release of catecholamines (Azzaro and Smith, 1975; Fischer and Cho,
1979; Fuller and Snoddy, 1979; Paton, 1973; Raiteri *et al.,* 1979), and a similar
mechanism should be considered for the chemical release of 5-HT. Indeed, an exchange diffusion mechanism would reconcile the different explanations for the
effect of uptake blockers on p-chloroamphetamine-induced release of 5-HT: i.e.,
the one hypothesis which argues that uptake blockers prevent the active accumulation of p-chloroamphetamine into nerve terminals (see for example, Fuller, 1980)
and the other which argues that the blockers prevent the carrier-mediated efflux of
the released amine (see for example, Ross and Ask, 1980).

Miscellaneous drugs

Morphine and physostigmine have been found to increase the release of 5-HT into a
push-pull cannula perfusate of the cerebral cortex (Aiello-Malmberg *et al.,* 1979). It
is not entirely clear that these effects are mediated by a direct effect on the 5-HT
nerve terminal. Physostigmine does release 5-HT when added to synaptosomes *in
vitro* (Hery *et al.,* 1977) but the concentrations required are much higher than
those active *in vivo.* Certainly a release of 5-HT should be considered in the pharmacological actions of these drugs. Tryptamine, α-methyltryptamine and related compounds release 5-HT from synaptosomes with IC_{50} values in the μM range (Baker *et
al.,* 1977). The monoamine oxidase inhibitor, α-ethyltryptamine, has similar
potency, while other monamine oxidase inhibitors have only weak or insignificant
effects (Baker *er al.,* 1980). The 5-HT uptake blocker, chlorimipramine (Gallager
and Aghajanian, 1975), and the 5-HT receptor blocker, mianserin (Raiteri *et al.,*
1976), may also have direct effects on 5-HT release. The doses required are well
above those which initiate the primary actions of these drugs. However, one should
be aware of a possible 5-HT releasing action when using these drugs as tools.

FUTURE TRENDS

Considering the evidence for compartmentalization of stored 5-HT and non-uniform release from the different compartments, the method of manipulating 5-HT storage sites (synaptosomes) for subsequent release studies may yield different results. For example, some investigators utilize synaptosomes prepared from reserpine-treated animals or routinely add a monoamine oxidase inhibitor to the incubation medium, treatments which obviously perturb the usual compartmentalization of the amine. In the future, it would seem important to investigate the validity of these approaches in careful systematic studies.

The precise mechanism of the control of 5-HT release remains to be determined. 5-HT autoreceptors on raphe cell bodies have a well-established role in the regulation of cell firing rate and consequently transmitter release. However, there is a need for additional studies of the possible involvement of interneuronal loops and autoreceptors located on serotonergic nerve terminals. Further, the relative importance of these three possible control points in regulating 5-HT release is as yet unknown.

Histological studies have suggested that two neurotransmitters can co-exist in neurones in the mammalian CNS. Of particular interest in the present context is the recent demonstration that substance P is found in some 5-HT neurones (Chan-Palay *et al.*, 1978; Hökfelt *et al.*, 1978). Considering this apparent co-existence in the same neurone, questions about storage and release of these co-transmitters are numerous and will certainly be examined in the future. The observation that substance P increases 5-HT release *in vitro* is interesting in this regard. If this effect were mediated by a specific peptide receptor (an interpretation still open to question) the local control of 5-HT release may depend on the proportion of these two co-transmitters released from nerve endings.

Finally, with regard to 5-HT releasing drugs, additional studies of their mechanisms of actions are needed. It is interesting to note that two classes of CNS stimulants with different mechanisms of DA release have been described (Shore, 1976). Although amphetamine and non-amphetamine stimulants, such as amfonelic acid and methylphenidate, both increase DA release, their actions on the DA neurone are clearly different perhaps reflecting release from different intraneuronal pools. Whether similar differences will be found with the 5-HT releasing drugs remains to be determined.

REFERENCES

Aghajanian, G. K., Haigler, H. J., and Bloom, F. E. (1972). Lysergic acid diethylamide and serotonin: direct actions on serotonin-containing neurons in rat brain. *Life Sci.* **11**, pt 1, 615–622.

Aghajanian, G. K. and Wang, R. Y. (1978). Physiology and pharmacology of central serotonergic neurons, in *Psychopharmacology: A Generation of Progress* (Eds M. A. Lipton, A. D. Mascio and K. F. Killam), pp. 171–183, Raven, New York.

Aiello-Malmberg, P., Bartolini, A., Bartolini, R., and Galli, A. (1979). Effects of morphine, physostigmine and raphe nuclei stimulation on 5-hydroxytryptamine release from the cerebral cortex of the cat, *Br. J. Pharmac.* **65**, 547-555.

Anderson, S. D., Basbaum, A. I., and Fields, H. L. (1977). Response of medullary raphe neurons to peripheral stimulation and to systemic opiates, *Brain Res.* **123**, 363-368.

Arnold, E. B., Molinoff, P. B., and Rutledge, C. O. (1977). The release of endogenous norepinephrine and dopamine from cerebral cortex by amphetamine, *J. Pharmac. Exp. Therap.* **202**, 544-557.

Ashkenazi, R., Holman, R. B., and Vogt, M. (1972). Release of transmitters on stimulation of the nucleus linearis raphe in the cat, *J. Physiol.* **223**, 255-259.

Ashkenazi, R., Holman, R. B., and Vogt, M. (1973). Release of transmitters into the perfused third cerebral ventricle of the cat, *J. Physiol.* **233**, 195-209.

Axelrod, J. (1972). Dopamine-β-hydroxylase: regulation of its synthesis and release from nerve terminals,, *Pharmac. Rev.* **24**, 233-243.

Azzaro, A. J. and Smith, D. J. (1975). The role of storage and catabolism in the accumulation of norepinephrine after short and long incubations, *J. Neurochem.* **24**, 811-813.

Azzaro, A. J., Ziance, R. J., and Rutledge, C. O. (1974). The importance of neuronal uptake of amines for amphetamine-induced release of ^{3}H-norepinephrine from isolated brain tissue, *J. Pharmac. Exp. Therap.* **189**, 110-118.

Baker, G. B., Hiob, L. E., and Dewhurst, W. G. (1980). Effects of monamine oxidase inhibitors on release of dopamine and 5-hydroxytryptamine from rat striatum *in vitro, Cell. Mol. Biol.* **26**, 82-186.

Baker, G. B., Martin, I. L., and Mitchell, P. R. (1977). The effects of some indolalkylamines on the uptake and release of 5-hydroxytryptamine in rat striatum, *Br. J. Pharmac.* **61**, 151P-152P.

Baldessarini, R. J. (1975). Release of catecholamines, in *Handbook of Psychopharmacology*, Vol. 3 (Eds L. L. Iversen, S. D. Iversen and S. H. Snyder), pp. 37-137, Plenum, New York.

Baumann, P. A. and Maitre, L. (1976). Is drug inhibition of dopamine uptake a misinterpretation of *in vitro* experiments? *Nature* (London), **264**, 789-790.

Blaustein, M. P. (1975). Effects of potassium, veratridine and scorpion venom on calcium accumulation of transmitter release by nerve terminals *in vitro, J. Physiol.* **247**, 617-655.

Blaustein, M. P., Johnson, E. M., and Needleman, P. (1972). Calcium-dependent norepinephrine release from presynaptic nerve endings *in vitro. Proc. Nat. Acad. Sci. USA*, **69**, 2237-2240.

Blaustein, M. P. and Weisman, W. P. (1970). Potassium ions and calcium ion fluxes in isolated nerve terminals, in *Drugs and Cholinergic Mechanisms in the CNS* (Eds. E. Heilbronn and A. Winter), pp. 291-307, Forsvarets Forskningsansta Research Institute of National Defense, Stockholm.

Bourgoin, S., Artaud, F., Enjalbert, A., Hery, A., Glowinski, J., and Hamon, M. (1977). Acute changes in central serotonin metabolism induced by the blockade of stimulation of serotonergic receptors during ontogenesis in the rat, *J. Pharmac. Exp. Therap.* **202**, 519-531.

Cerrito, F. and Raiteri, M. (1979). Serotonin release is modulated by presynaptic autoreceptors, *Eur. J. Pharmac.* **57**, 427-430.

Chan-Palay, V., Jonsson, G. and Palay, S. L. (1978). Serotonin and substance P coexist in neurons of the rat's central nervous system, *Proc. Nat. Acad. Sci.* **75**, 1582-1586.

Chase, T. N., Breese, G. R., and Kopin, I. J. (1967). Serotonin release from brain slices by electrical stimulation: regional differences and effect of LSD, *Science*, **157**, 1461–1463.

Chase, T. N., Katz, R. I., and Kopin, I. J. (1969). Release of [³H]serotonin from brain slices, *J. Neurochem.* **16**, 607–615.

Chiueh, C. C. and Moore, K. E. (1976). Effects of dopaminergic agonists and electrical stimulation of the midbrain raphe on the release of 5-hydroxytryptamine from the cat brain *in vivo*, *J. Neurochem.* **26**, 319–324.

Chun-Hwang, E. and Van Woert, M. H. (1979). Behavioral and biochemical effect of para-methoxyphenylethylamine, *Res. Commun. Chem. Path. Pharmac.* **23**, 419–431.

Chun-Hwang, E. and Van Woert, M. H. (1980). Comparative effects of substituted phenylethlamines on brain serotonergic mechanisms, *J. Pharmac. Exp. Therap.* **213**, 254–260.

Clineschmidt, B. V. and McGuffin, J. C. (1978). Pharmacological differentiation of the central 5-hydroxytryptamine-like actions of MK212 (6-chloro-2-[1-piperozenyl]-pyrazine), *p*-methoxyamphetamine and fenfluramine in an *in vivo* model system, *Eur. J. Pharmac.* **50**, 369–375.

Da Prada, M. and Pletscher, A. (1974). Mechanisms of 5-hydroxytryptamine storage in subcellular organelles of blood platelets, *Adv. Biochem. Psychopharmac.* **10**, 311–320.

Dismukes, R. K. and Mulder, A. H. (1976). Cyclic AMP and α-receptor-mediated modulation of noradrenaline release from rat brain slices, *Eur. J. Pharmac.* **39**, 383–388.

Elks, M. L., Youngblood, W. W., and Kizer, J. S. (1979a). Synthesis and release of serotonin by brain slices: effect of ionic manipulations and cationic ionophores, *Brain Res.* **172**, 461–469.

Elks, M. L., Youngblood, W. W., and Kizer, J. S. (1979b). Serotonin synthesis and release in brain slices: independence of tryptophan, *Brain Res.* **172**, 471–486.

Farnebo, L. O. and Hamberger, B. (1971). Drug-induced changes in the release of ³H-monoamines from field stimulated rat brain slices, *Acta Physiol. Scand.* Suppl. 371, 35–44.

Farnebo, L. O. and Hamberger, B. (1974). Regulation of [³H]5-hydroxytryptamine release from rat brain slices, *J. Pharm. Pharmac.* **26**, 642–644.

Fischer, J. F. and Cho, A. K. (1979). Chemical release of dopamine from striatal homogenates: Evidence for an exchange diffusion model, *J. Pharmac. Exp. Therap.* **208**, 203–209.

Frankhuyzen, A. L. and Mulder, A. H. (1980). Noradrenaline inhibits depolarization-induced ³H-serotonin release from slices of rat hippocampus, *Eur. J. Pharmac.* **63**, 179–182.

Fuller, R. W. (1980). Mechanism by which uptake inhibitors antagonize *p*-chloroamphetamine-induced depletion of brain serotonin, *Neurochem. Res.* **5**, 241–245.

Fuller, R. W. and Snoddy, H. D. (1979). Inability of methylphenidate or mazindol to prevent the lowering of 3,4-dihydroxyphenylacetic acid in rat brain by amphetamine, *J. Pharm. Pharmac.* **31**, 183–184.

Fuller, R. W., Perry, K. W., and Molloy, B. B. (1975). Reversible and irreversible phases of serotonin depletion by 4-chloroamphetamine, *Eur. J. Pharmac.* **33**, 119–124.

Gallagher, D. W. and Aghajanian, G. K. (1975). Effects of chlorimipramine and lysergic acid diethylamide on efflux of precursor-formed ³H-serotonin: correlations with serotonergic impulse flow, *J. Pharmac. Exp. Therap.* **193**, 785–795.

Glowinski, J. (1975). Properties and functions of intraneuronal monoamine compartments in central aminergic neurons, in *Handbook of Psychopharmacology* (Eds L. L. Iversen, S. D. Iversen and S. H. Snyder), pp. 136–167, Plenum, New York.

Gothert, M. and Huth, H. (1980). Alpha-adrenoceptor-mediated modulation of 5-hydroxytryptamine release from rat brain cortex slices, *Naunyn-Schmiedeberg's Arch. Pharmac.* **313**, 21–26.

Gothert, M. and Weinheimer, G. (1979). Extracellular 5-hydroxytryptamine release from rat brain cortex slices, *Naunyn-Schmiedeberg's Arch. Pharmac.* **310**, 93–96.

Green, A. R. and Grahame-Smith, D. G. (1975). 5-Hydroxytryptamine and other indoles in the central nervous system, in *Handbook of Psychopharmacology*, Vol. 3 (Eds. L. L. Iversen, S. D. Iversen, and S. H. Snyder), pp. 169–245, Plenum, New York.

Halaris, A. E. and Freedman, D. X. (1977). Vesicular and juxtavesicular serotonin: Effect of lysergic acid diethylamide and reserpine, *J. Pharmac. Exp. Therap.* **203**, 575–586.

Hamon, M., Bourgoin, S., Jagger, J., and Glowinski, J. (1974). Effects of LSD on synthesis and release of 5-HT in rat brain slices, *Brain Res.* **69**, 265–280.

Harvey, J. A., McMaster, S. E., and Yunger, L. M. (1975). *p*-Chloroamphetamine: Selective neurotoxic action in brain, *Science*, **187**, 841–843.

Heikkila, R. E., Orlansky, H., and Cohen, G. (1975). Studies on the distinction between uptake inhibition and release of ^{3}H-dopamine in rat brain tissue slices, *Biochem, Pharmac.* **24**, 847–852.

Hery, F., Bourgoin, S., Hamon, M. Ternaux, J. P., and Glowinski, J. (1977). Control of the release of newly synthetized ^{3}H-5-hydroxytryptamine by nicotinic and muscarinic receptors in rat brain hypothalamic slices, *Naunyn-Schmiedeberg's Arch. Pharmac.* **296**, 91–97.

Hery, F., Simonnet, G., Bourgoin, S., Soubrie, P., Artaud, F., Hamon, M., and Glowinski, J. (1979). Effect of nerve activity on the *in vivo* release of [^{3}H]-serotonin continuously formed from L-[^{3}H] tryptophan in the caudate nucleus of the cat, *Brain Res.* **169**, 317–334.

Hery, F., Soubrie, P., Bourgoin, S., Motastruc, J. L., Artaud, F., and Glowinski, J. (1980). Dopamine released from dendrites in the substantia nigra controls the nigral and striatal release of serotonin, *Brain Res.* **193**, 143–151.

Hökfelt, T., Ljungdahl, A., Steinbusch, H., Verhofstd, A., Nilsson, G., Brodin, E., Pernow, B., and Goldstein, M. (1978). Immunohistochemical evidence of substance P-like immunoreactivity in some 5-hydroxytryptamine-containing neurons in rat central nervous system, *Neuroscience*, **3**, 517–538.

Holman, R. B. and Vogt, M. (1972). Release of 5-hydroxytryptamine from caudate nucleus and septum, *J. Physiol.* **223**, 243–254.

Jonakait, G. M., Tamir, H., Gintzler, A. R., and Gershon, M. D. (1979). Release of ^{3}H-serotonin and its binding protein from enteric neurons, *Brain Res.* **174**, 55–69.

Kantak, K. M., Wayner, M. J., Tilson, H. A., Dwoskin, L. P., and Stein, J. M. (1978). Synthesis and turnover of ^{3}H-5-hydroxytryptamine in the lateral cerebroventricle, *Pharmac. Biochem. Behav.* **8**, 153–161.

Katz, R. I. and Kopin, I. J. (1969). Effect of *d*-LSD and related compounds on release of norepinephrine-H^3 and serotonin-H^3 evoked from brain slices by electrical stimulation, *Pharmac. Res. Commun.* **1**, 54–61.

Kerwin, R. W. and Pycock, C. J. (1979). The effect of some putative neurotransmitters on the release of 5-hydroxytryptamine and γ-aminobutyric acid from slices of the rat midbrain raphe area, *Neuroscience*, **4**, 1359–1365.

Koe, B. K. and Corkey, R. F. (1976). Inhibition of rat brain tryptophan hydroxylase with p-chloroamphetamine, *Biochem. Pharmac.* **25**, 31–36.

Lane, J. D. and Aprison, M. H. (1977). Calcium-dependent release of endogenous serotonin, dopamine and norepinephrine from nerve endings, *Life Sci.* **20**, 665–672.

Langer, S. Z. (1977). Presynaptic receptors and their role in the regulation of transmitter release, *Br. J. Pharmac.* **60**, 481–497.

Langer, S. Z., Briley, M. S., and Raisman, R. (1980). Regulation of neurotransmission through presynaptic receptors and other mechanisms: possible clinical relevance and therapeutic potential, in *Receptors for Neurotransmitters and Peptide Hormones* (Eds G. Pepeu, M. J. Kuhar, and S. J. Enna), pp. 203–212, Raven, New York.

Lees, A. J., Fernando, J. C. R., and Curzon, G. (1979). Serotonergic involvement in behavioural responses to amphetamine at high dosage, *Neuropharmacology*, **18**, 153–158.

Lewis, J. C. and Moertel, C. G. (1978). Platelet 5-hydroxytryptamine storage in the carcinoid syndrome: An electron microscopic autoradiographic study, *Amer. J. Clin. Pathol.* **70**, 628–631.

Lorez, H., Saner, A., Richards, J. G., and DaPrada, M. (1976). Accumulation of 5HT in non-terminal axons after p-chloro-N-methyl-amphetamine without degeneration of identified 5HT nerve terminals. *Eur. J. Pharmac.* **38**, 79–88.

Loullis, C. C., Hingtgen, J. N., Shea, P. A., and Aprison, M. H. (1980). *In vivo* determination of endogenous biogenic amines in rat brain using HPLC and push-pull cannula, *Pharmac. Biochem. Behav.* **12**, 959–963.

Marsden, C. A., Conti, J., Strope, E., Curzon, G., and Adams, R. N. (1979). Monitoring 5-hydroxytryptamine release in the brain of the freely moving unanesthetized rat using *in vivo* voltammetry, *Brain Res.* **171**, 85–99.

Massari, V. J., Tizabi, Y., Gottsfield, Z., and Jacobowitz, D. M. (1978a). A fluorescence histochemical and biochemical evaluation of the effect of p-chloroamphetamine on individual serotonergic nuclei in the rat brain, *Neuroscience*, **3**, 339–344.

Massari, V. J., Tizabi, Y., and Sanders-Bush, E. (1978b). Evaluation of the neurotoxic effects of p-chloroamphetamine: A histological and biochemical study, *Neuropharmacology*, **17**, 541–548.

Maynert, E. W., Levi, R., and deLorenzo, A. J. D. (1964). The presence of norepinephrine and 5HT in vesicles from disrupted nerve-ending particles, *J. Pharmac. Exp. Therap.* **144**, 385–392.

Meek, J. L., Fuxe, K., and Carlsson, A. (1971). Blockade of p-chloromethamphetamine induced 5-hydroxytryptamine depletion by chlorimipramine, chlorpheniramine and meperidine, *Biochem. Pharmac.* **20**, 707–709.

Menon, M. K., Tseng, L. F., and Loh, H. H. (1976). Pharmacological evidence for the central serotonergic effects of monomethoxyamphetamines, *J. Pharmac. Exp. Therap.* **197**, 272–279.

Mosko, S. S. and Jacobs, B. L. (1977). Electrophysiological evidence against negative neuronal feedback from the forebrain controlling midbrain raphe unit activity, *Brain Res.* **119**, 291–303.

Mulder, A. H., deLangen, C. D. J., deRegt, V., and Hogenboom, F. (1978). Alpha-receptor-mediated modulation of ^{3}H-noradrenaline release from rat brain cortex synaptosomes, *Naunyn-Schmiedeberg's Arch. Pharmac.* **303**, 193–196.

Mulder, A. H., van den Berg, W. B., and Stoof, J. C. (1975). Calcium-dependent release of radiolabelled catecholamines and serotonin from rat brain synaptosomes in a superfusion system, *Brain Res.* **99**, 419–424.

Neckers, L. M., Neff, N. H., and Wyatt, R. J. (1979). Increased serotonin turnover in corpus striatum following an injection of kainic acid: evidence for neuronal feedback regulation of synthesis, *Naunyn-Schmiedeberg's Arch. Pharmac.* **306**, 173–177.

Ögren, S. O. and Ross, S. B. (1977). Substituted amphetamine derivatives. II. Behavioural effects in mice related to monoaminergic neurones, *Acta Pharmac. et Toxic.* **41**, 353–368.

Paton, D. M. (1973). Mechanism of efflux of noradrenaline from adrenergic nerves in rabbit atria, *Br. J. Pharmac.* **49**, 614–627.

Pletscher, A., DaPrada, M., Burkard, W. P., and Tranzer, J. P. (1968). Effects of benzoquinolizines and ring-substituted aralkylamines on serotonin metabolism, *Adv. Pharmac.* **6B**, 55–69.

Powers, J. M., Mann, G. T., Jones, R., Ward, J. W., Elsea, J. R., and Smith, H. M. (1979). A reassessment of the significance of dark neurons in serotonergic cell groups, *Neuropharmacology,* **18**, 383–389.

Pressman, B. C. (1976). Biological application of ionophores, *Ann. Rev. Biochem.* **45**, 501–527.

Raiteri, M., Angelini, F., and Bertollini, A. (1976). Comparative study of the effects of mianserin, a tetracyclic antidepressant, and of imipramine on uptake and release of neurotransmitters in synaptosomes, *J. Pharm. Pharmac.* **28**, 483–488.

Raiteri, M. Angelini, F., and Levi, G. (1974). A simple apparatus for studying the release of neurotransmitters from synaptosomes, *Eur. J. Pharmac.* **25**, 411–414.

Raiteri, M., Bertollini, A., Angelini, F., and Levi, G. (1975). d-Amphetamine as a releaser or reuptake inhibitor of biogenic amines in synaptosomes, *Eur. J. Pharmac.* **34**, 189–195.

Raiteri, M., Carmine, R. D., and Bertollini, A., (1977). Effect of desmethylimipramine on the release of ^{3}H-norepinephrine induced by various agents in hypothalamic synaptosomes, *Mol. Pharmac.* **13**, 746–758.

Raiteri, M., Cerrito, F., Cervoni, A. M., and Levi, G. (1979). Dopamine can be released by two mechanisms differentially affected by the dopamine transport inhibitor nomifensine, *J. Pharmac. Exp. Therap.* **208**, 195–202.

Reinhard, J. F., Jr, and Wurtman, R. J. (1977). Relation between brain 5-HIAA levels and the release of serotonin into brain synapses, *Life Sci.* **21**, 1741–1746.

Reubi, J. C. (1980). *In vitro* release of 5-[^{3}H]hydroxytryptamine from rat spinal trigeminal nucleus, *Neurosci. Lett.* **16**, 263–267.

Reubi, J. C. and Emson, P. C. (1978). Release and distribution of endogenous 5HT in rat substantia nigra, *Brain Res.* **139**, 164–168.

Reubi, J. C., Emson, P. C., Jessell, T. M., and Iversen, L. L. (1978). Effects of GABA, dopamine, and substance P on the release of newly synthesized ^{3}H-5-hydroxytryptamine from rat substantia nigra *in vitro, Naunyn-Schmiedeberg's Arch. Pharmac.* **304**, 271–275.

Ross, S. B. (1976). Antagonism of the acute and long-term biochemical effects of 4-chloroamphetamine on the 5-HT neurons in the rat brain by inhibitors of the 5-hydroxytryptamine uptake, *Acta Pharmac. et Toxic.* **39**, 456–476.

Ross, S. B. and Ask, A. L. (1980). Structural requirements for uptake into serotonergic neurons, *Acta Pharmac et Toxic.* **46**, 270–277.

Ross, S. B. and Fröden, Ö. (1977). On the mechanism of the acute decrease of rat brain tryptophan hydroxylase activity by 4-chloroamphetamine, *Neurosci. Lett.* **5**, 215–220.

Ross, S. B. and Kelder, D. (1977). Efflux of 5-hydroxytryptamine from synaptosomes of rat cerebral cortex, *Acta Physiol. Scand.* **99**, 27–36.

Ross, S. B., Ögren, S. O., and Renyi, A. L. (1977). Substituted amphetamine derivatives. I. Effect on uptake and release of biogenic monoamines and on mono-amine oxidase in the mouse brain, *Acta Pharmac. et Toxic.* **4**, 337-352.

Saavedra, J. M., Brownstein, M., and Axelrod, J. A. (1973). A specific and sensitive enzymatic microassay for serotonin in tissues, *J. Pharmac. Exp. Therap.* **186**, 508-515.

Sanders-Bush, E., Bushing, J. A., and Sulser, F. (1972). *p*-Chloroamphetamine: Inhibition of cerebral tryptophan hydroxylase, *Biochem. Pharmac.* **21**, 1501-1510.

Sanders-Bush, E., Bushing, J. A., and Sulser, F. (1975). Long-term effects of *p*-chloroamphetamine and related drugs on central serotonergic mechanisms, *J. Pharmac. Exp. Therap.* **192**, 33-41.

Sanders-Bush, E., Gallagher, D. A., and Sulser, F. (1974). On the mechanism of brain 5-hydroxytryptamine depletion by *p*-chloroamphetamine and related drugs and the specificity of their action, *Adv. Biochem. Psychopharm.* **10**, 185-194.

Sanders-Bush, E. and Steranka, L. R. (1978) Immediate and long-term effects of *p*-chloroamphetamine on brain amines, *Ann. N.Y. Acad. Sci.* **305**, 208-221.

Schwarz, R. D., Uretsky, N. J., and Bianchine, J. R. (1980). The relationship between the stimulation of dopamine synthesis and release produced by amphetamine and high potassium in striatal slices, *J. Neurochem.* **35**, 1120-1127.

Shaskan, E. G. and Snyder, S. H. (1970). Kinetics of serotonin accumulation into slices from rat brain: relationship to catecholamine uptake, *J. Pharmac. Exp. Therap.* **175**, 404-418.

Sheard, M. H., Zolovick, A., and Aghajanian, G. K. (1972). Raphe neurons: effect of tricyclic antidepressant drugs, *Brain Res.* **43**, 690-694.

Shelanski, M. L. and Taylor, E. W. (1967). Isolation of a protein subunit from microtubules, *J. Cell. Biol.* **34**, 549-554.

Shields, P. J. and Eccleston, D. (1973) Evidence for the synthesis and storage of 5-hydroxytryptamine in two separate pools in the brain, *J. Neurochem.* **20**, 881-888.

Shore, P. A. (1972). Transport and storage of biogenic amines, *Ann. Rev. Pharmac.* **12**, 209-222.

Shore, P. A. (1976). Actions of amfonelic acid and other nonamphetamine stimulants on the dopamine neuron, *J. Pharm. Pharmac.* **28**, 855-857.

Shore, P. A. and Giachetti, A. (1978). Reserpine: Basic and clinical pharmacology in *Handbook of Psychopharmacology* Vol. 10 (Eds L. L. Iversen, S. D. Iversen, and S. H. Snyder), pp. 197-219, Plenum, New York.

Sloviter, R. S., Connor, J. D., Damiano, B. P., and Drust, E. G. (1980a). Parahalogenated phenylethylamines: Similar serotonergic effects in rats by different mechanisms, *Pharmac. Biochem. Behav.* **13**, 283-286.

Sloviter, R. S., Connor, J. D., and Drust, E. G. (1980b). Serotonergic properties of β-phenylethylamine in rats, *Neuropharmacology*, **19**, 1071-1074.

Sloviter, R. S., Drust, E. G., and Connor, J. D. (1978). Evidence that serotonin mediates some behavioral effects of amphetamine, *J. Pharmac. Exp. Therap.* **206**, 348-352.

Snodgrass, S. F. and Iversen, L. L. (1974). Formation and release of ^{3}H-tryptamine from ^{3}H-tryptophan in rat spinal cord slices, *Adv. Biochem. Psychopharmac.* **10**, 141-150.

Starke, K. and Montel, H. (1973). Involvement of α-receptors in clonidine-induced inhibition of transmitter release from central monoamine neurones, *Neuropharmacology*, **12**, 1073-1080.

Sulser, F. Owens, M., Strada, S., and Dingell, J. (1969). Modification by desipramine of the availability of norepinephrine released by reserpine in the hypothalamus of rat brain *in vivo*, *J. Pharmac. Exp. Therap.* **163**, 272-282.

Tamir, H., Bebirian, R., Muller, F., and Casper, D. (1980). Differences between intracellular platelet and brain proteins that bind serotonin, *J. Neurochem.*, **35**, 1033–1044.

Tamir, H. and Gershon, M. D. (1979). Storage of serotonin and serotonin binding protein in synaptic vesicles, *J. Neurochem.* **33**, 35–44.

Tamir, H. and Huang, Y. L. (1974). Binding of serotonin to soluble binding protein from synaptosomes, *Life Sci.* **14**, 83–93.

Tamir, H., Klein, A., and Rapport, M. M. (1976). Serotonin binding protein: Enhancement of binding by Fe^{++} and inhibition of binding by drugs, *J. Neurochem.*, **26**, 871–878.

Tamir, H. and Kuhar, M. J. (1975). Association of serotonin binding protein with projections of the midbrain raphe nuclei, *Brain Res.* **83**, 169–172.

Ternaux, J. P., Boireau, A., Bourgoin, S., Hamon, M., Hery, F., and Glowinski, J. (1976). *In vivo* release of 5-HT in the lateral ventricle of the rat: effects of 5-hydroxytryptamine and tryptophan, *Brain Res.* **101**, 533–548.

Ternaux, J. P., Hery, F., Hamon, M., Bourgoin, S., and Glowinski, J. (1977). 5-HT release from ependymal surface of the caudate nucleus in 'encephale isole' cats, *Brain Res.* **132**, 575–579.

Tessell, R. E. and Rutledge, C. O. (1976). Specificity of release of biogenic amines from isolated rat brain tissue as a function of the meta substituent of N-ethylamphetamine derivatives, *J. Pharmac. Exp. Therap.* **197**, 253–262.

Tseng, L. F., Harris, R. A., and Loh, H. H. (1978). Blockade of para-methoxyamphetamine-induced serotonergic effects by chlorimipramine, *J. Pharmac. Exp. Therap.* **204**, 27–38.

Tseng, L. F., Menon, M. K., and Loh, H. H. (1976). Comparative actions of monomethoxyamphetamines on the release and uptake of biogenic amines in brain tissue, *J. Pharmac. Exp. Therap.* **197**, 263–271.

Tuomisto, J. (1977). Nomifensine and its derivatives as possible tools for studying amine uptake, *Eur. J. Pharmac.* **42**, 101–106.

Vane, J. R. (1957). A sensitive method for the assay of 5-hydroxytryptamine, *Br. J. Pharmac.* **12**, 344–349.

Wemer, J., van der Lugt, J. C., deLangen, C. D. J., and Mulder, A. H. (1979). On the capacity of presynaptic alpha receptors to modulate norepinephrine release from slices of rat neocortex and the affinity of some agonists and antagonists for these receptors, *J. Pharmac. Exp. Therap.* **211**, 445–451.

Wong, D. T., Horng, J. S., and Fuller, R. W. (1973). Kinetics of serotonin accumulation into synaptosomes of rat brain: Effects of amphetamine and chloroamphetamines, *Biochem. Pharmac.* **22**, 311–322.

Yaksh, T. L. and Tyce, G. M. (1979). Microinjection of morphine into the periaqueductal gray evokes the release of serotonin from spinal cord, *Brain Res.* **171**, 176–181.

Yaksh, T. L. and Tyce, G. M. (1980). Resting and K^+-evoked release of serotonin and norepinephrine *in vivo* from the rat and cat spinal cord, *Brain Res.* **192**, 133–146.

Biology of Serotonergic Transmission
Edited by N. N. Osborne
© 1982, John Wiley & Sons Ltd.

Chapter 5

Axonal Transport of Serotonin

D. J. GOLDBERG and J. H. SCHWARTZ

Departments of Pharmacology, Physiology and Neurology and Centre for
Neurobiology and Behavior, College of Physicians and Surgeons,
Columbia University, New York, NY 10032, USA

INTRODUCTION

Fast axonal transport is the rapid movement of organelles along the axon of a nerve cell. During the past decade, axonal transport engaged the attention of most neural scientists primarily because of its usefulness as a neuroanatomical tracing technique. Recently, however, there is a heightened interest in axonal transport from the perspective of the cell biology of the neurone, which has led to the characterization of organelles that move in the anterograde direction, from the cell body to terminals, and in the retrograde direction, from terminals back to the cell body. Transmitter substances were among the first chemical markers used in the analysis of fast transport, and they have permitted the identification of a chief type of organelle that is transported in the anterograde direction, the transmitter storage vesicle. Thus, Dahlström and Häggendal (1966) showed that norepinephrine is carried along fibres in the rat sciatic nerve within dense core granules.

From cell biological studies, it has become apparent that fast transport is part of the universal mechanism of secretion in a highly differentiated neuronal form. Anterograde fast transport would be homologous to the mechanism by which secretory vesicles and other membranous components move from their somatic sites of synthesis and assembly in the endoplasmic reticulum and Golgi apparatus to the plasma membrane, and retrograde transport would be homologous to the mechanism by which elements of plasma membrane are engulfed by lysomoses and moved to the interior of the cell. In neurones, fast transport selectively aims at a highly interesting region of the plasma membrane—the synapse.

In this chapter, we shall review the evidence showing that serotonin moves by fast transport and leading to the identification of its organellar carriers, and we shall discuss the physiological significance of the transport.

ANTEROGRADE TRANSPORT

Molluscs

Axonal transport of serotonin has been demonstrated most clearly in the central nervous system of molluscs. Initial observations were made in whole nerves. Osborne and Cottrell (1970) showed that accumulation of serotonin, measured fluorometrically and by bio-assay, was much greater proximal than distal to a constriction of the visceral nerve in the land snail, *Helix*. A few years later Howes *et al.* (1974) described fast transport in the cerebro-visceral connective in the swan mussel, *Anodonta*. The rapidly-moving material was labelled by injection into the cerebral ganglion of the radioactive precursor of serotonin, [^{3}H] 5-hydroxytryptophan, but was not conclusively identified as [^{3}H] serotonin.

Definitive evidence for fast transport of serotonin has come from studies using a pair of identified giant neurones in the sea hare, *Aplysia californica*, the giant cerebral neurones (GCN; also called metacerebral cell (MCC), giant serotonergic cell (GSC) and Cl). Identified molluscan neurones offer important experimental advantages. In vertebrates, neurones are relatively small and nerves like the sciatic contain the axons of many thousands of neurones whose properties are varied, differing for example in type of transmitter and type of vesicles. In contrast, with an identified cell, transport can be studied in a single neurone whose transmitter type has been characterized. In addition, many molluscan neurones are extraordinarily large, and therefore permit intrasomatic or even intra-axonal injection of a variety of substances, for example, radiolabelled transmitters, drugs, and even protein molecules that are too large to penetrate through the neuronal membrane if administered extracellularly.

There is extensive evidence that GCN and homologous cells in other opisthobranch molluscs (Weiss and Kupfermann, 1976) are serotonergic. The cell body of

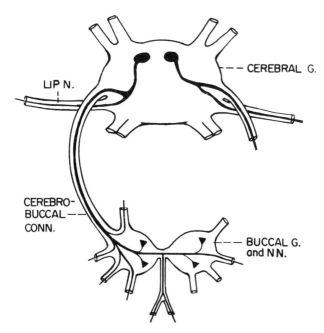

Fig. 1 Diagram of the dorsal surface of the cerebral ganglion of *Aplysia*, showing the pair of giant cerebral neurones and their axon distributions. The distributions of the axons in the lip nerve and cerebro-buccal connective (conn) were drawn after specimens injected with $CoCl_2$ and horseradish peroxidase. The distribution of axons and terminals in the buccal ganglion is based on radioautographic studies of nervous systems 15 h after intrasomatic injection of [^{3}H]N-acetylgalactosamine, an amino sugar precursor of membrane glycoproteins. Membranous organelles, including serotonergic storage vesicles, labelled in the cell body, are rapidly transported along the axon by fast axonal transport, and serve to identify the distant processes of the injected neurone in both the light and the electron microscope (Shkolnik and Schwartz, 1980)

GCN has been shown to contain high concentrations of serotonin (Weinreich *et al.*, 1973) and to convert tryptophan to serotonin (Eisenstadt *et al.*, 1973). Application of serotonin has been shown to mimic both the central (Gerschenfeld and Paupardin-Tritsch, 1974) and the peripheral (Weiss *et al.*, 1978) synaptic actions of GCN. These actions are blocked by serotonin antagonists (Gerschenfeld and Paupardin-Tritsch, 1974). Finally, Gerschenfeld *et al.* (1978) have shown that serotonin is released from the terminals of GCN upon stimulation.

The cell body of GCN is 200–300 μm in diameter and is near the surface of the cerebral ganglion (Fig. 1). Typical of invertebrate neurones, it gives rise to only one axon, which bifurcates within the neuropil of the ganglion into branches of similar diameter. One branch enters the ipsilateral cerebrobuccal connective and synapses on neurones in the paired buccal ganglia as well as on muscle fibres in the underlying buccal mass (Weiss *et al.*, 1978). The other branch runs in the ipsilateral

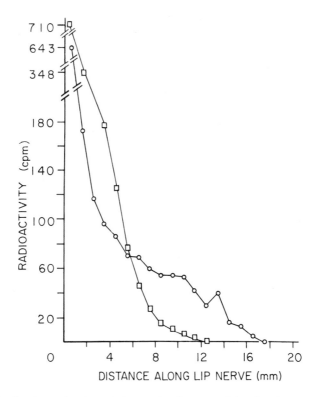

Fig. 2   Distribution of radioactivity in the lip nerve 3 h after intrasomatic injection of [³H]serotonin (○) and 5 h after intrasomatic injection of [³H]choline (□) (Goldberg and Schwartz, 1980). The excised nervous systems were injected and maintained at 23 °C in culture for 3 or 5 h; they were then frozen on a brass block. The lip nerve was cut into sequential millimeter segments and analysed for radioactivity by scintillation counting. High voltage electrophoretic analysis revealed that essentially all of the radioactivity in the neurone injected with [³H]serotonin was in the form of serotonin; all of the radioactivity in the axon of the cell injected with [³H]choline was in the form of metabolites: most of the material was betaine, and the rest was phosphorylcholine

posterior lip nerve and presumably synapses on lip musculature. The cell is large enough to tolerate impalement with a relatively large double-barrelled microelectrode (resistance = 1-4 MΩ) through which [³H]serotonin can be injected rapidly with pressure. Usually, an amount of [³H]serotonin equal to 10-50 per cent of the endogenous somatic content of serotonin is injected.

Serotonin has been shown to move from cell body to synapse by fast axonal transport in GCN. [³H]Serotonin injected into GCN is exported at a constant rate into the cerebrobuccal connective and lip nerve (Goldman *et al.*, 1976). The distribution of serotonin along the length of the nerve is quite different from the distribution of substances moving by diffusion, such as [³H]betaine (Goldberg and Schwartz, 1980). After intrasomatic injection, [³H]serotonin progresses much farther along the nerve in 3 h than [³H]betaine does in 5 h, and the profile of the spatial distribution of [³H]serotonin has a plateau region, in contrast to the smoothly decreasing profile with [³H]betaine (Fig. 2). In addition, exposure of a length of the nerve to colchicine (Goldman *et al.*, 1976) or to low temperature (1-3 °C) (Goldberg *et al.*, 1978), treatments known to block fast transport, prevents further movement of most of the [³H]serotonin, and results in a local accumulation.

Also diagnostic of fast transport is the constant rate of movement of [³H]serotonin in the axon of GCN. Exact measurement of this rate is difficult when efflux from the cell body is continuous throughout an experiment, because the front of radioactivity in the axon is often broad and ill-defined. The constant velocity is seen most clearly by analysing pulses of [³H]serotonin, which move as discrete patches of radioactivity with easily discernable peaks whose position in the axon can be determined precisely (Fig. 3). A pulse is produced by ligating the origin of the lip nerve at the cerebral ganglion just after the first substantial amounts of [³H]serotonin appear in the nerve. The pulse of radioactivity migrates at constant velocities of 130 mm per day at 23 °C and 48 mm per day at 14 °C (Goldberg *et al.*, 1978). These rates are similar to velocities of fast transport measured at those temperatures in nerves from other animals (Edström and Hanson, 1973; Ochs and Smith, 1975). The strong dependence of velocity on temperature is also characteristic of fast transport.

Mammals

The mammalian central nervous system is the only other site where evidence, albeit sketchy, of fast transport of serotonin has been obtained. The cell bodies of serotonergic neurones in the central nervous system occur mainly in the raphe nuclei of the brain stem and midbrain, with axons radiating widely to other regions of the brain and spinal cord. One radiation consists of axons that descend in the anterior and lateral funiculi of the spinal cord to innervate grey matter at all levels of the cord. Transport can be evaluated by transecting the cord and measuring the accumulation of serotonin or associated materials rostral to the cut.

Serotonin does accumulate above a transection (Dahlström *et al.*, 1973) as well as above a segment treated with colchicine or vinblastine (Dahlström, 1971). The rate of accumulation is slow, however: in 24 h, the content of serotonin in the 1 cm length of white matter just rostral to a transection was found to double (Dahlström *et al.*, 1973). Without correction for non-mobile serotonin, this would suggest that the transmitter moves at a rate of 10 mm per day. This rate cannot itself be taken as evidence of fast transport. It is an order of magnitude slower than established rates of fast transport in mammalian nerves (Schwartz, 1979). Although velocities of fast transport are often found to be slower in central than peripheral neurones, the most accurately measured rates in the central nervous system are still about

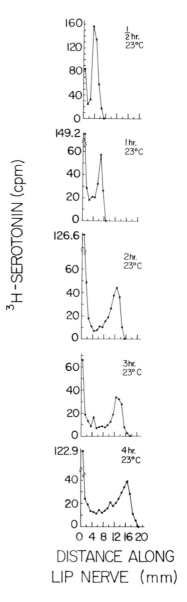

Fig. 3   Transport of [³H] serotonin along the lip nerve at 23 °C. The nerve was ligated at its point of exit from the cerebral ganglion 70 min after intrasomatic injection of [³H] serotonin in order to create a discrete pulse of radioactivity in the proximal end of the nerve. The nerve was frozen at the indicated times after ligation, and cut into millimetre pieces. The distribution of [³H] serotonin was then determined. Each graph is from a different representative experiment (Goldberg *et al.*, 1978)

100 mm per day, as for [³H]glycoprotein in axons projecting from the locus coerules to the hypothalamus (Levin, 1977).

A rate of 10 mm per day is almost certainly an underestimate; much of the serotonin probably is not capable of moving, either because it is not in axons or, if within axons, not in vesicles, but free in the axoplasm. Serotonin-binding protein, which is thought to be complexed to serotonin and therefore to be transported with it, was found to accumulate above a spinal transection more rapidly. Tamir and Gershon (1979) estimated that the binding protein is transported at a rate of at least 78 mm per day.

CARRIER OF SEROTONIN IN ANTEROGRADE FAST TRANSPORT

Molluscs

There is a great deal of evidence that serotonin is transported within vesicles, and little indication for the transport of transmitter unassociated with membrane-bound organelles. The most direct evidence for identifying vesicles as the carriers of serotonin comes from the analysis of the distribution of silver grains in electron microscopic radioautographs of GCN after injection of [³H]serotonin. As already discussed, [³H]serotonin that is injected into the cell body of this *Aplysia* neurone is exported into the axon at a constant rate and moves by fast transport. Having shown by biochemical analysis that all of the axonal radioactivity was in the form of [³H]serotonin, Goldman *et al.* (1976) found that a characteristic vesicle is the only labelled membranous organelle in the axon. The much more abundant axoplasmic reticulum was not specifically labelled nor were mitochondria and the axolemma. Association of the moving transmitter with vesicles was also supported by subcellular fractionation experiments. About half the [³H]serotonin in the axon of GCN was found to sediment with particulate material and the rest was recovered in the soluble fraction (Goldman *et al.*, 1976). In contrast, only 2 per cent of [³H]betaine, which moves in the axon of GCN by diffusion rather than fast transport (Goldberg and Schwartz, 1980), was particulate (Goldman *et al.*, 1976). The substantial proportion of [³H]serotonin found in soluble form in these subcellular fractionation experiments coincides with the observation in radioautographs that, in addition to vesicles, axoplasm was also significantly labelled by [³H]serotonin (Goldman *et al.*, 1976). Thus, while both subcellular fractionation and radioautography show that [³H]serotonin moving by fast transport is carried in storage vesicles, they raised the possibility that the transmitter is also transported as a soluble molecule.

We think it unlikely that serotonin is transported in soluble form, because soluble molecules are generally found to move by slow axoplasmic flow (Schwartz, 1979). The extravesicular location of [³H]serotonin in both types of experiments can be readily explained by assuming that some of the transmitter can escape from vesicles. Even in intact axons, discrete moving pulses of [³H]serotonin leave trails of radioactivity behind, suggesting that vesicles in living axons lose some transmitter as they move along the axon (Fig. 3; ignore the most proximal few millimetres; local immobilization of some radioactivity is due to the injury caused by the ligation used to produce the pulse). Escape of transmitter is likely to be artefactually enhanced both by cell breakage during subcellular fractionation and by fixation procedures for electron microscopy. Surely leakage occurs during subcellular fractionation of mammalian brain, since Tamir and Gershon (1979) showed that the amount of serotonin sedimenting with vesicles could be greatly enhanced by modifying the fractionation protocol. Leakage from vesicles might also occur in preparing tissue for electron microscopy, since we found that 30 per cent of the [^{3}H] serotonin in GCN escaped from the tissue during fixation with glutaraldehyde and osmium (Schwartz *et al.*, 1979). In addition, some of the silver grains that were assigned to axoplasm in the experiments of Goldman *et al.* (1976) might actually have obscured underlying vesicles, since the grains often were larger than the vesicles.

Electron microscopic analyses of serotonin in nerves of two other molluscs also support the idea that the transmitter is packaged in vesicles. An accumulation of vesicles accompanied the build up of serotonin proximal to a constriction in the visceral nerve of *Helix*; when chromium salts were added to the fixative, these vesicles exhibited the dense reaction product that indicates a high content of monoamine (Osborne and Cottrell, 1970). In the cerebro-visceral connective of *Anodonta*, analyses of the distribution of silver grains over axons revealed that the radioactivity in transported material was most heavily associated with dense cored vesicles, although mitochondria and other organelles were also labelled to a significant extent (Howes *et al.*, 1974). As mentioned above, however, it was not shown that all of the radioactivity in the tissue was in the form of [^{3}H] serotonin.

Mammals

Because there is scant direct evidence for fast transport of serotonin in mammalian neurones, little can be concluded about the vehicle in which it is moved except by analogy with adrenergic nerves, in which transport has been analysed more thoroughly. In these nerves, norepinephrine is transported in dense cored vesicles. When the sciatic nerve, which contains many unmyelinated adrenergic axons, was crushed, norepinephrine accumulated proximal to the crush at a rate indicative of a transport velocity of about 190 mm per day (Dahlström and Häggendal, 1966; Häggendal *et al.*, 1975). Dense cored vesicles, morphologically similar to those in axon terminals (Hökfelt, 1969) that are known to contain norepinephrine (Bisby and Fillenz, 1971) also accumulated (Kapeller and Mayor, 1966; Banks *et al.*, 1969a; Geffen and Ostberg, 1969). Accumulations of transmitter and vesicles were both prevented by colchicine (Banks *et al.*, 1971). Reserpine, which depletes aminergic vesicles of their stores of transmitter (Cooper *et al.*, 1978), also blocked the accumulation of norepinephrine (Dahlström, 1967); the vesicles continued to collect behind the crush, but were visibly depleted of dense cores (Banks *et al.*, 1969b). Reserpine also prevented the increase of serotonin rostral to a transection of the spinal cord (Dahlström *et al.*, 1973). This action of reserpine is the only direct evidence that serotonin is transported within vesicles in mammalian nerve.

MATURATION OF VESICLES DURING ANTEROGRADE TRANSPORT

The organelle that carries serotonin by fast transport along the axon of GCN is a 74 nm vesicle with a core (Fig. 4A). The core is bounded by a complete, trilaminar membrane of the same thickness as the outer vesicle membrane, and thus constitutes a vesicle within a vesicle (Shkolnik and Schwartz, 1980). This arrangement could be seen in vesicles in the proximal regions of the axon of GCN because the contents were not electron-dense enough to obscure the inner membrane when a glutaraldehyde-osmium fixative was used. Thus, we refer to these vesicles in the axon of GCN as compound rather than dense cored. Histochemical procedures for biogenic amines demonstrated that these organelles contain endogenous serotonin, however. When potassium permanganate or chromium salts were added to the fixative, 80-90 per cent of the vesicles in the axon had dense cores (Fig. 4B).

In contrast, special histochemical procedures were not needed to visualize the dense cores of serotonergic vesicles in the synaptic terminals of GCN, where 96 per cent of the vesicles with cores have dense cores after fixation with glutaraldehyde and osmium (Shkolnik and Schwartz, 1980) (Fig. 5A). These vesicles are also larger than the compound vesicles in the proximal regions of the axon, averaging 95 nm in diameter.

Does the serotonergic storage vesicle change during axonal transport? In GCN, the vesicles in terminals evidently have more serotonin, or other core substance, than do the axonal vesicles. The extra material could be added gradually during axonal transport. Alternatively, it might all be added at the nerve ending, perhaps because of differences in the cytoplasmic environment.

Interestingly, if the axon of GCN is severed in its proximal region, the vesicles that accumulate above the transection begin to resemble the vesicles in terminals, becoming dense cored after fixation with glutaraldehyde and osmium (Fig. 5B). This apparent maturation of vesicles at the transection can be explained by either a time-dependent or an environment-dependent addition of material. The vesicles trapped above the block not only are older than normal axonal vesicles by virtue of having been delayed in transit, but also are exposed to a new environment because of the local nerve injury. For example, nerve injury causes an increased influx of Ca^{2+} (Schlaepfer and Bunge, 1973); in this regard, the axon just proximal to the interruption might resemble the synaptic terminal, where there is repeated influx of Ca^{2+} associated with synaptic transmission (Katz, 1969). It is also possible that these dense cored vesicles are not serotonergic storage granules, but serve some other function in the injured axon. In several animals, dense cored vesicles have been found to appear rapidly in injured axons of unidentified transmitter type (Lanners and Grafstein, 1980).

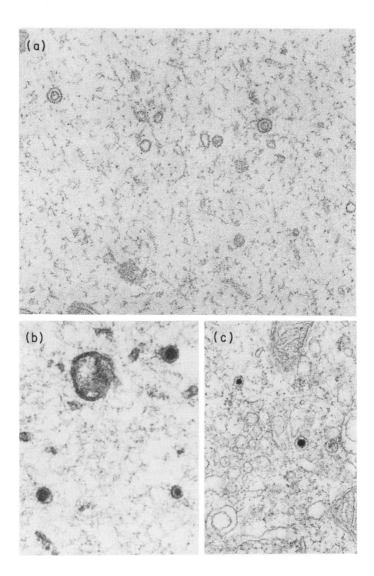

Fig. 4   The appearance of serotonergic vesicles in cross sections of the proximal axon of GCN in the cerebrobuccal connective prepared for electron microscopy: (a) after standard fixation with glutaraldehyde and osmium; (b) after fixation with potassium permanganate; (c) after treatment with chromium salts. (Shkolnik and Schwartz, 1980). Note that the electron-lucent compound vesicle seen after standard fixation becomes dense cored after cytochemical procedures previously used in vertebrates to enhance the density of aminergic vesicles. Magnification X 67 500

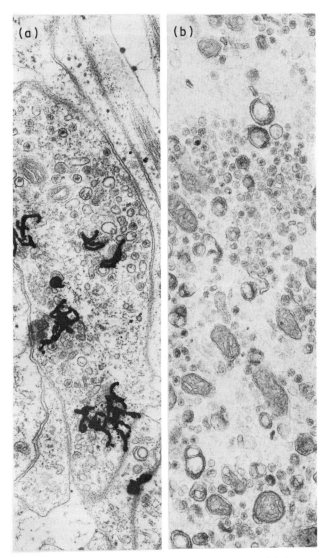

Fig. 5　Vesicles have a similar morphological appearance, (a) in an axon terminal of GCN within the buccal ganglion identified by radioautography (see caption to Fig. 1) and (b) in a 6-h accumulation above a ligature of the proximal axon in the cerebrobuccal connective (Shkolnik and Schwartz, unpublished experiments). Magnification X 44 000

Less ambiguous evidence for the gradual addition of transmitter during transport has been obtained in adrenergic nerves. Lagercrantz *et al.* (1974) measured the content of norepinephrine in large dense cored vesicles in three regions of the splenic nerve as well as in the synaptic terminals in the spleen. They found that the ratio of transmitter to total vesicle protein increases along a proximo-distal gradient in the nerve, though more modestly than at the terminals where there is a three-fold increase in vesicular norepinephrine (Lagercrantz, 1976). These results with adrenergic nerves favour the idea that some of the increase in vesicular serotonin suggested by the micrographs of the axon terminals of GCN might occur gradually during the course of axonal transport. This would not be surprising. The enzymes catalysing serotonin synthesis, tryptophan hydroxylase and aromatic amino acid decarboxylase, are in axons (Dahlström and Jonason, 1968; Pickel *et al.*, 1976) on the way from their site of synthesis in the cell body to the synapse (Dahlström and Jonason, 1968; Meek and Neff, 1972), where they help to maintain transmitter stores. Some enzyme may even be associated with the storage vesicles (Pickel *et al.*, 1976; Starkey and Brimijoin, 1979). Thus, accumulation of serotonin by axonal vesicles in transit probably does not play a functional role locally, but happens because vesicles and the synthetic enzymes are both present in the axon.

RETROGRADE TRANSPORT

Retrograde transport of serotonin has so far been described only in the mammalian central nervous system, where the impetus has been its potential usefulness as a neuroanatomical method for tracing serotonergic tracts. Stereotaxic injection of [^{3}H] serotonin into areas densely innervated by serotonergic terminals (neostriatum, substantia nigra and olfactory bulb) results in the appearance of radioactivity in the raphe nucleus within several hours (Léger *et al.*, 1977; Streit *et al.*, 1979; Araneda *et al.*, 1980a; Araneda *et al.*, 1980b). Araneda *et al.* (1980a, 1980b) have clearly shown that this radioactivity is due to retrograde transport of [^{3}H] serotonin. They injected [^{3}H] serotonin into the rat olfactory bulb and measured a sub-stantial, time-dependent increase in radioactivity only in the dorsal raphe nucleus. Much of this radioactive material was shown to be serotonin by chromatography, and no accumulation was observed when the serotonergic nerve terminals in the olfactory bulb were damaged by administration of 5,6-dihydroxytryptamine, a drug that is specifically taken up by serotonergic terminals and kills them.

In the experiments of Araneda *et al.* (1980a) the first significant amount of radioactivity appeared in the raphe nuclei 8 h after injection. Because these nuclei are about 16 mm distant from the bulb, the [^{3}H] serotonin must have travelled at a rate of at least 48 mm per day and this rate is likely to be underestimated since no assays were performed between 4 and 8 h after the injection. Moreover, application of colchicine prevented the accumulation of radioactivity in the raphe nuclei. These results strongly imply that [^{3}H] serotonin was taken up by serotonergic nerve terminals in the bulb and moved by retrograde fast axonal transport to the cell bodies in the raphe nuclei.

CARRIER OF SEROTONIN IN RETROGRADE TRANSPORT

It is likely that serotonin moving by retrograde fast axonal transport is carried in membranous organelles. Retrograde transport of [^{3}H] serotonin in axons of serotonergic raphe neurones persists in the absence of monoamine oxidase inhibition, implying that the transmitter moves in a protected form (Araneda *et al.*, 1980a). One function of retrograde transport presumably is the return to the cell body of worn-out constituents of the synaptic terminals for degradation by lysosomes and, possibly, reuse of the breakdown products as precursors for new macromolecules. Lysosomes and pre-lysosomal organelles such as multivesicular bodies form the predominant class of organelles moving in the retrograde direction (Smith, 1980; Tsukita and Ishikawa, 1980). Although the carrier for serotonin in the retrograde direction has not been directly identified, lysosomal organelles are the best candidates.

Only Pentreath (1976) has observed serotonin in axonal lysosomes. He iontophoretically injected [^{3}H] serotonin into the cell body of an identified serotonergic cell in *Helix*, the homologue of GCN in *Aplysia*, fixed the tissue many hours later, and then prepared electron microscopic radioautographs of distal regions of the axon. Silver grains appeared to be associated with lysosomes, although statistical analysis was not performed. It is unlikely that this [^{3}H] serotonin was actually moving by fast transport, because the nervous tissue was kept at 4 °C after injection, a temperature that blocks transport in *Aplysia* (Goldberg *et al.*, 1978). But this result suggests that lysosomes in the axon can take up serotonin.

There is no doubt that lysosomes in the cell body accumulate serotonin, perhaps even serving as storage depots for the transmitter rather than just sites of degradation. Large lysosomes in the cell body of a serotonergic cell in the land slug, *Limax*, exhibit an electron-dense reaction product when chromium salts are included in the fixative, suggesting the presence of a monoamine (Cottrell and Osborne, 1970). A striking association of [^{3}H] serotonin with large somatic lysosomes can be observed after the radioactive amine is injected into the cell body of *Aplysia's* GCN (Fig. 6) (Schwartz *et al.*, 1979). Quantititave electron microscopic analysis of the distribution of silver grains in the cell body at a time when essentially all of the radioactivity was in the form of [^{3}H] serotonin shows that storage vesicles and large lysosomes were the only significantly labelled organelles. Because the lysosomes occupy a much greater volume than the vesicles, they contained most of the somatic [^{3}H] serotonin.

Lysosomal uptake was not due to a non-specific process such as ion trapping, because injected [^{3}H] dopamine and [^{3}H] histamine did not accumulate in the lysosomes of GCN. Moreover, [^{3}H] serotonin was not sequestered by the structurally similar large lysosomes of R2, a cholinergic *Aplysia* neurone. The specificity with which the lysosomes of GCN take up [^{3}H] serotonin suggests that these organelles might contain functional binding proteins for serotonin deriving from storage vesicles that had been phagocytosed. It is unlikely that the lysosomes phagocytose newly synthesized vesicles, because the sugar [^{3}H] *N*-acetylgalactosamine, which is rapidly incorporated into newly synthesized storage vesicles (Ambron *et al.*, 1980), did not label the lysosomes (Schwartz *et al.*, 1979). We think that these binding proteins ultimately derive from old storage vesicles in synaptic terminals, which were phagocytosed and returned to the cell body in lysosomes. Transport of serotonin in the retrograde direction would therefore reflect the movement of lysosomal

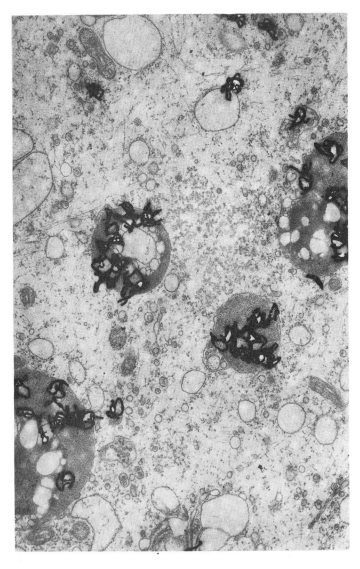

Fig. 6   Electron microscopic radioautographs showing that large, end-stage lysosomes in the cell body of GCN were labelled 3 h after injection of [^{3}H]sero-tonin. Quantitative analyses of the distribution of silver grains showed that somatic compound vesicles were also labelled (Schwartz *et al.*, 1979). Magnification X 30 000

elements, with the transmitter being taken up into lysosomes in the terminals as a constituent of phagocytosed vesicles and perhaps directly from the cytoplasm, as demonstrated in the cell body. Until serotonin moving in the retrograde direction in axons has been localized directly, however, this remains a suggestion rather than a firm conclusion.

Uptake of transmitter into lysosomes may be a characteristic feature of monoaminergic neurones. Bailey *et al.* (1981) have found similar labelling of a putative serotonergic facilitator neurone, L29, in the abdominal ganglion of *Aplysia*. Gotoh and Schwartz (1980) also showed that [^{3}H]histamine is taken up into end-stage lysosomes of C2, a histaminergic neurone in the cerebral ganglion of *Aplysia* (Weinreich, 1978). These histaminergic lysosomes were also labelled after injection of [^{3}H]serotonin into the cell body of C2, indicating that the lysosomal uptake in the histaminergic neurone does not discriminate between the two monoamine transmitter substances. Binding of monoamine transmitters to lysosomes may be a universal phenomenon that is not limited to all molluscan neurones. The colour of the human substantia nigra is due to pigmentation within lysosomal neuromelanin granules (Duffy and Tennyson, 1965; Barden, 1975) that results from degradation of dopamine (see, for example, Das *et al.*, 1978). Lysosomal binding of the transmitter presumably precedes its oxidation and polymerization into neuromelanin.

SPECIFICITY OF TRANSPORT IN SEROTONERGIC NEURONES

Since serotonin is transported along the axon within vesicles and lysosomes, the specificity of transport is likely to be a reflection of the selectivity of uptake into those organelles. Fast transport might therefore be used to assay the specificity of these uptake processes. Uptake of transmitter from the extracellular space into serotonergic terminals is not completely specific and uptake into vesicles is even less so. For example, dopamine, but not norepinephrine, is taken into serotonergic terminals in mammalian brain (Berger and Glowinski, 1978); both are sequestered by isolated serotonergic storage vesicles (Da Prada and Pletscher, 1969). It is therefore not surprising that dopamine applied exogenously is transported in the retrograde direction along serotonergic axons (Streit *et al.*, 1979) while norepinephrine (Araneda *et al.*, 1980a) and the structurally unrelated amino acid, γ-aminobutyric acid (Streit *et al.*, 1979), are not.

Because uptake across the external membrane appears to be more selective than uptake into vesicles, fast transport should carry an even wider array of foreign transmitters if the filter represented by the plasma membrane is circumvented by injecting the transmitters directly into the cell. With this procedure, anterograde fast transport could be used as an asaay of the *in vivo* specificity of uptake into storage vesicles, a process studied previously only with purified cell-free vesicles.

We have injected radioactive foreign transmitters into the cell body of GCN and have found that the specificity of fast transport resembles the specificity of uptake into purified aminergic vesicles (Goldberg and Schwartz, 1980). When injected, [^{3}H]dopamine, [^{3}H]octopamine and [^{3}H]histamine all moved along the nerve as rapidly as [^{3}H]serotonin (Fig. 7). In contrast, when [^{3}H]choline, [^{3}H]γ-aminobutyric acid or [^{3}H]putrescine were injected, radioactivity moved much more slowly along the nerve and at rates that diminished with distance from the cell

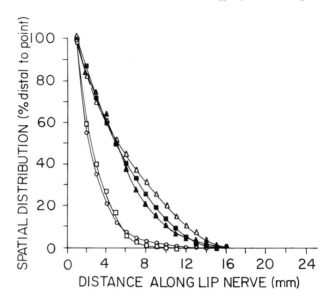

Fig. 7   Spatial distributions of radioactivity in the lip nerve after intrasomatic injections of a variety of foreign labelled transmitter substances and of [³H] choline. In order to normalize the distributions along axons from the different specimens, we constructed the curves by plotting the percentage of total axonal radioactivity distal to a given point along the lip nerve as a function of distance from the cell body 3 h after injection of [³H] serotonin (△), [³H] dopamine (■), [³H] histamine (▲), [³H] choline (○) and [³H] γ-aminobutyric acid (□) (Goldberg and Schwartz, 1980)

body. Unlike the biogenic amines, which appeared to be protected, choline, γ-amino-butyric acid and putrescine were broken down during passage along the axon. These slowly-moving compounds evidently diffuse in axoplasm, while the biogenic amines are transported within a carrier, presumed to be the serotonergic storage vesicle. Further evidence that the biogenic amines were all carried in the same organelle is provided by experiments in which injection of large amounts of unlabelled foreign monoamines (or serotonin) was shown to reduce the transport of [³H] serotonin that was injected at the same time. This result is most easily explained by assuming that the unlabelled material competitively inhibited uptake of [³H] serotonin into the vesicles.

IS TRANSPORTED SEROTONIN PHYSIOLOGICALLY IMPORTANT?

In summary, serotonin moves by anterograde fast axonal transport, carried by storage vesicles that are precursors of synaptic vesicles. Some serotonin also moves in the retrograde direction, possibly within lysosomal organelles. In cell bodies, the transmitter can readily be identified in characteristic end-stage lysosomes, which are absent from the axon. Presumably serotonin is carried in less-readily identified lysosomes or in pre-lysosomal structures such as multivesicular bodies.

Why is serotonin transported? We suggest that the serotonin transported either in the anterograde or the retrograde direction not needed for any essential physiological role. It just goes along for the ride, because it has been sequestered by moving organelles that take up serotonin. There would seem to be little need for serotonin to be supplied from the cell body to terminals, because nerve endings avidly recover serotonin released into the synaptic cleft and have the capability to synthesize enormous amounts of new transmitter (Cooper *et al.*, 1978). Synthesis at terminals can probably even be adjusted to conform to alterations in synaptic activity, for example, through changes in affinity of tryptophan hydroxylase for substrate or cofactor (Boadle-Biber, 1978).

Limiting to synaptic transmssion, however, are the macromolecular components of synaptic vesicles. Unlike the transmitter, these cannot be synthesized in the terminals, but must be formed in the pericaryon. Thus, the serotonergic vesicle itself, needed to replace its worn-out antecedents at the terminal, presumably constitutes the essential part of the package. Axonal transport of vesicle *contents* might be very important if neuroactive peptides co-exist with serotonin within storage vesicles, however (Chan-Palay *et al.*, 1978; Hökfelt *et al.*, 1978), because these macromolecules are formed only in the cell body, and no mechanism has been described for their synthesis or recapture in nerve terminals.

Serotonin that moves by fast transport is thus probably not of direct physiological importance to the neurone, being transported because of the requirements of the synapse for serotonergic vesicles. Even though the transport of serotonin appears to have no direct physiological function, it is still quite useful to the neurobiologist, as a tool for mapping the projections of serotonergic neurones in the brain and for studying the mechanism of fast transport and the selectivity of transmitter uptake into storage vesicles.

REFERENCES

Ambron, R. T., Goldman, J. E., Shkolnik, L. J., and Schwartz, J. H. (1980). Synthesis and axonal transport of membrane glycoproteins in an identified serotonergic neuron of *Aplysia, J. Neurophysiol.* **43**, 924–944.

Araneda, S., Bobillier, P., Buda, M., and Pujol, J. -F. (1980a). Retrograde axonal transport following injection of ^{3}H-serotonin in the olfactory bulb. I. Biochemical study, *Brain Res.* **196**, 405–415.

Araneda, S., Gamrani, H., Font, C., Calas, A., Pujol, J. -F., and Bobillier, P. (1980b). Retrograde axonal transport following injection of ^{3}H-serotonin into the olfactory bulb. II. Radioautographic study, *Brain Res.* **196**, 417–427.

Bailey, C. H., Hawkins, R. D., Chen, M. C., and Kandel E. R. (1981). Interneurons involved in mediation and modulation of gill-withdrawal reflex in *Aplysia*. IV. Morphological basis of presynaptic facilitation, *J. Neurophysiol* **45**, 358–378.

Banks, P., Kapeller, K., and Mayor, D. (1969a). The effects of iproniazid and reserpine on the accumulation of granular vesicles and noradrenaline in constricted adrenergic nerves, *Br. J. Pharm.* **37**, 10–18.

Banks, P., Mangnall, D., and Mayor, D. (1969b). The redistribution of cytochrome oxidase, noradrenaline and adenosine triphosphate in adrenergic nerves constricted at two points, *J. Physiol.* **200**, 745–762.

Banks, P., Mayor, D., Mitchell, M., and Tomlinson, D. (1971). Studies on the translocation of noradrenaline-containing vesicles in post-ganglionic sympathetic neurons *in vitro*. Inhibition of movement by colchicine and vinblastine and evidence for the involvement of axonal microtubules, *J. Physiol.* **216**, 625–639.

Barden, H. (1975). The histochemical relationships and the nature of neuromelanin, *Aging,* **1**, 79–117.

Berger, B. and Glowinski, J. (1978). Dopamine uptake in serotonergic terminals *in vitro*: A valuable tool for the histochemical differentiation of catecholaminergic and serotonergic terminals in rat cerebral structures, *Brain Res.* **147**, 29–45.

Bisby, M. A. and Fillenz, M. (1971). The storage of noradrenaline in sympathetic nerve terminals, *J. Physiol.* **215**, 163–179.

Boadle-Biber, M. C. (1978). Activation of tryptophan hydroxylase from central serotonergic neurones by calcium and depolarization, *Biochem Pharmacol.* **27**, 1069–1079.

Chan-Palay, V., Jonsson, G., and Palay, S. L. (1978). Serotonin and substance P coexist in neurons of the rat's central nervous system, *Proc. nat. Acad. Sci. USA,* **75**, 1582–1586.

Cooper, J. R., Bloom, F. E., and Roth, R. H. (1978). *The Biochemical Basis of Neuropharmacology*, Oxford University Press, New York.

Cottrell, G. A. and Osborne, N. N. (1970). Subcellular localization of serotonin in an identified serotonin-containing neurone, *Nature (Lond.),* **225**, 470–472.

Dahlström, A. (1967). The effect of reserpine and tetrabenazine on the accumulation of noradrenaline in the rat sciatic nerve after ligation, *Acta Physiol. Scand.* **69**, 167–179.

Dahlström, A. (1971). Effects of vinblastine and colchicine on monoamine-containing neurons of the rat, with special regard to the axoplasmic transport of amine granules, *Acta Neuropath. (Berl.)* Suppl. V, 226–237.

Dahlström, A. and Häggendal, J. (1966). Studies on the transport and life-span of amine storage granules in a peripheral adrenergic neuron system, *Acta Physiol. Scand.* **67**, 278–288.

Dahlström, A., Häggendal, J., and Atack, C. (1973). Localization and transport of serotonin, in *Serotonin and Behavior* (Eds J. Barchas and E. Usdin), pp. 87–96, Academic Press, New York.

Dahlström, A. and Jonason, J. (1968). DOPA-decarboxylase activity in sciatic nerves of the rat after constriction, *Eur. J. Pharm.* **4**, 377–383.

Da Prada, M. and Pletscher, A. (1969). Differential uptake of biogenic amines by isolated serotonin organelles of blood platelets, *Life Sci.* **8**, 65–72.

Das, K. C., Abramson, M. B., and Katzman, R. (1978). Neuronal pigments: Spectroscopic characterization of human brain melanin, *J. Neurochem.* **30**, 601–605.

Duffy, P. E., and Tennyson, V. M. (1965). Phase and electron microscopic observations of Lewy bodies and melanin granules in the substantia nigra and locus coerleus in Parkinson's disease. *J. Neuropathol. Exp. Neurol.* **24**, 398–414.

Edström, A., and Hanson, M. (1973). Temperature effects on fast axonal transport of proteins *in vitro* in frog sciatic nerves. *Brain Res.* **58**, 345–354.

Eisenstadt, M. L., Goldman, J. E., Kandel, E. R., Koike, H., Koester, J., and Schwartz, J. H. (1973). Intrasomatic injection of radioactive precursors for studying transmitter synthesis in identified neurons of *Aplysia californica, Proc. nat. Acad. Sci. USA.* **70**, 3371–3375.

Geffen, L. B. and Ostberg, A. (1969). Distribution of granular vesicles in normal and constricted sympathetic neurons, *J. Physiol.* **204**, 583–592.

Gerschenfeld, H. M., Hamon, M., and Paupardin-Tritsch, D. (1978). Release of endogenous serotonin from two identified serotonin-containing neurones and the physiological role of serotonin reuptake, *J. Physiol.* **274**, 265–278.

Gerschenfeld, H. M. and Paupardin-Tritsch, D. (1974). On the transmitter function of 5-hydroxytryptamine at excitatory and inhibitory monosynaptic junctions, *J. Physiol.* **243**, 457–481.

Goldberg, D. J. and Schwartz, J. H. (1980). Fast axonal transport of foreign transmitters in an identified neurone of *Aplysia californica, J. Physiol.* **307**, 259–273.

Goldberg, D. J., Schwartz, J. H., and Sherbany, A. A. (1978). Kinetic properties of normal and perturbed axonal transport of serotonin in a single identified axon, *J. Physiol.* **281**, 559–579.

Goldman, J. E., Kim, K. S., and Schwartz, J. H. (1976). Axonal transport of ^{3}H-serotonin in an identified neuron of *Aplysia californica, J. Cell Biol.* **70**, 304–318.

Gotoh, H. and Schwartz, J. H. (1980). Specific axonal transport of ^{3}H-histamine after intrasomatic injection of C2, an identified *Aplysia* neuron, *Neurosci. Abstr.* **6**, 502.

Häggendal, J., Dahlström, A., and Larsson, P. -A. (1975). Rapid transport of noradrenaline in adrenergic axons of rat sciatic nerve distal to a crush. *Acta Physiol. Scand.* **94**, 386–392.

Hökfelt, T. (1969). Distribution of noradrenaline storage particles in peripheral adrenergic neurons as revealed by electron microscopy, *Acta Physiol. Scand.* **76**, 427–440.

Hökfelt, T., Ljungdahl, Å., Steinbusch, H., Verhofstad, A., Nilsson, G., Brodin, E., Pernow, B., and Goldstein, M. (1978). Immunohistochemical evidence of substance P-like immunoreactivity in some 5-hydroxytryptamine-containing neurons in the rat central nervous system. *Neuroscience*, **3**, 517–538.

Howes, E. A., McLaughlin, B. J., and Heslop, J. P. (1974). The autoradiographic association of fast transported material with dense core vesicles in the central nervous system of *Anodonta cygnea* (L.). *Cell Tiss. Res.* **153**, 545–558.

Kapeller, K. and Mayor, D. (1966). Ultrastructural changes proximal to a constriction in sympathetic axons during first 24 hours after operation. *J. Anat.* **100**, 439–441.

Katz, B. (1969). *The Release of Neural Transmitter Substances.* Liverpool University Press, Liverpool.

Lagercrantz, H. (1976). On the composition and function of large dense cored vesicles in sympathetic nerves, *Neuroscience*, **1**, 81–92.

Lagercrantz, H., Kirksey, D. F., and Klein, R. L. (1974). On the development of sympathetic nerve vesicles during axonal transport. *J. Neurochem.* **23**, 769–773.

Lanners, H. N. and Grafstein, B. (1980). Early stages of axonal regeneration in the goldfish optic tract: an electron microscopic study. *J. Neurocytol.* **9**, 733–751.

Léger, L., Pujol, J. -F., Bobillier, P., and Jouvet, M. (1977). Transport axoplasmique de la sérotonin par voie rétrograde dans les neurones mono-aminergiques centraux. *C.R. Acad. Sci. Ser. D.* **285**, 1179–1182.

Levin, B. E. (1977). Axonal transport of ^{3}H-fucosyl glycoproteins in noradrenergic neurons in the rat brain. *Brain Res.* **130**, 421–432.

Meek. J. L. and Neff, N. H. (1972). Tryptophan-5-hydroxylase: Approximation of half-life and rate of axonal transport, *J. Neurochem.* **19**, 1519–1525.

Ochs, S. and Smith, C. (1975). Low temperature slowing and cold-block of fast axoplasmic transport in mammalian nerves *in vitro, J. Neurobiol.* **6**, 85–102.

Osborne, N. N. and Cottrell, G. A. (1970). Transport of amines along the visceral nerve of *Helix pomatia*, *Z. Zellforsch. Mikrosk. Anat.* **109**, 171–179.

Pentreath, V. W. (1976). Ultrastructure of the terminals of an identified 5-hydroxytryptamine-containing neurone marked by intracellular injection of radioactive 5-hydroxytryptamine. *J. Neurocytol.* **5**, 43–61.

Pickel, V. M., Joh, T. H., and Reis, D. J. (1976). Monoamine synthesizing enzymes in central dopaminergic, noradrenergic and serotonergic neurons: Immunocytochemical localization by light and electron microscopy, *J. Histochem. Cytochem.* **24**, 792–806.

Schlaepfer, W. W. and Bunge, R. P. (1973). Effects of calcium ion concentration on the degeneration of amputated axons in tissue culture, *J. Cell Biol.* **59**, 456–470.

Schwartz, J. H. (1979). Axonal transport: components, mechanism and specificity, *Ann. Rev. Neurosci.* **2**, 467–504.

Schwartz, J. H., Shkolnik, L. J., and Goldberg, D. J. (1979). Specific association of neurotransmitter with somatic lysosomes in an identified serotonergic neuron of *Aplysia californica. Proc. nat. Acad. Sci. USA.* **76**, 5967–5971.

Shkolnik, L. J. and Schwartz, J. H. (1980). Genesis and maturation of serotonergic vesicles in identified giant cerebral neuron of *Aplysia*, *J. Neurophysiol.* **43**, 945–967.

Smith, R. S. (1980). The short term accumulation of axonally transported organelles in the region of localized lesions of single myelinated axons, *J. Neurocytol.* **9**, 39–65.

Starkey, R. R. and Brimijoin, S. (1979). Stop-flow analysis of the axonal transport of DOPA decarboxylase (EC 4.1.1.26) in rabbit sciatic nerves. *J. Neurochem.* **32**, 437–441.

Streit, P., Knecht, E., and Cuénod, M. (1979). Transmitter specific retrograde labelling in the striato-nigral and raphe-nigral pathways. *Science.* **205**, 306–308.

Tamir, H. and Gershon, M. D. (1979). Storage of serotonin and serotonin binding protein in synaptic vesicles, *J. Neurochem.* **33**, 35–44.

Tsukita, S. and Ishikawa, H. (1980). The movement of membranous organelles in axons. Electron microscopic identification of anterogradely and retrogradely transported organelles, *J. Cell Biol.* **84**, 513–530.

Weinreich, D. (1978). Histamine-containing neurons in *Aplysia*, in *Biochemistry of Characterised Neurons* (Ed N. N. Osborne), pp. 153–175, Pergamon, Oxford.

Weinreich, D., McCaman, M. W., McCaman, R. E., and Vaughn, J. E. (1973). Chemical, enzymatic, and ultrastructural characterization of 5-hydroxytryptamine-containing neurons from the ganglia of *Aplysia californica* and *Tritonia diomedia*, *J. Neurochem.* **20**, 969–976.

Weiss, K. R., Cohen, J. L., and Kupfermann, I. (1978). Modulatory control of buccal musculature by a serotonergic neuron (metacerebral cell) in *Aplysia*, *J. Neurophysiol.* **41**, 181–203.

Weiss, K. R. and Kupfermann, I. (1976). Homology of the giant serotonergic neurons (metacerebral cells) in *Aplysia* and pulmonate molluscs, *Brain Res.* **117**, 33–49.

Biology of Serotonergic Transmission
Edited by N. N. Osborne
© 1982, John Wiley & Sons Ltd.

Chapter 6

Serotonin Turnover and Regulation

LEONARD M. NECKERS

*National Cancer Institute, Laboratory of Pathology, National Institute of Health,
Building 10, Room 2N-115, Bethesda, Maryland 20205, USA*

INTRODUCTION

Measurements of the turnover rate of serotonin (5-HT) have been widely used to study the functional changes in the serotonergic system which occur *in vivo* as a result of drug action on physiological modification of 5-HT neurones. The concept of turnover, although flawed, has been useful in that it allows biochemical investigation of the functional state of neurones. When one measures the absolute level of serotonin, one is measuring a static event from which the dynamics of the system cannot be inferred. The term turnover refers to the process of renewal of a substance in a tissue. This renewal can be accomplished in two different ways: (1) synthesis of new material within the tissue; or (2) synthesis of new material elsewhere followed by transport to the tissue in question via the circulation. Since little of the serotonin found in brain originates in peripheral tissues, we can assume that brain serotonin synthesis is equivalent to brain serotonin turnover. It is a measure of the dynamic state of the neurone and, since it has been shown that the turnover rate of forebrain 5-HT increases when the raphe nuclei are stimulated, it appears to be a measure of the functional state of the nuclei.

In this chapter, we will review and assess the most common methods available for determination of serotonin turnover in animals and man. We will discuss several current theories about what regulates serotonin turnover. Finally, we will briefly discuss possible future directions in this area of serotonin research.

SEROTONIN SYNTHESIS AND DEGRADATION

Since these subjects are to be described in detail in other chapters of this book, we will only briefly layout the metabolic pathway of serotonin, schematically shown in Fig. 1. Tryptophan, the amino acid precursor of brain serotonin, is found in high concentration in the plasma. It is a dietary amino acid and is not produced by the body. This compound is transferred by a transport system from plasma to brain where it is converted to 5-hydroxytryptophan (5-HTP) by the enzyme tryptophan hydroxylase. This hydroxylation appears to be the rate-limiting step in serotonin synthesis. 5-Hydroxytryptophan is converted to serotonin by aromatic amino acid decarboxylase. After serotonin is released presynaptically and bound to its receptor, it appears that it is returned to the serotonergic nerve terminal via an active re-uptake mechanism. After re-uptake, it is converted to 5-hydroxyindoleacetic acid (5-HIAA) by monoamine oxidase (MAO), a mitochondrial enzyme. Finally, 5-HIAA is cleared from the brain via an acid transport system.

METHODS FOR MEASURING TURNOVER IN THE BRAIN

The various methods of estimating serotonin turnover *in vivo* may be divided into two broad categories (Table 1): the steady-state and non-steady-state techniques. Steady-state methods offer the theoretical advantage that analysis does not involve a perturbation of the system, while non-steady-state methods rely on the blockade

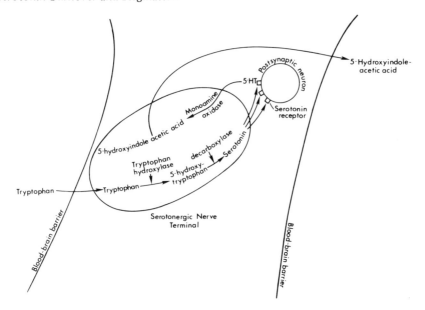

Fig. 1   A schematic representation of serotonin synthesis and degradation in brain. Reprinted with permission from *Life Sciences*, **19**, L. M. Neckers and J. L. Meek, (Copyright 1976, Pergamon Press, Ltd)

of serotonin metabolism at one point or another. Under 'steady-state methods' we will deal with tracer techniques of several types as well as measurement of serotonin pathway metabolite concentrations in CSF and urine. The 'non-steady-state methods' include inhibition of MAO, aromatic amino acid decarboxylase, tryptophan hydroxylase, or 5-HIAA transport out of the brain and following the resultant rise or fall in amine or metabolite concentrations.

Table 1. Summary of turnover methods

Steady-state methods	Non-steady-state methods
Radioactive tracer techniques using either 5-HT, 5-HTP or tryptophan	Inhibition of MAO
Non-radioactive tracer techniques using $^{18}O_2$	(a) Measurement of 5-HT increase (b) Measurement of 5-HIAA decline
Urinary or CSF 5-HIAA concentration	Inhibition of 5-HIAA efflux from brain
	Inhibition of 5-HTP decarboxylation
	Inhibition of tryptophan hydroxylation

Steady-state methods

Labelling with [³H] 5-HT (Bloom and Costa, 1971)

This technique assumes that brain 5-HT stores can be trace-labelled with very small quantities of high specific activity [³H] 5-HT. The decline in brain 5-HT specific activity would then be a measure of turnover. There are many problems with this technique which render it not very useful. Since serotonin does not cross the blood-brain barrier very well, labelled serotonin must be administered intraventricularly. When this is done, most of the label concentrates in periventricular structures, leaving serotonergic neurones and nerve terminals at a distance from the ventricles unlabelled or less labelled than terminals abutting the ventricles. Thus, if one interprets data obtained by this technique as a measure of whole brain 5-HT turnover, one's results would be highly skewed.

Using this technique, one must also assume that the radioactive serotonin equilibrates rapidly with the endogenous amine. There is much evidence in the literature that this is not the case (Weiner, 1974).

Labelling with [³H] 5-hydroxytryptophan (Bloom and Costa, 1971)

Another type of tracer analysis with many drawbacks utilizes labelling of the 5-HTP pool in brain with high specific activity [³H] 5-HTP. Since 5-HTP crosses the blood-brain barrier with ease, a peripheral route of administration can be used and uniform labelling of central stores might be expected. The endogenous 5-HTP pool is so small, however, that labelling with tracer significantly increases it, while at the same time bypassing the rate-limiting hydroxylation of tryptophan. Furthermore, because L-amino acid decarboxylase is present in catecholaminergic, as well as serotonergic neurones, [³H] 5-HTP may be taken up and converted to a false transmitter by non-serotonergic neurones, further complicating the interpretation of data. Lastly, because serotonin is produced in the periphery and stored in blood platelets, where it is protected from degradation, entrapment of [³H] 5-HT within blood vessels may interfere significantly with the determination of brain 5-HT specific activity and resultant turnover calculations.

Labelling with [³H] Tryptophan (Lin *et al.*, 1969)

In this method, labelled tryptophan is administered by either pulse injection or intravenous infusion. The rate of formation of labelled biogenic amine is determined in the brain by following the change in 5-HT specific activity. The specific activity of the labelled tryptophan is followed in either plasma or tissue. From these data the rate of turnover of serotonin can be determined. This technique assumes that the system is not perturbed by the amount of trace label. In view of the large endogenous levels of plasma and brain tryptophan, this is probably a valid assumption. One also must assume that the specific activity of the tryptophan at the site of hydroxylation inside the serotonergic neurone is the same as its specific activity in plasma or whole brain. This may not be the case, since evidence for differential rates of tryptophan uptake by various brain nuclei has recently been collected by us and others (Neckers *et al.*, 1977). Since amino acids are utilized by all cells in the brain—neurones and glia alike—for protein synthesis, it is highly likely that total tissue amino acid specific activity does not accurately reflect the specific activity of tryptophan in serotonergic neurones. Investigators have in fact demonstrated that after a pulse injection of labelled tyrosine, the specific activity of this substance in adrenergic neurones is considerably greater than that in the remainder of the tissue (Costa *et al.*, 1972). One can assume an analogous situation for tryptophan in serotonergic neurones. If plasma tryptophan specific activity is used for the calculations, one obtains a minimal value for amine synthesis, since, theoretically, amino acid specific activity in the neurone cannot exceed that of the plasma. Another assumption required of this method is that the newly synthesized radioactive serotonin is not metabolized or lost from the tissue. This assumption is not very likely, particularly if the newly synthesized serotonin is more labile and is more readily released from the tissue or more readily metabolized. This has been reported to be the case for catecholamines (Theirry *et al.*, 1973). Again, to the degree that the loss or metabolism of labelled serotonin is not considered, one will underestimate the rate of turnover.

Labelling with $^{18}O_2$ (Galli *et al.*, 1977, 1978)

The procedures mentioned to this point are unsuitable for human studies due to their common use of either ^3H or ^{14}C isotopes. A new and promising tracer method utilizing a stable isotope of $^{16}O_2$ can potentially be applied to human studies. The initial biosynthetic event for the formation of serotonin involves the utilization of oxygen during the enzymatic hydroxylation of tryptophan. In animal studies, $^{18}O_2$ has been substituted for $^{16}O_2$ in the breathing air, resulting in the *in vivo* labelling of endogenous 5-HT with $^{18}O_2$. A gas chromatograph-mass spectrometer has then been used to isolate 5-HT and distinguish between molecules containing ^{18}O and ^{16}O. Turnover rate estimates can be made from the relative abundance of ^{18}O incorporated into 5-HT and 5-HIAA in blood, urine and CSF. Unfortunately, methods for determining $[^{18}O]$5-HIAA have yet to be developed. This technique also has similar problems as the previous methods in terms of assumptions which must be made.

Measurement of urinary or CSF 5-HIAA

Since 5-HIAA is the major metabolite of 5-HT and occurs in brain with the same distribution as 5-HT, its concentration in CSF and urine has been used as a measure of central 5-HT turnover. High levels of 5-HIAA in CSF have been found in subgroups of depressed individuals where 5-HT turnover appears to be increased (Traskman *et al.*, 1979). Conversely, low 5-HIAA CSF concentration would be interpreted as signifying a decreased serotonin turnover. There are, however, several problems with this approach. Steady-state levels of 5-HIAA can only be used to estimate turnover rate of serotonin if the concentration of the metabolite in lumbar CSF (where taps are usually made) corresponds to that in the cerebral ventricles. Secondly, one would have to assume that the 5-HT/5-HIAA ratio is constant in all brain areas, otherwise the ratio in the periventricular areas would be determining the 'whole brain' turnover rate. In fact, the 5-HT/5-HIAA ratio is not constant in all brain regions. The turnover rate of 5-HT in nuclei containing serotonergic cell bodies is 10–15 times greater than the turnover in areas with only 5-HT terminals. Thus, CSF monitoring of 5-HIAA steady-state levels yields an underestimated turnover rate, probably reflecting the relatively low turnover seen in 5-HT terminals. Another problem in using steady-state CSF metabolite levels lies in the mechanism of the acid transport system. One must assume that the efficiency of the transport of 5-HIAA out of the brain (and CSF) is independent of its concentration; i.e. that the acid transport is not rate-limiting. One has to rule out the possibility that a decreased or increased level of 5-HIAA in CSF is not the result of a perturbation in the mechanism removing the metabolite from brain. Although the many assumptions make interpretation of CSF steady-state metabolite data quite difficult, this technique is still used quite fruitfully in some cases.

A prolonged elevation in the urinary excretion of 5-HIAA is the best measurement of an increased synthesis of these neurotransmitters in man (Costa, 1972). However, study of the change in urinary metabolite concentration does not yield information on the turnover rate of serotonin in individual structures such as the brain, let alone in individual nuclei within these structures.

Since serotonin is synthesized in the gut and perhaps other peripheral tissues, the presence of 5-HIAA in the urine represents a whole-body measure of serotonin metabolism. None the less, a major and prolonged derangement in central serotonin metabolism would probably be noticeable by monitoring urinary 5-HIAA levels.

Non-steady-state methods

All of the methods have the theoretical disadvantage of disturbing a finely-tuned system by blocking a part of it. The assumption is made that, at least initially, the turnover will not be affected by this intervention. As we shall see, in some instances this is not the case.

Pargyline inhibition of monoamine oxidase (Neff *et al.*, 1969a)

Referring to Fig. 1, we see that there are several steps along the serotonin pathway which can be blocked.

If we assume that for every molecule of 5-HT formed one is degraded, then turnover can be measured by blocking the destruction of 5-HT and measuring its

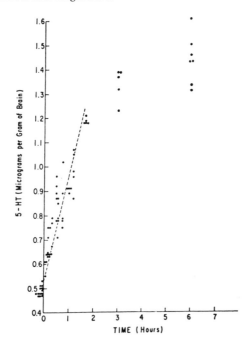

Fig. 2   The rise of serotonin levels in brain after MAO inhibition. (From Neff *et al.*, 1969a; reproduced by permission of Raven Press, New York)

accumulation with time. Thus, when parglyine hydrochloride, a non-reversible monoamine oxidase inhibitor, is administered to rats, brain 5-HT concentration increases linearly for about 1 h and then approaches a horizontal asymptote. Turnover rate is measured from the initial accumulation of 5-HT (Fig. 2). For this procedure to be valid, one must assume complete and immediate inhibition of MAO as well as the absence of any 5-HT diffusion out of brain. In fact, MAO is completely inhibited within several minutes after pargyline administration. It is not known, however, if the MAO in brain nuclei containing serotonin terminals or cell bodies is inhibited more or less quickly. Although 5-HT in normal concentrations does not leave or enter the brain by diffusion, it can find its way into the brain when administered peripherally in large doses. Thus, it might also diffuse out from the brain if intracerebral concentrations rise high enough after MAO inhibition. It is important when using this technique to monitor 5-HT levels during the first 90 min after MAO inhibition and to interpret only that data which fits a straight line. When this is done, the rate of accumulation of 5-HT after MAO inhibition is proportional to its rate of synthesis.

If the rate of formation and elimination of 5-HIAA is equal to the rate of formation of 5-HT, then the rate of elimination of 5-HIAA under steady-state conditions will be a direct measure of 5-HT synthesis. After intraperitoneal injection of

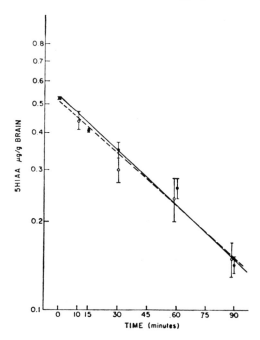

Fig. 3   The decline of 5-HIAA levels in brain after MAO inhibition. (From Neff *et al.*, 1969a; reproduced by permission of Raven Press, New York

MAO inhibitors, brain concentrations of 5-HIAA decline exponentially (Fig. 3). By measuring 5-HIAA concentrations at various times after pargyline administration, one can calculate the rate of decline in brain 5-HIAA concentration after MAO inhibition. Since we are assuming that the rate of 5-HIAA efflux from the brain is equal to its rate of formation which in turn is equal to the rate of 5-HT turnover, we have a method for estimating 5-HT turnover. Again this method is accurate only to the degree that the assumptions listed above are valid. Since turnover rates determined using either pargyline procedure are quire similar to each other and to those determined using other methods, these assumptions are probably valid most of the time. However, their validity should not be taken as an axiom under all experimental conditions.

Probenecid inhibition of 5-HIAA efflux from brain (Neff *et al.*, 1969a)

Since we have already assumed that the rate of efflux of 5-HIAA from brain (and CSF) is equivalent to the rate of formation of 5-HT, if we block 5-HIAA efflux from brain and measure its accumulation with time we should have another method of determining 5-HT turnover. Probenecid is a drug which blocks the transport of acids in the kidney and liver and is also used to block 5-HIAA efflux from brain.

Turnover is determined from the initial linear rise in 5-HIAA in brain or CSF. This technique, although theoretically as feasible as previous ones, has some major drawbacks. First, one must completely block the 5-HIAA efflux for at least 90 min to get enough data to analyse. In some rat strains, the doses of probenecid required for this are so high they prove to be toxic. In those strains where complete blockade can be achieved at non-toxic doses of drug, the method appears valid for whole brain turnover rates but not for rates determined in various brain nuclei (Meek and Neckers, 1977). Another serious problem with this method is the propensity of probenecid to elevate brain tryptophan concentrations (van Wijk *et al.*, 1979). If any of these non-steady-state methods disturbs the equilibrium of any step other than the one intended, interpretations of the data are no longer valid. If tryptophan hydroxylase is not saturated with substrate, then increasing brain tryptophan concentration would be expected to elevate 5-HT. This increased amount of 5-HT would then be converted to more 5-HIAA whose efflux out of the brain has been blocked. Thus overestimations of turnover may occur. For these reasons, acid transport blockade is no longer widely used in 5-HT turnover determinations.

Inhibition of aromatic amino acid decarboxylase (Carlsson *et al.*, 1972)

The hydroxylation of tryptophan to 5-hydroxytryptophan limits the rate of 5-HT synthesis. This is the bottleneck in the pathway. 5-HTP is rapidly converted to 5-HT so that the endogenous pool of 5-HTP is at or below the limits of detection of even very sophisticated monitoring devices. However, by blocking the decarboxylation of 5-HTP to 5-HT, one allows 5-HTP to accumulate. By measuring the initial rate of accumulation of 5-HTP one can determine turnover values in an analogous manner to the pargyline-5-HT procedure. The validity of this method depends on the usual assumptions of complete and essentially immediate inhibition of decarboxylation, lack of diffusion of 5-HTP out of brain and lack of feedback inhibition of 5-HTP on tryptophan hydroxylation. Since 5-HTP readily diffuses into brain it may just as easily diffuse out if it is allowed to accumulate. The linearity of the 5-HTP increase is apparent for only about 30 min before bending of the curve occurs. Turnover data calculated from the initial part of the curve (within the first 30 min) give results similar to those determined using other methods.

Inhibition of tryptophan hydroxylase

The final method we will mention is the measurement of the decline in 5-HT concentration after inhibition of tryptophan hydroxylase. This technique is analogous to the measurement of the decline of 5-HIAA levels after MAO inhibition. Using either P-chlorophenylalanine or α-propyldopacetmide to inhibit the enzyme, one finds an exponential decline in brain 5-HT concentration occurring over the next several hours. One must again assume complete, irreversible and immediate inhibition of the enzyme tryptophan hydroxylase throughout the brain. Although this appears to be the case in the pineal gland (Neff *et al.*, 1969b), these assumptions do not hold when applied to whole brain or brain regions (Meek and Neckers, 1977).

HOW VALUABLE ARE TURNOVER STUDIES?

In the previous section we have reviewed all the commonly used methods for estimating brain 5-HT turnover. The validity of each method depends on several assumptions which we have described in detail. The basic premise on which these techniques are based is that serotonin synthesis and degradation is an open system with each occurring at equal rates. That is, serotonin enters the system by synthesis and is lost by metabolism with a steady-state existing in which the rate of formation of 5-HT equals its rate of degredation. However, it has not been established that studies of 5-HT turnover by any of these methods provide an accurate estimate of absolute turnover or synthesis rates. Most of the possible errors would tend to lead to underestimation of the true value. These methods are of use in determining comparative overall differences in turnover rate under different experimental conditions.

It is imperative that one does not rely on just one method when interpreting experimentally-induced changes in turnover rate. Because of the problems all of the methods possess, similar results found with several of these procedures make one's data much stronger and less open to criticism. Large changes in 5-HT turnover in specific neuronal pathways or nuclei might be missed when one measures turnover in whole brain or large brain regions. This may be very important when one is studying particular physiological processes in brain which are regulated by an individual 5-HT neuronal pathway. Can these techniques for measuring gross turnover be applied to individual brain nuclei containing either serotonin nerve terminals or cell bodies?

APPLICATION OF TURNOVER STUDIES TO BRAIN NUCLEI

Measurement of serotonin turnover rate in discrete nuclei can provide information on the anatomical site of action of drugs affecting the serotonergic system. Turnover studies in individual nuclei can be used to examine the physiological regulation of serotonin metabolism in unique neuronal pathways. The small amounts of tissue available make turnover calculations quite difficult. Use of the 'steady-state' methods previously discussed becomes technically difficult and prohibitively expensive when one attempts to get adequate amounts of radioactivity incorporated into milligram amounts of tissue. Several of the 'non-steady-state' methods we have described can and have been used successfully in estimating turnover in individual brain nuclei (Table 2). The successful utilization of these techniques requires measurement of tryptophan, 5-HTP, 5-HT or 5-HIAA in minute pieces of brain tissue. Appropriate detection methods have recently been published by ourselves and others (Neckers and Meek, 1976; Meek and Lofstrandh, 1976; Neckers *et al.*, 1977; Koch and Kissinger, 1980; Lyness *et al.*, 1980). Most methods require the use of high-pressure

Table 2. 5-HT Turnover rate in brain nuclei

	(ng 5-HT mg^{-1} protein hr^{-1})			
Calculation method	Dorsal raphe	Median raphe	Caudate	Hippocampus
5-HT increase after pargyline	45	27	2.8	2.7
5-HIAA decline after pargyline	34	20	2.3	2.8
5-HIAA increase after probenecid	16	20	3.1	3.3
5-HT decline after α-PDA	–	–	1.1	1.6
5-HTP increase after RO4 4602	33	25	1.4	0.8

From Neckers and Meek, 1976; reproduced by permission of Pergamon Press, Ltd

liquid chromatography with either fluorescence detection or electrochemical detection. Such procedures have been reviewed previously (Neckers and Bohlen, 1980). In this section we will review the use of several of these methods, some results obtained with them and what these results imply for understanding the significance of serotonin turnover in brain.

Of the five non-steady-state methods we have described for use in whole brain, all have been tried in attempts to measure 5-HT turnover in brain nuclei (Meek and Neckers, 1977). The probenecid method, which measures accumulation of 5-HIAA in brain after blocking its efflux is effective in some brain areas but not in others. It is clear that this procedure will be effective only if it reaches the brain area in question at a sufficient concentration to block acid transport. The method appears to work (in that its results compared well to other methods) in hippocampus, caudate nucleus and median raphe nucleus, but not in dorsal raphe nucleus. Since the dorsal raphe nucleus is periventricular, it is possible that 5-HIAA diffuses into the ventricle even though acid transport out of the nucleus is blocked. This would result in a low turnover estimation for the dorsal raphe using the probenecid technique, a result which we have obtained in previous experiments.

Measurements of 5-HT disappearance after inhibiting tryptophan hydroxylase with α-proplydopacetimide also prove ineffective in brain nuclei. The decline in 5-HT is not exponential as is required. Apparently, the drug does not inhibit tryptophan hydroxylase completely in the nuclei that contain 5-HT cell bodies.

The two methods involving pargyline inhibition of MAO appear to work well in all nuclei studied to date. These include nuclei-containing serotonin cell bodies, the median and dorsal raphe, and nuclei that contain 5-HT nerve terminals such as septum, caudate nucleus and hippocampus. Measuring either the rate of 5-HT accumulation or 5-HIAA decline, one obtains values that are similar to each other and to those obtained with other methods.

The final method used with some success involves measurement of 5-HTP accumulation after blockade of decarboxylase activity. This procedure yields values for the raphe nuclei comparable to other methods. Its use in nerve terminals is hindered by the relatively low levels of 5-HTP seen, even after decarboxylase inhibition. With new and more sensitive detection devices now available for 5-HTP, the utility of this procedure in nerve terminals should also be quite good.

Previously, serotonin turnover in whole brain was reported to be about 2 nmol g^{-1} tissue h^{-1} (Neff *et al.*, 1969a). In brain stem, where serotonergic cell bodies are located, it was reported to be about 4 nmol g^{-1} tissue h^{-1} (Neff *et al.*, 1969a). Other brain areas had a 5-HT turnover rate of about 2 nmol g^{-1} h^{-1} (Neff *et al.*, 1969a).

When we measured 5-HT turnover in brain regions containing serotonergic nerve terminals, such as caudate nucleus and hippocampus, we obtained values very close to those reported for whole brain and 'other brain areas ' (besides brain stem) (Neckers and Meek, 1976). However, when we measured the turnover directly in median and dorsal raphe nuclei (where the 5-HT cell bodies are located) we found a rate 5-10 times that reported previously for the brain stem (Neckers and Meek, 1976). Since the raphe nuclei make up only a very small portion of the brain stem, all the nonserotonergic tissue (and serotonergic nerve terminals) were diluting the actual 5-HT cell body turnover rate. By looking at the raphe nuclei directly we could find a turnover rate 10-15 times that seen in 5-HT nerve terminals. The ratio of turnover rate to 5-HT content can be used to normalize for the density of serotonergic innervation. The ratio in the cell body-containing nuclei is nearly double that seen in terminal-containing nuclei. The very rapid turnover in serotonergic cell bodies was not recognized in whole brain determinations or even brain stem determinations. Because serotonergic cell bodies innervating the forebrain are uniquely concentrated in a very small part of the brain, one must get at these nuclei directly to correctly perceive their metabolism. The utility of measuring turnover in individual 5-HT terminal-containing nuclei is also readily demonstrated in the next section where we show that drugs and other treatments which affect serotonin metabolism do not have equal actions in all brain regions. Even the dorsal and medium raphe nuclei respond differently to drugs. Injection of the 5-HT re-uptake blocker chlorimipramine to rats decreases serotonin turnover in median raphe nucleus but not in dorsal raphe nucleus (Meek and Lofstrandh, 1976). Our ability to study turnover in discrete brain nuclei should open new approaches to studying how 5-HT turnover in discrete neuronal pathways is modulated.

REGULATION OF SEROTONIN TURNOVER

The processes that modulate 5-HT turnover *in vivo* have not been clearly defined (Costa and Meek, 1974). Until recently, most studies have focused on whole brain serotonin turnover and consequently could come to only general conclusions if any at all. Many drugs have been reported to affect serotonin synthesis but their mechanisms of action are not always well-understood. In this section we will refer only to those drugs whose use has shed some light on regulatory mechanisms. Remembering that we are interested in the *in vivo* regulation of serotonin turnover, let us list the possible regulatory sites along the 5-HT synthetic pathway. Since the hydroxylation of tryptophan is the rate-limiting step in 5-HT synthesis, this would be an obvious place to exert regulatory control. One could do this by affecting either brain tryptophan concentration or tryptophan hydroxylase activity (or by doing both). If serotonin receptor occupation in some way is involved with 5-HT synthesis, then affecting the occupancy of serotonin receptors might alter the turnover of the amine. These are the possible regulating mechanisms that we will now discuss in some detail.

Regulation of tryptophan hydroxylation

Tryptophan hydroxylation requires two prerequisites: the availability of tryptophan and the activity of tryptophan hydroxylase.

Regulation via brain tryptophan concentration

Since there is not enough tryptophan in the brain to saturate tryptophan hydroxylase, brain tryptophan content itself may partially regulate serotonin synthesis. Intraperitoneal injections of tryptophan increase the brain 5-HT level (Fernstrom and Wurtman, 1971), and even in discrete nuclei 5-HT synthesis is increased after tryptophan administration (Meek and Lofstrandh, 1976). Dietary fluctuation in tryptophan content, or in the content of other neutral amino acids which compete with it for uptake into brain, can alter CNS serotonin concentration (Fernstrom and Wurtman, 1972).

Even changing the ratio of free tryptophan to albumin-bound tryptophan in plasma will affect brain 5-HT levels (Knott and Curzon, 1972). However, it is also apparent that different brain nuclei are not all equally affected by elevated tryptophan levels (Neckers *et al.*, 1977). When challenged with the same intraperitoneal injection of tryptophan, different nuclei have a wide variation (4–5 fold) in the level of amino acid reached within that nucleus (and thus available for hydroxylation). Thus, it is not surprising that different brain areas have differential sensitivity to physiological and pharmacological manipulations of tryptophan content.

Brain 5-HT and 5-HIAA concentrations correlate fairly well with brain tryptophan and with free tryptophan in plasma, but the correlations are not perfect (Costa and Meek, 1974). There are conditions when brain or plasma tryptophan levels are high, while 5-HT, the concentration of 5-HIAA or of both are below normal, and vice versa. This incomplete correlation means that there are other parameters that can control brain 5-HT synthesis, even though the tryptophan concentration probably does regulate the turnover of the amine in some brain areas. It seems unlikely that a neurotransmitter such as 5-HT, which is possibly involved in sleep, sexual activity, motor behaviour and emotional states, would be entirely under the control of dietary composition. It seems likely that the diverse brain functions of serotonergic neurones are not all controlled by gross changes in plasma tryptophan content.

Regulation of tryptophan hydroxylase

Two co-factors required for hydroxylation of tryptophan are molecular oxygen and reduced pterin. Not only is tryptophan hydroxylase unsaturated with substrate, it also appears to be unsaturated with respect to oxygen and pterin concentration as well (Cost and Meek, 1974). The Km of tryptophan hydroxylase (from rabbit brain) for oxygen is about 2.5 per cent, near the oxygen concentration reported for brain. Where rats inhale 100 per cent oxygen, oxygen pressure in various brain areas nearly doubles with a concomitant increase in brain 5-HT turnover. Anoxia causes a decrease in *in vivo* tryptophan hydroxylation which parallels the decline in arterial oxygen pressure. Although these effects can be experimentally induced by altering oxygen pressure, one wonders what effect normal physiological fluctuation in oxygen pressure has on serotonin synthesis.

It appears that brain pterin concentration is also similar to the Km of the hydroxylase, about 30 μM (Costa and Meek, 1974). That is, fluctuations in pterin levels could affect enzyme activity. Since endogenous pterins have only recently been examined in the CNS it is still too early to detail their regulatory role.

Feedback regulation of serotonin synthesis

In brain, the short-term regulation of monoamine synthesis involves feedback control of the firing rate of monoaminergic neurones, which is linked in some unknown way to the regulation of monoamine synthesis rate. This theory holds that, when serotonin is occupying its post-synaptic receptor, the receptive neurone is somehow transmitting this information to the serotonergic neurone, perhaps via a neuronal feedback loop. If the receptor is over-occupied (i.e. there is too much serotonin available), then a signal is sent to reduce the firing rate of, and 5-HT synthesis in, the serotonergic neurone. If the receptor is under-occupied (i.e. there is too little serotonin available), then a signal is given to increase the synthesis rate and firing rate of 5-HT neurones. Much evidence has accumulated over the years to support this model of short-term regulation (for review see Costa and Meek, 1974).

As we have previously mentioned, serotonin is removed from the vicinity of its receptor by re-uptake into the serotonergic terminal. Drugs which block the re-uptake mechanism and lead to increased serotonin concentrations at the receptive post-synaptic neurone decrease the firing rate of 5-HT neurones and their synthesis of serotonin (Meek and Werdinius, 1970; Fuller and Wang, 1977; Hyttel, 1977).

Various serotonin receptor blockers elevate 5-HT turnover, presumably by preventing serotonin from occupying its receptor (Bhargava and Kasabdji, 1977; Golombiowska-Nikitin *et al.*, 1977). The 5-HT receptive neurone, sensing no serotonin on its receptors, alerts the 5-HT neurone of this fact and synthesis and release are concomitantly increased. We recently reported evidence supporting the existence of a negative neuronal feedback loop acting as a sort of breaking device on 5-HT neurones which project to the caudate nucleus (Neckers *et al.*, 1979). After using kainic acid to destroy striatal serotonergic receptors we found increased serotonin turnover in the striatum. Whether or not the feedback loop is strictly inhibitory, or both excitatory or inhibitory, is not clear. It is apparent, however, that the 5-HT-receptive neurone can participate in the regulation of serotonin turnover.

FUTURE DIRECTIONS IN THE STUDY OF SEROTONIN TURNOVER

In this final section I would like to mention two areas of research involving serotonin turnover and metabolism which I think are quite exciting and rich in undiscovered information.

Independent regulation of neuronal cell body and terminal turnover

We have already touched upon the fallacy of regarding brain serotonin metabolism as a unified whole when responding to various physiological and pharmacological manipulations. The first obvious distinction must be made between cell bodies and nerve terminals. This becomes quite clear when one examines the effects of *para*-chloramphetamine (PCA) on serotonin metabolism. This drug decreases tryptophan hydroxylase activity and 5-HT turnover rate in brain, but all brain areas are not equally susceptible to its effects (Neckers *et al.*, 1976). Serotonin metabolism in the dorsal and median raphe nuclei remains little affected after PCA treatment, although serotonin metabolism in dorsal raphe is transiently depressed (Fig. 4). However, serotonin metabolism in nerve terminals is severely depressed, in hippocampus and caudate for as long as 2 months after treatment but in septum for only 3 weeks. Even here one can see the different response of these 5-HT terminal-containing nuclei to the drug. What is quite remarkable is that although 5-HT metabolism in neuronal cell bodies (in the median raphe) is normal, the metabolism at the end of the neurone (the terminals in the hippocampus) is only about 25 per cent of normal. This finding led us to consider the hypothesis that 5-HT metabolism in a nerve terminal and cell body of the same neurone could be independently regulated.

When kainic acid is injected into rats intrastriatally to destroy 5-HT receptors, serotonin turnover increases in the treated striatum but not in the contralateral one nor in the dorsal raphe nucleus (the location of serotonergic cell bodies projecting to the striatum) (Neckers *et al.*, 1979). Here then is another example where nerve terminal 5-HT metabolism is specifically changed independent of neuronal

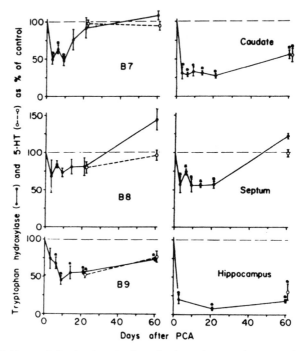

Fig. 4 Depletion of brain tryptophan hydroxylase and serotonin by PCA in 5-HT cell-body-containing nuclei (B7, B8 and B9) and 5-HT terminal-containing nuclei (caudate, septum and hippocampus). (From Neckers *et al.*, 1976; reproduced by permission of Williams and Wilkins, Baltimore, Maryland)

cell body metabolism. In an exhaustive investigation of the effects of tricyclic anti-depressants on serotonin turnover in several brain regions, Meek and his group has supplied further evidence of this phenomenon (Marco and Meek, 1979). Chlorimi-pramine, a drug which blocks serotonin re-uptake and causes a general reduction in whole brain 5-HT turnover, decreases amine turnover in many 5-HT terminal-containing areas of the brain including the caudate nucleus, a dorsal raphe projection area. However, when the effects of chlorimipramine on the dorsal and median raphe nuclei are examined, only median raphe 5-HT turnover is depressed. Serotonin turnover in dorsal raphe remains unchanged following drug treatment (Meek and Lofstrandh, 1976).

The reverse has also been found to occur; 5-HT metabolism in the cell body can change without producing a concomitant change in the respective nerve terminals. In rats, when lesions of the lateral habenula are performed, serotonin turnover increases in the dorsal raphe nucleus for as long as 1 week post lesion. At no time, however, are significant changes observed in caudate nucleus 5-HT metabolism (Speciale *et al.* 1980).

These results leave one with the impression that the two ends of a serotonergic neurone can be regulated independently of each other. Whether or not this occurs normally under physiological regulation is not yet certain but continued research in this area should supply provocative new information that might change the way we think of neuronal function.

Physiological and pharmacological regulation of individual serotonergic neuronal pathways in CNS

In the past several years, anatomical studies have been reported which delineate the major ascending serotonergic neuronal pathways in mammalian brain (Lorens and Guldberg, 1974; Bobillier *et al.*, 1975). The 5-HT neuronal cell bodies are clustered in two nuclei of the brainstem, the median and dorsal raphe nuclei. These nuclei project to several anatomically defined areas of the forebrain with surprisingly little overlapping innervation.

Because our ability to measure turnover in these brain nuclei is quite recent, studies on 5-HT turnover in discrete nuclei in the central nervous system are still scarce. What is emerging to date is the realization that dorsal and median raphe are independently regulated—both physiologically and pharmacologically. Meek and Lofstrandh (1976) have reported that median raphe 5-HT turnover is depressed by intraperitoneal administration of chlorimipramine while dorsal raphe metabolism remains unchanged. Why the median raphe but not the dorsal raphe appears sensitive to this anti-depressant is not known. We have recently reported the presence of a peptidergic substance P-containing pathway from lateral habenula innervating the dorsal raphe, but not the median raphe (Neckers *et al.*, 1980). When one injures the lateral habenula, destroying this connection, 5-HT turnover in the dorsal raphe is increased for as long as 1 week post lesion. No effect is seen in the median raphe (Speciale *et al.*, 1980).

Here we have one pharmacological and one physiological example of dissimilar responsivity of these two raphe nuclei. What is responsible for these differences? Certainly differential innervation of the nuclei by other structures is possible. Does a different neuronal and glial milieu in the two nuclei explain their differential responsivity to drugs? Questions like these must remain unanswered until the complex interlocking neuronal network of each nucleus is uncovered.

These two general areas of serotonin research may contain the answer to the question of how serotonin metabolism in the CNS is so finely regulated that various 5-HT neuronal networks can be implicated in so many diverse physiological functions.

SUMMARY

In this chapter we have reviewed the most commonly used methods of measuring brain serotonin turnover. These include several steady-state and non-steady-state procedures. All of the methods are dependent on several simplifying assumptions which may not always be true under all experimental conditions. For this reason, at least two, and if possible three, different methods should be used in a given experiment. Recent advances in methodology have permitted us to measure 5-HT turnover in discrete brain nuclei. These studies have demonstrated the intricate complexity of the serotonergic system in brain. By trying to understand the diverse nature of 5-HT neuronal pathways we may begin to perceive the true role of serotonin in CNS function.

REFERENCES

Bhargava, H. N. and Kasabdji, D. (1977). Effect of mianserin on brain serotonin turnover in mice, *Res. Comm. Chem. Pathol. Pharmacol.* **17**, 735–738.

Bloom, F. E. and Costa, E. (1971). The effects of drugs on serotonergic nerve terminals, *Adv. in Cytopharmacology*, **1**, 379–395.

Bobillier, P., Petitjean, F., Salvert, D., Ligier, M., and Seguin, S. (1975). Differential projections of the nucleus raphe dorsalis and nucleus raphe centralis as revealed by auto-radiography, *Brain Res.* **85**, 205–210.

Carlsson, A., Davis, J. N., Keht, W., Lindqvist, M., and Atack, C. V. (1972). Simultaneous measurement of tyrosine and tryptophan hydroxylase activities in brain *in vivo* using an inhibitor of the aromatic amino acid decarboxylase, *Naunyn-Schmiedebergs Arch. Pharmacol.* **275**, 153–168.

Costa, E. (1972). Appraisal of current methods to estimate the turnover rate of serotonin and catecholamines in human brain, *Adv. Biochem. Psychopharmacol.* **4**, 171–183.

Costa, E., Green. A. R., Koslow, S. H., Lefevre, H. F., Revuelta, A. V., and Wang, C. (1972). Dopamine and norepinephrine in noradrenergic axons: A study *in vivo* of their precursor product relationships by mass fragmentography and radiochemistry, *Physiol. Rev.* **24**, 167–190.

Costa, E. and Meek, J. L. (1974). Regulation of biosynthesis of catecholamines and serotonin in the CNS, *Annu. Rev. Pharmacol.* **14**, 491–511.

Fernstrom, J. D. and Wurtman, R. J. (1971). Brain serotonin content: Physiological dependence on plasma tryptophan levels, *Science*, **173**, 149–150.

Fernstrom, J. D. and Wurtman, R. J. (1972). Brain serotonin content: Physiological regulation by plasma neutral amino acids, *Science*, **178**, 414–416.

Fuller, R. W. and Wang, D. T. (1977). Inhibition of serotonin reuptake, *Fed. Proc.* **36**, 2154–2158.

Galli, C., Commissiong, J. W., Costa, E., and Neff, N. H. (1978). Incorporation of $^{18}O_2$ into brain serotonin *in vivo* as a procedure for estimating turnover: A feasibility study in animals, *Life Sci.* **22**, 473–478.

Galli, C., Commissiong, J. W., and Neff, N. H. (1977). Measuring the formation of biogenic amines utilizing $^{18}O_2$, *Biochem. Pharmacol.* **26**, 1271–1273.

Golombiowska-Nikitin, K., Wiszeniowska, G., and Marchasj, J. (1977). The effect of serotonin receptor blocking agents cyprohepradine and danitracen on serotonin turnover in the rat brain, *Pol. J. Pharmcol. Pharm.* **29**, 485–495.

Hyttel, J. (1977). Effect of selective 5-HT reuptake inhibitor—LU 10-171—on rat brain 5-HT turnover, *Acta. Pharmacol. Toxicol.* **40**, 439–446.

Knott, P. J. and Curzon, G. (1972). Free tryptophan in plasma and brain tryptophan metabolism, *Nature (Lond.)*, **239**, 452–453.

Koch, D. D. and Kissinger, P. T. (1980). Liquid chromatography with pre-column sample enrichment and electrochemical detection. Regional determination of serotonin and 5-hydroxyindoleacetic acid in brain tissue, *Life Sci.* **26**, 1099–1107.

Lin, R. C., Costa, E., Neff, N. H., Wang, T. C., and Ngai, S. H. (1969). *In vivo* measurement of 5-hydroxytryptamine turnover rate in the rat brain from the conversion of L-^{14}C-tryptophan to ^{14}C-5-hydroxytryptamine, *J. Pharmacol. Exp. Therap.* **170**, 232–238.

Lorens, S. A. and Guldberg, H. C. (1974). Regional 5-hydroxytryptamine following selective midbrain raphe lesions in the rat. *Brain Res.* **78**, 45–56.

Lyness, W. H., Fliedle, N. M., and Moore, K. E. (1980). Measurement of 5-hydroxytryptamine and 5-hydroxyindoleacetic acid in discrete brain nuclei using reverse phase liquid chromatogrpahy with electrochemical detection, *Life Sci.* **26**, 1109–1114.

Marco, E. and Meek, J. L. (1979). The effects of antidepressants on serotonin turnover in discrete regions of rat brain, *Naunyn-Schmiedebergs Arch. Pharmacol.* **306**, 75–79.

Meek, J. L. and Loftstrandh, S. (1976). Tryptophan hydroxylase in discrete brain nuclei: Comparison of activity *in vitro* and *in vivo*, *Europ. J. Pharmacol.* **37**, 377–380.

Meek, J. L. and Neckers, L. M. (1977). Studies of serotonin turnover in discrete nuclei using HPLC, in *Structure and Function of Monoamine Enzymes* (Eds E. Usdin, N. Weiner, and M. Youdim), pp. 799–809, Marcel Dekker, New York.

Meek, J. L. and Werdinius, B. (1970). Hydroxytryptamine turnover decreased by the antidepressant drug chlorimipramine, *J. Pharm. Pharmacol.* **22**, 141–143.

Neckers, L. M., Bertilsson, L., Koslow, S. H. and Meek, J. L. (1976). Reduction of tryptophan hydroxylase activity and 5-hydroxytryptamine concentration in certain rat brain nuclei after p-chloroamphetamine, *J. Pharmacol. Exp. Ther.* **196**, 333–338.

Neckers, L. M., Biggio, G., Moja, E., and Meek, J. L. (1977). Modulation of brain tryptophan hydroxylase activity by brain tryptophan content, *J. Pharmacol. Exp. Ther.* **201**, 110–116.

Neckers, L. M. and Bohlen, P. (1980). High performance liquid chromatography in the neurosciences: Applications, in *Physico-Chemical Methodologies in Psychiatric Research* (Eds I. Hanin and S. Koslow), pp. 23–26, Raven Press, New York.

Neckers, L. M., and Meek, J. L. (1976). Measurement of 5-HT turnover rate in discrete nuclei of rat brain, *Life Sci.* **19**, 1579–1584.

Neckers, L. M., Neff, N. H., and Wyatt, R. J. (1979). Increased serotonin turnover in corpus striatum following an injection of kainic acid: Evidence for neuronal feedback regulation of synthesis. *Naunyn-Schmiedebergs Arch. Pharmacol.* **306**, 173–177.

Neckers, L. M., Schwartz, J. P., Wyatt, R. J., and Speciale, S. G. (1980). Substance P afferents from the habenula innervate the dorsal raphe nucleus, *Exp. Brain Res.* **37**, 619–623.

Neff, N. H., Barrett, R. E., and Costa, E. (1969b). Kinetic and fluorescent histochemical analysis of the serotonin compartments in rat pineal gland, *Eur. J. Pharmacol.* **5**, 348–356.

Neff, N. H., Lin, R. C., Nagai, S. H., and Costa, E. (1969a). Turnover rate measurements of brain serotonin in unanesthetized rats, *Adv. Biochem. Psychopharmacol.* **1**, 91–109.

Speciale, S. G., Neckers, L. M., and Wyatt, R. J. (1980). Habenular modulation of raphe indoleamine metabolism, *Life Sci.* **27**, 2367–2372.

Thierry, A. M., Blanc, G., and Glowinski, J. (1973). Further evidence for the heterogeneous storage of noradrenalin in central noradrenergic terminals, *Naunyn-Schmiedebergs Arch. Pharmacol.* **279**, 255–266.

Traskman, L., Asberg, M., Bertilsson, L., Cronholm, B., Mellstrom, B., Neckers, L. M., Sjoqvist, F., and Tybring, G. (1979). Plasma levels of chlorimipramine and its dimethyl metabolite during treatment of depression, *Clin. Pharmacol. Thera.* **26**, 600–610.

Weiner, N. (1974). A critical assessment of methods for the determination of monoamine synthesis turnover rates *in vivo*, in *Neuropsychopharmacology of Monoamines and Their Regulatory Enzymes* (Ed E. Usdin), pp. 143–159, Raven Press, New York.

van Wijk, M., Sebens, J. B., and Korf, J. (1979). Probenecid-induced increase of 5-hydroxytryptamine synthesis in rat brain, as measured by formation of 5-hydroxytryptophan, *Pharmacology*, **60**, 229–235.

Biology of Serotonergic Transmission
Edited by N. N. Osborne
© 1982, John Wiley & Sons Ltd.

Chapter 7

The Characteristics of Serotonin Uptake Systems

SVANTE B. ROSS

Research and Development Laboratories,
Astra Läkemedel AB, S-151 85 Södertälje, Sweden

INTRODUCTION

The considerable physiological importance of neuronal re-uptake mechanisms for terminating action of neurotransmitters released from the nerve terminals was first established for noradrenaline (Hertting and Axelrod, 1961). Soon after, it was found that similar uptake mechanisms existed for several other transmitters including serotonin (Blackburn *et al.*, 1967; Ross and Renyi, 1967). All these neuronal re-uptake systems are characterized by their high affinity for the transmitters, their low transport capacity, their relatively high specificity and their requirement for external Na^+.

In common with the corresponding transport systems for noradrenaline (reviewed by Paton, 1976) and dopamine (reviewed by Horn, 1979) the accumulation of serotonin in serotonergic neurones is dependent upon two transport mechanisms. One uptake site is localized at the neurone membrane and catalyses the transport of serotonin across this membrane. The other uptake mechanism facilitates the uptake of serotonin into the intraneuronal storage vesicles. The distinction between these two uptake sites is ascertained by their sensitivity to inhibitors. The membranal serotonin uptake is inhibited by cocaine, several antidepressant agents and ouabain. The uptake of serotonin into storage vesicles is blocked by reserpine and similar drugs.

CRITERIA FOR UPTAKE INTO SEROTONERGIC NEURONES

The observation that serotonin is accumulated in brain tissue is not in itself proof that serotonin is actively transported by a specific carrier into serotonergic neurones. However, many observations clearly indicate a specific active uptake mechanism for serotonin in serotonergic neurones (Blackburn *et al.*, 1967; Ross and Renyi, 1967, 1969; Bogdanski *et al.*, 1968, 1970a; Fuxe *et al.*, 1968; Tissari *et al.*, 1969; Carlsson, 1970; Shaskan and Snyder, 1970; Kuhar *et al.*, 1972a,b; Kuhar and Aghajanian, 1973; Björklund *et al.*, 1975; Fuller *et al.*, 1975; Aghanjanian and Bloom, 1976, Ross *et al.*, 1976). Thus, the serotonin uptake in brain tissues, (1) is saturable following Michaelis-Menten kinetics, (2) shows structural specificity, (3) is temperature dependent, (4) is inhibited by metabolic inhibitors, (5) is inhibited by ouabain, (6) is Na^+ dependent, (7) is inhibited by selectively acting compounds,

(8) is reduced by lesions in the raphe nuclei, which contain serotonergic cell bodies, (9) is reduced by selectively acting neurotoxins, e.g. 5,7 dihydroxytryptamine and *p*-chloroamphetamine, (10) is regionally distributed similar to endogenous serotonin.

SEROTONIN UPTAKE IN VARIOUS TISSUES

Nervous tissues

Various rat brain regions have been shown to accumulate labelled serotonin when studied *in vitro* with slices or synaptosomes. The relative amount of [^{3}H] serotonin accumulated in the homogenates of ten rat-brain regions according to Köhler *et al.* (1978) is shown in Table 1. Intraventricularly injected [^{3}H] serotonin is accumulated in nerve terminals and unmyelinated axons and has a similar distribution to that of endogenous serotonin (Fuxe *et al.*, 1968; Aghajanian and Bloom, 1976).

Serotonin is specifically accumulated in pinealocytes via a mechanism very similar to the neuronal serotonin uptake mechanisms (Ducis and Di Stefano, 1980).

Elements in the myenteric plexus of the small intestine of the guinea-pig take up [^{3}H] serotonin utilizing a mechanism which fulfills the criteria for a selective high affinity uptake (Gershon and Altman, 1971; Gershon *et al.*, 1976; Gershon and Jonakait, 1979; Gershon *et al.*, 1980).

Chick, rabbit and bovine retina accumulate serotonin specifically with a high affinity uptake mechanism (Suzuki *et al.*, 1978; Ehinger and Florén, 1978; Florén, 1979 Thomas and Redburn, 1979, 1980; Osborne, 1980). This accumulation is saturable, temperature and Na^+ dependent, ouabain and reserpine sensitive and inhibited by specific serotonin uptake inhibitors. Tryptaminergic neurones in the rabbit retina have been demonstrated by fluorescence histochemistry (Ehinger and Florén, 1978). Low affinity uptake of serotonin was also demonstrated in this tissue (Ehinger and Florén, 1978; Osborne, 1980).

Serotonin is accumulated in adrenergically innervated tissues, e.g. vas deferens (Thoa *et al.*, 1969), most probably by utilizing the noradrenaline transport system.

Non-nervous tissues

Platelets contain a high affinity membranal transport mechanism for serotonin (Born and Gillison, 1959) which is very similar to that of serotonergic neurones (for review see Sneddon, 1973).

Suddith *et al.* (1978) reported that cultured glia cells (rat C6 astrocytoma cells) take up serotonin via a high affinity system (Km = 1 μM), which was temperature and Na^+ dependent and inhibited by serotonin uptake inhibitors. Similar results were reported by Bakhanashvili *et al.* (1979) from a study with rabbit glia cells. Unfortunately only high concentrations (10^{-5} M) of the uptake inhibitors were used in

Table 1. Accumulation of $[^{14}C]$ serotonin in ten rat brain regions, expressed in pmol $[^{14}C]$ serotonin per mg protein per min. Homogenates of regions were incubated with 5×10^{-8} M $[^{14}C]$ 5-HT for 4 min at 37 °C

Olfactory bulb	Frontal cortex	Entorhinal cortex	Hippocampus		Septum	Striatum	Hypothalamus	Colliculi	
			ventral	dorsal				superior	inferior
1.132	1.890	2.276	1.281	0.836	1.762	1.485	1.939	3.216	1.295

Values taken from Köhler et al. (1978).

these studies, which makes a comparison to the neuronal serotonin uptake system difficult. Nevertheless, these observations indicate the presence of a membranal serotonin transport system in glia cells.

Serotonin enters rabbit erythrocytes *in vitro* with a saturable facilitated diffusion mechanism with low affinity (Km = 5.6 mM) (Blakeley and Nicol, 1978).

Mouse pancreatic β-cells accumulate serotonin with a high affinity mechanism (Km = 1.6 μM) which is temperature and Na$^+$ dependent and inhibited by metabolic inhibitors (Lindström *et al.*, 1980).

In vitro methods for the determination of the neuronal accumulation of serotonin

The rate of accumulation of serotonin in serotonergic neurones is dependent on several factors, e.g. transport across the neurone membrane, uptake and binding in the storage vesicles, monoamine oxidase (MAO) activity and outward diffusion or release of serotonin taken up. Furthermore, serotonin can also be non-specifically taken up or bound in other tissue compartments. In order to ascertain that the accumulation of serotonin in serotonergic neurones is that measured some precautions need to be taken. By using low concentrations (10-100 nM), non-specific uptake and binding is largely avoided. Addition of a MAO inhibitor to the incubation medium, e.g. pargyline at the concentration 50-100 μM (higher concentrations are inhibitory on the serotonin uptake) diminishes the oxidative deamination of serotonin. Brief incubations are recommended in studies of drug effects on the serotonin uptake. The initial rate of serotonin uptake measures the undirected inflow of serotonin. This can be obtained by using very brief incubation or by extrapolation of time curves of the log rate of accumulation to zero time. It must be emphasized that even although a time curve is linear, the vesicular and possibly extravesicular binding of serotonin taken up becomes rapidly involved in the over all mechanism. For example, if a compound, which antagonizes the intraneuronal binding of serotonin, is studied, it may shift the rate limiting step from the membranal uptake to these intraneuronal binding sites. The results thus obtained may therefore be misinterpreted as uptake inhibition.

The active accumulation of serotonin into serotonergic neurones is often calculated from the difference between the accumulation obtained at 37 °C and 0 °C or in absence or presence of an uptake inhibitor, e.g. cocaine, or, preferably, a selective serotonin uptake inhibitor.

Two different brain tissue preparations have generally been used in the determination of [^{14}C] or [^{3}H] serotonin, brain slices and synaptosomes.

Brain slices (Shaskan and Snyder, 1970; Ross and Renyi, 1975a)

Slices from the brain region examined are made by means of a tissue chopper, which provides small slices, e.g. 0.2 × 0.2 × 0.5 mm, or by hand with a razor blade, which yields larger slices. The slices (10–40 mg) are preincubated at 37 °C in a saline-bicarbonate buffer (e.g. Krebs-Henseleit's buffer) in the presence of a MAO inhibitor, ascorbic acid (1.1 mM), EDTA Na$_2$ (0.13 mM) and glucose (6.5 mM) in an atmosphere of 6.5 per cent CO_2 in O_2. The labelled serotonin (10–100 nM final concentration) is then added and the incubation is continued for the required time. The slices are separated from the medium, disintegrated and the radioactivity is measured.

Synaptosomes (Bogdanski *et al.*, 1968; Kuhar *et al.*, 1972a; Kannengiesser *et al.*, 1973; Ross and Renyi, 1975a)

The use of synaptosomes has some advantage compared with the slice technique, particularly in kinetic experiments, since a homogeneous suspension is obtained and the nerve endings are freely exposed to the medium. A rather crude preparation can be used, e.g. a homogenate in 0.32 M sucrose centrifuged at 800 × g for 10 min. The tedious isolation of synaptosomes is therefore not necessary in most experiments. Synaptosomes corresponding to 5–10 mg brain tissue are incubated in 2 ml of a saline–bicarbonate buffer, pH 7.4, containing a MAO inhibitor, ascorbic acid, EDTA-Na$_2$ and glucose as described above. The reaction is interrupted either by adding ice-chilled buffer or chilling on an ice bath. The synaptosomes are collected either by centrifugation or by Millipore filtration. The latter method is most suitable for brief incubations. The pellets or the filters are washed with ice chilled buffer or saline solution. The organic material is disintegrated and the radioactivity measured.

For both the techniques described above it is possible to measure simultaneously the accumulation of serotonin and noradrenaline or dopamine by the double-labelling technique (Ross and Renyi, 1975a).

The results obtained with the two techniques described differ in some respects. For example, the inhibitory potencies of most compounds are about 10 times higher when using synaptosomes compared with slices (Ross and Renyi, 1975a, Heikkila *et al.*, 1976). The reason for these discrepancies is yet unclear but may be due to factors such as non-specific binding and uptake of the compounds in other cellular compartments, diffusion barriers in the slices and disruption of the cell organization by homogenization. Whether uptake of serotonin into glia cells by a similar transport mechanism as that in serotonergic neurones, which was shown in a recent study (Suddith *et al.*, 1978), contributes to the serotonin accumulation in brain slices has to be further elucidated. One disadvantage with the synaptosomes is that they are relatively unstable upon prolonged incubation at 37 °C. This is especially noticable in experiments when synaptosomes are pre-loaded with labelled serotonin for release studies (Heikkila *et al.*, 1976; Ross and Kelder, 1977).

In vivo methods

In studies of serotonin uptake inhibitors an *in vivo* method is needed. Since serotonin passes the blood brain barrier very poorly, the serotonin uptake can not be measured directly after injection of labelled serotonin. An indirect method has been devised by Carlsson *et al.* (1969), in which the depletion of serotonin in the brain evoked by α-ethyl-4-methyl-*m*-tyramine (H 75/12) is measured. This depletion of serotonin is antagonized by serotonin uptake inhibitors, probably due to the fact that H 75/12 is taken up in the serotonergic neurones by the membrane carrier. *p*-Chloroamphetamine is sometimes used instead of H 75/12 (Meek *et al.*, 1971; Fuller *et al.*, 1975).

Ex vivo determination of the inhibition of the serotonin uptake is another method in which the accumulation of labelled serotonin in brain slices or synaptosomes is determined after injection of the test compound (Ross and Renyi, 1975b; Wong *et al.*, 1975).

Kinetic experiments

The methods used for calculation of kinetic constants for the serotonin uptake have recently been discussed by Stahl and Meltzer (1978). As pointed out by these authors, several investigators have neglected the contribution of the uptake of passive diffussion and non-specific uptake or binding of serotonin and thereby obtained erroneous constants. Figure 1 shows an experiment, in which the 60 sec uptake of $[^{14}C]$ serotonin by partially purified synaptosomes from rat hypothalamus was measured at 37 °C and 0 °C using the Millipore filtration technique for the termination of the reaction. The concentration range of $[^{14}C]$ serotonin was 0.01–500 μM. The difference between the uptake at 37 °C and 0 °C are shown in the figure. The broken line shows the contribution of the non-specific uptake of serotonin as determined from the highest concentration examined. Double reciprocal plots of the difference between the observed uptake and the non-specific uptake at various serotonin concentration (Fig. 2) indicate three uptake mechanisms with different affinities and maximal uptake velocities. However, the kinetic constants obtained from these plots are not numerically correct, since the sum of the inverted velocities is not equal to the inverted sum of the velocities. The correct kinetic constants may be found by computerized trial and error fitting of the kinetic constants of the three uptake systems according to the Michaelis–Menten equations

$$v = \frac{V_1}{1 + Km_1 \cdot [S]^{-1}} + \frac{V_2}{1 + Km_2 \cdot [S]^{-1}} + \frac{V_3}{1 + Km_3 \cdot [S]^{-1}}$$

in which v is the observed uptake at each concentration [S] of serotonin examined, V_1, V_2 and V_3 and Km_1, Km_2 and Km_3 are the maximal uptake velocities and the apparent Km values, respectively, of the three uptake systems.

Graphically it is possible to resolve the three uptake systems from Eadi-Hofstee plots v versus $v \cdot [S]^{-1}$ according to the method of Rosenthal (1967) as shown in Fig. 3.

KINETICS OF THE NEURONAL SEROTONIN UPTAKE

The uptake of serotonin in serotonergic neurones is characterized by its high affinity and low capacity. Examples on apparent Km and V values obtained in various

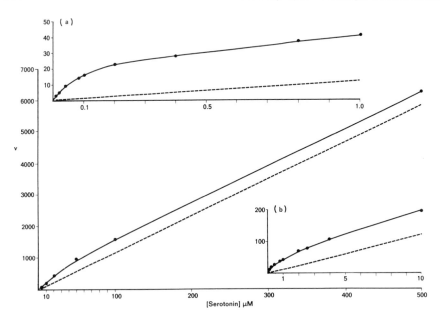

Fig. 1   Initial (60 sec) uptake of [^{14}C]serotonin in rat hypothalamic synapto-somes. Pooled hypothalamic tissues were homogenized in 10 vol 0.25 M sucrose, centrifuged at 800 × g for 10 min, the supernatant was centrifuged at 20 000 × g for 20 min and the pellet was resuspended in the original volume of Krebs–Henseleit's buffer. The uptake of [^{14}C]serotonin (0.01–500 μM) by synaptosomes (corres-ponding to 10 mg wet weight) for 60 sec was determined at 37 °C and 0 °C. The reaction was stopped by addition of 5 ml ice chilled saline solution containing 10^{-4} M serotonin as carrier and filtration through Whatman GF/B glass filter which was washed with 2 × 5 ml of the saline solution. The radioactivity was extracted from the filter with 1 ml Soluene-350 in a counting vial, 10 ml of Econofluor was added and the radioactivity measured. Each point represents the mean of four determinations at 37 °C minus that at 0 °C. The broken line denotes the contribu-tion of non-specific uptake as determined from 500 μM [^{14}C]serotonin. The uptake (v) is expressed as pmol mg^{-1} protein min^{-1}. The inserted figures (a) and (b) show the uptake at the [^{14}C]serotonin concentrations 0.01–1.0 μM and 0.1–10 μM, respectively

laboratories are given in Table 2. As shown, the values vary considerably. One cause of this variation appears to be the technique employed. Thus, the slice technique generally gives higher apparent Km values than the synaptosome technique. This can be due to diffussion barriers and non-specific uptake and binding in other cells. This non-specific uptake or binding of serotonin contributes largely also in experi-ments with synaptosomes as shown in Fig. 1. As previously discussed, the methods used for calculation of the kinetic constants can give different values (Stahl and Meltzer, 1978).

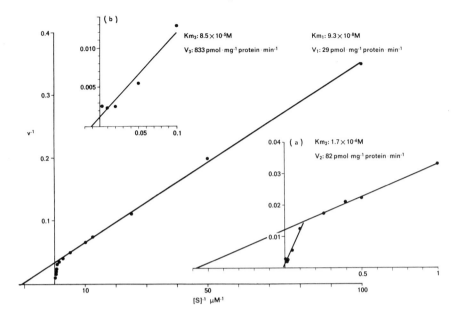

Fig. 2   Double reciprocal plots of the uptake of [^{14}C] serotonin shown in Fig. 1.
The estimated non-specific uptake is substracted from the measured uptake. v =
pmol mg^{-1} protein min^{-1} ; [S] = serotonin concentration in μM

Shaskan and Snyder (1970) examined the kinetics of the serotonin accumulation
in rat hypothalamic and striatal slices. Two uptake mechanisms were reported, one
high-affinity uptake into serotonergic neurones (Table 2) and a second low-affinity
uptake (Km = 8 μM, V = 8.0-9.1 nmol/min/g in hypothalamus and 22.6-33.0
nmol/min/g in striatum). Since the low-affinity uptake was inhibited by noradrenaline and dopamine to a larger degree than was the high-affinity uptake of serotonin,
the authors discussed the possibility that the low-affinity uptake occurs in catecholaminergic neurones. However, they pointed out that the maximal uptake was
much larger than could be expected from the corresponding values of the catecholamine uptake, since the maximal velocity of transport with an uptake system is independent of which compound is transported. The high V value obtained in this study
may be due to a large contribution of non-specific uptake and binding of serotonin.
However, uptake of serotonin into catecholaminergic neurones is clearly indicated
by the observation of Iversen (1970) that pretreatment of rats with intraventric-
larly injected 6-hydroxydopamine, which selectively destroys catecholaminergic
nerve terminals, reduced the serotonin accumulation in synaptosomes considerably
more at high than at low serotonin concentrations. Moreover, 5,6- and 5,7-dihy-
droxytryptamine injected intraventricularly not only destroy serotonergic nerve
terminals but also, although less potently, catecholaminergic nerve terminals
(Björklund *et al.*, 1975; Baumgarten *et al.*, 1978). The effect of these neurotoxins

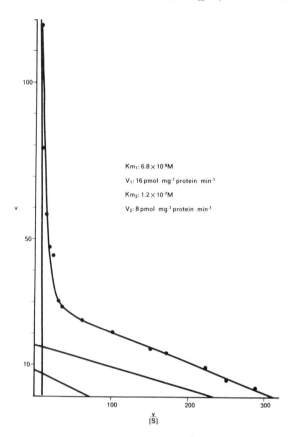

Fig. 3   An Eadi–Hofstee plot of the same data shown in Figs. 1 and 2. The experimental curve was divided in three lines according to Rosentahl (1967). v = pmol mg^{-1} protein min^{-1} ; [S] = serotonin concentration in μM

on the noradrenergic system is antagonized by desipramine, which indicates that hydroxylated tryptamines are transported via the noradrenaline carrier. Uptake of serotonin in peripheral noradrenergic nerve terminals has also been demonstrated (Thoa *et al.*, 1969).

In order to examine whether a low-affinity uptake of serotonin in the μM range can be demonstrated, the experiment shown in Figs 1–3 was performed. As already discussed three uptake systems appeared of which the very low-affinity uptake may be non-specific uptake or binding of serotonin. The characteristic high-affinity uptake was easily demonstrated. However, between these two uptake systems an intermediate uptake of serotonin seems to occur which may possibly represent uptake into catecholaminergic neurones. It must be emphasized that the kinetic constants estimated for this uptake system are only rough approximations due to the slight contribution of this uptake to the total uptake.

Table 2. Apparent Km and V values for the high affinity uptake of serotonin in various tissue preparations

Tissue	Concentration range (μM)	Km (μM)	V	Reference
Brain slices				
Mouse midbrain	0.05–1.0	0.7	0.23[a]	Ross and Renyi (1969)
Whole rat brain	0.02–0.57	0.57	0.88[a]	Blackburn et al. (1967)
Rat hypothalamus	0.4–4	0.14	0.66–0.82[a]	Shaskan and Snyder (1970)
Rat striatum	0.4–4	0.17	1.12–1.40[a]	Shaskan and Snyder (1970)
Synpatosomes				
Rabbit brain stem	0.057–1.4	0.071	18.3[b]	Bogdanski et al. (1970a)
Whole rat brain	0.07–2.0	0.10	1130[b]	Wong et al. (1973)
Rat brain stem	0.04–10	0.076	0.33[a]	Tuomisto and Tuomisto (1973)
Whole rat brain	0.01–0.11	0.049	0.15[a]	Hyttel (1977)
Rat hypothalamus	0.01–500	0.068	16[b]	Fig. 3
Guinea-pig small intestine				
Myenteric plexus	0.1–6.5	0.74	0.22[a]	Gershon and Altman (1971)
Snail (Helix pomatia)				
Suboesophagal ganglion	0.01–100	0.085	0.077	Osborne et al. (1975)
Platelets				
Rat	0.25–2.0	0.48	4500[c]	Wielotz et al. (1976)
Human	0.25–2.0	0.63	870[c]	Wielotz et al. (1976)
Human	0.05–50	0.096	20[c]	Stahl and Meltzer (1978)

[a] nmol g^{-1} wt tissue min^{-1}, [b] pmol mg^{-1} protein min^{-1}, [c] pmol 10^{-9} platelets min^{-1}.

DRIVING FORCE OF THE SEROTONIN TRANSPORT

The role of (Na^+ + K^+) activated ATPase

The observations that ouabain inhibits the accumulation of serotonin in brain tissue (Blackburn *et al.*, 1967; Ross and Renyi, 1967; Bogdanski *et al.*, 1968) indicates that the transport of serotonin across the neurone membrane is a secondary transport mechanism which requires a functional sodium pump. Tissari *et al.* (1969) and Tissari and Bogdanski (1971) analysed the action of ouabain on the accumulation of serotonin in cerebral cortical synaptosomes. They found that ouabain inhibits serotonin accumulation following a time delay of a few min. High K^+ concentrations in the medium partially antagonized the effect of ouabain. Lowering the medium concentration of Na^+ from 143 mM to 50 mM decreased the inhibitory effect of ouabain. The authors concluded that ouabain inhibits the serotonin transport by blocking a K^+ dependent factor in the transport, probably (Na^+ + K^+) ATPase.

Na^+ gradient hypothesis

The absolute requirement of Na^+ for the serotonin transport indicates that the assymetrical distribution of Na^+ across the neurone membrane is the driving force of the serotonin transport. Since the sodium pump ([Na^+ + K^+]-activated ATPase) produces the Na^+ gradient, the inhibitory action of ouabain appears to be the result of the breakdown of this gradient. Furthermore, the finding by Bogdanski *et al.* (1970b) that omission of K^+ in the incubation medium after a time lapse of 15 min markedly decreased the serotonin accumulation compared with that at optimum K^+ concentration (6 mM) can be explained by a decreased activity in the sodium pump due to lack of activating concentrations of K^+.

Ouabain (0.1 mM) inhibited the Na^+ + K^+ ATPase maximally and almost doubled the intracellular Na^+ concentration but had no effect on serotonin uptake in snail ganglia (Stahl *et al.*, 1977). This observation appears to be contradictory to the view that the Na^+ gradient is the energy source for the serotonin uptake. These authors discussed the possibility that the inhibitory effect of ouabain at higher doses may be due to reduction of high energy phosphates in the cells.

Kinetic experiments on the accumulation of serotonin in synaptosomes with various concentrations of external Na^+ reveal that Na^+ facilitates the serotonin transport mainly by increasing the affinity of serotonin for the carrier, leaving the maximal velocity of transport almost unchanged (Bogdanski *et al.*, 1970a; Ross and Kelder, 1977). High external K^+ concentrations decrease the serotonin transport, possibly by competing with Na^+ and thereby reducing the affinity of serotonin for the carrier (Bogdanski *et al.*, 1970a, Tissari and Bogdanski, 1971). These authors proposed that serotonin, Na^+ and the carrier form a ternary complex on the outside of the neurone membrane. This complex is translocated to the inside where it is dissociated presumably by a conformation change in the carrier, induced by the high

K^+ concentration. The carrier possibly in a complex with K^+ is translocated back to the outside and changes the conformation to that ready for binding serotonin. The translocation step may be a real movement of the complex or a conformation change in the carrier protein opening pores for the passage of serotonin and Na^+.

Working with plasma membrane vesicles of platelets, which have membrane serotonin transport system very similar to that in serotonergic neurones, Rudnick and Nelson, in a series of studies have analysed the serotonin transport mechanism in more detail. These plasma membrane vesicles have the great advantage in that both the inner and the outer milieu can be experimentally controlled. Thus, intracellular binding compartments and metabolizing enzymes do not interfere with the uptake of serotonin in this preparation. The plasma membrane vesicles accumulate serotonin 80-fold if the appropriate Na^+ and K^+ gradients are constructed (Rudnick, 1977). The uptake of serotonin is inhibited by antidepressant agents, e.g. imipramine and ionophores, e.g. gramicidine, whereas reserpine and ouabain have no or little effect. There is an absolute requirement of external Na^+ for serotonin transport in this system. In accordance with the findings of Lingjaerde (1971) external Cl^- is also necessary for transport. The apparent Km of the initial uptake of serotonin was found to be 0.6 μM.

Subsequent studies (Rudnick and Nelson, 1978a; Nelson and Rudnick, 1979) showed that serotonin transport appears to be independent on a measurable electrical potential across the membrane when K^+ is present internally. In absence of internal K^+ the serotonin transport is apparently electrogenic with net positive charge crossing the membrane with serotonin. Internal K^+ stimulates, whereas high external concentrations of K^+ inhibits the serotonin transport.

With this model system it is also possible to study the outwardly directed transport of serotonin. It was observed that a Na^+ gradient (in $\leqslant$ out) stimulates the efflux of serotonin and is further stimulated by high external K^+ or unlabelled serotonin. The latter requires, however, presence of external Na^+.

From these and other observations Nelson and Rudnick (1979) proposed a model for the serotonin transport in the platelet (Fig. 4) which is an extension of the transport model proposed by Bogdanski *et al.* (1970a) for the neuronal transport of serotonin. The first step in the translocation is the binding of Na^+ and then of protonated serotonin to the carrier. Cl^- is not required for the binding but is necessary for the net transport. The translocation involves a conformation change of the carrier whereby Na^+ serotonin and Cl^- are dissociated at the interior side of the membrane. In the presence of internal K^+ this ion binds to the carrier and the complex is translocated back to the exterior side. In the absence of K^+ the carrier is also translocated to the outside, but more slowly. In the latter case net positive charge enters the vesicles.

Although it has not been possible to examine the neuronal transport of serotonin in such detail as the platelet serotonin uptake, several observations indicate that the characteristics are similar. For example, the neuronal serotonin transport requires external Cl^- (Kuhar and Zarbin, 1978). The efflux of labelled serotonin

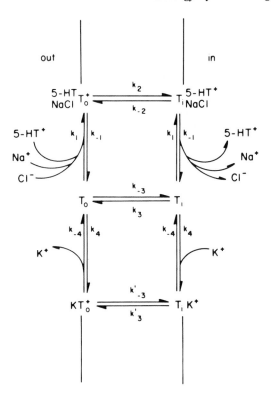

Fig. 4   A model of the serotonin transport across the platelets cell membranes according to Nelson and Rudbnick (1979). (Reproduced by permission of *Journal of Biological Chemistry*)

from preloaded cerebral cortical synaptosomes obtained from reserpinized rats is stimulated by agents which increase the internal Na^+ concentration (Ross and Kelder, 1977). This evoked increase in the serotonin efflux is inhibited by serotonin uptake inhibitors. These observations indicate a facilitated outward transport of serotonin with the uptake carrier.

Intracellular binding

Although the experiments by Rudnick and co-workers clearly show that the Na^+ gradient provides energy for the accumulation of serotonin in platelet vesicle membranes, the rapid binding of serotonin in storage vesicles and possibly at extravesicular binding sites significantly contributes to the high degree of accumulation of serotonin in serotonergic neurones and platelets.

STRUCTURAL REQUIREMENTS FOR UPTAKE INTO
SEROTONERGIC NEURONES

Carrier mediated transport mechanisms are characterized by their structural specificity for the compounds transported. Although no systematic study of the structural requirements for the serotonin transport mechanisms has yet been performed, there are some observations which indicate that the serotonin carrier has a relatively high specificity. For example, besides serotonin itself only 5,6-dihydroxytryptamine, 5,7-dihydroxytryptamine and their N-methylated derivatives have shown to be accumulated in serotonergic neurones (Björklund *et al.*, 1975; Baumgarten *et al.*, 1978). Judging from the irreversible damage of the serotonergic neurones and their inhibitory effect on the serotonin accumulation the N-methylated dihydroxytryptamines appear to have higher affinity for the serotonin carrier than the corresponding primary amines (Horn *et al.*, 1973; Björklund *et al.*, 1975). Whether or not hydroxytryptamines, with the hydroxyl group placed in other than 5 position, are transported has not been examined.

Non-hydroxylated tryptamines do not appear to be transported by the serotonin carrier in mouse and rat brain tissue (Ross and Ask, 1980), in human platelets (Born *et al.*, 1972), or in snail (*Helix pomatia*) suboesophagal ganglia (Schröder *et al.*, 1979).

Although Shaskan and Snyder (1970) did not find any evidence that catecholamines are transported by the neuronal serotonin carrier, Berger and Glowinski (1978) observed, with a histochemical technique, that dopamine, but not noradrenaline, is accumulated in serotonergic nerve terminals. Since this accumulation of dopamine was antagonized by fluoxetine, a selective serotonin uptake inhibitor, and by lesion with 5,7-dihydroxytryptamine but not by 6-hydroxydopamine, the dopamine uptake appears to occur with the serotonin carrier.

It has been suggested that α-ethyl-4-methyl-*m*-tyramine, *p*-chloromethamphetamine, *p*-chloroamphetamine, *p*-methoxyamphetamine, fenfluramine and similar drugs are transported by the serotonin carrier, since the release of brain 5-HT induced by these compounds is antagonized by selective serotonin uptake inhibitors (Carlsson *et al.*, 1969; Meek *et al.*, 1971; Fuller *et al.*, 1975 Clineschmidt *et al.*, 1978b). Attempts to verify this hypothesis by measuring the accumulation of labelled *p*-chloroamphetamine and fenfluramine in rat synaptosomes have failed to show any specific carrier mediated accumulation of these compounds (Belin *et al.*, 1976; Ross 1976). This failure might have been due to non-specific uptake into other synaptosomes, which are in   great surplus, or as a result of rapid diffusion of the substance out from the synaptosomes. It is also possible that these compounds are simply not transported by the carrier.

No evidence that *p*-chloroamphetamine is accumulated in the serotonergic neurones by the membrane carrier was obtained in a study by Ross and Ask (1980), in which the *in vivo* antagonism of *p*-chloroamphetamine, which is a relatively potent MAO-A inhibitor, on the irreversible MAO inhibition produced by pheniprazine in serotonergic neurones was determined. In contrast, the MAO protecting effect produced by α-ethyl-4-methyl-*m*-tyramine, although only partial, was antagonized by norzimelidine, a selective serotonin uptake inhibitor, which indicates that this compound unlike *p*-chloroamphetamine and *p*-methoxyamphetamine (unpublished observation) is transported by the serotonin carrier.

INHIBITION OF THE SEROTONIN UPTAKE

Compounds can cause inhibition of the accumulation of serotonin by different mechanisms.

1. Inhibition of the membrane transport causes an inhibition of the initial uptake. Competitive inhibition is the result of binding of the compound to the serotonin binding site of the carrier.
2. Transport of the compound itself by the membrane carrier.
3. Inhibition of the vesicular binding, i.e. a reserpine-like effect. The initial uptake, e.g. determined from a time curve extrapolated to zero (Ross and Renyi, 1969) is not inhibited.
4. Release of serotonin newly taken up by interaction with vesicular or extravesicular binding sites, e.g. tryptamines and para-substituted amphetamines.

The classification of any particular inhibitor of the serotonin accumulation requires the examination of the compound in more than one experimental situation. The releasing effect may be tested by incubating or superfusing slices or synaptosomes preloaded with labelled serotonin (Heikkila *et al.*, 1976; Mulder *et al.*, 1975; Raiteri *et al.*, 1977; Ross *et al.*, 1977). The superfusion technique has the advantage that it diminishes the re-uptake of released serotonin. A compound possessing an equal serotonin-releasing activity to its accumulation-inhibiting activity has been suggested as primarily a releasing agent (Heikkila *et al.*, 1976). However, this method does not resolve the question regarding how much more potent the compound is in releasing serotonin than in inhibiting the uptake. Ross (1979) observed that serotonin-releasing compounds, e.g. tryptamine derivatives and *p*-chloroamphetamine, are, to a varying degree, more potent in inhibiting the serotonin accumulation by synaptosomes from reserpinized rats than by those from non-treated animals, whereas serotonin-uptake inhibitors are equally active in both preparations. Although the mechanism underlying the reserpine potentiation is not fully understood, this method appears to have a value for classifying a compound as an uptake inhibitor or a serotonin releaser. For example, α-methyltryptamine was found to be ten times more active in inhibiting the serotonin accumulation in the reserpinized tissue than in the control tissue, which indicates that it is primarily a serotonin releasing compound.

Two possible explanations for the mechanism by which serotonin accumulation is inhibited by releasing agents have been considered.

1. An accelerative exchange diffusion which induces an increased efflux of the

labelled serotonin taken up (Paton, 1974; Rutledge, 1978). A necessary prerequisite is that the compounds are transported by the serotonin carrier. The potentiating effect of reserpine is then caused by the elevated concentration of free serotonin in the neurones.

2. The compounds block some hypothetical extravesicular binding sites for serotonin (Ross, 1976; Ross and Kelder, 1977). The explanation to the potentiating effect of reserpine is the same as that mentioned above.

More detailed studies on the mode of action of the serotonin releasing agents are obviously necessary.

Tryptamine derivatives

Horn and co-workers have examined the structural activity relationships for tryptamine derivative in inhibiting the accumulation of serotonin in rat hypothalamic homogenates (Horn, 1973; Horn *et al.*, 1973; Björklund *et al.*, (1975). They found that the α-methyl, α-ethyl and *N,N*-dimethyl derivatives were more potent than tryptamine itself (Table 3). However, these compounds probably act via a release mechanism as discussed above (Baker *et al.*, 1977; Ross, 1979). Prolongation of the side chain with one methyl group also increases the inhibitory potency.

The aromatic hydroxylated tryptamine derivatives have about the same inhibitory activities as the non-hydroxylated tryptamines (Horn *et al.*, 1973 Björklund *et al.*, 1975). The order of potency being 5-OH > 5,6-diOH > 7-OH > 4,5-diOH > 6-OH > 4-OH ≫ 5,6,7-triOH. *N*-Methylation of the 5,6-diOH derivative tends to increase the activity whereas α-methylation of the 5,7-diOH derivative does not change the potency. 5-Methoxytryptamine has a poor effect. *N,N*-Dimethylation also decreased the potency. At variance with this structure activity relationship is

Table 3 Structure activity relationships for tryptamine derivatives in inhibiting the serotonin accumulation in rat hypothalamic homogenate

Substitution	IC_{50} (μM)	Substitution	IC_{50} (μM)
None	2.8[a]	5-hydroxy-*N,N*-dimethyl	1.5[a]
α-Methyl	0.7[a]	5-methoxy	25[a]
α-Ethyl	0.34[a]	4,5-dihydroxy	1.7[d]
N,N-dimethyl	0.85[a]	5,6-dihydroxy	0.6[d]
7,*N,N*-trimethyl	0.4[b]	5,7-dihydroxy	4.0[d]
4-Hydroxy	4.5[a]	6,7-dihydroxy	4.4[a]
6-Hydroxy	0.8[c]	5,6-dihydroxy-*N*-methyl	0.37[e]
7-Hydroxy	2.8[d]	5,7-dihydroxy-α-methyl	4.4[e]
		5,6,7-trihydroxy	>10[e]

Values taken from: [a]Horn (1973), [b]Glennon *et al.* (1978), [c]Baumgarten *et al.* (1972), [d]Horn *et al.* (1973), [e]Björklund *et al.* (1975).

the finding by Uebelhack *et al.* (1978) that 5-methoxy-*N,N*-dimethyltryptamine is quite a potent inhibitor. These authors also reported high potency of LSD on inhibiting the serotonin accumulation in an enriched synaptosome preparation of the whole rat brain with significant inhibition at 10^{-11} M. This result contrasts with that obtained by Ross and Renyi (1967), who found LSD to be a poor inhibitor of the serotonin accumulation in slices of the mouse cerebral cortex.

Phenethylamine derivatives

The para substituted phenethylamines and amphetamines are potent inhibitors of the serotonin accumulation in brain sices and synaptosomes (Carlsson, 1970; Ross *et al.*, 1972, 1977; Wong *et al.*, 1973; Chung Hwang and Van Woert, 1980). For the phenethylamine derivatives the following order of potency was observed: Br > J > Cl > CH_3 > CH_3O ≫ F (Chung Hwang and Van Woert, 1980). The *meta* and *orto* substituted phenethylamines have a lower activity. Since these compounds are potent releasers of serotonin, at least a part of the inhibition of the serotonin accumulation appears to be exerted by this release mechanism. In accordance with this view Ross ((1979) found that *p*-chloroamphetamine was four times more potent in inhibiting the serotonin accumulation in hypothalamic synaptosomes from reserpinized rats than a control preparation.

Like the tryptamine derivatives, the para substituted amphetamines (Meek *et al.*, 1971; Fuller *et al.*, 1975; Ross, 1976) and phenethylamines (if combined with a MAO inhibitor) Chung Hwang and Van Woert, 1980) cause a release of brain serotonin *in vivo* and produce a behaviour syndrome (e.g. head twitches, lateral head moving, 'wet dog shake', abduction of hind legs, tremor), which is characteristic for stimulation of serotonin receptors in the brain. The serotonin release and the behaviour syndrome are blocked by serotonin uptake inhibitors (Meek *et al.*, 1971; Fuller *et al.*, 1975; Ross, 1976; Ross *et al.*, 1976; Chung Hwang and Van Woert, 1980). Thus, these compounds are dependent upon a functional serotonin transport for exerting their central effects on the serotonin neurones. A plausible explanation of these observations is that the phenethylamines and amphetamines are transported by the serotonin carrier and thereby release serotonin (Meek *et al.*, 1971; Fuller *et al.*, 1975; Rutledge, 1978). As already discussed above, this transport has not, however, been experimentally established for most of these compounds. An alternative explanation of these observations is therefore that these lipophilic compounds enter the serotonin neurones by diffusion, release serotonin from binding sites and the free serotonin is then transported in an outward direction via the carrier by facilitated diffusion across the nerve membrane (Ross, 1976; Ross and Kelder, 1977).

Tricyclic antidepressant agents

The structure activity relationships for tricyclic antidepressant and related agents have been reviewed by Maxwell and White (1978). The possible involvement of

Table 4 Inhibition of serotonin and noradrenaline uptake in rat brain synaptosomes and human platelets by some tricyclic antidepressant agents

Compound	IC_{50} Synaptosomes		Human platelets serotonin
	Serotonin (μM)	Noradrenaline (μM)	
Desipramine	1.1^a	0.0033^a	1.0^b
Imipramine	0.14^a	0.028^a	0.08^b
Chlorimipramine	0.018^a	0.060^a	0.09^b
3-Cyanoimipramine	0.0016^c	–	–
Nortriptyline	0.87^a	0.054^a	0.9^b
Amitriptyline	0.18^a	0.024^a	0.3^b
Maprotiline	20^d	0.1^d	$>100^d$

[a]Ross and Renyi (1975a), [b]Tuomisto (1974), [c]Keller *et al.* (1980), [d]Baumann and Maitre (1979).

serotonin in affective disorders (Coppen, 1967; Carlsson *et al.*, 1969; Lapin and Oxenkrug, 1969) has inititated much research into the development of selective inhibitors of serotonin uptake. The early findings that (a) desipramine is much more potent in inhibiting the uptake of noradrenaline than of serotonin (Ross and Renyi, 1967; Carlsson *et al.*, 1969) and that (b) chlorimipramine is *in vitro* a considerably more potent inhibitor of the uptake of serotonin than that of noradrenaline (Carlsson, 1970; Shaskan and Snyder, 1970) clearly indicate different structural requirements for the inhibition of the two amine uptake systems.

The inhibitory potencies of some tricyclic antidepressant agents are given in Table 4. They are true uptake inhibitors, since much higher concentrations are needed to induce serotonin release than inhibition of uptake (Heikkila *et al.*, 1976). In general the tertiary amino derivatives are more potent than their secondary analogues (Carlsson *et al.*, 1969; Ross and Renyi, 1969). As shown, substituents at the 3 position, increase the potency and the selectivity on the serotonin uptake. Thus, 3-cyanoimipramine appears to be the most potent of all serotonin uptake inhibitors hitherto synthetized. The quaternary methiodide derivative of imipramine is *in vitro* equipotent with imipramine (Horn and Trace, 1974; Ross and Renyi, 1976) showing that the protonated form of the molecule binds to the carrier. The inhibitory potency is reduced by increasing or decreasing the length of the side chain or introducing a methylgroup at α- or β-position (Horn and Trace, 1974). Aromatic hydroxylation of imipramine makes little difference in activity (Heikkila *et al.*, 1976; Javaid *et al.*, 1979). However, Bertilsson *et al.* (1979) found that the 10-hydroxylated amitriptyline loses the inhibitory activity on the serotonin uptake. A non-coplanar structure of the tricyclic ring is essential for optimal inhibitory activity. Thus, replacement of the dimethylene bridge by a sulphur atom (e.g. as in

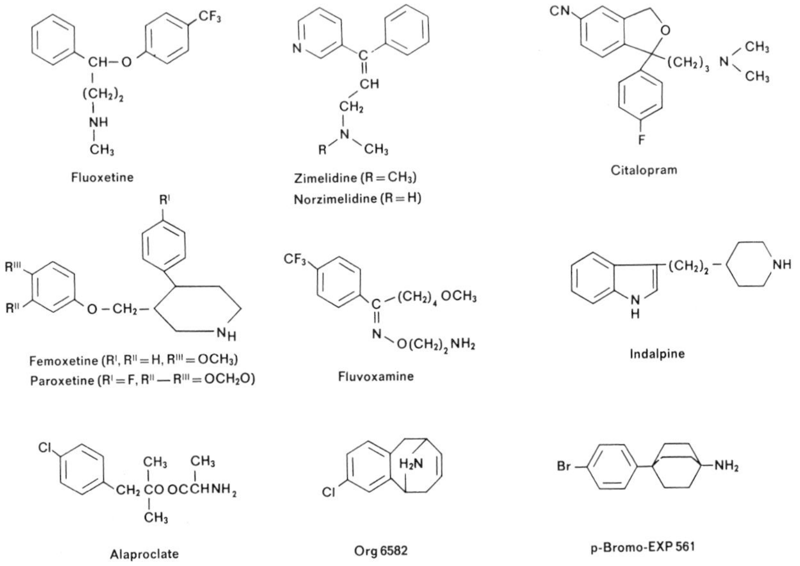

Fig. 5   Chemical structures of some selective inhibitors of serotonin uptake

chlorpromazine) or replacement of the N=C bond in the side chain of imipramine by a C=C double bond (e.g. as in amitriptyline) decreases the inhibitory potency (Horn and Trace 1974; Horn, 1976; Ross and Renyi, 1976). Grunewald *et al.* (1979) reported that reduction of one or both aromatic rings of imipramine and desipramine to cyclohexane rings do not change the inhibitory activity on the serotonin accumulation in chopped rat cerebral cortex. The location of the side chain amino group in relation to the tricyclic ring system appears to be essential, since rigid spiro analogues of amitriptyline are poor inhibitors of the serotonin uptake but are potent inhibitors of the noradrenaline uptake (Carnmalm *et al.*, 1974).

Selective inhibitors of the serotonin uptake

Several new classes of selective inhibitors of serotonin uptake have been developed during recent years. The chemical structures of some of these new compounds are shown in Fig. 5 and their *in vivo* inhibitory activities are summarized in Table 5. The conclusion drawn from the tricyclic agents that the tertiary amino derivatives are more potent than their secondary amino analogues can not be generalized for all classes of compounds. Thus, several secondary amines are very potent inhibitors of serotonin uptake, e.g. fluoxetine (Wong *et al.,* 1974, 1975), norzimelidine (Ross and Renyi, 1977), femoxtine (Buus Lassen *et al.,* 1975), paroxetine (Petersen *et al.,* 1977) and indalpine (LeFur and Uzan, 1977). In other classes the primary amine derivatives are potent inhibitors of serotonin uptake, e.g. Org 6582 (Mirey-

Table 5 The activity of some selective inhibitors on the monoamine uptake in rat synaptosomes. The concentrations of the labelled amines employed in the various experiments are given below the table

Compound	IC$_{50}$			Reference
	Serotonin (μM)	Noradrenaline (μM)	Dopamine (μM)	
Alaproclate	0.17[a]	25[g]	24[g,i]	Lindberg *et al.* (1978a)
p-Bromo-EXP561	0.68[j]	54[j]	–	Fuller *et al.* (1978)
CGP6085A	0.0055[b]	1.4[c]	7.5[d]	Waldmeier *et al.* (1979)
Citalopram	0.0055[b]	4.4[c]	$\geqslant 10^d$	Waldmeier *et al.* (1979)
Femoxetine	0.019[b]	0.36[c]	3.7[d]	Waldmeier *et al.* (1979)
Fluoxetine	0.027[b]	0.55[c]	7.5[d]	Waldmeier *et al.* (1979)
Fluvoxamine	0.084[a]	30[j,i]	63[j,i]	Claassen *et al.* (1977)
Indalpine	0.04[h]	2[h]	4[h]	LeFur and Uzan (1977)
Ly 125180	0.06[a]	2.2[a]	2.5[a]	Wong *et al.* (1980)
Norzimelidine	0.04[a]	0.076[g]	12[g]	Ross and Renyi (1977)
Org 6582	0.089[a]	2.9[e]	13[e]	Mireylees *et al.* (1978)
Paroxetine	0.03	–	–	Petersen *et al.* (1977)
Trazodone	0.57[d]	35[g]	53[f]	Riblet *et al.* (1979)
Zimelidine	0.13[a]	2.7[g]	12[g]	Ross and Renyi (1977)

The concentration of labelled amine: [a]K_i value, [b]2.5 nM, [c]5 nM, [d]10 nM, [e]25 nM, [f]33nM, [g]50 nM, [h]100 nM, [i]unpublished observation, [j]not stated.

lees *et al.*, 1978), *p*-bromo-EXP 561 (Fuller *et al.*, 1978) and alaproclate (Lindberg *et al.*, 1978a).

Compounds with rigid structures can give valuable information regarding the topography of the serotonin carrier. From inspections of models of the structures of the very rigid compounds Org 6582 and *p*-bromo-EXP 561, it is obvious that it is impossible to superimpose both the aromatic rings and the nitrogen atom of these two compounds. Thus, the distance between the centre of the aromatic rings and the amino group is about 2 Å longer for *p*-bromo-EXP 561. Since the corresponding distance for the fully staggered transform of serotonin is 1.3 Å shorter than that for EXP 561 (Horn, 1977), it is also impossible to superimpose EXP 561 and serotonin. Since both Org 6582 (Mireylees *et al.*, 1978) and *p*-chloro-EXP 561 (Fuller *et al.*, 1978) appear to be competitive inhibitors, they are not able to be bound to the same elements of the serotonin carrier at the same conformation stage. One possible explanation of these conflicting sturctural activity relationships is that some inhibitors are attached by structures adjacent to the serotonin carrier (Koe, 1976). Alternatively, the serotonin carrier may exist in different conformation states.

As a result of a conformation study of the R-enantiomer of alaproclate by a single-crystal X-ray determination and energy calculation (Lindberg *et al.*, 1978b) it was suggested that the favoured conformation of serotonin for binding to the uptake sites is the gauche form, in which the side chain is approximately perpendicular to the indole plane and the terminal nitrogen is bent towards 2 position of the indole in a staggered conformation (Lindberg *et al.*, 1978a). A hypothetical model of the serotonin carrier was proposed in which the geometry and electronic properties of the site binding serotonin were defined by a centre of negative charge capable of ionic binding, a centre of the aromatic nucleus capable of van der Vaal's interactions and a centre of partial positive charge capable of localized charge transfer. Several other serotonin uptake inhibitors, e.g. fluoxetine, norzimelidine and Org 6582, are able to adapt to this model. However, *p*-chloro-EXP 561, as discussed above, and indalpine (Gueremy *et al.*, 1980) deviate from this model.

Inhibition of serotonin uptake by other compounds

Various other types of compounds have been examined for their *in vitro* inhibitory actions on the serotonin uptake.

1. Anaesthetics, e.g. ketamine (Azzaro and Smith, 1977).
2. Anoretic drugs, e.g. chlorphentermine (Ross *et al.*, 1972), fenfluaramine (Belin *et al.*, 1976; Clineschmidt *et al.*, 1978b; Kannengiesser *et al.*, 1976), mazindol (Heikkila *et al.*, 1977; Sugrue and Mireylees, 1978), dita (Heikkila *et al.*, 1977).
3. Anticholinergics; e.g. orphenadrine (Van der Zee and Hespe, 1978).
4. Antihistaminics, e.g. chlorpheniramine and bromopheniramine (Carlsson, 1970).
5. Benzhydrylethers (Van der Zee and Hespe, 1978).
6. Cannabinoids (Sofia *et al.*, 1971; Banerjee *et al.*, 1975).
7. β-Carbolines (Lessin *et al.*, 1967; Tuomisto and Tuomisto, 1973; Rommelspacher *et al.*, 1978; Airaksinen *et al.*, 1980; Komulainen *et al.*, 1980).
8. 6-Chloro-2-(1-piperazinyl)pyrazine (MK-212) (Clineschmidt *et al.*, 1978a).
9. 1-(3-Dimethylamino)propyl)-5-methyl-3-phenyl-1H-imidazole (FS 3) and its secondary amino analogue (FS 97) (Koide and Uyemura, 1980).
10. Diphenylcycloalkylamines and related compounds (Carnmalm *et al.*, 1975, 1977).
11. Methylxanthines (Cardinali, 1977).
12. Narcotic analgesics (Ciafalo, 1974; Moffat and Jhamandas, 1976; Warwick *et al.*, 1977; Mennini *et al.*, 1978).
13. Tetrahydronapthylamine (Bruinvels, 1971).

UPTAKE OF SEROTONIN IN BLOOD PLATELETS

Since the mechanisms for uptake, binding and storage of serotonin in blood platelets are similar to those in serotonergic neurones (for review see Sneddon, 1973), platelets are frequently used as the model for the serotonergic neurones. In several respects the use of platelets have great experimental advantages compared with synaptosomes or brain slices, e.g. they are easily obtained even from humans; they can be isolated from other cells; as mentioned previously, plasma membrane vesicles can be prepared and allow detailed analysis of the serotonin transport; and storage vesicles can be isolated and studied.

The serotonin uptake capacity of the platelets is high, a factor which has to be taken into account in experiments with platelet rich plasma (PRP). Thus, the serotonin uptake will already, after a brief incubation period, become non-linear when low concentrations of serotonin are used (Stahl and Meltzer, 1978).

The serotonin uptake by human platelets is inhibited by antidepressant drugs with a similar order of potency as in rat synaptosomes (Table 4). Wielotz *et al.*, (1976) compared the potencies of imipramine, chlorimipramine and fenfluramine in inhibiting the serotonin uptake in human and rat platelets. They observed that imipramine and chlorimipramine were seven and ten times, respectively, more potent in human than in rat platelets. This species difference in the effect of inhibitors on serotonin uptake is notable, since the rat is frequently used in animal experiments with antidepressant and related agents. Thus, results obtained in rats with this type of drugs can not be validly generalized to be valid to humans. For instance, desipramine which is a weak inhibitor of the serotonin uptake in the rat brain (Ross and Renyi, 1975b) and in rat platelets (Ross, unpublished observation) causes a marked inhibition of the serotonin uptake in human platelets when examined in PRP from desipramine-treated patients (Ross *et al.*, 1980)). In accordance with this species difference, it was also observed that the plasma concentrations of norzimelidine producing the same degree of inhibition of serotonin uptake in rat and human platelets *in vivo* was five times higher in rats (Ross *et al.*, 1981). The inhibition of the rat platelet serotonin uptake was obtained in the same dose range as that of the brain serotonin uptake, which indicates that results obtained with platelets can be extrapolated to the brain at a steady-state equilibrium of the drug treatment.

In addition to the usefulness of platelets in studies of the efficacy of serotonin uptake inhibitors under clinical conditions, potential relationships between various psychiatric and other disorders and the serotonin uptake mechanism can be examined with the platelet model. Several groups of investigators have reported a reduced serotonin uptake in platelets from patients with endogenous depression (Coppen *et al.*, 1978; Hallström *et al.*, 1976; Tuomisto and Tukianen, 1976; Tuomisto *et al.*, 1979). It has also been reported that the platelets serotonin uptake is decreased in schizophrenic patients (Rotman *et al.*, 1979).

SEROTONIN UPTAKE IN STORAGE VESICLES

Serotonin transported into serotonergic neurones is bound in vesicles through a similar mechanism as that of the catecholamines (Slotkin and Bareis, 1980). Thus, reserpine produces a pronounced depletion of serotonin (Pletscher *et al.*, 1956) and reduces the accumulation of labelled serotonin in brain tissue (Blackburn *et al.*, 1967; Ross and Renyi, 1967). However, it has not yet been possible to measure the serotonin uptake into brain serotonergic vesicles separated from the catecholaminergic vesicles (Slotkin and Bareis, 1980). Chemical lesions of serotonergic neurones do not reduce the vesicle uptake of serotonin whereas lesions of catecholaminergic neurones do reduce this uptake (Slotkin *et al.,* 1978). It appears therefore that the vesicle uptake is much less specific than is the membrane transport systems and that serotonin uptake in catecholaminergic vesicles is predominant when measured under *in vitro* conditions.

Studies with storage vesicles from isolated platelets have revealed that the mechanism by which serotonin is taken up into these organelles is very similar to that by which catecholamines are taken up into chromaffin granules and synaptic vesicles. Thus, the vesicle membrane contains a carrier for the transport of serotonin into the vesicles (Wilkins *et al.,* 1978; Rudnick *et al.,* 1980). This transport of serotonin is saturable and is blocked by reserpine and related compounds (Pletscher *et al.,* 1971; Rudnick *et al.,* 1980). Kinetic experiment with dense granule from porcine platelets revealed two uptake mechanisms, one saturable component with Km = 3.3 μM and V = 0.79 nmol mg^{-1} protein mg^{-1} and a saturable uptake which appeared to be diffusion-limited (Wilkins *et al.,* 1978). The uptake of serotonin in isolated vesicles requires ATP and Mg^{2+} (Rudnick *et al.,* 1980). A Mg^{2+} dependent APTase in the vesicle membrane generates a H$^+$ gradient (in $\geqslant$ out) across the membrane which produces the main driving force for the serotonin transport (Rudnick *et al.,* 1980). The electrochemical gradient across the membrane may also contribute as an energy source. ATP and Mg^{2+} have a dual role in the uptake and binding of serotonin in the storage vesicles, since they form an osmotically inactive aggregate with serotonin in the vesicles (Pletscher *et al.,* 1971).

The mechanisms for the uptake of binding of biogenic monoamines in storage vesicles from various sources (adrenal medulla, nerve terminals and platelets) are remarkably similar. It seems therefore likely that the uptake and storage of serotonin in serotonergic neuronal vesicles occurs with essentially the same mechanism as described for the platelets.

ONTOGENY OF SEROTONIN UPTAKE

The transport and storage of [^{14}C]serotonin in synaptosomes of rat brain tissue mature simultaneously with the endogenous serotonin content (Tissari, 1975). In the 21 day foetus the active transport of [^{14}C]serotonin in synaptosomes from rat brain stem was only 12 per cent of that in the adult rats. The post-natal development of the serotonin accumulation occurred successively during the first 5 weeks, at which time the adult capacity to accumulate serotonin was reached (Tissari, 1975).

HORMONAL EFFECTS ON THE SEROTONIN UPTAKE

The influence of hormones on serotonin uptake has been examined in a few studies. No clear pattern of results seems, however, to be forthcoming. Ovariectomized rats treated with oestradiol on two consecutive days and killed on the third day had significantly higher serotonin uptake in slices of anterior and posterior hypothalamus than had vehicle treated rats (Cardinali and Gomez, 1977). Contrastingly, Endersby and Wilson (1973) failed to find any effect of 3-day treatment of ovariectomized rats with oestradiol, progesterone and oestradiol + progesterone on the serotonin uptake in hypothalamic slices. Two hours after an acute injection of oestradiol to ovariectomized rats Wirz-Justice *et al.* (1974) did not find any effect on the serotonin uptake in slices of various brain regions whereas progesterone significantly increased the serotonin uptake in the pre-optic + septum regions. Meyer and Quay (1976) observed a pronounced increase of the serotonin uptake in slices of suprachiasmatic region from pro-estreous female rats at the late afternoon, just prior to the peak of plasma luteinizing hormone. Thus, the exact effect of the sexual hormones upon serotonin uptake remains largely undecided. Nor do the glucocorticoids seem implicated. Thus, Lieberman *et al.* (1980) did not find any direct effect of hydrocortisone (10^{-9} M–10^{-5} M) on the serotonin accumulation in rat hypothalamic slices.

The influence of neonatal hypo- and hyperthyroidism in rats upon the accumulation of serotonin in synaptosomes from basal ganglia, brain stem and hypothalamus was examined by Schwark and Keesy (1978). They found an elevated serotonin accumulation in cretinous rats of 30 days age but no change in the hyperthyroid rats. The treatment of cretinous rats with L-triiodothyronine from the birth counteracted the increase in serotonin accumulation.

CIRCADIAN AND SEASONAL RHYTHMS

Serotonin uptake is to some extent dependent upon the diurnal rhythm. Wirz-Justice (1974) found a significantly higher accumulation of serotonin in hippocampal slices from male rats at 8 p.m. than at 8 a.m. The serotonin uptake in slices and homogenates of hypothalamus and suprachiasmatic nuclear region of male rats showed a daily rhythm with a peak at the end of light period (Meyer and Quay, 1976). In the hypothalamus additional peaks were observed during the dark period.

Oxenkrug *et al.* (1978) observed a circadian rhythm affecting the uptake of serotonin in human blood platelets with a significantly higher uptake (150 per cent) at 9 p.m. than at 9 a.m. (100 per cent). Interestingly, these authors found that patients with endogenous depression did not have this circadian rhythm.

In ovariectomized rats a seasonal variation of serotonin uptake in striatal slices has been indicated with the highest activity in October (Wirz-Justice, 1974).

STRESS

In rats exposed to ether stress, the serotonin uptake in slices of hypothalamus and mesencephalon was significantly reduced 20 min after stress but had normalized 60 min after stress (Vermes and Telegdy, 1977).

THIAMINE DEFICIENCY

Thiamine deficiency in rats decreased the serotonin uptake in synaptosomes from the cerebellum by 50 per cent and also tended to decrease the uptake in synaptosomes from hypothalamus (Plaitakis *et al.*, 1978). The uptakes of GABA, glutamic acid, choline and noradrenaline were not affected.

TOPOGRAPHY OF THE SEROTONIN CARRIER

The possible role of gangliosides and glucoproteins containing sialic acid in the terminal position in the binding and transport of serotonin was examined in synaptosomes from rat cerebral cortex (Dette and Wesemann, 1978). Treatment of the synaptosomes with neuraminidase caused a 50 per cent reduction of the maximal uptake velocity without changing the affinity (apparent Km). The results obtained indicated that sialic acid may take part in the serotonin transport. This is in accordance with the observation that the enzymatically catalysed incorporation of *N*-acetylneuraminic acid into human platelets accelerated the uptake of serotonin in the platelets (Szabados *et al.*, 1975).

Concanavalin A and lectin isolated from *Lens culinaris* did not reduce the accumulation of serotonin by synaptosomes from rat cerebral cortex (Wang *et al.*, 1975). Trypsin treatment of synaptosomes from the cerebral cortex did not change the accumulation of serotonin (Hitzemann *et al.*, 1975; Wang *et al.*, 1975).

[^{3}H] IMIPRAMINE BINDING

The high affinity binding of [^{3}H] imipramine on the human platelet membrane was first demonstrated by Rudnick and co-workers (Talvenheimo *et al.*, 1979). They observed that plasma membrane vesicles require external Na^+ in order to bind imipramine and that Cl^- enhanced the binding. Serotonin and fluoxetine reversed the imipramine binding. The apparent K_D value was in this preparation 23 nM, which was in good agreement with the K_i value 17 nM for the competitive inhibition of the initial serotonin uptake in the same preparation. The authors concluded that imipramine is bound to the serotonin carrier of the platelet membrane but is not transported by the carrier. The high affinity binding of imipramine to human platelet membranes has been confirmed by others (Briley *et al.*, 1979; Paul *et al.*, 1980; Langer *et al.*, 1980a). The apparent K_D values reported in these studies was about 1.5 nM at 0 °C but decreased to 7 nM at 37 ° C (Paul *et al.*, 1980). Imipramine binding is antagonized by antidepressant agents and other compounds which inhibit serotonin uptake (Paul *et al.*, 1980, Langer *et al.*, 1980a). It is interesting to note that the maximal binding of imipramine to platelets was significantly reduced in depressed drug-free patients when compared to healthy controls (Briley *et al.*, 1980). The affinity was unchanged.

The high affinity binding of [^{3}H] imipramine has also been observed in membranes from rat and human brains (Raisman *et al.*, 1979a,b 1980; Rehavi *et al.*,

1980). This binding is very similar to that in platelets. The K_D value (1.7 nM) for this binding in human brain is almost identical with that in human platelets but three to five times lower than that in rat brain (Rehavi *et al.*, 1980). This species difference agrees with that previously observed for the inhibitory activity of imipramine upon the serotonin uptake (Wielotz *et al.*, 1976). The [^{3}H] imipramine binding is stereoselectively antagonized by Z-forms of zimelidine and norzimelidine (Langer *et al.*, 1980b), which are also considerably more potent than their E-analogues as serotonin uptake inhibitors (Ross and Renyi, 1977). These and other observations clearly indicate that the [^{3}H] imipramine binding in platelets and brain membranes occurs at the site of the serotonin uptake carrier. The high affinity binding of [^{3}H] imipramine may therefore become a valuable experimental method in studies of the serotonin uptake mechanism.

FUTURE TRENDS

The ultimate goal concerning the characterization of the serotonin transport systems is their description in molecular terms, e.g. identification of the biochemical components. How does the carrier bind and transport serotonin across the membrane, the role of Na^+, K^+ and Cl^- in this process, and how is the transport coupled to the energy sources utilized?

A first step toward this characterization is to solublize the carrier protein and to reconstitute it with lipids in an artification system. Rudnick and Nelson (1978b) were not able to solubilize the serotonin carrier in platelets membrane by treatment with bile acid cholate but the membranes were disrupted and the transport capacity was inactivated. The carrier appeared to be associated with other membrane components. Removal of cholate did not restore the transport capacity but did so in presence of soybean phopholipids. The characteristics of the serotonin transport in these proteioliposomes were the same as in the intact platelets. These findings will certainly stimulate further attempts to solubilize the platelet membrane serotonin carrier. The possible use of the [^{3}H] imipramine binding technique to identify the serotonin carrier may become helpful in the procedures of solibilization and purification of the carrier protein.

The development of techniques to prepare synaptosome membrane vesicles without the intracellular components should be of great advantage in studies of the neuronal serotonin transport.

The [^{3}H] imipramine-binding technique will give valuable information on the serotonin carrier, particularly in the human brain, since post-mortem tissue can be used in this assay technique. The question of whether psychiatric disorders may directly or indirectly be coupled with changes in the serotonin uptake system as indicated for endogenous depression could be answered by such studies.

There remains much to learn about the structural requirement for transport with the serotonin carrier. For instance, are such compounds as non-hydroxylated tryptamines and para-substituted amphetamines transported by the serotonin carrier? Is it possible to develop selectively acting drugs (e.g. MAO inhibitors) by utilizing the membrane carrier to accumulate compounds into serotonergic neurones?

The clinical application of the serotonin uptake system is already evidenced by the antidepressive action of selective serotonin uptake inhibitors. These compounds are also valuable tools for studies of the physiological role of serotonin systems in the brain.

REFERENCES

Aghajanian, G. K. and Bloom. F. E. (1976). Localization of tritiated serotonin in rat brain by electron-microscopic autoradiography, *J. Pharmacol. exp. Ther.* **156**, 23–30.

Airkasinen, M. M., Svensk, H., Tuomisto, J., and Komulainen, H. (1980). Tetrahydro-β-carbolines and corresponding tryptamines: *in vitro* inhibition of serotonin and dopamine uptake by human blood platelets, *Acta pharmacol. toxicol.* **46**, 308–313.

Azzaro, A. J. and Smith, D. J. (1977). The inhibitory action of ketamine HCl on [^{3}H] 5-hydroxy-tryptamine accumulation by rat brain synaptosomal-rich fractions: comparison with [^{3}H] catecholamine and [^{3}H] aminobutyric acid uptake, *Neuropharmacology*, **16**, 349–356.

Bakhanashvili, T. A., Maisov, N. I., and Zharikova, A. D. (1979). Inhibition of serotonin uptake in synaptosomes and glia cells by some pharmacological substances, *Biull. Eksp. Biol. Med.* **88**, 564–566.

Baker, G. B., Martin, I. L., and Mitchell, P. R. (1977). The effects of some indolealkylamines on the uptake and release of 5-hydroxytryptamine in rat striatum, *Br. J. Pharmacol.* **61**, 151P–152P.

Banerjee, S. P., Snyder, S. H., and Mechoulam, R. (1975). Cannabinoids: Influence on neurotransmitter uptake in rat brain synaptosomes, *J. Pharmacol. Exp. Ther.* **194**, 74–81.

Baumann, P. A. and Maitre, L. (1979). Neurobiochemical aspects of maprotiline (Ludiomil[R]) action, *J. Int. Med. Res.* **7**, 391–400.

Baumgarten, H. G., Evetts, K. D., Holman, R. B., Iversen, L. L., Vogt, M., and Wilson, G. (1972). Effects of 5,6-dihydroxytryptamine on monoaminergic neurones in the central nervous system of the rat, *J. Neurochem.* **19**, 1587–1597.

Baumgarten, H. G., Klemm, H. P., Lachenmayer, L., Björklund, A., Lovenberg, W., and Schlossberger, H. G. (1978). Mode and mechanism of action of neurotoxic indoleamines: a review and a progress report, *Ann. NY. Acad. Sci.* **305**, 3–24.

Belin, M.-F., Kouyoumidjian, J.-C., Bardakdjian, J., Duhault, J., and Gonnard, P. (1976). Effects of fenfluramine on accumulation of 5-hydroxytryptamine and other neurotransmitters into synaptosomes of rat brain, *Neuropharmacology*, **15**, 613–617.

Berger, B. and Glowinski, J. (1978). Dopamine uptake in serotoninergic terminas *in vitro*. A valuable tool for the histochemical differentiation of catecholaminergic and serotoninergic terminals in rat cerebral structures, *Brain Res.* **147**, 29–45.

Bertilsson, L., Mellström, B., and Sjöqvist, F. (1979). Pronounced inhibition of noradrenaline uptake by 10-hydroxymetabolites of nortriptyline, *Life Sci.* **25**, 1285–1292.

Björklund, A., Horn, A. S., Baumgarten, H. G., Nobin, A., and Schlossberger, H. G. (1975). Neurotoxicity of hydroxylated tryptamines: structure-activity relationships. II. *In vitro* studies on monoamine uptake inhibition and uptake impairment, *Acta physiol. scand.* suppl. 429, pp. 30–60.

Blackburn, K. J., French, P. C., and Merrills, R. J. (1967). 5-Hydroxytryptamine uptake by rat brain *in vitro*, *Life Sci.* **6**, 1653–1663.

Blakeley, A. G. H. and Nicol. C. J. M. (1978). Accumulation of amines by rabbit erythrocytes *in vitro*, *J. Physiol.* (Lond.), **277**, 77–90.

Bogdanski, D. F., Tissari, A., and Brodie, B. B. (1968). Role of sodium, potassium, ouabain and reserpine in uptake, storage and metabolism of biogenic amines in synaptosomes, *Life Sci.* **7**, 419–428.

Bogdanski, D. F., Tissari, A. H., and Brodie, B. B. (1970a). Mechanism of transport and storage of biogenic amines. III. Effects of sodium and potassium on kinetics of 5-hydroxytryptamine and norepinephrine transport by rabbit synaptosomes, *Biochim. biophys. Acta,* **219**, 189–199.

Bogdanski, D. F., Blaszkowiski, T. P., and Tissari, A. H. (1970b). Mechanisms of biogenic amine transport and storage. IV. Relationship between K^+ and the Na^+ requirement for transport and storage of 5-hydroxytryptamine and norepinephrine in synaptosomes. *Biochim. biophys. Acta,* **211**, 521–532.

Born, F. V. R. and Gillison, R. E. (1959). Studies on the uptake of 5-HT by blood platelets, *J. Pharmacol. (Lond),* **146**, 472–491.

Born, G. V. R., Juengjaroen, K., and Michal, E. (1972). Relative activities on and uptake by human blood platelets of 5-hydroxytryptamine and several analogues, *Br. J. Pharmacol.* **44**, 117–139.

Briley, M. S., Raisman, R., and Langer, S. Z. (1979). Human platelets possess high-affinity binding sites for [^{3}H]imipramine, *Eur. J. Pharmacol.* **58**, 347–348.

Briley, M. S., Langer, S. Z., Raisman, R., Sechter, D., and Zarifian, E. (1980). Tritiated imipramine binding sites are decreased in platelets of untreated depressed patients, *Science,* **209**, 303–305.

Bruinvels, J. (1971). Evidence for inhibition of the reuptake of 5-hydroxytryptamine and noradrenaline by tetrahydronaphthylamine in rat brain, *Br. J. Pharmacol.* **42**, 281–286.

Buus Lassen, J., Squires, R. F., Christensen, J. A., and Molander, L. (1975). Neurochemical and pharmacological studies of a new 5-HT-uptake inhibitor, FG 4963, with potential antidepressant properties, *Psychopharmacologia,* **42**, 21–26.

Cardinali, D. P. (1977). Effects of pentoxifylline and theophylline on neurotransmitter uptake and release by synaptosome-rich homogenates of the rat hypothalamus, *Neuropharmacology,* **16**, 785–790.

Cardinali, D. P. and Gomez, E. (1977). Changes in hypothalamic noradrenaline, dopamine and serotonin uptake after oestradiol administration to rats, *J. Endocrinol.* **73**, 181–182.

Carlsson, A. (1970). Structural specificity for inhibition of [^{14}C] 5-hydroxytryptamine uptake by cerebral slices, *J. Pharm. Pharmacol.* **22**, 729–732.

Carlsson, A., Corrodi, H., Fuxe, K., and Hökfelt, T. (1969). Effect of antidepressant drugs on the depletion of intraneuronal brain 5-hydroxytryptamine stores caused by 4-methyl-a-ethyl-meta-tyramine, *Eur. J. Pharmacol.* **5**, 357–366.

Carnmalm, G., Jacupvic, E., Johansson, L., dePaulis, T., Rämsby, S., Stjernström, N. E., Renyi, A. L., Ross, S. B., and Ogren, S. O. (1974). Antidepressant agents. 1. Chemistry and pharmacology of amino-substituted spiro-(5H-dibenzo[a,d]cycloheptene-5,1-cycloalkanes), *J. Med. Chem.* **17**, 65–72.

Carnmalm, B., dePaulis, T., Jacupovic, E., Johansson, L., Lindberg, U. L., Ulff, B., Stjernström, N. E., Renyi, A. L., Ross, S. B., and Ögren, S. O. (1975). Antidepressant agents. IV. Phenylcycloalkylamines, *Acta Pharm. Suecica,* **12**, 149–172.

Carnmalm, B., dePaulis, T., Jacupovic, E., Renyi, A. L., Ross, S. B., Ögren, S. O., and Stjernström, N. E. (1977). Antidepressant agents. VIII. Chemistry and pharmacology of some aminomethyldiphenylcycloalkanes, *Acta Pharm. Suecica,* **14,** 377–390.

Chung, Hwang, E. and Van Woert, M. H. (1980). Comparative effects of substituted phenylethylamines on brain serotonergic mechanisms. *J. Pharmacol. Exp. Ther.* **213,** 254–260.

Ciafolo, L. (1974). Methadone inhibition of ^{3}H-5-hydroxytryptamine uptake by synaptosomes, *J. Pharmacol. Exp. Ther.* **189,** 82–89.

Claassen, V., Davies, J. E., Hertting, G., and Placheta, P. (1977). Fluvoxamine, a specific 5-hydroxytryptamine uptake inhibitor, *Br. J. Pharmacol.* **60,** 505–516.

Clineschmidt, B. V., Totaro, J. A., Pflueger, A. B., and McGuffin, J. C. (1978a). Inhibition of the serotoninergic uptake system by MK-212 (6-chloro-2-(1-piperazinyl)-pyrazine), *Pharmacol. Res. Commun.* **10,** 219–228.

Clineschmidt, B. V., Zacchei, A. G., Totaro, J. A., Pflueger, A. B., Guffin, J. C., and Wishousky, T. I. (1978b). Fenfluramine and brain serotonin, *Ann. N.Y. Acad. Sci.* **305,** 222–242.

Coppen, A. (1967). The biochemistry of affective disorders, *Br. J. Psychiat.* **113,** 1237–1264.

Coppen, A., Swade, C., and Wood, K. (1978). Platelet 5-hydroxytryptamine accumulation in depressive illness, *Clin. Chem. Acta,* **87,** 165–168.

Dette, B. A. and Wesemann, W. (1978). On the significance of sialic acid in high affinity 5-hydroxytryptamine uptake by synaptosomes, *Hoppe-Seyler's Z. Physiol. Chem.* **359,** 399–406.

Ducis, I. and Di Stefano, V. (1980). Characterization of serotonin uptake in isolated pinealocyte suspensions, *Molec. Pharmacol.,* **18,** 447–454.

Ehinger, B. and Florén, I. (1978). Quantitation of the uptake of indoleamines and dopamine in the rabbit retina, *Exp. Eye Res.* **26,** 1–11.

Endersby, C. A. and Wilson, C. (1973). The effect of ovarian steroids on the uptake of ^{3}H-noradrenaline, ^{3}H-dopamine and ^{3}H-5-hydroxytryptamine by hypothalamic tissue *in vitro, Br. J. Pharmacol.* **47,** 647P–648P.

Florén, I. (1979). Arguments against 5-hydroxytryptamine as neurotransmitter in the rabbit retina, *J. Neurol. Transmission,* **46,** 1–15.

Fuller, R. W., Perry, K. W., and Molloy, B. B. (1975). Reversible and irreversible phases of serotonin depletion by 4-chloroamphetamine, *Eur. J. Pharmacol.* **33,** 119–124.

Fuller, R. W., Snoddy, H. D., Perry, K. W., Bymaster, F. P., and Wong, D. T. (1978). Studies on 4-(p-bromophenyl)-bicyclo (2,2,2)octan-1-amine as an inhibitor of uptake into serotonin neurones. *Neuropharmacology,* **17,** 815–818.

Fuxe, K. Hökfelt, T., Ritzén, M., and Ungerstedt, U. (1968). Studies on uptake of intraventricularly administered tritiated noradrenaline and 5-hydroxytryptamine with combined fluorescence histochemical and autoradiographic techniques, *Histochemie,* **16,** 186–194.

Gershon, M. D. and Altman, R. F. (1971). An analysis of the uptake of 5-hydroxytryptamine by the myenteric plexus of the small intestine of the guinea-pig, *J. Pharmacol. Exp. Ther.* **179,** 29–41.

Gershon, M. D. and Jonakait, G. M. (1979). Uptake and release of 5-hydroxytryptamine by enteric 5-hydroxytryptaminergic neurones: effects of fluoxetine (Lilly 110140) and chlorimipramine, *Br. J. Pharmacol.* **66,** 7–9.

Gershon, M. D., Robinson, R. G., and Ross, L. L. (1976). Serotonin accumulation in the guinea-pig myenteric plexus: ion dependence, structure-activity relationship and the effect of drugs, *J. Pharmacol. Exp. Ther.* **198,** 548–561.

Gershon, M. D., Sherman, D. L., and Dreyfus, C. F. (1980). Effects of indolic neurotoxins on enteric serotonergic neurons, *J. Comp. Neurol.* **190,** 581–596.

Glennon, R. A., Martin, B., Johnson, K. M., and End, D. (1978). 7,*N*,*N*-trimethyltryptamine: a selective inhibitor of synaptosomal serotonin uptake, *Res. Commun. Chem. Pathol. Pharmacol.* **19**, 161–164.

Grunewald, G. L., Reitz, T. J., Ruth, J. A., Vollmer, S., Eiden, L. E., and Rutledge, C. O. (1979). Inhibition of neuronal uptake of ^{3}H-biogenic amines into rat cerebral cortex by partially and fully saturated derivatives of imipramine and desipramine, *Biochem. Pharmacol.* **28**, 417–421.

Gueremy, C., Audian, F., Champseix, A., Uzan, A., LeFur, G., and Rataud, J. (1980). 3-(4-Piperidinylalkyl)indoles, selective inhibitors of neuronal 5-hydroxytryptamine uptake, *J. Med. Chem.*, **23**, 1306–1310.

Hallström, C. O. S., Pare, C. M. B., Rees, W. L., Trendrand, A., and Turner, P. (1976). Platelet uptake of 5-hydroxytryptamine and dopamine in depression. *Postgrad. Med. J.* (Suppl.) **52**, 40–44.

Heikkila, R. E., Cabbat, F. S., and Mytilineou, C. (1977). Studies on the capacity of mazindol and dita to act as uptake inhibitors or releasing agents for ^{3}H-biogenic amines in rat brain tissue slices, *Eur. J. Pharmacol.* **45**, 329–333.

Heikkila, R. E., Goldinger, S. S., and Orlansky, H. (1976). The effect of various phenothiazines and tricyclic antidepressants on the accumulation and release of [3H]norepinephrine and [3H]5-hydroxytryptamine in slices of rat occipital cortex, *Res. Commun. Chem. Pathol. Pharmacol.* **13**, 237–250.

Hertting, G. and Axelrod, J. (1961). Fate of tritiated noradrenaline at the sympathetic nerve-endings, *Nature (Lond.)*, **192**, 172–173.

Hitzemann, B. A., Hitzemann, R. J., and Loh, H. H. (1975). On the specificity of trypsin (EC 3.4.4.4.) of nerve endings particles to inhibit norepinephrine transport, *J. Neurochem.* **24**, 323–330.

Horn, A. S. (1973). Structure activity relations for the inhibition of 5-HT uptake into rat hypothalamic homogenates by serotonin and tryptamine analogues, *J. Neurochem.* **21**, 883–888.

Horn, A. S. (1976). The interaction of tricyclic antidepressants with the biogenic amine uptake systems in the central neurones system, *Postgraduate Med. J. (Suppl. 3)*, **52**, 25–30.

Horn, A. S. (1977). The binding of inhibitors of serotonin uptake to biological receptors, *Post-grad. med. J.* (Suppl. 4), **53**, 9–13.

Horn, A. S. (1979). Characteristics of dopamine uptake, in *The neurobiology of dopamine*, (Eds A. S. Horn, J. Korf, and B. H. C. Westerink) pp. 217–235, Academic Press, New York.

Horn, A. S. and Trace, R. C. A. M. (1974). Structure-activity relations for the inhibition of 5-hydroxytryptamine uptake by tricyclic antidepressants into synaptosomes from serotoninergic neurones in rat brain homogenates, *Br. J. Pharmacol.* **51**, 399–403.

Horn, A. S., Baumgarten, H. G., and Schlossberger, H. G. (1973). Inhibition of the uptake of 5-hydroxtryptamine, noradrenaline and dopamine into rat brain homogenates by various hydroxylated tryptamines, *J. Neurochem.* **21**, 233–236.

Hyttel, J. (1977). Neurochemical characterization of a new potent and selective serotonin uptake inhibitor, Lu 10-171, *Psychopharmacol.* **51**, 225–233.

Iversen, L. L. (1970). Neuronal uptake processes for amines and amino acids, in *Advances in Biochemical Psychopharmacology*, Vol. 2 *Biochemistry of simple neuronal model* (Eds E. Costa and E. Giacobini), pp. 109–132, Raven Press, New York.

Javaid, J. I., Perel, J. M., and Davis, J. M. (1979). Inhibition of biogenic amines uptake by imipramine, desipramine, 2-OH-imipramine and 2-OH-desipramine in rat brain, *Life Sci.* **24**, 21–28.

Kannengiesser, M. H., Hunt, P., and Raynaud, J. P. (1976). An *in vitro* model for the study of psychotropic drugs and as a criterion of antidepressant activity, *Biochem. Pharmacol.* **22**, 73–84.

Kannengiesser, M.-H., Hunt, P., and Raynaud, J.-P. Comparative action of fenflur-mine on the uptake and release of serotonin and dopamine, *Eur. J. Pharmacol.* **35**, 35–43.

Keller, H. H., Burkard, W. P., and DaPrada, M. (1980). Dopamine receptor blockade in rat brain after acute and subchronic treatment with tricyclic antidepressants, in *Long-term Effects of Neuroleptics, (Advances in Biochemical Psychopharma-cology,* Vol. 24, Eds. F. Cattabeni, G. Racagni, P. F. Spano, and E. Costa), pp. 175–179, Raven Press, New York.

Koe, B. K. (1976). Molecular geometry of inhibitors of the uptake of catechola-mines and serotonin in synaptosomal preparations of rat brain, *J. Pharmacol. Exp. Ther.* **199**, 649–661.

Köhler, C., Ross, S. B., Srebo, B., and Ögren, S.-O. (1978). Long-term biochemical and behavioural effects of *p*-chloroamphetamine in the rat, *Ann. N.Y. Acad. Sci.* **305**, 645–663.

Koide, T. and Uyemura, K. (1980). A comparison of the inhibitory effects of new non-tricyclic amine uptake inhibitors on the uptake of norepinephrine and 5-hydroxytryptamine into synaptosomes of the rat brain, *Neuropharmacol.* **19**, 349–354.

Komulainen, H., Tuomisto, J., Airaksinen, M. M., Kari, I., Peura, P., and Pollari, L. (1980). Tetrahydro-β-carbolines and corresponding tryptamines: *in vitro* inhibi-tion of serotonin, dopamine and noradrenaline uptake in rat brain synapto-somes, *Acta pharmacol. toxicol.* **46**, 299–307.

Kuhar, M. J. and Aghajanian, G. K. (1973). Selective accumulation of ^{3}H-serotonin by nerve terminals of raphe neurones: and autoradiographic study, *Nature (Lond.),* **241**, 187–189.

Kuhar, M. J. and Zarbin, M. A. (1978). Synaptosomal transport: a chloride depen-dence for choline, GABA, glycine and several other compounds, *J. Neurochem.* **31**, 251–256.

Kuhar, M. J., Roth, R. H., and Aghajanian, G. K. (1972a). Synaptosomes from fore-brains of rats with midbrain raphe lesions: selective reduction of serotonin uptake, *J. Pharmacol. Exp. Ther.* **181**, 36–45.

Kuhar, M. J., Aghajanian, G. K., and Roth, R. H. (1972b). Tryptophan hydroxylase activity and synaptosomal uptake of serotonin in discrete brain regions after midbrain raphe lesions: Correlations with serotonin levels and histochemical fluorescence, *Brain Res.* **44**, 165–176.

Langer, S. Z., Briley, M. S., Raisman, R., Henry, J.-F., and Morselli, P. L. (1980a). Specific ^{3}H-imipramine binding in human platelets. Influence of age and sex, *Archiv. Pharmacol.* **313**, 189–194.

Langer, S. Z., Raisman, R., and Briley, M. S. (1980b). Stereoselective inhibition of ^{3}H-imipramine binding by antidepressant drugs and their derivatives, *Eur. J. Pharmacol.* **64**, 89–90.

Lapin, I. P. and Oxenkrug, G. F. (1969). Intensification of the central serotonergic processes as a possible determinant of the thymoleptic effects. *Lancet,* **1**, 132–136.

Lessin, A. W., Long, R. F., and Parkes, M. W. (1967). The central stimulant proper-ties of some substituted indolealkylamines and β-carbolines and their activities as inhibitors of monoamine oxidase and the uptake of 5-hydroxytryptamine, *Br. J. Pharmacol. Chemother.* **29**, 70–79.

LeFur, G. and Uzan, A. (1977). Effect of 4-(3-indolealkyl)piperidine derivatives on uptake and release of noradrenaline, dopamine and 5-hydroxytryptamine in rat brain synaptosomes, rat heart and human blood platelets, *Biochem. Pharmacol.* **26**, 497-503.

Lieberman, K. W., Stokes, P. E., Fanelli, C. J., and Klevan, T. (1980). Reuptake of biogenic amines by brain slices: effect of hydrocortisone. *Psychopharmacology,* **70**, 59-61.

Lindberg, U. H., Thorberg, S.-O., Bengtsson, S., Renyi, A. L., Ross, S. B., and Ögren, S.-O. (1978a). Inhibitors of neuronal monoamine uptake. 2. Selective inhibition of 5-hydroxytryptamine uptake by α-amino acid esters of phenethyl alcohols, *J. Med. Chem.* **21**, 448-456.

Lindberg, U. H., Ross, S. B., Thorberg, S.-O., Ögren, S.-O., Malmros, G., and Wägner, A. (1978b). A conformational study of R-alaproclate, a new selective inhibitor of neuronal 5-hydroxytryptamine uptake, *Tetrahedron Lett.* **20**, 1779-1782.

Lindström, P., Sehlin, J., and Täljedal, I.-B. (1980). Characteristics of 5-hydroxytryptamine transport in pancreatic islets, *Br. J. Pharmacol.* **68**, 773-778.

Lingjaerde, O. (1971). Uptake of serotonin in blood platelets *in vitro.* I. The effect of chloride, *Acta physiol. scand.* **81**, 75-83.

Maxwell, R. A. and White, H. L. (1978). Tricyclic and monoamine oxidase inhibitor antidepressants: structure-activity relationships, *in Handbook of psychopharmacology,* Vol. 14 (Eds. L. L. Iversen, S. D. Iversen, and S. H. Snyder), pp. 83-155, Plenum, New York.

Meek, J. L., Fuxe, K., and Carlsson, A. (1971). Blockade of *p*-chloromethamphetamine induced 5-hydroxytryptamine depletion by chlorimipramine, chlorpheniramine and meperidine, *Biochem. Pharmacol.* **20**, 707-709.

Mennini, T., Pataccini, R., and Samanin, R. (1978). Effects of narcotic analgesics on the uptake and release of 5-hydroxytryptamine in rat synaptosomal preparations, *Br. J. Pharmacol.* **64**, 75-82.

Meyer, D. C. and Quay, W. B. (1976). Hypothalamic and suprachiasmatic uptake of serotonin *in vitro:* twenty-four hour changes in male and proestrous female rats, *Endocrinol.* **98**, 1160-1165.

Mireylees, S. E., Goolet, I. and Sugrue, M. F. (1978). Effects of Org. 6582 on monoamine uptake *in vitro, Biochem. Pharmacol.* **27**, 1023-1027.

Moffat, J. A. and Jhamandas, K. (1976). Effects of acute and chronic methadone treatment on the uptake of ^{3}H-5-hydroxytryptamine in rat hypothalamus slices, *Eur. J. Pharmacol.* **36**, 289-297.

Mulder, A. H., Berg, W. B., and van der Stoof, J. C. (1975). Calcium-dependent release of radiolabelled catecholamines and serotonin from rat brain synaptosomes in a superfusion system, *Brain Res.* **99**, 419-424.

Nelson, P. J. and Rudnick, G. (1979). Coupling between platelets 5-hydroxytryptamine and potassium transport, *J. Biol. Chem.* **254**, 10084-10089.

Osborne, N. N. (1980). *In vitro* experiments of the metabolism, uptake and release of 5-hydroxytryptamine in bovine retina, *Brain Res.* **184**, 283-297.

Osborne, N. N., Hiripi, L., and Neuhoff, V. (1975). The *in vitro* uptake of biogenic amines by snail (*Helix pomatia*) nervous system, *Biochem. Pharmacol.* **24**, 2141-2148.

Oxenkrug, G. F., Prakhje, I., and Mikhalenko, I. N. (1978). Disturbed circadian rhythm of 5-HT uptake by blood platelets in depressive psychosis, *Activ. nerv. sup. (Praha),* **20**, 66-67.

Paton, D. M. (1974). Mechanism of inhibition by cocaine of action of indirectly sympathomimetic amines, *Am. Heart. J.* **88**, 128-129.

Paton, D. M. (ed.) (1976). The mechanism of neuronal and extraneuronal transport of catecholamines. Raven Press, New York.

Paul, S. M., Rehavi, M., Skolnick, P., and Goodwin, F. K. (1980). Demonstration of specific 'high affinity' binding sites for [^{3}H]imipramine on human platelets, *Life Sci.* **26**, 953–959.

Petersen, E. N., Olsson, S.-O., and Squires, R. (1977). Effects of 5-HT uptake inhibitors on the pressor response to 5-HT in the pithed rat. The significance of the 5-HT blocking property, *Eur. J. Pharmacol.* **43**, 209–215.

Plaitakis, A., Nicklas, W. J., and Berl, S. (1978). Thiamine deficiency: selective impairment of the cerebellar serotonergic system, *Neurology*, **28**, 691–698.

Pletscher, A., Shore, P. A., and Brodie, B. B. (1956). Serotonin as a mediator of reserpine action in brain, *J. Pharmacol. Exp. Ther.* **116**, 84–89.

Pletscher, A., DaPrada, M., Berneis, K. H., and Tranzer, J. P. (1971). New aspects on the storage of 5-hydroxytryptamine in blood platelets, *Experientia* **27**, 993–1120.

Raisman, R., Briley, M. S., and Langer, S. Z. (1979a). Specific tricyclic antidepressant binding sites in rat brain, *Nature (Lond.)*, **281**, 148–150.

Raisman, R., Briley, M. S., and Langer, S. Z. (1979b). High affinity ^{3}H-imipramine binding in rat cerebral cortex, *Eur. J. Pharmacol.* **54**, 307–308.

Raisman, R., Briley, M. S., and Langer, S. Z. (1980). Specific tricyclic antidepressant binding sites in rat brain characterized by high affinity ^{3}H-imipramine binding, *Eur. J. Pharmacol.* **61**, 373–380.

Raiteri, M., Del Carmine, R., Bertollini, A., and Levi, G. (1977). Effect of sympathomimetic amines on the synaptosomal transport of noradrenaline, dopamine and 5-hydroxytryptamine, *Eur. J. Pharmacol.* **41**, 133–143.

Rehavi, M., Paul, S. M., Skolnick, P. and Goodwin, F. K. (1980). Demonstration of specific high affinity binding sites for [^{3}H]imipramine in human brain, *Life Sci.* **26**, 2273–2279.

Riblet, L. A., Gatewood, C. F., and Mayol, R. F. (1979). Comparative effects of trazadone and tricyclic antidepressants on uptake of selected neurotransmitters by isolated rat brain synaptosomes, *Psychopharmacology*, **63**, 99–101.

Rommelspacher, H., Strauss, S. M., and Rehse, K. (1978). β-Carbolines: a tool for investigating structure-activity relationships of the high affinity uptake of serotonin, noradrenaline, dopamine, GABA and choline into a synaptosome-rich fraction of various regions from rat brain, *J. Neurochem.* **30**, 1573–1578.

Rosenthal, H. E. (1967). A graphic method for the determination and presentation of binding parameters in a complex system, *Anal. Biochem.* **20**, 525–532.

Ross, S. B. (1976). Antagonism of the acute and long-term biochemical effects of 4-chloroamphetamine on the 5-HT neurones in the rat brain by inhibition of the 5-hydroxytramine uptake, *Acta pharmacol. toxicol.* **39**, 456–476.

Ross, S. B. (1979). Interactions between reserpine and various compounds on the accumulation of [^{14}C]5-hydroxytryptamine and [^{3}H]noradrenaline in homogenates from rat hypothalamus, *Biochem. Pharmacol.* **28**, 1085–1088.

Ross, S. B. and Ask, A.-L. (1980). Structural requirements for uptake into serotoninergic neurones, *Acta pharmacol. toxicol.* **46**, 270–277.

Ross, S. B. and Kelder, D. (1977). Efflux of 5-hydroxytryptamine from synaptosomes of rat cerebral cortex. *Acta physiol. scand.* **99**, 27–36.

Ross, S. B. and Renyi, A. L. (1967). Accumulation of tritiated 5-hydroxytryptamine in brain slices, *Life Sci.* **6**, 1407–1415.

Ross, S. B. and Renyi, A. L. (1969). Inhibition of the uptake of tritiated 5-hydroxytryptamine in brain tissue, *Eur. J. Pharmacol.* **7**, 270–277.

Ross, S. B. and Renyi, A. L. (1975a). Tricyclic antidepressants agents. I. Comparison of the inhibition of the uptake of ^{3}H-noradrenaline and ^{14}C-5-hydroxytryptamine in slices and crude synaptosome preparations of the midbrain-hypothalamus region of the rat brain, *Acta pharmacol. toxicol.* **36**, 382–394.

Ross, S. B. and Renyi, A. L. (1975b). Tricyclic antidepressants. II. Effects of oral administration on the uptake of ^{3}H-noradrenaline and ^{14}C-5-hydroxytryptamine in slices from the midbrain-hypothalamus region of the rat brain, *Acta phamacol. toxicol.* **36**, 395–408.

Ross, S. B. and Renyi, A. L. (1976). Structural requirements for inhibition of the uptake of noradrenaline and 5-hydroxytryptamine by tricyclic antidepressants and related compounds, in *Symposium on pharmacology of catecholaminergic and serotoninergic mechanisms* (Eds. J. Knoll and K. Magyar) pp. 1–8, Akadémiai Kiadò, Budapest.

Ross, S. B. and Renyi, A. L. (1977). Inhibition of the neuronal uptake of 5-hydroxytryptamine and noradrenaline in rat brain by (Z)- and (E)-3-(4-bromophenyl)-*N,N*-dimethyl-3-(3-pyridyl) allylamines and their secondary analogues, *Neuropharmacology*, **16**, 57–63.

Ross, S. B., Renyi, A. L., and Ögren, S.-O. (1972). Inhibition of the uptake of noradrenaline and 5-hydroxytryptamine by chlorphentermine and chlorimipramine, *Eur. J. Pharmacol.* **17**, 107–112.

Ross, S. B. Ögren, S.-O., and Renyi, A. L. (1976). Z-Dimethylamino-1-(4-bromophenyl)-1-(3-pyridyl)propene (H 102/09), a new selective inhibitor of the neuronal 5-hydroxytryptamine uptake, *Acta pharmacol. toxicol.* **39**, 152–166.

Ross, S. B., Ögren, S.-O., and Renyi, A. L. (1977). Substituted amphetamine derivatives. I. Effect on uptake and release of biogenic monoamines and on monoamine oxidase in the mouse brain, *Acta pharmacol. toxicol.* **41**, 337–352.

Ross, S. B., Aperia, B., Beck-Friis, J., Jansa, S., Wetterberg, L., and Åberg, A. (1980). Inhibition of 5-hydroxytryptamine uptake in human platelets by antidepressive agents *in vivo, Psychopharmacology*, **67**, 1–7.

Ross, S. B., Hall, H., Renyi, A. L., and Westerlund, D. (1981). Effects of zimelidine on serotonergic and noradrenergic neurones after repeated administration in the rat, *Psychopharmacology*, **72**, 219–225.

Rotman, A., Mondai, I., Munitz, H., and Wijsenbeek, H. (1979). Active uptake of serotonin by blood platelets of schizophrenic patients, *FEBS Letters*, **101**, 134–136.

Rudnick, G. (1977). Active transport of 5-hydroxytryptamine by plasma membrane vesicles isolated from human blood platelets, *J. Biol. Chem.* **252**, 2170–2174.

Rudnick, G. and Nelson, P. J. (1978a). Platelet 5-hydroxytryptamine transport, an electroneutral mechanism coupled to potassium, *Biochemistry*, **17**, 4739–4742.

Rudnick, G. and Nelson, P. J. (1978b). Reconstitution of 5-hydroxytryptamine transport from cholate-disrupted platelet plasma membrane vesicles, *Biochemistry*, **17**, 5300–5303.

Rudnick, G., Fishkes, H., Nelson, P. J., and Schuldiner, S. (1980). Evidence for two distinct serotonin transport systems in platelets, *J. Biol. Chem.* **255**, 3638–3641.

Rutledge, C. O. (1978). Effect of metabolic inhibitors and ouabain on amphetamine- and potassium-induced release of biogenic amines from isolated brain tissue, *Biochem. Pharmacol.* **27**, 511–516.

Schröder, H. U., Neuhoff, V., Priggemeier, E. and Osborne, N. N. (1979). The influx of tryptamine into snail (*Helix pomatia*) ganglia: comparison with 5-hydroxytryptamine, *Malacologia*, **18**, 517–525.

Schwark, W. S. and Keesey, R. R. (1978). Altered thyroid function and synaptosomal uptake by serotonin in developing rat brain, *J. Neurochem.* **30**, 1583–1586.

Shaskan, E. G. and Snyder, S. H. (1970). Kinetics of serotonin accumulation into slices from rat brain. Relationship to catecholamine uptake, *J. Pharmacol. Exp. Ther.* **175**, 404–418.

Slotkin, T. A. and Bareis, D. L. (1980). Uptake of catecholamines by storage vesicles, *Pharmacology*, 21, 109–122.

Slotkin, T. A., Seidler, F. J., Withmore, W. L., Lau, C., Salvaggio, M., and Kirksey, D. F. (1978). Rat brain synaptic vesicles. Uptake and specificities of [^{3}H] norepinephrine and [^{3}H] serotonin in preparation from whole brain and brain regions, *J. Neurochem.* 31, 961–968.

Sneddon, J. M. (1973). Blood platelets as a model for monoamine-containing neurones, *Prog. Neurobiol.* 1, 151–198.

Sofia, D. R., Ertel, R. J., Dixit, B. N., and Barry III, H. (1971). The effect of Δ-tetrahydrocannabinol on the uptake of serotonin by rat brain homogenates, *Eur. J. Pharmacol.* 16, 257–259.

Stahl, S. M. and Meltzer, H. Y. (1978). A kinetic and pharmacological analysis of 5-hydroxytryptamine transport by human platelets and platelet storage granules: comparison with central serotonergic neurons, *J. Pharmacol. Exp. Ther.* 205, 118–132.

Stahl, W. L., Neukoff, V., and Osborne, N. N. (1977). Role of sodium in uptake of 5-hydroxytryptamine by *Helix* ganglia, *Comp. Biochem. Physiol.* 56C, 13–18.

Suddith, R. L., Hutchinson, H. T., and Haber, B. (1978). Uptake of biogenic amines by glial cells in culture. I. A neuronal-like transport of serotonin, *Life Sci.* 22, 2179–2188.

Sugrue, M. F. and Mireylees, S. E. (1978). Effects of mazindol on rat brain synaptosomal monoamine uptake, *Biochem. Pharmacol.* 27, 1843–1847.

Suzuki, O., Noguchi, E., and Yagi, K. (1978). Uptake of 5-hydroxytryptamine by chick retina, *J. Neurochem.* 30, 295–296.

Szabados, L., Mester, L., Michal, F., and Born, G. V. R. (1975). Accelerated uptake of 5-hydroxytryptamine by human platelets enriched in a sialic acid, *Biochem. J.* 148, 335–336.

Talvenheimo, J. T., Nelson, P. J., and Rudnick, G. (1979). Mechanism of imipramine inhibition of platelet 5-hydroxytryptamine transport, *J. Biol. Chem.* 245, 4631–4635.

Thoa, N. B., Ecclestone, D., and Axelrod, J. (1969). The accumulation of ^{14}C-serotonin in the guinea-pig vas deferens, *J. Pharmacol. Exp. Ther.* 169, 68–73.

Thomas, T. N. and Redburn, D. A. (1979). 5-Hydroxytryptamine—a neurotransmitter of bovine retina, *Exp. Eye Res.* 28, 55–61.

Thomas. T. N. and Redburn, D. A. (1980). Serotonin uptake and release by subcellular fractions of bovine retina, *Vision Res.* 20, 1–8.

Tissari, A. H. (1975). Pharmacological and ultrastructural maturation of serotonergic synapses during ontogeny, *Med. Biology*, 53, 1–14.

Tissari, A. H. and Bogdanski, D. F. (1971). Biogenic amine transport. VI. Comparison of effects of ouabain and K$^+$ deficiency on the transport of 5-hydroxytryptamine and norepinephrine by synaptosomes, *Pharmacology*, 5, 225–234.

Tissari, A. H., Schönhöfer, P. S., Bogdanski, D. F., and Brodie, B. B. (1969). Mechanism of biogenic amine transport. II. Relationship between sodium and the mechanism of ouabain blockade of the accumulation of serotonin and norepinephrine by synaptosomes, *Molec. Pharmacol.* 5, 593–604.

Tuomisto, J. (1974). A new modification for studying 5-HT uptake by blood platelets: a re-evaluation of tricyclic antidepressants as uptake inhibitors, *J. Pharm. Pharmacol.* 26, 96–100.

Tuomisto, J. and Tukianen, E. (1976). Decreased uptake of 5-hydroxytryptamine in blood platelets from depressed patients, *Nature (Lond.)*, 262, 596–598.

Tuomisto, L. and Tuomisto, J. (1973). Inhibition of monoamine uptake in synaptosomes by tetrahydroharmane and tetrahydroisoquinoline compounds, *Arch. Pharmacol.* 279, 371–380.

Tuomisto, J., Tukianen, E., and Ahlfors, V. G. (1979). Decreased uptake of 5-hydroxytryptamine in blood platelets from patients with endogenous depression, *Psychopharmacology,* **65,** 141–147.

Uebelhack, R., Franke, L., and Seidel, K. (1978). Wirking von LSD und 5-Methoxy-*N,N*-dimethyltryptamine auf die aktive Aufnahme von [^{3}H]-Serotonin in Synaptosomenfraktionen aus Rattenhirn, *Acta biol. med. germ.* **37,** 1611–1614.

Van der Zee, P. and Hespe, W. (1978). A comparison of the inhibitory effects of aromatic substituted benzhydryl ethers on the uptake of catecholamines and serotonin into synaptosomal preparations of the rat brain, *Neuropharmacol.* **17,** 483–490.

Vermes, I. and Telegdy, G. (1977). Effect of stress on activity of the serotoninergic System in limbic brain structures and its correlation with pituitary-adrenal function in the rat, *Acta physiol. acad. scient. Hung.* **49,** 37–44.

Waldmeier, P. C., Baumann, P. A., and Maître, L. (1979). CGP 6085A, a new, specific inhibitor on serotonin uptake: neurochemical characterization and comparison with other serotonin uptake blockers, *J. Pharmacol. Exp. Ther.* **211,** 42–49.

Wang, Y.-J., Gurd, J. W., and Mahler, H. R. (1975). Topography of synaptosomal high affinity uptake systems, *Life Sci.* **17,** 725–734.

Warwick, R. O., Bousquet, W. F., and Schnell, R. C. (1977). Effect of acute and chronic morphine treatment on serotonin uptake into rat hypothalamic synaptosomes, *Pharmacology,* **15,** 415–427.

Wielotz, M., Salmona, M., de Gaetano, G., and Garattini, S. (1976). Uptake of ^{14}C-5-hydroxytryptamine by human and rat platelets and its pharmacological inhibition. A comparative kinetic analysis, *Arch. Pharmacol.* **296,** 59–65.

Wilkins, J. A., Greenawalt, J. W., and Huang, L. (1978). Transport of 5-hydroxytryptamine by dense granules from porcine platelets, *J. Biol. Chem.* **253,** 6260–6265.

Wirz-Justice, A. (1974). Possible circadian and seasonal rhythmicity in an *in vitro* model: monoamine uptake in rat brain slices, *Experientia,* **30,** 1240–1241.

Wirz-Justice, A., Hackmann, E., and Lichtsteiner, M. (1974). The effect of oestradiol dipropionate and progesterone on monoamine uptake in rat brain, *J. Neurochem.* **22,** 187–189.

Wong, D. T., Horng, J.-S., and Fuller, R. W. (1973). Kinetics of serotonin accumulation into synaptosomes of rat brain—effects of amphetamine and chloroamphetamines, *Biochem. Pharmacol.* **22,** 311–322.

Wong, D. T., Horng, J.-S., Bymaster, F. P., Hauser, K. L., and Molloy, B. B. (1974). A selective inhibitor of serotonin uptake: Lilly 110140, 3-(*p*-trifluoromethylphenoxy)-*N*-methyl-3-phenylpropylamine, *Life Sci.* **15,** 471–479.

Wong, D. T., Bymaster, F. P., Horng, J.-S., and Molloy, B. B. (1975). A new selective inhibitor for uptake of serotonin into synaptosomes of rat brain: 3-(*p*-trifluoromethylphenoxy)-*N*-methyl-3-phenylpropylamine, *J. Pharmacol. Exp. Ther.* **193,** 804–811.

Wong, D. T., Bymaster, F. P., Chen, S., and Molloy, B. B. (1980). *N,N*-Dimethyl-α-(2-(*p*-tolyloxy)ethyl)benzylamine hydrochloride (LY 125180). Effects on serotonin uptake and serotonin synthesis in rat brain *in vitro* and *in vivo, Biochem. Pharmacol.* **29,** 935–941.

Biology of Serotonergic Transmission
Edited by N. N. Osborne
© 1982, John Wiley & Sons Ltd.

Chapter 8

Characteristics of 5-HT Metabolism and Function in the Developing Brain

M. HAMON and S. BOURGOIN

Groupe NB, INSERM U.114, Collège de France,
11, Place Marcelin Berthelot, F-75231 Paris cedex 05, France

INTRODUCTION

Serotonin (5-HT)-containing neuroblasts first appear in the central nervous system (CNS) as early as the twelfth day of gestation in the rat (Olson and Seiger, 1972). Although the presence of 5-HT inside cells undoubtedly means that they are differentiating into serotonergic neurones, this does not allow to conclude that the two preceding stages of neuronal development, i.e. neuroblast proliferation and migration, are achieved. Indeed, in the case of serotonergic systems, evidence has been obtained which indicates that cell proliferation, migration and differentiation are overlapping during ontogenesis. Thus Lauder and Bloom (1974) observed that cell proliferation in the raphe nuclei takes place till the sixteenth day of gestation, i.e. 4 days after the raphe cells have started to differentiate (i.e. to store 5-HT). In addition, Cadilhac and Pons (1976) mentioned that the migration of 5-HT-containing cells from the lateral parts of the brain stem to the midline, thus forming the dorsal raphe nucleus, occurs on the twentieth day of gestation, i.e. only 1 day before birth. In this respect, serotonergic neurones in the CNS markedly differ from those at the periphery since Gershon *et al.* (1980) recently noted that 5-HT-containing neuroblasts in the (chick) gut undergo cell proliferation and migration well before expressing typical characters of well-differentiated serotonergic neurones.

Accordingly, it can be concluded that, in contrast to those in the peripheral nervous system, central serotonergic neurones do not follow the classical pattern of neuronal maturation. This is not the only particularity of developing serotonergic neurones in brain. Others concern the biochemical characteristics of 5-HT metabolism, the possible function of 5-HT-containing neurones in the fetal brain and the remarkable ability of these neurones to build up new processes and terminals following axotomy particularly during the early life period. The present review is an attempt to summarize the original characteristics of developing serotonergic neurones and to propose some concepts for their possible physiological significance during brain maturation.

CHARACTERISTICS OF 5-HT METABOLISM IN THE DEVELOPING BRAIN

Although serotonergic neurones are among the first categories of neurones which mature in the CNS, their differentiation is far from being complete at birth in the rat. This is illustrated by the large differences existing between the density or activity of presynaptic markers in the new born versus the adult rat.

Endogenous levels of 5-HT, tryptophan and 5-hydroxyindoleacetic acid (5-HIAA) in the brain of developing rats

At birth, 5-HT levels in the brain stem are only 32 per cent of those found in the same region in adult rats (Fig. 1). They increase progressively thereafter till the end of the third postnatal week thus reaching a value slightly higher than that found in adult rats.

In the forebrain, 5-HT levels at birth are even less (22 per cent) as compared to adults. As illustrated in Fig. 1, they increase much more slowly than in the brain stem since they are still only 75 per cent of those found in the forebrain of adult rats at the end of the fifth postnatal week.

The presence of 5-HT in tissues is in fact the consequence of the capacity of

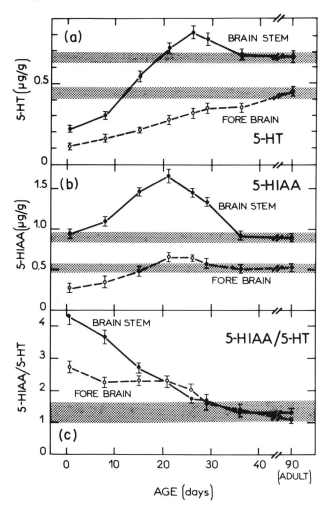

Fig. 1   Developmental changes in the levels of (a) 5-HT, (b) 5-HIAA and (c) in the 5-HIAA/5-HT ratio in the brain stem (●) and in the forebrain (○) of rats. Time 0 corresponds to birth. Each point is the mean ± SEM of six to eight determinations. (From Bourgoin *et al.*, 1977a; reproduced by permission of Pergamon Press)

serotonergic neurones to store the indoleamine. Therefore, these observations indicate that the 5-HT-storage capacity develops faster in the brain stem than in forebrain areas. In this respect, serotonergic neurones behave as other neurones in brain since the differentiation of neuronal cells generally proceeds from the lower to the upper brain parts.

    In the CNS, 5-HT is synthesized from tryptophan, an essential amino acid in mammals. Under normal conditions, the concentration of tryptophan in the brain of adult rats is relatively low since it does not saturate tryptophan hydroxylase, the enzyme catalyzing the rate-limiting step in 5-HT biosynthesis (Hamon and Glowinski, 1974). In contrast, brain tryptophan levels are high during the early postnatal period. Thus, Bourgoin *et al.* (1974) have noted that the concentration of the precursor amino acid in the CNS is five to ten times higher than in adults for the first 2 days following birth. During this short period, tryptophan hydroxylase is saturated by its amino acid substrate and the rate of 5-HT synthesis does not depend on fluctuations in tryptophan levels (of central or peripheral origin, Bourgoin *et al.*, 1974), in contrast to what has been repeatedly observed in adults (see Hamon and Glowinski, 1974, for a review).

Like the precursor, the metabolite of 5-HT, 5-HIAA is relatively abundant in the brain of developing rats (Bourgoin *et al.*, 1977a). In the brain stem for instance, 5-HIAA levels are already as high at birth as in adults (Fig. 1). As illustrated in Fig. 1, the concentration of 5-HIAA increases progressively both in the brain stem and the fore brain for the first three postnatal weeks; at the end of this period, they are significantly higher than in adults, particularly in the brain stem. Later on, the concentration of the 5-HT metabolite slowly decreases thus reaching adult value at the end of the fifth postnatal week (Fig. 1).

In adult rats, the ratio of 5-HIAA to 5-HT levels is generally considered as a good index of 5-HT turnover in brain. In this connection, one can conclude that the turnover of the indole amine is quite rapid in young animals since this ratio remains three to four times higher than in control adults for the first 2 weeks following birth. The turnover of 5-HT depends on the electrical activity of serotonergic neurones in adults (Aghajanian *et al.*, 1967) and, in fact, an increase in the 5-HIAA/5-HT ratio may be the consequence of an enhanced rate of neuronal firing. The direct application of this reasoning to young rats would lead to the proposal that nerve impulse flow in 5-HT-containing cells is very rapid during the early life period. Although the firing rate of 5-HT-containing cells has never been measured directly in the brain of neonates, indirect evidence strongly suggests that it should be extremely low (Hoff *et al.*, 1977; Hamon and Bourgoin, 1979). Therefore, as it will be discussed, the high 5-HIAA/5-HT ratio in newborn animals is more likely the consequence of some biochemical particularities of developing serotonergic neurones.

Biochemical particularities of serotonergic systems during ontogenesis in the rat

Brain and serum tryptophan

Numerous studies in adult rats have shown that fluctuations in brain tryptophan are very often, if not always, correlated with those in brain 5-HIAA levels (Gessa and Tagliamonte, 1974) simply because the rates of 5-HT synthesis and turnover directly depend on the availability of the precursor amino acid in tissues. Accordingly, the high 5-HIAA/5-HT ratio found in the brain of neonates may be—at least partly—the consequence of the elevated concentration of tryptophan (Bourgoin *et al.*, 1974).

The accumulation of tryptophan in brain depends on:

1. the intrinsic activity of the specific carrier in neuronal membranes,
2. the concentration of free tryptophan (the only form entering tissues) in plasma,
3. the levels of circulating substances interacting more or less directly with the tryptophan carrier in membranes such as hormones (notably hydrocortisone and corticosterone, Neckers and Sze, 1975) and neutral amino acids (Fernstrom and Wurtman, 1972).

Systematic examination of these factors has shown that at least two major particularities are involved in favouring the entry of tryptophan into the brain of newborn animals. First, the binding of circulating tryptophan to serum albumin is extremely low during the early postnatal period (Bourgoin *et al.*, 1974, 1977b). Therefore, serum tryptophan is almost totally available for the tryptophan carrier in neuronal membranes and then for 5-HT synthesis. This situation is quite different than that of adult rats in which more than 90 per cent of circulating tryptophan is bound to serum albumin. Consequently, the actual concentration of the amino acid available for 5-HT synthesis is limited to only 10 per cent of total tryptophan in serum (Gessa and Tagliamonte, 1974; Hamon and Glowinski, 1974). At least three causes are responsible for the very poor binding of peripheral tryptophan in neonates:

1. The purified and defatted serum albumin extracted from the serum of newborn rats exhibits much less capacity to bind tryptophan than that extracted from the serum of adult rats (Bourgoin *et al.*, 1977b). This does not allow to conclude that neonatal and adult serum albumins are different proteins; in fact, they exhibit the same apparent immunoreactivity when tested with a specific rabbit antibody against serum albumin from adult rats.
2. During the first three postnatal weeks and particularly for the first 7 days following birth, the concentration of serum albumin in serum is significantly lower than in adult rats (Bourgoin *et al.*, 1977b). This results in a marked reduction in the number of specific binding sites for tryptophan in the serum of neonates.
3. The concentration of unesterified fatty acids in serum is higher during the lactating period, i.e. during the first three postnatal weeks, than later on, in adult rats. Since unesterified fatty acids are competitive inhibitors of tryptophan binding onto serum albumin, this particularity also converges to make the essential amino acid largely unbound in the serum of newborn rats. Indeed, defatting the serum albumin extracted from the serum of newborn animals results in the appearance of a highly significant tryptophan binding capacity (Bourgoin *et al.*, 1977b).

Not only peripheral tryptophan but also the tryptophan carrier in neuronal membranes seems to be adapted for facilitating the entry of the amino acid in the brain of neonates. Thus, using slices and synaptosomes, we observed that the accumulation of tryptophan in the brain stem is about twice as high in 0-3-week-old rats as in adults (Hamon and Bourgoin, 1979). Similar findings have been obtained with slices or synaptosomes from the cerebellum, the raphe area and the hypothalamus (Bourgoin and Hamon, unpublished observations). Kinetic analyses of tryptophan uptake in brain stem synaptosomes indicate that the increased activity of the tryptophan carrier is mainly associated with an increased V_{max} during the neonatal period (Hamon and Bourgoin, 1979). The apparent affinity (Km^{-1}) of the uptake process for tryptophan remains essentially as in adults for the whole developmental period. This strongly suggests that the same carrier molecule exists in central neurones throughout life time. Indeed, the specificity of the tryptophan carrier in membranes exhibits identical characteristics in new born and in adult rats: large neutral aminoacids (tyr, phe, val, leu, ileu) compete to the same extent with the uptake of tryptophan in synaptosomes from young or adult rats (Hamon and Bourgoin, unpublished observations).

As emphasized by Fernstrom and Wurtman (1972), the entry of tryptophan into brain also depends on the ratio of the concentration of tryptophan to those of large neutral amino acids in plasma. In light of the efficient competition exerted by these amino acids on the tryptophan carrier in young rats, this ratio would also have to be considered when studying the characteristics of tryptophan entry into brain of neonates. Indeed, Roux and Jachan (1974) have shown that the plasmatic concentrations of large neutral amino acids significantly increase during the first 3 days after birth; therefore, the rapid decay in brain tryptophan occurring for this period might well involve the increased competition of these amino acids at the level of the tryptophan carrier in neuronal membranes.

In conclusion, both peripheral and central mechanisms are adapted to favour tryptophan accumulation in brain during the early life period. This results not only in maintaining a high rate of 5-HT synthesis but also in fulfilling the great demand in tryptophan due to the elevated protein synthesis in the developing brain. Recently, Chanez *et al.* (1981) have observed that intrauterine growth retardation (IUGR) resulting from uterine artery ligation on the seventeenth day of gestation is associated with a significant delay in the appearance of tryptophan binding capacity in serum. As a consequence, the concentration of tryptophan in brain remains at a higher level than in adults for a longer period in IUGR rats than in normal developing animals. Not only the synthesis of 5-HT but also that of proteins are, therefore, very active in the brain of IUGR rats for a rather long period during ontogenesis. This might explain why the brain is so slightly affected in IUGR animals. In contrast, the growth of peripheral organs (liver, kidney, lung) is markedly restricted in IUGR rats.

Tryptophan hydroxylase

In the rat brain, tryptophan hydroxylase is first detected on the sixteenth day of gestation (Renson, 1973). Since this enzyme catalyzes the rate limiting step in 5-HT

Table 1. Relative changes in 5-HT levels and tryptophan hydroxylase activity (TRP-OH) in the brain stem of developing rats. 5-HT levels are expressed in $\mu g\ g^{-1}$ of fresh tissue. The activity of tryptophan hydroxylase was measured with 0.2 mM of tryptophan and 0.16 mM of 6-methyl-5,6,7,8-tetrahydropterin (6-MPH$_4$) as the co-factor. It is expressed in nmol of 5-HTP synthesized mg prot^{-1} and per 15 min. Each value is the mean ± SEM of six to ten determinations

Age (days)	5-HT	TRP-OH	TRP-OH/5-HT
~ 0.5 (new born)	0.22 ± 0.01	0.91 ± 0.02	4.1 ± 0.2*
8	0.30 ± 0.02	1.55 ± 0.04	5.2 ± 0.3*
15	0.54 ± 0.03	2.72 ± 0.06	5.0 ± 0.3*
22	0.71 ± 0.03	3.04 ± 0.06	4.3 ± 0.2*
28	0.78 ± 0.03	2.81 ± 0.03	3.6 ± 0.2*
35	0.69 ± 0.02	2.20 ± 0.05	3.2 ± 0.1
~90 (adult)	0.67 ± 0.02	1.88 ± 0.04	2.8 ± 0.2

*$p < 0.05$ when compared to the ratio calculated with data obtained in adult (~ 90 day-old) rats.

biosynthesis, its activity might well play a key role in maintaining a high 5-HIAA/ 5-HT ratio in the CNS of developing rats; in particular, if it exceeds the 5-HT storage capacity in serotonergic neurones, a large portion of newly synthesized 5-HT would be catabolized thus achieving a high 5-HIAA/5-HT ratio in brain. Indeed, tryptophan hydroxylase activity in the brain stem of new born rats is already half that found in the same region in adult rats (Table 1). This enzymatic activity increases rapidly thereafter thus reaching the adult level on the tenth day after birth and peaking at the end of the third postnatal week. At this time, it is 80 per cent higher than in adult rats (Table 1). Data in Table 1 also indicate that the ratio of tryptophan hydroxylase activity to the endogenous 5-HT content (an index of the storage capacity of serotonergic neurones) in the brain stem remains significantly higher in young than in adult rats for the first postnatal month. Similar observations have been made in the forebrain (unpublished observations). However, data in Table 1 concern the activity of tryptophan hydroxylase measured *in vitro* using a synthetic pterin co-factor (6-methyl-5,6,7,8-tetrahydropterin, 6-MPH$_4$), i.e. under different conditions than those occurring *in vivo* with the natural pterin co-factor and cellular environment. This led us to measure the rate of 5-HT synthesis *in vivo* by following the accumulation of 5-HT in tissues after the blockade of monoamine oxidase (MAO) by pargyline (75 mg kg^{-1} ip). As illustrated in Fig. 2, the rate of 5-HT synthesis in the brain stem (and in the fore brain) exhibits developmental changes similar to those already described for tryptophan hydroxylase activity. Furthermore, the ratio of the rate of 5-HT synthesis to the endogenous 5-HT tissue content is significantly higher (+ 43–100 per cent) in young (less than 1 month-old) than in adult rats. Accordingly, similar observations are made whether the 5-HT biosynthetic capacity is estimated *in vivo* or *in vitro* by the only measure-

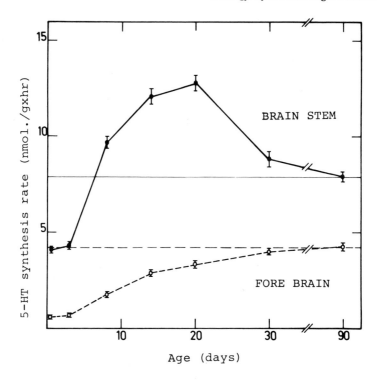

Fig. 2   Developmental changes in the rate of 5-HT synthesis in the brain stem (●) and the forebrain (○) of rats. The rate of 5-HT synthesis was estimated by measuring the accumulation of 5-HT in tissues for the first 15 min following the administration of pargyline (75 mg/kg ip, MAOI). Data are expressed in nmol of 5-HT synthesized per g of fresh tissue and per h. Each point is the mean ± SEM of six independent determinations

ment of tryptophan hydroxylase activity. This further illustrates that tryptophan hydroxylase is, as in adult rats, the enzyme catalizing the rate-limiting step in the biosynthesis of 5-HT in the developing brain. Since the ratio of the rate of 5-HT synthesis (or tryptophan hydroxylase activity) to the 5-HT content in tissues is higher in young than in adult rats, this suggests that the 5-HT biosynthetic capacity develops faster than the 5-HT storage capacity in serotonergic neurones. For the first postnatal month, this might well result in the leakage of 5-HT synthesized in excess to catabolizing pathways, notably oxidative deamination, thus leading to a high 5-HIAA/5-HT ratio in brain.

The relative excess of 5-HT biosynthetic versus storage capacity in tissues of young when compared to adult rats may well derive from the respective anatomy of serotonergic systems in both groups of animals. In a serotonergic neurone, the storage capacity mainly involves nerve terminals whereas tryptophan hydroxylase is first synthesized in cell bodies before migrating to nerve terminals. Since the density of nerve terminals relative to cell bodies is lower in young than in adult rats, it is in fact not surprising to find a lower storage capacity relative to bio-synthetic activity in young animals. Indeed, when the density of nerve terminals in the brain stem is higher than normal (i.e. after the neonatal administration of 5,7-dihydroxytryptamine for instance, Jonsson *et al.*, 1978; Hamon and Bourgoin, 1979), the 5-HT storage capacity is markedly higher than expected when con-sidering the 5-HT biosynthetic capacity in the same region (Hamon *et al.*, 1981).

Control of 5-HT synthesis during the neonatal period

In addition to the control of 5-HT synthesis exerted by the peripheral and central systems regulating the entry of tryptophan in brain (see above), other regulatory mechanisms exist which affect the presynaptic metabolism of 5-HT. These mechan-isms are apparently triggered by presynaptic receptors. The stimulation of these receptors by 5-HT itself or by other agonists like LSD and quipazine induces the reduction of both synthesis and release of the indoleamine neurotransmitter in adult rats (Bourgoin *et al.*, 1977c). Conversely, the blockade of these receptors by an antagonist like methiothepin produces a marked acceleration of 5-HT synthesis and release in brain (Bourgoin *et al.*, 1977c).

Under normal conditions in adult rats, these regulatory mechanisms exert a tonic inhibitory control on 5-HT synthesis since their reinforcement (with the administration of a 5-HT agonist) reduces the rate of 5-HT synthesis whereas their blockade (by a treatment with a 5-HT antagonist) produces the reverse effect. If such mechanisms are lacking in the developing brain, no negative feedback would control 5-HT synthesis and turnover thus also contributing to the elevated 5-HIAA/5-HT ratio in tissues (Fig. 1). This led us to analyse the effects of some 5-HT agonists (quipazine, LSD) and antagonist (methiothepin) on the rate of 5-HT synthesis in the developing rat brain. Such studies have been engaged since Enjal-bert *et al.* (1978a, b) and Nelson *et al.* (1980a, b) have shown that the primary target of these drugs, i.e. specific receptors for 5-HT, are present in the rat brain very early during ontogenesis.

As confirmed in Table 2, the administration of LSD to adult rats produces a marked reduction in the rate of 5-HT synthesis in brain. A similar effect is detected in 2-week-old rats. In contrast, the rate of 5-HT synthesis remains unaltered by the administration of LSD to 0.5- or 7-day-old rats. Similar findings have been obtained using another 5-HT agonist, quipazine. Thus, the peripheral administration of quipazine (10 mg kg^{-1} ip) produces a significant decrease in the rate of 5-HT synthesis in adults (-25 per cent) but fails to alter the indoleamine synthesis in rats younger than 2 week-old (Bourgoin *et al.*, 1977c). These results strongly suggest that negative feedback processes controlling 5-HT synthesis via the stimu-lation of 5-HT receptors are not operating during the early life period. As other

Table 2. Effects of LSD treatment on the rate of 5-HT synthesis in the brain of rats at various ages. LSD (1 mg kg^{-1} ip) or saline ('control') was administered 10 min before benserazid (800 mg kg^{-1} ip), a potent inhibitor of central 5-HTP decarboxylase activity. Animals were killed 30 min after benserazid treatment and the rate of 5-HT synthesis in the whole brain was calculated from the slope of 5-HTP accumulation in tissues. Results are expressed in nmol 5-HT (in fact 5-HTP) synthesized g^{-1} of fresh tissue and h^{-1}. Each value is the mean ± SEM of 5–7 independent determinations. Figures in parentheses are the percent reductions due to LSD treatment

	5-HT synthesis rate (nmol $g^{-1} \times h^{-1}$)		
Age (days)	Control	LSD	(%)
~ 0.5 (new born)	0.49 ± 0.08	0.47 ± 0.04	(4)
7	0.59 ± 0.05	0.64 ± 0.06	(−8)
14	1.12 ± 0.07	0.56 ± 0.05*	(50)
~90 (adult)	1.89 ± 0.16	0.90 ± 0.04*	(52)

*$p < 0.05$ when compared to the respective control value.

particularities already described, this may also contribute to maintain a high turnover rate of 5-HT in newborn rats.

If such a negative feedback is lacking, the administration of a 5-HT antagonist should not accelerate 5-HT synthesis in the brain of neonates. In contrast Bourgoin *et al.* (1977c) observed that methiothepin treatment (200 mg kg^{-1} ip) produces a significant increase in the rate of 5-HT synthesis not only in adults (35 per cent) but also in 24-h-old (+34 per cent) and 8-day-old rats (+39 per cent). These findings suggest that the exact mechanism of action of methiothepin might be more complex than the only blockade of central 5-HT receptors. In fact, several studies have shown that this drug is also a potent antagonist of histamine, noradrenaline and dopamine in the CNS (see Nelson *et al.*, 1979a).

Using brain stem slices, Bourgoin *et al.* (1977c) noted that the K^+-induced release of [³H] 5-HT newly synthesized from [³H] tryptophan is increased in the presence of methiothepin (10 μM). Conversely, LSD (10 μM) produces a marked reduction in the K^+-evoked release of the indoleamine neurotransmitter. Both effects are observed with tissues from adult or newborn rats (Bourgoin *et al.*, 1977c). Therefore, the negative control of 5-HT release via the stimulation of 5-HT receptors is apparently already effective at birth.

In conclusion, feedback processes controlling 5-HT release but not those controlling 5-HT synthesis seem to be acting very early during development. Such an uncomplete control of 5-HT metabolism via negative mechanisms triggered by 5-HT receptors might also contribute, but to a limited extent only, to maintaining a high rate of 5-HT turnover during the early life period in the rat brain.

5-HT inactivation

5-HT reuptake process Since the carrier involved in the reuptake of 5-HT only exists in the membrane of serotonergic neurones, it is considered as one of the most specific presynaptic markers of these neurones in the CNS. Apparently the same 5-HT carrier is present in adult and in new born rats (Nomura *et al.*, 1975, 1976) so that the measurement of [^{3}H] 5-HT uptake is quite appropriate to follow the development of serotonergic fibres and terminals in a given brain region.

For the first 2 weeks following birth, the synaptosomal uptake of [^{3}H] 5-HT increases rapidly in various brain regions (Kirksey and Slotkin, 1979). In the brain stem, the synaptosomal uptake of [^{3}H] 5-HT is thus slightly higher in 15–21-day-old rats than in adults. Both in the brain stem and the forebrain, the developmental pattern of [^{3}H] 5-HT uptake closely resembles that of endogenous 5-HT content. Thence, the ratio of the 5-HT storage capacity to that of the 5-HT reuptake process remains rather stable in brain from birth until adulthood (unpublished observation). This indirectly confirms that these two specific capacities of serotonergic neurones (storage and reuptake) concern the same neuronal compartment, i.e. the terminal arborization. In addition, the stability of this ratio (storage/re-uptake) throughout the life time indicates that the re-uptake process plays no role in maintaining a high rate of 5-HT turnover in brain during the early life period.

5-HT catabolism by MAO A In addition to the reuptake process, the enzymatic conversion of 5-HT into 5-HIAA by MAO type A plays a key role in the inactivation of the neurotransmitter in central serotonergic synapses. A high rate of enzymatic degradation would also explain the high 5-HIAA/5-HT ratio in the CNS of developing rats (Fig. 1). In fact, *in vitro* experiments with brain stem slices demonstrated that the conversion of [^{3}H] 5-HT into [^{3}H] 5-HIAA is faster in new born than in adult rats (Bourgoin *et al.*, 1977a). This might result from two different factors possibly acting together; (a) a poor storage capacity of [^{3}H] 5-HT in tissues from young rats so that high concentrations of unprotected [^{3}H] 5-HT are available to MAO A, (b) a high intrinsic activity of this enzyme in brain during the neonatal period. In order to test the first hypothesis, both adult and newborn rats were treated with reserpine (5 mg kg^{-1}, ip, 5 h before death) under conditions producing the complete disappearance of 5-HT storage capacity in tissues (Bourgoin *et al.*, 1977a). Although this treatment induces a marked acceleration (> 100 per cent) in the conversion of [^{3}H] 5-HT into [^{3}H] 5-HIAA in tissues of newborn or adult rats, it does not reduce the difference previously noted between the two groups: as observed with brain stem slices from control animals, the catabolism of [^{3}H] 5-HT into [^{3}H] 5-HIAA is still faster in tissues from young rats after reserpine treatment (Bourgoin *et al.*, 1977a). This strongly suggests that a possible defect in the 5-HT storage capacity is apparently not involved in maintaining a high rate of 5-HT turnover in the developing brain. Therefore, we considered the second hypothesis and measured MAO A activity in the rat brain during ontogenesis.

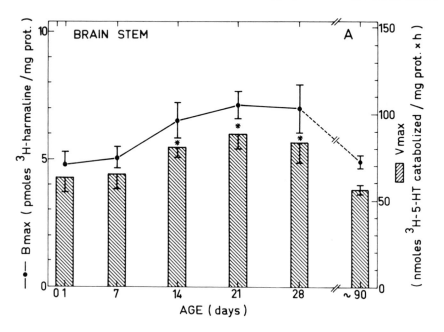

Fig. 3   Developmental changes in MAO A activity in the brain stem of rats. MAO
A activity was measured with [³H] 5-HT as the substrate (Nelson *et al.*, 1979b).
Each bar is the mean ± SEM of five to ten independent determinations of the
V_{max} of MAO A activity (in nmol [³H] 5-HT catabolized per mg prot and per h).

    The total number (B_{max}) of MAO A molecules was determined by measuring the
specific binding of [³H] harmaline to crude membranes (see Nelson *et al.*, 1979b).
Each point is the mean ± SEM of five to ten independent determinations of the
total number of MAO A molecules (i.e. the B_{max} of [³H] harmaline specific
binding sites) expressed as pmol per mg prot. *$p < 0.05$ when compared to the
respective V_{max} (MAO A activity) and B_{max} ([³H] harmaline binding) of adult
(~ 90 day-old) rats. (From Nelson *et al.*, 1979c; reproduced by pemission of
Pergamon Press)

    As illustrated in Fig. 3, the activity of MAO A in the brain stem is already as
high as in adults on the first postnatal day. It increases for the first 3 weeks after
birth thus reaching a peak corresponding to 158 per cent of the adult activity.
Analyses of the kinetic characteristics of MAO A in adult and in developing rats
have demonstrated that the higher enzymatic activity found in young animals is
associated with a higher V_{max} (Fig. 3, Bourgoin *et al.*, 1977a). The apparent
affinity of MAO A for 5-HT (Km = 0.1–0.2 mM) does not change significantly
throughout the life time. Recently Nelson *et al.* (1979b) have developed a simple
assay to determine the concentration of active sites of MAO A in tissues. It consists of measuring the specific binding of [³H] harmaline, a reversible selective
inhibitor of MAO A, to brain membranes under well-defined conditions. This led
us to use this method for further characterizing MAO A during ontogenesis in the

rat brain. We thus observed that the turnover number of MAO A does not change significantly during brain growth further confirming that the same enzyme molecule is present in adults and in newborn rats. The high MAO A activity detected in the brain stem (Fig. 3) and the forebrain (Nelson *et al.*, 1979c) of developing rats for the first postnatal month results only from the presence of a higher density of MAO A molecules in tissues (Fig. 3, Nelson *et al.*, 1979c).

Further studies on MAO A labelled with [^{3}H]harmaline indicate that the synthesis and degradation of this enzyme are faster in young animals (Nelson *et al.*, 1979c).

Concluding remarks

On the basis of the respective values of the 5-HIAA/5-HT ratio in brain (Fig. 1), it can be deducted that the turnover of 5-HT is three to four times faster in neonates than in adult rats. *In vivo* as well as *in vitro* experiments indicate that this high turnover rate of the indoleamine neurotransmitter is not dependent on the nerve impulse flow within serotonergic neurones. In fact, it is entirely due to biochemical particularities in the new-born organism such as the poor binding of peripheral tryptophan, the high intrinsic activity of the tryptophan carrier in membranes and the high concentration of MAO A in brain. Since these biochemical characteristics are quite different than those observed in adult rats (see Hamon and Glowinski, 1974), it can be asked whether 5-HT also plays a neurotransmitter function in the brain of neonates. Attempts to answer—indirectly—this question have been made by analysing the effects of the degeneration of serotonergic neurones on sleep both in new born and in adult animals. In adult rats and cats, the destruction of serotonergic neurones induces a significant decrease in the duration of slow wave and REM sleeps during the day time (Jouvet, 1972). In contrast, the electrical or chemical lesion of the central serotonergic systems in newborn animals (Adrien, 1976; Adrien *et al.*, 1977, 1980, 1981) effectively produces a marked depletion of 5-HT in brain (Adrien *et al.*, 1981; Bourgoin *et al.*, 1977d) but fails to alter the sleep pattern in developing animals. Therefore, 5-HT is not involved in the control of sleep during the early life period. In fact, the indoleamine might well exert quite different functions in neonates and in adults.

IS 5-HT A TROPHIC FACTOR STIMULATING THE MATURATION OF THE CNS DURING DEVELOPMENT?

Among functions possibly exerted by 5-HT during development, one has been repeatedly advanced which suggests that the indoleamine may stimulate the brain growth (Baker and Quay, 1969). Several observations support this hypothesis: first, as already mentioned, serotonergic neurones mature very early during brain development, well before other non-monoaminergic neuronal systems; secondly, 5-HT can stimulate the growth and differentiation of various embryonic tissues not only in invertebrates (Franquinet, 1979) but also in mammals (Leonov *et al.*, 1969); thirdly, 5-HT receptors coupled to adenylate cyclase are present in the CNS before birth (Enjalbert *et al.*, 1978a and unpublished observations). In light of the observations made by McMahon (1974) that the second messenger of trophic factors stimulating the growth and differentiation of tissues is generally cyclic AMP, such a rapid development of this particular type of 5-HT receptors has to be emphasized.

In addition to these indirect arguments, some experimental evidence exists which supports the hypothesis of 5-HT being a growth factor during brain development. Thus, the depletion of brain 5-HT resulting from either the blockade of tryptophan hydroxylase by *p*-chlorophenylalanine (Hole, 1972a) or the destruction of storage sites by reserpine (Patel *et al.*, 1977) is associated with a significant retardation in brain growth. A careful analysis of the effect of *p*-chlorophenylalanine has been made by Lauder and Krebs (1978) which further supports that 5-HT is selectively involved in brain maturation. These authors administered *p*-chlorophenylalanine to pregnant rats for various times (2-8 days) starting on the eighth day of gestation. At the end of this treatment, i.e. when the 5-HT concentration in the fetal brain was markedly reduced, $[^3H]$ thymidine was injected into the mother. After birth, pups were allowed to survive until 30 days of age. At that time, cells heavily labelled with tritium were counted in various brain regions. These cells were those ceasing to divide on the time of $[^3H]$ thymidine injection. Therefore, the time course for cell differentiation could be established. Using this approach, Lauder and Krebs (1978) have concluded that the 5-HT depleting drug, *p*-chlorophenylalanine, retards the onset of neuronal differentiation (cessation of germinal cell proliferation) specifically in those fetal brain regions which will, in the adults, receive serotonergic terminals or have a high content of 5-HT (future target cells of serotonergic neurones).

If 5-HT depletion delays the brain growth, conversely, one can expect that increased 5-HT levels may result in some acceleration in brain development. By injecting 5-HT directly into the albumen of incubating eggs, Ahmad and Zamenhof (1978) were able to increase the rate of protein synthesis in the brain of chick embryos. Similarly, Chumasov *et al.* (1978) reported that the addition of 5-HT to the medium of cultured dorsal hippocampal fragments taken from newborn rats (1-5 day-old) results in an increased growth and accelerated cytodifferentiation of neuronal and glial cells. Furthermore, the formation of interneuronal synaptic connections is also markedly stimulated by exogenous 5-HT under these conditions. However, in these two situations (Ahmad and Zamenhof, 1978; Chumasov *et al.*, 1978), it has to be emphasized that relatively large doses of 5-HT (1.0 mg egg^{-1} and 3.0 mg ml^{-1} of culture medium) were necessary to produce a significant acceleration of growth.

In vivo, the possible trophic effect of 5-HT is likely to occur very early during brain development. Indeed, the chronic blockade of central 5-HT receptors by various antagonists (cyproheptadine, methysergide, BC105) between the seventh and fifteenth postnatal days does not alter the brain development in rats (Hole, 1972b). Similarly, we observed that the chronic administration of methiothepin (20 mg kg^{-1} ip each day for the first postnatal week), a powerful long-lasting antagonist blocking all types of 5-HT receptors in the CNS (Nelson *et al.*, 1979a), has no significant consequence on the brain weight and on the maturation of central serotonergic neurones during at least the first month following birth. However, such a treatment resulted in a marked delay in body weight gain for the same period.

In conclusion, the demonstration of a growth promoting effect of 5-HT on brain development *in vivo* is far from being clearly established at present. Apparently, this function would occur before birth and is therefore very difficult to assess. Pharmacological treatments to foetus in fact consist of drug administrations to the pregnant rat (see Lauder and Krebs, 1978) so that the interpretation of data obtained is always open to criticism. The best approach would consist of directly treating developing rats at a time when serotonergic and other neuronal systems are quite immature. Following the injection of 5,6- or 5,7-dihydroxytryptamine (5,6- or 5,7-HT) to newborn rats, serotonergic systems in the brain stem behave as immature neurones since they subsequently develop numerous fibres and terminals like during the normal ontogenesis (Adrien *et al.*, 1981; Hamon and Bourgoin, 1979; Jonsson *et al.*, 1978). This may offer the opportunity to study the growth and possible functions of serotonergic systems—other than those associated with synaptic neurotransmission—even after birth. The characteristics of such 5,6- or 5,7-HT-induced outgrowth of serotonergic systems during the postnatal period are analysed in the following section.

REGENERATION AND SPROUTING OF SEROTONERGIC NEURONES IN THE CNS OF YOUNG AND ADULT RATS

Following the selective destruction of serotonergic terminals by 5,6- or 5,7-HT, the cell bodies in raphe nuclei can develop new processes and terminals possibly going up the same pathways as normally (Björklund and Wilklund, 1980; Wiklund and Björklund, 1980). This occurs following the peripheral (in newborn rats) or central (in newborn or adult rats) administration of these neurotoxins. In adults, the growth of new serotonergic fibres and terminals is rather slow since Björklund and Wik-lund (1980) observed that 8 months elapse between the time of intraventricular injection of 5,6-HT and the reappearance of serotonergic terminals in the lower (thoracic and lumbar) spinal cord. In regions close to the cell bodies, i.e. within the medulla oblongata, the proliferation of new serotonergic terminals is faster since a supranormal density of serotonergic terminals is found only 2 months after the neurotoxin administration (Björklund and Wiklund, 1980). The time course of these events is markedly shorter in young animals since Jonsson *et al.* (1978) noted that the sprouting of new serotonergic terminals in the mesencephalon-pons medulla occurs within 2 weeks after the peripheral administration of 5,7-HT. In other regions such as the cerebral cortex and the spinal cord, only half of the normal density of serotonergic terminals in these areas is recovered within 2 months after the injection of the neurotoxin (Jonsson *et al.*, 1978). Apparently, the variations in the time course of regeneration from one region to the other, in neonates and in adult rats, are not due to possible differences in the mechanisms involved but are simply related to the various distances serotonergic processes have to cover for reaching the terminal areas.

Like normal serotonergic terminals, those proliferating in the brain stem after neonatal 5,7-HT administration contain tryptophan hydroxylase and are able to synthesize $[^3H]$-5-HT from $[^3H]$ tryptophan (Hamon *et al.*, 1981); they also have the capacity to release $[^3H]$ 5-HT when exposed to depolarizing stimuli (Jonsson *et al.*, 1978) and to accumulate $[^3H]$ 5-HT by a specific reuptake process involved in the inactivation of the released neurotransmitter (Hamon and Bourgoin, 1979; Jonsson *et al.*, 1978). However, the detailed comparison of these presynaptic capacities with those of serotonergic terminals in developing rats reveals several differences. In particular, the rate of 5-HT turnover in the brain stem seems to be significantly lower in 5,7-HT-treated than in control rats (Ponzio and Jonsson, 1978; Hamon *et al.*, 1981). In addition, the serotonergic terminals which proliferate after 5,7-HT treatment apparently contain less tryptophan hydroxylase than normally (Hamon *et al.*, 1981). However, such differences do not allow to conclude that the serotonergic innervation developing after 5,7-HT administration is abnormal with a reduced metabolic activity since the reverse picture is seen in regions such as the spinal cord and the cerebral cortex. Indeed, the turnover of 5-HT in such distal regions is faster in 5,7-HT-treated than in control rats (Ponzio and Jonsson, 1978; Hamon *et al.*, 1981). This led to the conclusion that following the administration of a dihydroxytryptamine neurotoxin in newborn (Ponzio and Jonsson, 1978; Hamon *et al.*, 1981) as well as in adult (Björklund and Wiklund, 1980) rats, the turnover of 5-HT is increased in areas supplied with an infranormal amount of serotonergic fibres (spinal cord, cerebral cortex) and conversely is decreased in regions hyperinnervated by serotonergic terminals (brain stem). Accordingly, the differences in the metabolic characteristics of growing serotonergic terminals would not be of intrinsic nature but likely correspond to adaptive regulations of all serotonergic terminals.

Among the various problems inherent in the 5,6-HT or 5,7-HT-induced sprouting of serotonergic fibres in rats after birth, that dealing with the mechanism of growth initiation has been, and is still, the matter of debate. According to Jonsson *et al.* (1978), the developing serotonergic neurones seem to be programmed to produce a certain quantity of nerve terminals which they try to conserve after 5,7-HT-induced injury. This is based on the observation that the decrease in synaptosomal $[^3H]$ 5-HT uptake in areas distal from the cell bodies is compensated by the increase due to the terminal sprouting in the mesencephalon-pons medulla of rats treated with 5,7-HT at birth. According to this view, the development of serotonergic neurones would occur independently of the environment, notably of target cells. In support

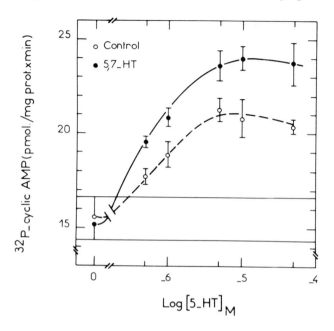

Fig. 4   Stimulating effects of 5-HT on adenylate cyclase activity in the brain stem of control and 5,7-HT-treated rats. 5,7-HT (100 mg/kg sc) was injected twice on the first and second postnatal days. Control and 5,7-HT-treated rats were killed on the seventh postnatal day and the activity of 5-HT-sensitive adenylate cyclase was measured in brain stem homogenates as described by Enjalbert *et al.* (1978a). Each point is the mean of triplicate determinations (in two separate experiments) of [^{32}P] cyclic AMP (in pmoles) synthesized per mg protein and per min at 30 °C. (From Hamon *et al.*, 1981; reproduced by permission of Pergamon Press)

with their statement, Jonsson *et al.* (1978) mentioned that the ontogenic evolution of presynaptic markers has apparently no link with that of postsynaptic 5-HT receptors. However, more recent studies challenge this conclusion since Hamon *et al.* (1981) observed that the V_{max} of the 5-HT-sensitive adenylate cyclase increases in various brain areas (notably hippocampus and brain stem) quite rapidly after 5,7-HT administration to newborn rats (see Fig. 4). Furthermore, Wiklund and Björklund (1980) recently obtained evidence strongly suggesting that the growth of new serotonergic processes following 5,6-HT-induced axotomy depends in fact on target cells. In particular, they observed that the 5,6-HT-induced degeneration of serotonergic systems in the inferior olive is followed by the reappearance within 1-4 months of a dense network of new 5-HT-containing fibres and terminals only when the olive cells are intact. When rats are pretreated with 3-acetylpyridine, a drug that induces the destruction of neurones in the inferior olive, the reinnervation of this area by serotonergic systems (which normally occurs after 5,6-HT-induced axotomy) is markedly impaired. Moreover, when some olivary neurones survive to

3-acetylpyridine treatment, the growth of new serotonergic fibres is roughly proportional to the number of remaining olivary neurones in the area. These data suggest that target (olive) cells may release some trophic factors stimulating the growth of afferent serotonergic fibres. The recent observation by Ebbott and Hendry (1978) that neurones of the raphe nuclei have the ability to take up nerve growth factor (NGF) in their terminal regions for retrograde transport to the cell body may be of significance in this respect. In fact, Bjerre *et al* (1973) have mentioned that NGF can stimulate the growth of new axonal sprouts from lesioned central catecholamine- and serotonin-containing neurones in adult rats. Subsequent studies challenged this view and indicated that in fact NGF is not involved in the normal or 6-hydroxydopamine (or 6-hydroxydopa)-induced growth of noradrenergic systems in the rat brain (Dreyfus *et al.*, 1980; Konkol *et al.*, 1978). In light of these results concerning noradrenergic neurones, the possible role of NGF as a growth promoting factor for central serotonergic neurones is quite unlikely.

As illustrated by the conflicting reports of Jonsson *et al* (1978) and Wiklund and Björklund (1980), even the possible role of target cells in the growth of new serotonergic terminals after dihydroxytryptamine-induced axotomy is still far from being elucidated. This led us to reinvestigate this problem by analysing possible changes in the 5,7-HT-induced increased outgrowth of serotonergic terminals in the brain sten when central 5-HT receptors are chronically blocked in neonatal rats (Hamon *et al.*, 1981). If target cells are involved in the sprouting of serotonergic terminals, the blockade of 5-HT receptors would change these cells into non-target cells and therefore alter their influence on the 5,7-HT-induced growth of presynaptic elements. Metergoline, a potent and selective 5-HT antagonist in the CNS (Bourgoin *et al.*, 1978), was administered daily (5 mg kg^{-1}, ip) for 3 weeks starting immediately after the neonatal treatment with 5,7-HT. Neither the increase in endogenous 5-HT levels, the enhancement of tryptophan hydroxylase activity nor the elevation of [^{3}H] 5-HT synaptosomal uptake normally seen in the brain stem after neonatal 5,7-HT administration was altered in metergoline-treated rats (Hamon *et al.*, 1981). Accordingly, the 5,7-HT-induced sprouting of serotonergic terminals does not depend—apparently—on some functional relationships between presynaptic fibres and postsynaptic target cells.

In conclusion, serotonergic neurones have the capacity to regenerate after axotomy in the rat brain. Although numerous hypothetical mechanisms have been proposed to explain this capacity, neither the intrinsic genetic programming of serotonergic neurones nor the influence of target cells via the release of a putative trophic factor has yet been firmly established.

FUTURE TRENDS

Although the sequence of events taking place from the proliferation of neuroblasts to the differentiation of serotonergic neurones is well known in the rat brain, the mechanisms governing the cell multiplication, their migration to precise areas in the CNS, the growth of fibres and the formation of synaptic (or non-synaptic) contacts are still poorly understood. This mainly results from the fact that the maturation of serotonergic neurones occurs very early *in vivo* (almost entirely during the fetal life in the rat), thus impeding appropriate experimental investigations.

Two original approaches have been proposed recently to study the ontogenesis of central neurones. They consist of examining the survival and growth of neurones in neonatal tissues implanted into various regions of the adult brain (Kromer *et al.*, 1979) on one hand, and to culture neurones of foetuses under well defined conditions *in vitro* (Prochiantz *et al.*, 1979) on the other. In the case of catecholaminergic neurones, these two techniques have been already very fruitful to explore the possible role of target cells in the control of the growth of terminal arborization (Prochiantz *et al.*, 1979). Similar approaches applied to serotonergic neurones would be very helpful to answer the following important questions:

1. Are 5-HT-containing neurones spontaneously dying during the normal onto-genesis in order to realize the final number of serotonergic neurones in the mature brain? Alternatively, is the cell proliferation perfectly controlled and ceasing immediately when the whole population of serotonergic neurones is set up?
2. Is the differentiation of neuroblasts into serotonergic neurones irreversible?
3. What governs the migration of neuroblasts which will be serotonergic neurones into the right places (raphe nuclei) in the adult brain?
4. What are the possible trophic factors involved in the differentiation of neuro-blasts into serotonergic neurones in brain?
5. Why do some serotonergic terminals make true synaptic contacts whereas others do not, having apparently no synaptic membrane differentiation with post-synaptic elements (see Wiklund and Björklund, 1980)? Alternatively, can the same serotonergic terminal make first non-synaptic contacts and then true synaptic connections in the course of its maturation? Are environmental influences of critical importance for the selection of synaptic or non-synaptic contacts? This question has to be asked since the proportion of serotonergic terminals with well-differentiated synapses is quite variable from one region to the other in the CNS of adult rats (see Wiklund and Björklund, 1980).
6. What is really the role of 5-HT in brain during the early life period, i.e. when it is not yet a neuroregulator (neuromodulator or neurotransmitter)? What would happen to the brain growth when 5-HT synthesis is permanently altered? In this connection, the use of albumin-deficient mutant rats (Nagase *et al.*, 1979) in which central 5-HT synthesis is not negatively controlled by peripheral trypto-phan binding would be quite appropriate.
7. In light of the presence of both 5-HT and a peptide like substance P (Hökfelt *et al.*, 1980) or enkephalin (Glazer *et al.*, 1981) in some serotonergic neurones, what can be the implication of this peptide in the maturation of serotonergic

neurones? Does the capacity of these neurones to synthesize this peptide occur simultaneously or not with their ability to synthesize 5-HT? Some data in the literature do suggest that peptides are involved in the growth of neurones. Thus Narumi and Fujita (1978) have shown that substance P can stimulate the neurite outgrowth in some embryonic chick ganglia. Harston *et al.* (1980) were able to enhance the sprouting of noradrenergic terminals in the pons medulla of 6-OH-dopa-treated rats by injecting morphine immediately after birth. Whether endomorphine compounds, notably enkephalins, have the same potency has not yet been explored.

These are only some of the numerous questions yet unsolved concerning the maturation of central serotonergic neurones. Developmental neurobiology is a so rapidly moving search field at present that one can be sure that most of these questions will be answered in the near future.

SUMMARY

The proliferation, migration and differentiation of neuroblasts into serotonergic neurones take place very early during ontogenesis. Thus, in the central nervous system of the rat, the first sign of differentiation of these cells, i.e. their capacity to store 5-HT, is already detected on the twelfth day of gestation. Then, the outgrowth of fibres and terminals from 5-HT-containing cell bodies in the raphe nuclei proceeds for several weeks following a postero-anterior progression. During this maturation period, the CNS exhibits several biochemical particularities (elevated concentration of the precursor, tryptophan, in brain; poor efficacy of the feedback processes controlling 5-HT synthesis; high intrinsic activity of the catabolizing enzyme, MAO A) all contributing to accelerate the turnover of 5-HT. The special characteristics of 5-HT metabolism during ontogenesis strongly suggest that the indoleamine may exert function(s) other than neurotransmission in the developing brain. In this connection, we have reanalysed the possible role of 5-HT as a growth promoting factor during early brain development in the rat.

Following 5,6- or 5,7-dihydroxytryptamine-induced axotomy, particularly in newborn rats, serotonergic neurones develop new fibres and terminals mainly in the mesencephalon-pons medulla. These new terminals apparently exhibit the same biochemical characteristics as those seen in untreated rats therefore indicating that the dihydroxytryptamine-induced sprouting may be considered as a useful model for studying the development of serotonergic systems after birth. Other techniques (*in vitro* culture of serotonergic neurones, *in vivo* implantation of fetal brain tissues into the CNS of adult rats) also seem very promising to better characterize the various regulatory mechanisms involved in setting up the serotonergic systems in brain.

ACKNOWLEDGEMENTS

This research has been supported by grants from INSERM, DGRST (No. 80.7.0184), CNRS (no. 4149), DRET (79.077) and Rhône-Poulenc SA.

REFERENCES

Adrien, J. (1976). Lesion of the anterior raphe nuclei in the new born kitten and the effects on sleep, *Brain Research*, **103**, 579–583.

Adrien, J., Bourgoin, S., and Hamon, M. (1977). Midbrain raphe lesion in the newborn rat. I. Neurophysiological aspects of sleep, *Brain Research*, **127**, 99–110.

Adrien, J., Bourgoin, S., Laguzzi, R., and Hamon, M. (1981). Effets de l'injection intraventriculaire de 5,7-dihydroxytryptamine sur le sommeil et les monoamines cérébrales du chaton, *J. Physiol.* (Paris), **77**, 425–430.

Adrien, J., Laguzzi, R., Bourgoin, S., and Hamon, M. (1980). Le sommeil du rat à lésion précoce du raphé: pharmacologie du système sérotoninergique, *Waking and Sleeping*, **4**, 119–129.

Aghajanian, G. K., Rosecrans, J. A. and Sheard, M. H. (1967). Serotonin: release in the forebrain by stimulation of midbrain raphe, *Science*, **156**, 402–403.

Ahmad, G. and Zamenhof, S. (1978). Serotonin as a growth factor for chick embryo brain, *Life Sci.* **22**, 963–970.

Baker, P. C. and Quay, W. B. (1969). 5-hydroxytryptamine metabolism in early embryogenesis and the development of brain and retinal tissues. A review, *Brain Research*, **12**, 273–295.

Bjerre, B., Björklund, A., and Stenevi, U. (1973). Stimulation of growth of new axonal sprouts from lesioned monoamine neurones in adult rat brain by nerve growth factor, *Brain Research*, **60**, 161–176.

Björklund, A. and Wiklund, L. (1980). Mechanisms of regrowth of the bulbospinal serotonin system following 5,6-dihydroxytryptamine induced axotomy. I. Biochemical correlates, *Brain Research*, **191**, 109–127.

Bourgoin, S., Artaud, F., Bockaert, J., Héry, F., Glowinski, J., and Hamon, M. (1978). Paradoxical decrease of brain 5-HT turnover by metergoline, a central 5-HT receptor blocker. *Naunyn-Schmiedeberg's Arch. Pharmacol.* **302**, 313–321.

Bourgoin, S., Artaud, F., Adrien, J., Héry, F., Glowinski, J., and Hamon, M. (1977a). 5-hydroxytryptamine catabolism in the rat brain during ontogenesis, *J. Neurochem.* **28**, 415–422.

Bourgoin, S., Faivre-Bauman, A., Héry, F., Ternaux, J. P., and Hamon, M. (1977b). Characteristics of tryptophan binding in the serum of the newborn rat, *Biol. Neonate*, **31**, 141–154.

Bourgoin, S., Artaud, F., Enjalbert, A., Héry, F., Glowinski, J., and Hamon, M. (1977c). Acute changes in central serotonin metabolism induced by the blockade or stimulation of serotoninergic receptors during ontogenesis in the rat, *J. Pharmacol. Exp. Ther.* **202**, 519–531.

Bourgoin, S., Enjalbert, A., Adrien, J., Héry, F., and Hamon, M. (1977d). Midbrain raphe lesion in the newborn rat. II. Biochemical alterations in serotoninergic innervation, *Brain Research*, **127**, 111–126.

Bourgoin, S., Faivre-Bauman, A., Benda, P., Glowinski, J., and Hamon, M. (1974). Plasma tryptophan and 5-HT metabolism in the CNS of the newborn rat, *J. Neurochem.* **23**, 319–327.

Cadilhac, J. and Pons. F. (1976). Le développment prénatal des neurones à monoamines chez le rat. *C.R. Soc. Biol.* **170**, 25–30.

Chanez, C., Priam, M., Flexor, M. A., Hamon, M., Bourgoin, S., Kordon, C., and Minkowski, A. (1981). Long lasting effects of intrauterine growth retardation on 5-HT metabolism in the brain of developing rats. *Brain Research*, **207**, 397–408.

Chumasov, E. I., Chubakov, A. R., Konovalov, G. V., and Gromova, E. A. (1978). Effect of serotonin on growth and differentiation of the hippocampal cellular elements in cultures, *Ann. Anat. Histol. Embryol.* LXXIV (1), 98–106.

Dreyfus, C. F., Peterson, E. R., and Crain, S. M. (1980). Failure of nerve growth factor to affect fetal mouse brain stem catecholaminergic neurons in culture, *Brain Research*, **194**, 540–547.

Ebbott, S. and Hendry, I. (1978). Retrograde transport of nerve growth factor in the rat central nervous system, *Brain Research*, **139**, 160–163.

Enjalbert, A., Bourgoin, S., Hamon, M., Adrien, J., and Bockaert, J. (1978a). Postsynaptic serotonin-sensitive adenylate cyclase in the central nervous system. I. Development and distribution of serotonin and dopamine-sensitive adenylate cyclases in rat and guinea pig brain, *Mol. Pharmacol.* **14**, 2–10.

Enjalbert, A., Hamon, M., Bourgoin, S., and Bockaert, J. (1978b). Postsynaptic serotonin-sensitive adenylate cyclase in the central nervous system. II. Comparison with dopamine- and isoproterenol-sensitive adenylate cyclases in rat brain, *Mol. Pharmacol.* **14**, 11–23.

Fernstrom, J. D. and Wurtman, R. J. (1972). Brain serotonin content: physiological regulation by plasma neutral amino acids, *Science*, **178**, 414–416.

Franquinet, R. (1979). Rôle de la sérotonine et des catecholamines dans la régénération de la planaire Polycelis tenuis, *J. Embryol. Exp. Morph.* **51**, 85–95.

Gershon, M. D., Epstein, M. L., and Hegstrand, L. (1980). Colonization of the chick gut by progenitors of enteric serotonergic neurons: distribution, differentiation, and maturation within the gut, *Develop. Biol.* **77**, 41–51.

Gessa, G. L. and Tagliamonte, A. (1974). Possible role of free serum tryptophan in the control of brain tryptophan level and serotonin synthesis, *Adv. Biochem. Psychopharmacol.* **11**, 119–131.

Glazer, E. J., Steinbusch, H., Verhofstad, A., and Basbaum, A. I. (1981). Serotonin neurons in nucleus raphe dorsalis and paragigantocellularis of the cat contain enkephalin, *J. Physiol.* (Paris) **77**, 241–245.

Hamon, M. and Bourgoin, S. (1979). Ontogenesis of tryptophan transport in the rat brain, *J. Neural Trans. Suppl.* **15**, 93–105.

Hamon, M. and Glowinski, J. (1974). Regulation of serotonin synthesis. *Life Sci.* **15**, 1533–1548.

Hamon, M., Nelson, D. L., Mallat, M., and Bourgoin, S. (1981). Are 5-HT receptors involved in the sprouting of serotoninergic terminals following neonatal 5,7-dihydroxytryptamine treatment in the rat? *Neurochem. International*, **3**, 69–79.

Harston, C. T., Morrow, A., and Kostrzewa, R. M. (1980). Enhancement of sprouting and putative regeneration of central noradrenergic fibres by morphine, *Brain Research Bull.* **5**, 421–424.

Hoff, K. M., Baker, P. C., and Buda, R. E. (1977). Effects of parachlorophenylalanine on indoleamines in maturing mouse brain. *Gen. Pharmacol.* **8**, 213–215.

Hökfelt, T., Johansson, O., Ljungdahl, Å., Lundberg, J. M., and Schultzberg, M. (1980). Peptidergic neurones, *Nature (Lond.)*, **284**, 515–521.

Hole, K. (1972a). Behavior and brain growth in rats treated with *p*-chlorophenylalanine in the first weeks of life, *Develop. Psychobiol.* **5**, 157–173.

Hole, K. (1972b). The effects of cyproheptadine, methysergide, BC 105 and reserpine on brain 5-hydroxytryptamine and brain growth, *Eur. J. Pharmacol.* **19**, 156–159.

Jonsson, G., Pollare, T., Hallman, H., and Sachs, C. (1978). Developmental plasticity of central serotonin neurons after 5,7-dihydroxytryptamine treatment, *Ann. N. Y. Acad. Sci.* **305**, 328–345.

Jouvet, M. (1972). The role of monoamine- and acetylcholine-containing neurons in the regulation of sleep-waking cycle, *Ergebn. Physiol.* **64**, 165–307.

Kirksey, D. F. and Slotkin, T. A. (1979). Concomitant development of [^{3}H] dopamine and [^{3}H] 5-hydroxytryptamine uptake systems in rat brain regions, *Brit. J. Pharmacol.* **67**, 387–391.

Konkol, R. J., Mailman, R. B., Bendeich, E. G., Garrison, A. M., Mueller, R. A., and Breese, G. R. (1978). Evaluation of the effects of nerve growth factor and anti-nerve growth factor on the development of central catecholamine-containing neurons, *Brain Research*, **144**, 277–285.

Kromer, L. F., Björklund, A., and Stenevi, U. (1979). Intracephalic implants: A technique for studying neuronal interactions, *Science*, **204**, 1117–1119.

Lauder, J. M. and Bloom, F. E, (1974). Ontogeny of monoamine neurons in the locus coeruleus, raphe nuclei and substantia nigra of the rat. I. Cell differentiation, *J. Comp. Neurol.* **155**, 469–482.

Lauder, J. M. and Krebs, H. (1978). Serotonin and early neurogenesis, in *Maturation of neurotransmission* (Eds A. Vernadakis, E. Giacobini, and G. Filogamo), pp. 171–180, Karger, Basel.

Leonov, B. V., Kamakhin, A. P., and Buznikov, G. A. (1969). Effects of serotonin and antiserotonin preparations on the development of early mouse embryos, cultured *in vitro*, *Dokl. Akad. Nauk. SSSR*, **188**, 958–960.

McMahon, D. (1974). Chemical messengers in development: a hypothesis, *Science*, **185**, 1012–1021.

Nagase, S., Shimamune, K., and Shumiya, S. (1979). Albumin-deficient rat mutant, *Science*, **205**, 590–591.

Narumi, S. and Fujita, T. (1978). Stimulatory effects of substance P and nerve growth factor (NGF) on neurite outgrowth in embryonic chick dorsal root ganglia, *Neuropharmacol.* **17**, 73–76.

Neckers, L. and Sze, P. Y. (1975). Regulation of 5-hydroxytryptamine metabolism in mouse brain by adrenal glucocorticoids, *Brain Research*, **93**, 123–132.

Nelson, D. L., Herbet, A., Adrien, J., Bockaert, J., and Hamon, M. (1980b). Serotonin-sensitive adenylate cyclase and [³H]serotonin binding sites in the CNS of the rat. II. Respective regional and subcellular distributions and ontogenetic developments. *Biochem. Pharmacol.* **29**, 2455–2463.

Nelson, D. L., Herbet, A., Enjalbert, A., Bockaert, J., and Hamon, M. (1980a). Serotonin-sensitive adenylate cyclase and [³H]serotonin binding sites in the CNS of the rat. I. Kinetic parameters and pharmacological properties. *Biochem. Pharmacol.* **29**, 2445–2453.

Nelson, D. L., Herbet, A., Glowinski, J., and Hamon, M. (1979c). [³H]harmaline as a specific ligand of MAO A. II. Measurement of the turnover rates of MAO A during ontogenesis in the rat brain, *J. Neurochem.* **32**, 1829–1836.

Nelson, D. L., Herbet, A., Pétillot, Y., Pichat, L., Glowinski, J., and Hamon, M. (1979b). [³H]harmaline as a specific ligand of MAO A. I. Properties of the active site of MAO A from rat and bovine brains, *J. Neurochem.* **32**, 1817–1827.

Nelson, D. L., Herbet, A., Pichat, L., Glowinski, J., and Hamon, M. (1979a). *In vitro* and *in vivo* disposition of ³H-methiothepin in brain tissues. Relationship to the effects of acute treatment with methiothepin on central serotoninergic receptors, *Naunyn-Schmiedeberg's Arch. Pharmacol.* **310**, 25–33.

Nomura, Y., Naitoh, F., and Segawa, T. (1976). Regional changes in monoamine content and uptake of the rat brain during postnatal development, *Brain Research*, **101**, 305–315.

Nomura, Y., Tanaka, Y., and Segawa, T. (1975). Development of the influences of sodium, catecholamine and tricyclic antidepressant drug on the uptake of [³H] 5-hydroxytryptamine by rat brain synaptosomes, *Brain Research*, **100**, 705–709.

Olson, L. and Sieger, Ä. (1972). Early prenatal ontogeny of central monoamine neurons in the rat: fluorescence histochemical observations, *Z. Anat. Entwickl. Gesch.* **137**, 301–316.

Patel, A. J., Béndek, G., Balàzs, R., and Lewis, P. D. (1977). Effect of reserpine on cell proliferation in the developing rat brain: a biochemical study, *Brain Research*, **129**, 283-297.

Ponzio, F. and Jonsson, G. (1978). Effects of neonatal 5,7-dihydroxytryptamine treatment on the development of serotonin neurons and their transmitter metabolism, *Dev. Neurosci.* **1**, 80-89.

Prochiantz, A., Di Porzio, U., Kato, A., Berger, B., and Glowinski, J. (1979). *In vitro* maturation of mesencephalic dopaminergic neurons from mouse embryos is enhanced in presence of their striatal target cells, *Proc. Natl. Acad. Sci. (USA)*, **76**, 5387-5391.

Renson, J. (1973). Assays and properties of tryptophan 5-hydroxylase, in *Serotonin and Behavior* (Eds J. Barchas and E. Usdin), pp. 19-32, Academic Press, New York.

Roux, J. M. and Jachan, T. (1974). Plasma level of amino acids in the developing young rat after intrauterine growth retardation, *Life Sci.* **14**, 1101-1107.

Wiklund, L. and Björklund, A. (1980). Mechanisms of regrowth in the bulbospinal serotonin system following 5,6-dihydroxytryptamine induced axotomy. II. Fluorescence histochemical observations, *Brain Research*, **191**, 129-160.

Biology of Serotonergic Transmission
Edited by N. N. Osborne
© 1982, John Wiley & Sons Ltd.

Chapter 9

Drugs Acting on Serotonergic Neuronal Systems

R. W. FULLER

The Lilly Research Laboratories,
Eli Lilly and Company, Indianapolis, Indiana 46285, USA

INTRODUCTION

Drugs acting on serotonergic neurones are of interest not only because they are presently or potentially useful in the treatment of various diseases, but also because they are valuable tools for elucidating mechanisms by which serotonergic neuronal systems function and physiological roles of serotonergic neurones. This chapter describes a number of drugs that act in known ways on serotonergic neuronal systems. Certain classes of these drugs, e.g. serotonin neurotoxins, are dealt with elsewhere in this book in more detail.

Serotonin neurones in brain are thought to function in the following way. The serotonin precursor that is derived from dietary nutrients is the amino acid L-tryptophan, which is converted by tryptophan 5-hydroxylase to 5-hydroxy-L-tryptophan. Serotonin formed by the action of the aromatic L-amino acid decarboxylase on 5-hydroxy-L-tryptophan is held in storage granules or vesicles within the serotonergic nerve terminals. This serotonin is released at nerve impulse into the synaptic cleft, where it acts on responsive sites called receptors on the neurone(s) on the other side of the synaptic junction. Serotonin in the synaptic cleft is inactivated by being transported back into the neurone that released it through the action of membrane carriers or uptake pumps on the serotonin neuronal membrane. Once inside the neurone, serotonin is already inactivated in so far as synaptic receptors are concerned. It may then be re-used in storage granules or degraded enzymatically through the action of mitochondrial monoamine oxidase.

Drug intervention is possible at several sites. Drugs may act directly on the synaptic receptors to mimic or antagonize the actions of serotonin. Drugs may inhibit the uptake pump, thereby prolonging the action of endogenously released serotonin on synaptic receptors. Drugs may release serotonin from storage granules into the synaptic cleft. Drugs may inhibit serotonin synthesis, resulting in decreased amounts of serotonin available for release. Serotonin precursors or inhibitors of monoamine oxidase increase the amounts of serotonin available for release. Drugs that act in each of these ways are known and have been studied in various ways. I will review some of the studies with these drugs, particularly in regard to their effects on mammalian brain function.

DIRECT SEROTONIN AGONISTS

Among compounds that mimic the action of serotonin on synaptic receptors, some more widely studied ones are listed in Table 1.

Various substitutions on the serotonin molecule can be made without loss of receptor activity. For instance, alkyl substituents can be added to the primary amine function, to the hydroxyl group, or to the alkyl side chain of serotonin with retention of pharmacologic activity on isolated peripheral smooth muscle or in respect to ability to compete with [^{3}H] serotonin binding to brain membrane receptors *in vitro*. Examples include *N,N*-dimethyl-5-methoxytryptamine and *N,N*-dimethyl-serotonin (bufotenin). In competing with [^{3}H] serotonin binding to calf brain synaptic membranes *in vitro*, these two compounds have IC_{50} values of 12 and 6nM, respectively, almost as low as that for serotonin itself (2.5 nM) (Whitaker and Seeman, 1978). Serotonin does not cross the blood–brain barrier and so does not stimulate brain receptors when injected systemically, except in newborn animals without a developed blood–brain barrier. The ring hydroxyl group is responsible for the inability of serotonin to penetrate into the brain, since tryptamine injected systemically does reach the brain. However, tryptamine and other compounds lacking the 5-hydroxy group are weak serotonin agonists based on *in vitro* binding studies (Bennett and Snyder, 1976; Whitaker and Seeman, 1978). Methylation of the hydroxyl group and of the amino group permits the molecule to penetrate the blood brain barrier and affords some protection against enzymic

Table 1. Direct-acting serotonin agonists

Compound	Source	Reference(s)
N,N-Dimethyl-5-methoxytryptamine	Aldrich Chemical, Regis	Whitaker and Seeman, 1978; Green and Grahame-Smith, 1975
Bufotenin	Aldrich Chemical	Whitaker and Seeman, 1978
Quipazine	Miles Laboratories	Rodriquez *et al.*, 1973; Green *et al.*, 1976
1-(*m*-Trifluoromethylphenyl)piperazine	Aldrich Chemical	Fuller *et al.*, 1978 Fuller *et al.*, 1980a
1-(*m*-Chlorophenyl)piperazine	Aldrich Chemical	Samanin *et al.*, 1979; Fuller *et al.*, 1980a; Rokosz-Pelc *et al.*, 1980
RU24969	Roussel	Euvrard and Boissier, 1980
MK-212	Merck	Clineschmidt *et al.*, 1977; Clineschmidt, 1979; Clineschmidt and McGuffin, 1978

destruction by monoamine oxidase. Thus *N,N*-dimethyl-5-methoxytryptamine is a serotonin analog that can be injected systemically to produce central serotonergic stimulation (Green and Grahame-Smith, 1976).

Several non-indole compounds that mimic the action of serotonin on synaptic receptors have also been identified. The first such compound to be extensively studied was quipazine, 1-(2-quinolyl)piperazine. This compound was observed several years ago to produce serotonin-like central effects in cats (behavioural alterations characterized by sham-rage reactions, lack of responsiveness to external stimuli and catatonic stereotyped postures as well as EEG synchronization) (Rodriguez *et al.*, 1973). More recently, quipazine has been shown to compete for [^{3}H] serotonin binding to brain membrane receptors (Whitaker and Seeman, 1978; Fuller *et al.*, 1978, 1980a), to reduce brain serotonin turnover (Fuller *et al.*, 1976), to cause a characteristic behavioural syndrome in rats (Green *et al.*, 1976), to cause head twitches in mice (Malick *et al.*, 1977), to decrease food intake in rats (Samanin *et al.*, 1977) and to have various other actions consistent with stimulation of central serotonergic receptors. Quipazine remains one of the most characterized and useful non-indole serotonin agonists, though its ability to inhibit monoamine oxidase competitively (Green *et al.*, 1976) and to inhibit the neuronal re-uptake of serotonin (Fuller *et al.*, 1976) suggests that some of its ability to enhance serotonergic function may result indirectly from presynaptic actions as well as direct stimulation of postsynaptic serotonin receptors.

Other more recently described substituted piperazines may have some potency and specificity advantages over quipazine. These include 1-(*m*-trifluoromethylphenyl)piperazine and 1-(*m*-chlorophenyl)piperazine, both of which are more potent than quipazine in competing for [^{3}H] serotonin binding to brain receptors *in vitro* and appear to act more purely on postsynaptic receptors with less presynaptic action. These compounds are potent appetite suppressants in animals (Samanin *et al.*, 1979; Fuller *et al.*, 1981).

Another compound, MK-212, though weaker in competing with [^{3}H] serotonin binding *in vitro*, has numerous effects *in vivo* resembling those of quipazine and thought to be due to direct stimulation of serotonin receptors (Clineschmidt, 1979). Quipazine is several hundred-fold more potent than MK-212 in stimulating serotonin receptors on the rat uterus *in vitro*, whereas it is less than 10-fold more potent than MK-212 in competing for [^{3}H] serotonin binding to rat brain membrane receptors *in vitro* and produces effects *in vivo* in approximately the same dosage range as MK-212. Although direct comparative data are lacking, the possibility has been suggested that MK-212 is a relatively selective agonist of central, as opposed to peripheral, serotonin receptors compared to quipazine (Clineschmidt, 1979). If central and peripheral serotonin receptors differ appreciably, complete selectivity may ultimately be achieved with agonists (and with antagonists).

A compound that acted as a serotonin agonist at brain but not peripheral serotonin receptors would not only be useful in studying physiological roles of serotonin experimentally but might have significant therapeutic advantages as well. As discussed elsewhere, a variety of possible therapeutic uses of serotonin agonists can be suggested. The only serotonin agonist on which clinical data appear to be in the literature is quipazine. Quipazine is reported to cause gastrointestinal side effects including nausea, vomiting, abdominal pain and diarrhoea in human subjects (Parati *et al.*, 1980). Presumably these side effects are due to direct stimulation of gastrointestinal serotonin receptors, and agents that stimulated only the brain serotonin receptors would lack these side effects. Additional studies are needed to ascertain if sufficient differences do exist between gut and brain serotonin receptors to permit completely selective agonist activity.

RU 24969 is a synthetic indole analogue which is a potent competitor of [^{3}H]-serotonin binding to brain membranes *in vitro* and which causes central effects *in vivo* thought to be due to stimulation of central serotonergic activity (Euvrard and Boissier, 1980).

Recently Peroutka and Snyder (1979) have postulated and defined two types of serotonin receptors in brain based on *in vitro* radioligand studies with rat frontal cerebral cortex. The 5-HT$_1$ receptor, preferentially labelled by [^{3}H] serotonin, may be linked to adenylate cyclase, since guanine nucleotides inhibited [^{3}H] serotonin binding much as they inhibited the binding of other neurotransmitters whose effects are known to be linked to an adenylate cyclase. A second type of serotonin receptor, the 5-HT$_2$ receptor, was defined as that labelled with [^{3}H] spiperone in frontal cortex, although relatively little evidence was presented to support the contention that the [^{3}H] spiperone binding site actually was a receptor for serotonin. Peroutka and Snyder (1979) cited electrophysiological results supporting the existence of multiple serotonin receptors in brain. If this concept is supported by further experimentation, it will be interesting to see if these different receptors can be linked to specific pharmacologic actions of various agonists, such as appetite suppression, the serotonin behavioural syndrome, body temperature changes, neuroendocrine effects, and so on, and also to see if completely selective agonists for the two different types of receptors can be found. Also the relationship of 5-HT$_1$ and 5-HT$_2$ receptors in the brain to serotonin receptors in various peripheral tissues needs to be examined.

One problem with the study of serotonin receptors in the brain is that functional effects are difficult to demonstrate *in vitro*. Radioligand binding studies are useful but give information only about affinity for the receptor, not necessarily whether a compound is an agonist or an antagonist. In peripheral smooth muscle, a contractile response can indicate agonist activity, whereas block of serotonin-induced contraction can indicate antagonist activity. Brain tissue does not contract in response to serotonin, and similar kinds of functional demonstration of agonist activity in brain have not usually been possible. An example of an experimental system in which this may be possible relates to a postulated role of serotonergic neurones in neuroendocrine regulation. Jones *et al.* (1976) and Buckingham and Hodges (1979) have shown that serotonin stimulates the release of corticotrophin-releasing factor from isolated rat hypothalamus *in vitro*. Their findings are compatible with a series of *in vivo* studies suggesting that serotonin neurones have a stimulatory influence on pituitary-adrenocortical function. Although results have not been reported with serotonin agonists other than serotonin itself, measuring corticotrophin-releasing factor secretion from the isolated rat hypothalamus may be a means of differentiating agonists from antagonists *in vitro*.

INDIRECT SEROTONIN AGONISTS

Inhibitors of serotonin uptake

Since neuronal re-uptake is thought to be the major means for inactivating serotonin that has been released into the synaptic cleft, inhibitors of the membrane uptake pump that takes up serotonin are expected to and apparently do enhance serotonergic activity. For many years, tricyclic antidepressant drugs and certain related compounds have been known to inhibit the neuronal uptake of serotonin as well as catecholamines. More recently, agents that inhibit the uptake of serotonin selectively have been identified and developed. These selective inhibitors of serotonin uptake have aided in establishing that neuronal re-uptake is an important means of inactivating serotonin released into the synaptic cleft by enabling investigators to associate various functional effects with blockade of serotonin uptake. Such association was not possible with uptake inhibitors that blocked the uptake of norepinephrine and other catecholamines as well as serotonin.

Fluoxetine (LY110140) was the first highly potent and selective inhibitor of serotonin uptake effective and selective *in vivo* to be described in the literature. Wong *et al.* (1974) reported that fluoxetine was selective *in vitro* and *in vivo*, in contrast to chlorimipramine, which was previously known to inhibit serotonin uptake selectively *in vitro* but which is not selective *in vivo*. The explanation for the differences between these two drugs *in vivo* despite their similarities *in vitro* lies in the fact that both drugs are metabolized by N-demethylation, the metabolite of fluoxetine being also a selective inhibitor of serotonin uptake whereas the metabolite of chlorimipramine is a more potent inhibitor of norepinephrine uptake (see Fuller *et al.*, 1977).

Soon after fluoxetine was described, various other selective serotonin uptake inhibitors were reported. Several of these are listed in Table 2. Although these compounds differ somewhat in potency in various test systems for evaluating serotonin uptake, they all appear to be capable of inhibiting serotonin uptake *in vivo* while having little or no effect on the uptake of norepinephrine or other catecholamines. As such, these compounds are useful experimental tools for studying serotonin neurones in animals. Some of these compounds are being evaluated in humans for the treatment of mental depression and other diseases. Presently there are published reports claiming that both zimelidine (Cox *et al.*, 1978) and fluvoxamine (Saletu *et al.*, 1977) are effective antidepressant drugs. Early clinical studies with fluoxetine also indicate antidepressant activity for this drug.

Table 2. Selective inhibitors of serotonin uptake

Compound	Source	Reference(s)
Alaproclate	Astra	Lindberg et al., 1978
CGP 6085 A	Ciba-Geigy	Waldmeier et al., 1979
Citalopram	Lundbeck	Hyttel, 1977, 1978
Femoxetine	Ferrosan	Buus Lassen et al., 1975
Fluoxetine	Lilly	Wong et al., 1974;
		Fuller et al., 1975
Fluvoxamine	Philips-Duphar	Claassen et al., 1977
LM5008	Groupe Pharmuka	LeFur et al., 1978
Org 6582	Organon	Sugrue et al., 1976
Paroxetine	Ferrosan	Buus Lassen, 1978
Pirandamine	Ayerst	Lippmann and Pugsley, 1976
Ro 11-2465	Roche	Burkard, 1980
Zimelidine	Astra	Ross et al., 1976;
		Ross and Renyi, 1977

There seems little doubt that uptake inhibitors acutely increase serotonin concentration in the synaptic cleft and enhance serotonergic function. However, the possibility has been suggested that chronic administration of uptake inhibitors may not be associated with increased serotonergic function (Hwang *et al.*, 1980). At least two types of adaptive changes may occur when serotonin uptake inhibitors are given. The first is a rapid compensatory decrease in serotonergic neurone firing and in serotonin release. This response may occur by increased stimulation of pre-synaptic autoreceptors that sense the concentration of serotonin in the neuronal synapse. The second is an adaptive down-regulation of post-junctional serotonin receptors that may occur over a period of several days or a few weeks. Some investigators have demonstrated these receptor changes (Segawa *et al.*, 1979); Fuxe *et al.*, 1979; Maggi *et al.*, 1980) by radioligand binding studies, whereas others have observed no effects after chronic treatment with serotonin uptake inhibitors (Wirz-Justice *et al.*, 1978; Savage *et al.*, 1979). Further investigation is necessary to resolve the discrepancies, which might be due to different durations of treatment or to other variables.

Both types of adaptive changes would serve to offset the direct effect of uptake inhibitors and may represent homeostatic mechanisms. The question of whether or not serotonergic function continues to be enhanced after prolonged treatment with uptake inhibitors is of obvious importance. Serotonergic function might even be decreased if the net effect of the adaptive changes were greater than that of continued uptake inhibition. This issue can only be clarified through further research and is mentioned here to alert the reader of the possible complexity associated with the use of serotonin- uptake-inhibiting drugs chronically.

Serotonin-releasing drugs

Increased concentrations of serotonin in the synaptic cleft can be produced by stimulating the release of serotonin as well as by inhibiting its re-uptake. Although agents like reserpine, Ro 4-1284, and tetrabenazine release serotonin, they also release other monoamines including catecholamines and hence are of little value as specific pharmacologic tools. Because of their non-specificity, they will not be discussed further here. Some compounds that release serotonin with a high degree of selectivity are listed in Table 3. *p*-Chloroamphetamine is perhaps the most studied of these.

Although *p*-chloroamphetamine has complex actions on serotonin neurones including neurotoxic effects leading to long-lasting depletion of serotonin levels, the acute effects after a single injection of *p*-chloroamphetamine appear to be mediated by serotonin release. The release of serotonin from neurones was demonstrated *in vivo* in rats through the use of push–pull cannulae implanted into the dorsal hippocampus (Gallager and Sanders-Bush, 1973). Various acute effects of *p*-chloroamphetamine have been associated with serotonin release, including production of the serotonin behavioural syndrome (Trulson and Jacobs, 1976), suppression of food intake (Kaergaard Nielsen *et al.*, 1967), and elevation of serum corticosterone and prolactin (Fuller and Snoddy, 1980; Fuller *et al.*, 1980b).

Table 3. Serotonin releasers

Compound	Source	Reference(s)
p-Chloroamphetamine	Regis	Gallager and Sanders-Bush, 1973; Marsden et al., 1979; Ross, 1979; Trulson and Jacobs, 1976
Fenfluramine	A. H. Robins	Clineschmidt and McGuffin, 1978; Trulson and Jacobs, 1976
α-Methyltryptamine	Aldrich Chemical	Marsden, 1979; Ross, 1979

A new technique for monitoring extraneuronal concentrations of serotonin *in vivo* affords a more elegant and immediate demonstration of serotonin release. This technique involves electrochemical detection of serotonin by means of electrodes implanted directly into serotonin-rich regions of brain (Adams *et al.*, 1978). Using a micrographite electrode stereotaxically placed into the striatum, Marsden *et al.* (1979) demonstrated that *p*-chloroamphetamine caused voltammetric changes indicative of increased extraneuronal serotonin concentrations in freely moving unanaesthetized rats. The electrochemical signal was increased within 10-20 min after the ip injection of *p*-chloroamphetamine at a dose of 5 mg/kg, and the increase lasted for about 90 min. These signal changes roughly paralleled the behavioural effects of *p*-chloroamphetamine attributed to serotonin release (lateral head weaving, forepaw treading, hind limb abduction and Straub tail). Similar results were obtained with electrodes placed in the dorsal hippocampus. Verification that the increased electrochemical signal was due to serotonin released by *p*-chloroamphetamine was made through cyclic voltammetry showing that the peak obtained after *p*-chloroamphetamine occurred at the proper voltage and by showing that *p*-chloroamphetamine pretreatment to deplete serotonin stores or fluoxetine pretreatment to block *p*-chloroamphetamine uptake into serotonin neurones prevented the increase in electrochemical signal produced by *p*-chloroamphetamine and attenuated or prevented the behavioural response as well.

Numerous structural analogues of *p*-chloroamphetamine also have been found to influence serotonin levels in a manner similar to *p*-chloroamphetamine itself and presumably release serotonin as well (see Fuller, 1978). Such analogues include amphetamines with substituents other than halogens in the para position, for example *p*-methoxyamphetamine (Clineschmidt and McGuffin, 1978).

Although *p*-chloroamphetamine has been reported to have appetite suppressant activity (Kaergaard Nielsen *et al.*, 1967) and antidepressant activity (Van Praag and Korf, 1973) in humans, fenfluramine is the only drug listed in Table 3 that is currently used in clinical therapy. Fenfluramine is marketed as an appetite suppressant drug for use in the treatment of obesity but has been reported to have antihypertensive effects in humans as well (Lake *et al.*, 1979). In animals, fenfluramine produces numerous effects resembling those of *p*-chloroamphetamine and attributed to release of serotonin, including behavioural (Trulson and Jacobs, 1976), appetite suppressant (Garattini *et al.*, 1975), and neuroendocrine (Quattrone *et al.*, 1978; Fuller and Snoddy, 1980) changes.

A structural analogue of serotonin that is included in Table 3 is α-methyltryptamine. This compound has relatively little affinity for the serotonin receptor but is a potent releaser of serotonin from brain synaptosomes *in vitro* (Ross, 1979). It also has been shown to release serotonin *in vivo* by electrochemical measurements *in situ* as discussed above for *p*-chloroamphetamine (Marsden, 1979).

Both *p*-chloroamphetamine and fenfluramine are reported to cause long-lasting neurotoxic effects on brain serotonin neurones in rats (Harvey, 1978). The extent to which these occur in other species and in particular in humans at the relatively low oral doses used therapeutically is still incompletely understood. Of the drugs listed in Table 3, α-methyltryptamine is the only one not known to be capable of producing neurotoxic effects on serotonin neurones, suggesting a possible advantage of this agent over the others. However, α-methyltryptamine has been studied less extensively as a serotonin releaser *in vitro* than have the other compounds.

Serotonin precursors

Serotonin is formed from L-tryptophan via the intermediate 5-hydroxy-L-tryptophan. The administration of either of these amino acids can increase the amount of serotonin formed and released into the synaptic cleft. The relative merits of each of these amino acids as a means of enhancing function have been debated.

5-Hydroxy-L-tryptophan is the immediate precursor to serotonin, and a higher percentage of an administered dose is converted to serotonin than is true with L-tryptophan. The only enzyme required for conversion of ·5-hydroxy-L-tryptophan to serotonin is the aromatic L-amino acid decarboxylase (EC 4.1.1.26). This enzyme is not restricted to serotonin neurones but appears to be relatively ubiquitous, playing a physiological role in the biosynthesis of catecholamines and perhaps various other amines. Thus when 5-hydroxy-L-tryptophan is administered, it can be converted to serotonin in cells other than serotonin neurones. The possibility of non-specific effects occurring through formation of serotonin at sites that normally do not form it is a criticism of 5-hydroxy-L-tryptophan as a means of influencing serotonergic function.

L-tryptophan requires the enzyme tryptophan 5-hydroxylase (EC 1.14.16.4) to be converted to serotonin. This enzyme apparently is present only within cells that normally synthesize serotonin, so the possibility of non-specific effects of tryptophan due to formation of serotonin at non-physiological sites is not of concern. There are, however, other concerns about tryptophan as a means of enhancing serotonergic function. Only a small percentage of an administered dose of tryptophan is converted to serotonin. Thus relatively large doses of tryptophan are required to produce relatively small increases in the amount of serotonin formed. Tryptophan is converted by numerous other enzymic pathways, for instance pyrrole ring cleavage leading to nicotinic acid and various other metabolites, decarboxylation to tryptamine and subsequent potential conversion to methylated or deaminated metabolites of tryptamine, and transamination. In addition, tryptophan is an essential amino acid required for protein synthesis. In the rat, tryptophan has a special role in protein synthesis, and injection of tryptophan increases polysome aggregation and the synthesis of various proteins (Smith *et al.,* 1979). Thus the possibility of non-specific effects of L-tryptophan is real, although those non-specific effects are different than would be obtained with 5-hydroxy-L-tryptophan.

In fact, histochemical studies have suggested that only at large doses of 5-hydroxy-L-tryptophan are substantial quantities of serotonin formed indiscriminately. When doses of 500–1000 mg/kg of 5-hydroxytryptophan were given, particularly in combination with a peripheral decarboxylase inhibitor, diffuse fluorescence attributed to serotonin was observed, suggestive of serotonin formation in catecholamine neurones and possibly other non-serotonin sites in the brain (Fuxe *et al.*, 1971). On the other hand, after lower doses of 5-hydroxytryptophan (20–100 mg/kg), the presence of serotonin derived from the amino acid could be detected only in serotonin neurones, suggesting a relative specificity of enhanced serotonin formation.

Two means are available for enhancing the specificity of injected 5-hydroxytryptophan. One is to co-administer an inhibitor of decarboxylase that does not penetrate the blood–brain barrier. Such an inhibitor prevents the decarboxylation of 5-hydroxytryptophan in peripheral tissues, reducing side effects due to formation of serotonin in the periphery and increasing the amount of 5-hydroxytryptophan available for uptake into the brain. Decarboxylase inhibitors that have been particularly useful in this regard are carbidopa (α-methyldopa hydrazine) and benserazide (Ro 4-4602). A second means of enhancing the specificity of 5-hydroxytryptophan that can be used in conjunction with peripheral decarboxylase inhibition is to pretreat with a selective inhibitor of serotonin uptake. Any effects of 5-hydroxytryptophan that are mediated by serotonergic synapses should be enhanced by pretreatment with the uptake inhibitor to potentiate the effects of serotonin formed from the precursor and released into the synaptic cleft, whereas non-specific effects not mediated by serotonergic synapses should not be potentiated since serotonin uptake pumps would not play a role in limiting those effects. Very low doses of 5-hydroxy-L-tryptophan are effective in various experimental paradigms when given in combination with an uptake inhibitor.

In general, I would advocate the use of low doses of 5-hydroxy-L-tryptophan in combination with a selective inhibitor of serotonin uptake (and perhaps also a peripheral decarboxylase inhibitor) as a means of enhancing central serotonergic function by precursor loading, although others feel that L-tryptophan may be a more physiological means of enhancing serotonin formation. Both approaches may be taken if a clear basis for choosing between them is not available in a given experimental situation.

Monoamine oxidase inhibitors

Since inhibitors of monoamine oxidase increase brain levels of other monoamines (such as dopamine, norepinephrine, and epinephrine) as well as serotonin, they have limited use in themselves as specific tools for altering serotonergic function. Monoamine oxidase inhibitors may be useful in conjunction with serotonin precursors to verify that effects of the precursors are mediated by an amine rather than the amino acid itself or some other metabolite.

Monoamine oxidase inhibitors have sometimes been used as tools for evaluating serotonin turnover, using the rate of accumulation of serotonin content after monoamine oxidase inhibition as an index of turnover. This procedure seems clearly to be inappropriate when applied to uptake inhibitors and thus must be questioned when used in conjunction with other kinds of drugs as well. Several groups (Fuller and Steinberg, 1976; Hyttel, 1977; Marco and Meek, 1979) have found that serotonin uptake inhibitors do not diminish the rate of serotonin accumulation after monoamine oxidase inhibition, although these uptake inhibitors decrease serotonin turnover as shown by several other methods for evaluating serotonin turnover. The basis of this unexpected finding is not understood, but Marco and Meek (1979) have suggested that measurement of changes in serotonin accumulation after monoamine oxidase inhibition is a poor index of serotonin turnover.

SEROTONIN RECEPTOR ANTAGONISTS

Some of the compounds considered to be antagonists of brain serotonin receptors are listed in Table 4. Among other effects, serotonin antagonists are reported to antagonize 5-hydroxytryptophan-induced head twitch in mice (Corne *et al.*, 1963), agonist or precursor induced neuroendocrine effects such as elevation of serum corticosterone (Fuller and Snoddy, 1979) and prolactin (Quattrone *et al.*, 1978), the serotonin behavioural syndrome (Deakin and Green, 1978), the appetite suppressant effects of direct and indirect acting agonists (Samanin *et al.*, 1977, 1979; Garattini *et al.*, 1979), the hind limb flexor reflex in spinal rats (Maj *et al.*, 1976), the convulsant effects of injected tryptamine (Przegalinski *et al.*, 1979; Vargaftig *et al.*, 1971), the hyperthermic effects of direct or indirect serotonin agonists (Maj *et al.*, 1978), and the antihypertensive effects of direct or indirect serotonin agonists (T. T. Yen and R. W. Fuller, data to be published). Thus there seems little doubt that serotonin antagonists are effective in blocking brain serotonin receptor *in vivo*.

Serotonin antagonists like methysergide, cyproheptadine, cinanserin, methiothepin and metergoline have failed to block the inhibitory effect of microiontophoresed serotonin applied to brain areas known to receive serotonergic input in electrophysiological studies (Haigler and Aghajanian, 1977). The excitatory effects of serotonin microiontophoresed onto some other brain regions are blocked by these antagonists. Recently, McCall and Aghajanian (1979) postulated based on results from electrophysiological studies that at least two types of postsynaptic serotonin receptors exist in the rat brain. The first type of receptor mediates depression of neuronal firing and is not blocked by classical serotonin antagonists, while the second type facilitates excitatory inputs and is antagonized by methysergide, at least. Whether other serotonin antagonists antagonize these latter receptors remains to be studied. The presynaptic receptors involved in serotonin-mediated inhibition of serotonergic neurones in the dorsal raphe nucleus are not blocked by peripheral serotonin antagonists, which may explain why these antagonists do not increase serotonin turnover (see below). Eventually it may be possible to associate these electrophysiologically distinguished receptor types with specific serotonergic functions or with receptors characterized by radioligand binding.

Although increased neurotransmitter turnover occurs in the brain when dopamine receptors and norepinephrine receptors are blocked, a similar compensatory increase in turnover may not occur with serotonin. In general, serotonin antagonists have failed to increase serotonin turnover. One exception is methiothepin, which does increase serotonin turnover (Jalfre *et al.*, 1974). However, there is some evidence that the effect of methiothepin may be mediated at least partly by an increase in tryptophan concentration in brain (Jacoby *et al.*, 1975). Also, methiothepin is known to be a relatively non-specific drug, affecting for example catecholamine receptors as well (Jalfre *et al.*, 1974). Thus the increase in serotonin turnover produced by methiothepin may not be simply a consequence of its blocking serotonin receptors. In any case, most other antagonists which do cause various pharmacologic effects thought to be due to blocking brain serotonin receptors do not increase brain serotonin turnover or block the decrease in turnover caused by an agonist. One explanation might be that serotonin turnover is not being tonically inhibited by serotonin receptor mediated mechanisms. Another is that the postsynaptic serotonin receptor antagonists do not antagonize presynaptic autoreceptors controlling serotonin turnover.

Originally, Snyder and his colleagues proposed that serotonin antagonists could be discriminated from agonists by radioligand binding studies *in vitro* using [^{3}H] serotonin and [^{3}H] LSD as radioligands (Bennett and Snyder, 1976). Agonists inhibited [^{3}H] serotonin binding more effectively than [^{3}H] LSD binding, whereas the converse was true with antagonists. [^{3}H] LSD was thought to label an antagonist state of the serotonin receptor, whereas [^{3}H] serotonin was thought to label an agonist state. Recently these workers (Peroutka and Snyder, 1979) have suggested a different basis for the dissimilarity between [^{3}H] serotonin and [^{3}H] LSD binding. [^{3}H] Serotonin is suggested to label a 5-HT$_1$ receptor, and [^{3}H] spiperone is suggested to label a 5-HT$_2$ receptor in brain. [^{3}H] LSD is thought to label both receptors. The 5-HT$_1$ and 5-HT$_2$ sites are suggested to be different receptor populations not having to do with agonist versus antagonist states.

Trazodone, which is listed in Table 4, is reported to have serotonin antagonist effects at low doses in rats but to have agonist effects at higher doses (Maj *et al.*, 1979). Trazodone is known to be metabolized to 1-(*m*-chlorophenyl)piperazine, a serotonin agonist listed in Table 1 (Melzacka *et al.*, 1979). A likely explanation for the biphasic pharmacologic effects of trazodone is that the antagonist effects at low doses are due to trazodone itself whereas the agonist effects at higher doses are due to the metabolite. We have found trazodone to be a very weak inhibitor of [^{3}H] -serotonin binding to rat brain membranes *in vitro* but a potent inhibitor of [^{3}H] -LSD binding, the degree of selectivity being higher than for any other compound we have studied to date (Fuller *et al.*, 1981). This selectivity may make trazodone a particularly useful compound in further understanding the nature of serotonin receptors in brain in association with particular physiologic functions.

A compound not included in Table 4 because the compounds listed there are serotonin antagonists effective in the brain is xylamidine. Xylamidine is a serotonin antagonist that apparently does not penetrate the blood-brain barrier so that when it is injected systemically it blocks serotonin receptors in the periphery but not in the brain (Copp *et al.*, 1967). Thus xylamidine has been used as a tool for deter-

Table 4. Serotonin receptor antagonists

Compound	Source	Reference(s)
Benzoctamine	Ciba/Geigy	Przegalinski et al., 1978
Cinanserin	Squibb	Clineschmidt and Lotti, 1974
Cyproheptadine	Merck	Corne et al., 1963;
		Maj et al., 1976;
		Clineschmidt and Lotti, 1974
Danitracen	Dr K. Thomae	Maj et al., 1976;
		Maj et al., 1978
Etoperidone	Angelini Francesco	Przegalinski and Lewandowska, 1979
Metergoline	Farmitalia	Ferrini and Glasser, 1965;
		Maj et al., 1978;
		Fuxe et al., 1978
Methiothepin	Roche	Jalfre et al., 1974
Methysergide	Sandoz	Corne et al., 1963;
		Clineschmidt and Anderson, 1970
Mianserin	Organon	Vargaftig e al., 1971;
		Maj et al., 1978
Pizotyline	Sandoz	Przegalinski et al., 1979
Trazodone	Angelini Francesco	Maj et al., 1979;
	Mead Johnson	Baran et al., 1979
LY53857	Lilly	Fuller and Snoddy, 1979

mining if an effect of direct or indirect serotonin agonists is centrally mediated (Clineschmidt and McGuffin, 1978).

SEROTONIN DEPLETORS

The lowering of serotonin concentration in brain can be brought about by inhibiting synthesis of serotonin or by impairing its storage in intraneuronal granules.

A widely used inhibitor of serotonin synthesis is p-chlorophenylalanine (Koe and Weissman, 1968). p-Chlorophenylalanine is only a weak inhibitor of tryptophan 5-hydroxylase *in vitro* but causes long-lasting irreversible inhibition of the enzyme *in vivo*. p-Chlorophenylalanine leads to a relatively rapid depletion of serotonin content, and approximately 1 week is required for tryptophan 5-hydroxylase activity and serotonin concentration to return to normal in rat brain. Thus p-chlorophenylalanine is useful when given alone to determine the functional consequences of depleting serotonin, and it is useful to give in advance of indirec-acting serotonin agonists like uptake inhibitors, releasers, or the precursor amino acid L-tryptophan to determine if effects produced by these agents are in fact mediated by serotonin. A disadvantage of p-chlorophenylalanine is that it is not absolutely specific, causing small and transient effects, for example, on catecholamines in brain and inhibiting phenylalanine hydroxylase in liver as well as tryptophan hydroxylase. An additional disadvantage of p-chlorophenylalanine is that high doses have to be given (usually 100–320 mg/kg), and these have effects on tissue uptake and perhaps utilization of other amino acids. p-Chlorophenylalanine is metabolized relatively slowly in the rat and persists for some time at high levels in blood and tissues. Despite these limitations, p-chlorophenylalanine has been and remains a useful tool in studies of serotonergic systems.

Various reversible inhibitors of tryptophan 5-hydroxylase are also known. One that has been used extensively is α-propyldopacetamide. This compound inhibits tryptophan hydroxylase by virtue of being a catechol. It therefore inhibits other tetrahydropteridine-requiring enzymes as well, including tyrosine hydroxylase, and thus is not specific. α-Propyldopacetamide is useful in evaluating serotonin turnover. Agents that accelerate serotonin turnover hasten the disappearance of serotonin after α-propyldopacetamide injection, for example. Other reversible inhibitors of tryptophan hydroxylation include 6-fluorotryptophan.

Because of the ubiquitous nature of the aromatic L-amino acid decarboxylase and the relatively high activity of this enzyme, it has not represented a promising target as a means of inhibiting serotonin synthesis completely and selectively. Inhibitors of 5-hydroxytryptophan decarboxylation are useful, however, as an alternative means of evaluating serotonin turnover. The rate of accumulation of 5-hydroxytryptophan after decarboxylase inhibition has been used to determine the influence of various types of drugs on serotonin turnover in brain.

A potent enzyme-activated irreversible inhibitor of the aromatic L-amino acid decarboxylase, α-monofluoromethyldopa, has recently been described to lower serotonin as well as catecholamine concentration in brain (Jung *et al.*, 1979).

Various agents deplete serotonin stores by causing their release or by impairing serotonin storage in granules. Many of these agents are non-specific, e.g., reserpine, tetrabenazine, Ro 4-1284, and so have limited use as pharmacologic tools because they affect catecholamine stores as well. Other compounds are reasonably specific depletors of serotonin. These include compounds already discussed in the section on serotonin releasers, namely *p*-chloroamphetamine, fenfluramine, and various structural analogues of these compounds.

p-Chloroamphetamine causes a rapid depletion of serotonin in rat brain. The effect is very long-lasting, serotonin concentration remaining decreased for at least several months after a single ip injection of *p*-chloroamphetamine (Sanders-Bush *et al.*, 1972). The depletion of serotonin is not immediately irreversible, however. When an uptake inhibitor is given prior to *p*-chloroamphetamine or soon after *p*-chloroamphetamine, the depletion of serotonin is either prevented or is reversed totally. However, if the uptake inhibitor is not given until 24–32 h after *p*-chloroamphetamine, the depletion of serotonin is no longer reversible (Fuller *et al.*, 1975). Between 4 and 24 h after *p*-chloroamphetamine, there is a progressive decline in the degree of reversibility of the serotonin depletion. Similar relationships hold for fenfluramine and other halogenated amphetamines (Fuller, 1978). Although the initial depletion of serotonin by these drugs appears to be due to release of serotonin, the mechanism of the persistent effects is still not understood. Despite that, the compounds can be used as pharmacologic tools for depleting serotonin, though investigators must bear in mind that the effect, at least in the rat, becomes irreversible.

The long-lasting effects of *p*-chloroamphetamine and related compounds resemble in several ways those of the serotonin neurotoxins, 5,6- and 5,7-dihydroxytryptamine. These compounds are covered in a separate chapter in this book and will not be discussed in detail here. The dihydroxytryptamines do not penetrate the blood-brain barrier in adult animals and so must be injected directly into the brain or cerebrospinal fluid, whereas the halogenated amphetamines can be injected systemically. There are also differences between these groups of compounds in their effects on serotonin in various brain regions.

POTENTIAL THERAPEUTIC USES OF DRUGS AFFECTING SEROTONERGIC FUNCTION

There are several potential therapeutic applications of drugs that affect serotonergic function, though it has not been proven that any drugs currently in clinical use actually act in this manner entirely. An agent that enhanced central serotonergic function might, however, be useful in the treatment of mental depression or other behavioural disorders, in the treatment of intention myoclonus, for appetite suppression, as an analgesic adjunct, to lower blood pressure in hypertensive patients, in the treatment of sleep disorders like narcolepsy/cataplexy, or in the treatment of alcoholism. Serotonin antagonists may be useful for treating migraine, hypersecretory endocrine disorders, or possibly other diseases. The rationale for these suggested potential uses is summarized briefly in the following paragraphs.

An involvement of brain serotonin in mental depression has long been suspected. Some investigators have reported decreased amounts of serotonin or of its principal metabolite 5-hydroxyindoleacetic acid in post-mortem studies on brains of depressed patients or of cerebrospinal fluid levels of 5-hydroxyindoleacetic acid in living depressed patients. Some investigators have also reported decreased concentrations of 5-hydroxyindoleacetic acid in the urine of at least some depressed patients. Findings of these sorts have remained controversial, and not all investigators have detected such differences between depressed and non-depressed patients. Most drugs used in the treatment of mental depression do influence brain serotonergic systems. These include monoamine oxidase inhibitors, which increase brain levels of serotonin as well as other monoamines (norepinephrine, dopamine, epinephrine). They also include various tricyclic drugs that inhibit neuronal reuptake, many of these affecting serotonin as well as norepinephrine and epinephrine or dopamine neurones. There have been claims that serotonin precursors (tryptophan or 5-hydroxytryptophan) have antidepressant effects on their own or enhance the effects of monoamine oxidase-inhibiting or uptake-inhibiting drugs. Some workers have found that inhibiting serotonin synthesis can counteract the effect of monoamine oxidase-inhibiting or uptake-inhibiting antidepressant drugs (Shopsin *et al.*, 1975, 1976). The possibility has been suggested that subgroups of depressed patients exist, some of whom are deficient in serotonergic function whereas others are not (Asberg *et al.*, 1976). For additional reading on serotonin involvement in mental depression, the reader is referred to recent reviews by Murphy *et al.* (1978) and by Burns and Mendels (1979).

As mentioned earlier in this chapter, two selective inhibitors of serotonin uptake, fluvoxamine and zimelidine, are reported to have antidepressant efficacy in humans. Interestingly, both compounds were claimed to produce significant improvement in depressive symptoms as early as the first week of treatment. Typically antidepressant drugs have required 2–3 weeks for onset of antidepressant activity, and this delay has been a problem in the management of severely depressed (especially suicidal) patients. If it is possible to speed up the onset of antidepressant activity by influencing serotonergic systems selectively, improvement in the drug therapy of depression would be realized.

The possibility of combining drugs that inhibit serotonin uptake with serotonin precursors or with monoamine oxidase inhibitors has been discussed in the literature, and apparently in both cases further experimental work is required to evaluate the potential benefits (and risks) of such combination therapy. In addition to serotonin uptake inhibitors, monoamine oxidase inhibitors, and serotonin precursors, direct-acting serotonin agonists would seem to have potential application as antidepressant drugs as well, a possibility that remains to be evaluated.

Besides mental depression, other types of behavioural disorders may involve serotonergic dysfunction or respond to drugs that alter serotonergic function. Yaryura-Tobias (1977) has theorized that obsessive-compulsive disorders may involve a serotonergic disturbance and that the reported efficacy of drugs like chlorimipramine in this disease involves enhancement of serotonergic activity through inhibition of the neuronal reuptake of serotonin.

Post-hypoxic intention myoclonus has been treated successfully by several investigators through the administration of 5-hydroxytryptophan alone or in conjunction with carbidopa, an inhibitor of decarboxylase that acts only in the periphery (Van Woert *et al.*, 1977). Recently the effects of 5-hydroxytryptophan have been found to be dramatically enhanced in a small number of patients studied by the co-administration of an inhibitor of serotonin uptake, fluoxetine (Van Woert *et al.*, 1980). Additional investigation is required to verify the synergistic effects of serotonin uptake inhibition and serotonin precursor loading, which would strongly support the idea that enhancement of central serotonergic function is of value in this disease.

Fenfluramine, a widely used appetite suppressant drug, appears to reduce food intake through stimulation of brain serotonin receptors by release of endogenous serotonin (Garattini *et al.*, 1979). In addition, several direct serotonin agonists are known to have appetite suppressant activity in animals (Samanin *et al.*, 1977, 1978, 1980; Clineschmidt, 1979, Fuller *et al.*, 1981). Thus appetite suppression in the management of obesity is one clearly defined use of agents that enhance central serotonergic activity. The recent findings of Wurtman and Wurtman (1979) that drugs acting in this way selectively reduce non-protein caloric intake in animals suggest that these agents may be a particularly attractive class of anti-obesity drugs.

Abundant evidence has accumulated over the past few years that the analgesic effect of morphine in several models of experimental pain in animals is enhanced by serotonin uptake inhibitors or direct-acting serotonergic receptor stimulants (see, for example, Ogren and Holm 1980; Malec and Langwinski, 1980). Whether such a drug combination would be useful clinically in the management of pain remains to to be evaluated, but recent findings that the potentiation of morphine analgesia by uptake inhibitors is particularly striking in morphine-tolerant animals and that the side effect of morphine that commonly limits the doses that can be used clinically, namely respiratory depression, is not enhanced by uptake inhibitors (M. D. Hynes and R. W. Fuller, data to be published) make attractive the possibility that a combination of a direct or indirect serotonin stimulant with morphine could be therapeutically useful.

Blood pressure in spontaneously hypertensive rats can be lowered by the administration of tryptophan (Sved *et al.*, 1979) or 5-hydroxytryptophan, which is active at a very low dose when combined with an inhibitor of serotonin uptake (Fuller *et al.*, 1979). Direct-acting serotonin agonists and fenfluramine, a serotonin-releasing drug, also lower blood pressure in spontaneously hypertensive rats (T. T. Yen and R. W. Fuller, data to be published). The observation that fenfluramine lowers blood pressure in hypertensive human patients (Lake *et al.*, 1979) suggests that drugs enhancing serotonergic function might have therapeutic application in the treatment of hypertension.

Cataplexy, sudden, and often total, muscle weakness usually triggered by emotion or muscular activity, often occurs in humans in association with narcolepsy. Although the sleep attacks in humans with narcolepsy can often be controlled with stimulant drugs, when cataplexy occurs in these patients it usually does not respond to those drugs. Recently drugs that selectively inhibit the neuronal re-uptake of serotonin have been found to be effective in the treatment of cataplexy in dogs (Babcock *et al.,* 1976) as well as in humans (see Schachter and Parkes, 1980).

Among numerous studies of drug effects on the voluntary ingestion of ethyl alcohol by rats, the serotonin precursor 5-hydroxytryptophan (Zabik *et al.,* 1978) and selective inhibitors of serotonin uptake (Rockman *et al.,* 1979) have been found to decrease alcohol consumption. Although the possibility that drugs which enhance serotonergic function might be useful clinically in the management of alcoholism is raised, no data are available presently to support this possibility.

Another potential use of drugs affecting serotonergic systems is in the treatment of migraine. A possible role of serotonin in the pathogenesis of migraine has long been suspected. Indeed, several compounds thought to be serotonin antagonists have been used in the prevention or treatment of migraine attacks. The mechanism by which these compounds act remains controversial, and if serotonin receptors are involved in the anti-migraine effects of these drugs the exact nature of the drug's interaction with these receptors remains to be elucidated (Hardebo *et al.,* 1978). For example, the possibilities have been suggested that these drugs mimic the contractile effect of serotonin on cerebral arteries, antagonize the contractile effect of serotonin, or sensitize the vessels to serotonin. If serotonin is involved in the pathogenesis of migraine, is it serotonin present in neurones that innervate the cranial vessels or serotonin present in and released from blood platelets, or both, that is involved? Until these issues are better understood, uncertainty remains about the required characteristics of serotonergic drugs that may have potential in the treatment of migraine.

A serotonin antagonist, cyproheptadine, is reported to be effective in lowering adrenocorticotrophic hormone secretion in patients with Cushing's disease or Nelson's syndrome (Krieger, 1980). Considering the evidence from animal studies that serotonin neurones have a stimulatory influence in the hypothalamic regulation of adrenocorticotrophic hormone release from the pituitary, one imagines that the effect of cyproheptadine is mediated by blockade of serotonin receptors in the brain, but this point remains unproven. Additional clinical data are required before the potential of cyproheptadine or other serotonin antagonists in treating these or other neuroendocrine disorders can be better understood.

SUMMARY

The function of serotonergic neurones can be altered by agents that mimic or antagonize serotonin action on synaptic receptors, by agents that inhibit the neuronal uptake of serotonin thereby prolonging the action of released serotonin on synaptic receptors, by agents that release serotonin or impair its storage in intraneuronal granules, by precursors or monoamine oxidase inhibitors that increase the intra-neuronal stores of serotonin available for release, or by inhibitors of serotonin synthesis or agents that otherwise deplete serotonin content. Agents that enhance serotonergic neuronal function have been found to cause behavioural changes, increase serum corticosterone and prolactin concentration, have antihypertensive effects, produce analgesia or enhance the action of analgesic drugs, suppress food intake, and to exert certain other actions in experimental animals. Several potential therapeutic uses of such drugs based on their properties in animals or postulates about disease aetiology are discussed. Agents that block serotonin receptors or deplete serotonin stores antagonize the actions of direct or indirect serotonin agonists and may have therapeutic utility as well.

REFERENCES

Adams, R. N., Conti, J., Marsden, C. A., and Strope, E. (1979). The measurement of dopamine and 5-hydroxytryptamine release in CNS of freely moving unan-aesthetised rats, *Brit. J. Pharmacol.* **64**, 470–471.

Asberg, M., Thorne, P., Traskman, L., Bertilsson, L., and Ringberger, V. (1976). Serotonin depression—A biochemical subgroup within the affective disorders, *Science,* **191**, 478–480.

Babcock, D. A., Narver, E. L., Dement, W. C., and Mitler, M. M. (1976). Effects of imipramine, chlorimipramine, and fluoxetine on cataplexy in dogs, *Pharmacol. Biochem. Behav.* **5**, 599–602.

Baran, L., Maj. J., Rogoz, Z., and Skuza, G. (1979). On the central antiserotonin action of trazodone, *Pol. J. Pharmacol. Pharm.* **31**, 25–33.

Bennett, Jr., J. P. and Snyder, S. H. (1976). Serotonin and lysergic acid diethy-lamide binding in rat brain membranes: relationship to postsynaptic serotonin receptors, *Mol. Pharmacol.,* **12**, 373–389.

Buckingham, J. C. and Hodges, J. R. (1979). Hypothalamic receptors influencing the secretion of corticotrophin releasing hormone in the rat, *J. Physiol.* **290**, 421–431.

Burkard, W. P. (1980). Specific binding sites in rat brain for a new and potent inhibitor of 5-hydroxytrytamine uptake, *Eur. J. Pharmacol.* **61**, 409–410.

Burns, D. D. and Mendels, J. (1979). Serotonin and affective disorders, in *Current Developments in Psychopharmacology* (Eds W. B. Essman and L. Valzelli), Vol. 5, pp. 293–359, Medical and Scientific Books, New York.

Buus Lassen, J. (1978). Potent and long-lasting potentiation of two 5-hydroxy-tryptophan-induced effects in mice by three selective 5-HT uptake inhibitors, *Eur. J. Pharmacol.* **47**, 351–358.

Buus Lassen, J., Squires, R. F., Christensen, J. A., and Molander, L. (1975). Neuro-chemical and pharmacological studies on a new 5HT-uptake inhibitor, FG4963, with potential antidepressant properties, *Psychopharmacologia,* **42**, 21–26.

Claassen, V., Davies, J. E., Hertting, G., and Placheta, P. (1977). Fluvoxamine, a specific 5-hydroxytryptamine uptake inhibitor, *Brit. J. Pharmacol.* **60**, 505–516.

Clineschmidt, B. V. (1979). MK-212: a serotonin-like agonist in the CNS, *Gen. Pharmac.* **10**, 287–290.

Clineschmidt, B. V. and Anderson, E. G. (1970). The blockade of bulbospinal inhibition by 5-hydroxytryptamine antagonists, *Exp. Brain Res.* **11**, 175–186.

Clineschmidt, B. V. and Lotti, V. J. (1974). Indoleamine antagonists: relative potencies as inhibitors of tryptamine and 5-hydroxytryptophan-evoked responses, *Brit. J. Pharmacol.* **50**, 311–313.

Clineschmidt, B. V. and McGuffin, J. C. (1978). Pharmacological differentiation of the central 5-hydroxytryptamine-like actions of MK-212 (6-chloro-2-[1-piperazinyl]-pyrazine), *p*-methoxyamphetamine and fenfluramine in an in vivo model system, *Eur. J. Pharmacol.* **50**, 369–375.

Clineschmidt, B. V., McGuffin, J. C., and Pflueger, A. B. (1977). Central serotonin-like activity of 6-chloro-2-[1-piperazinyl]-pyrazine (CPP; MK-212), *Eur. J. Pharmacol.* **44**, 65–74.

Copp, F. C., Green, A. F., Hodson, H. F., Randall, A. W., and Sim, M. F. (1967). New peripheral antagonists of 5-hydroxytryptamine, *Nature*, (Lond.) **214**, 200–201.

Corne, S. J., Pickering, R. W., and Warner, B. T. (1963). A method for assessing the effects of drugs on the central actions of 5-hydroxytryptamine, *Brit. J. Pharmacol.* **20**, 106–120.

Cox, J., Moore, G., and Evans, L. (1978). Zimelidine: a new antidepressant, *Prog. Neuro-Psychopharmacol.* **2**, 379–384.

Deakin, J. F. W. and Green, A. R. (1978). The effects of putative 5-hydroxytryptamine antagonists on the behaviour produced by administration of tranylcypromine and L-tryptophan or tranylcypromine and L-dopa to rats, *Brit. J. Pharmacol.* **64**, 201–209.

Euvrard, C. and Boissier, J. R. (1980). Biochemical assessment of the central 5-HT agonist activity of RU 24969, *Eur. J. Pharmacol.* **63**, 65–72.

Ferrini, R. and Glasser, A. (1965). Antagonism of central effects of tryptamine and 5-hydroxytryptophan by 1,6-dimethyl-8β-carbobenzyloxyaminomethyl-10α-ergoline, *Psychopharmacologia*, **8**, 271–276.

Fuller, R. W. (1978) Structure-activity relationships among the halogenated amphetamines, *Ann. N.Y. Acad. Sci.* **305**, 147–157.

Fuller, R. W., Holland, D. R., Yen, T. T., Bemis, K. G., and Stamm, N. B. (1979). Antihypertensive effects of fluoxetine and L-5-hydroxytryptophan in rats, *Life Sci.* **25**, 1237–1242.

Fuller, R. W., Mason, N. R., and Molloy, B. B. (1980a). Structural relationships in the inhibition of [^{3}H]-serotonin binding to rat brain membranes *in vitro* by 1-phenyl-piperazines, *Biochem Pharmacol.* **29**, 833–835.

Fuller, R. W., Perry, K. W., and Molloy, B. B. (1975). Reversible and irreversible phases of serotonin depletion by 4-chloroamphetamine, *Eur. J. Pharmacol.* **33**, 119–124.

Fuller, R. W. and Snoddy, H. D. (1979). The effects of metergoline and other serotonin receptor antagonists on serum corticosterone in rats, *Endocrinology*, **105**, 923–928.

Fuller, R. W. and Snoddy, H. D. (1980). The effect of serotonin-releasing drugs on serum corticosterone concentration in rats, *Neuroendocrinology*, **31**, 96–100.

Fuller, R. W., Snoddy, H. D., and Clemens, J. A. (1980b). Elevation of serum prolactin acutely after administration of *p*-chloroamphetamine in rats, *Endocr. Res. Commun.* **7**, 77–85.

Fuller, R. W., Snoddy, H. D., Mason, N. R., and Molloy, B. B. (1978). Effect of 1-(*m*-trifluoromethylphenyl)piperazine on ^{3}H-serotonin binding to membranes from rat brain in vitro and on serotonin turnover in rat brain *in vivo*, *Eur. J. Pharmacol.* **52**, 11–16.

Fuller, R. W., Snoddy, H. D., Mason, N. R., and Owen, J. E. (1981). Disposition and pharmacologic effects of *m*-chlorophenylpiperazine in rats, *Neuropharmacology*, **20**, 155–162.

Fuller, R. W., Snoddy, H. D., Perry, K. W., Bymaster, F. P., and Wong, D. T. (1977). Importance of duration of drug action in the antagonism of *p*-chloroamphetamine depletion of brain serotonin-comparison of fluoxetine and chlorimipramine, *Biochem. Pharmacol.* **27**, 193-198.

Fuller, R. W., Snoddy, H. D., Perry, K. W., Roush, B. W., Molloy, B. B., Bymaster, F. P., and Wong, D. T. (1976). The effects of quipazine on serotonin metabolism in rat brain, *Life Sci.* **18**, 925-934.

Fuller, R. W. and Steinberg, M. (1976). Regulation of enzymes that synthesize neurotransmitter monoamines, *Adv. Enz. Regul.* **14**, 347-390.

Fuxe, K., Butcher, L. L., and Engel, J. (1971). DL-5-Hydroxytryptophan-induced changes in central monoamine neurons after peripheral decarboxylase inhibition, *J. Pharm. Pharmacol.* **23**, 420-424.

Fuxe, K., Ogren, S.-O., and Agnati, L. F. (1979). The effects of chronic treatment with the 5-hydroxytryptamine uptake blocker zimelidine on central 5-hydroxytryptamine, *Neurosci. Letters,* **13**, 307-312.

Fuxe, K., Ogren, S.-O., Agnati, L. F., and Jonsson, G. (1978). Further evidence that methergoline is a central 5-hydroxytryptamine receptor blocking agent, *Neurosci. Letters,* **9**, 195-200.

Gallager, D. W. and Sanders-Bush, E. (1973). *In vivo* measurement of the release of 5-hydroxytryptamine (5HT) form the hippocampus of the rat: effect of Ro 4-1284, pargyline and *p*-chloroamphetamine (PCA), *Fed. Proc.* **32**, 303.

Garattini, S., Caccia, S., Mennini, T., Samanin, R., Consolo, S., and Ladinsky, H. (1979). Biochemical pharmacology of the anoretic drug fenfluramine: a review, *Curr. Med. Res. Suppl.* 1, **6**, 15-27.

Garattini, S., Jori, A., Buczko, W., and Samanin, R. (1975). The mechanism of action of fenfluramine, *Postgrad. Med. J.* Suppl. 1, **51**, 27-35.

Green, A. R. and Grahame-Smith, D. G. (1976). Effects of drugs on the processes regulating the functional activity of brain 5-hydroxytryptamine, *Nature,* (Lond.) **260**, 487-491.

Green, A. R., Youdim, M. B. H., and Grahame-Smith, D. G. (1976). Quipazine: its effects on rat brain 5-hydroxytryptamine metabolism, monoamine oxidase activity and behaviour, *Neuropharmacology,* **15**, 173-179.

Haigler, H. J. and Aghajanian, G. K. (1977). Serotonin receptors in the brain, *Fed. Proc.* **36**, 2159-2164.

Hardebo, J. E., Edvinsson, L., Owman, C., and Svendgaard, N.-A. (1978). Potentiation and antagonism of serotonin effects on intracranial and extracranial vessels. Possible implications in migraine, *Neurology,* **28**, 64-70.

Harvey, J. A. (1978). Neurotoxic action of halogenated amphetamines, *Ann. N.Y. Acad. Sci.* **305**, 289-302.

Hwang, E. C., Magnussen, I., and Van Woert, M. H. (1980). Effects of chronic fluoxetine administration on serotonin metabolism, *Res. Commun. Chem. Pathol. Pharmacol.* **29**, 79-98.

Hyttel, J. (1977). Effect of a selective 5-HT uptake inhibitor—Lu 10-171—on rat brain 5-HT turnover, *Acta pharmacol. toxicol.* **40**, 439-446.

Hyttel, J. (1978). Effect of a specific 5-HT uptake inhibitor, citalopram (Lu 10-171), on [3]H-5HT uptake in rat brain synaptosomes *in vitro, Psychopharmacology,* **60**, 13-18.

Jacoby, J. H., Shabshelowitz, H., Fernstrom, J. D., and Wurtman, R. J. (1975). The mechanisms by which methiothepin, a putative serotonin receptor antagonist, increases brain 5-hydroxyindole levels, *J. Pharmacol. Exp. Ther.* **195**, 257-264.

Jalfre, M., Ruch-Monachon, M-A., and Haefely, W. (1974). Methods for assessing the interaction of agents with 5-hydroxytryptamine neurons and receptors in the brain, *Adv. Biochem. Psychopharmacol.* **10**, 121-134.

Jones, M. T., Hillhouse, E. W., and Burden J. (1976). Effect of various putative neurotransmitters on the secretion of corticotrophin-releasing hormone from the rat hypothalamus *in vitro*—a model of the neurotransmitters involved, *J. Endocr.* **69**, 1-10.

Jung, M. J., Palfreyman, M. G., Wagner, J., Bey, P., Ribereau-Gayon, G., Zraika, M., and Koch-Weser, J. (1979). Inhibition of monoamine synthesis by irreversible blockade of aromatic aminoacid. decarboxylase with α-monofluoromethyldopa, *Life Sci.* **24**, 1037-1042.

Kaergaard Nielsen, C., Magnussen, M. P., Kampmann, E., and Frey, H.-H. (1967). Pharmacological properties of racemic and optically active *p*-chloroamphetamine, *Arch. int. Pharmacodyn.* **170**, 428-444.

Koe, B. K. and Weissman, A. (1968). The pharmacology of *para*-chlorophenylalanine, a selective depletor of serotonin stores, *Adv. in Pharmacol.* **6B**, 29-47.

Krieger, D. T. (1980). Cyproheptadine: drug therapy for Cushing's disease, *in Neuroactive Drugs in Endocrinology* (Eds E. E. Muller), pp. 361-370, Elsevier/ North Holland, Amsterdam.

Lake, C. R., Coleman, M. D., Ziegler, M. G., and Murphy, D. L. (1979). Fenfluramine and its effect on the sympathetic nervous system in man, *Curr. Med. Res. Opin.* Suppl. 1, **6**, 63-71.

Le Fur, G., Kabouche, M., and Uzan, A. (1978). On the regional and specific serotonin uptake inhibition by LM 5008, *Life Sci.* **23**, 1959-1966.

Lindberg, U. H., Ross, S. B., Thorberg, S.-O., Ogren, S.-O., Malmros, G., and Wagner, A. (1978). A conformational study of R-alaproclate, a new selective inhibitor of neuronal 5-hydroxytryptamine uptake, *Tetrahedron Letters,* **20**, 1779-1782.

Lippmann, W. and Pugsley, T. A. (1976). Pirandamine, a relatively selective 5-hydroxytryptamine uptake inhibitor, *Pharmacol. Res. Commun.* **9**, 387-405.

Maggi, A., U'Prichard, D. C., and Enna, S. J. (1980). Differential effects of antidepressant treatment on brain monoaminergic receptors, *Eur. J. Pharmacol.* **61**, 91-98.

Maj, J., Palider, W., and Baran, L. (1976). The effects of serotonergic and antiserotonergic drugs on the flexor reflex of spinal rat: a proposed model to evaluate the action on the central serotonin receptor, *J. Neural Transmission,* **38**, 131-147.

Maj, J., Palider, W., and Rawlow, A. (1979). Trazodone, a central serotonin antagonist and agonist, *J. Neural Transmission,* **44**, 237-248.

Maj, J., Sowinska, H., Baran, L., Gancarczyk, L., and Rawlow, A. (1978). The central antiserotonergic action of mianserin, *Psychopharmacology,* **59**, 79-84.

Malec, D. and Langwinski, R. (1980). Effect of quipazine and fluoxetine on analgesic-induced catalepsy and antinociception in the rat, *J. Pharm. Pharmacol.* **32**, 71073.

Malick, J. B., Doren, E., and Barnett, A. (1977). Quipazine-induced head-twitch mice, *Pharmacol. Biochem. Behav.* **6**, 325-329.

Marco, E. J. and Meek, J. L. (1979). The effects of antidepressant on serotonin turnover in discrete regions of rat brain, *Naunyn-Schmiedeberg's Arch. Pharmacol.* **306**, 75-79.

Marsden, C. A. (1979). Evidence for the release of hippocampal 5-hydroxytryptamine by α-methyltryptamine, *Brit. J. Pharmacol.* **67**, 438-439.

Marsden, C. A., Conti, J., Strope, E., Curzon, G., and Adams, R. N. (1979). Monitoring 5-hydroxytryptamine release in the brain of the freely moving unanaesthetized rat using *in vivo* voltammetry, *Brain. Res.* **171**, 85-99.

McCall, R. B. and Aghajanian, G. K. (1979). Serotonergic facilitation of facial motoneuron excitation, *Brain Res.* **169**, 11-27.

Melzacka, M., Boksa, J., and Maj, J. (1979). 1-(*m*-Chlorophenyl)-piperazine: a metabolite of trazodone isolated from rat urine, *J. Pharm. Pharmacol.* **31**, 855–856.

Murphy, D. L., Campbell, I. C., and Costa, J. L. (1978). The brain serotonergic system in the affective disorders, *Prog. Neuro-Psychopharmacol.* **2**, 1–31.

Ogren, S. O. and Holm, A.-C. (1980). Test-specific effects of the 5-HT reuptake inhibitors alaproclate and zimelidine on pain sensitivity and morphine analgesia, *J. Neural Transmission,* **47**, 253–271.

Parati, E. A., Zanardi, P., Cocchi, D., Caraceni, T., and Muller, E. E. (1980). Neuroendocrine effects of quipazine in man in health state or with neurological disorders, *J. Neural Transmission,* **47**, 273–297.

Peroutka, S. J. and Snyder, S. H. (1979). Multiple serotonin receptors: differential binding of [^{3}H]-5-hydroxytryptamine, [^{3}H]-lysergic acid diethylamide and [^{3}H]-spiroperidol, *Mol. Pharmacol.* **16**, 687–699.

Przegalinski, E., Baran, L., Palider, W., and Bigajska, K. (1978). On the central antiserotonin action of benzoctamine and opipramol, *Pol. J. Pharmacol. Pharm.* **30**, 781–790.

Przegalinski, E., Baran, L., Palider, W., and Siwanowicz, J. (1979). The central action of pizotifen, *Psychopharmacology,* **62**, 295–300.

Przegalinski, E. and Lewandowska, A. (1979). The effect of etoperidone, a new potential antidepressant drug, on the central serotonin system, *J. Neural Transmission,* **46**, 303–312.

Quattrone, A., Di Renzo, G., Schettini, G., Tedeschi, G., and Scopacasa, F. (1978). Increased plasma prolactin levels induced in rats by *d*-fenfluaramine: relation to central serotonergic stimulation, *Eur. J. Pharmacol.* **49**, 163–167.

Rockman, G. E., Amit, Z., Carr, G., Brown, Z. W. and Ogren, S.-O. (1979). Attenuation of ethanol intake by 5-hydroxytryptamine uptake blockade in laboratory rats. I. Involvement of brain 5-hydroxytryptamine in the mediation of the positive reinforcing properties of ethanol, *Arch. int. Pharmacodyn.* **241**, 245–259.

Rodriguez, R., Rojas-Ramirez, J. A., and Drucker-Colin, R. R. (1973). Serotoninlike actions of quipazine on the central nervous system, *Eur. J. Pharmacol.* **24**, 164–171.

Rokosz-Pelc, A., Antkiewicz-Michaluk, L., and Vetulani, J. (1980). 5-Hydroxytryptamine-like properties of *m*-cholorophenylpiperazine: comparison with quipazine, *J. Pharm. Pharmacol.* **32**, 220–222.

Ross, S. B. (1979). Interactions between reserpine and various compounds on the accumulation of [^{14}C]-5-hydroxytryptamine and ^{3}H-noradrenaline in homogenates from rat hypothalamus, *Biochem. Pharmacol.* **28**, 1085–1088.

Ross, S. B., Ogren, S. -O., and Renyi, A. L. (1976). (Z)-Dimethylamino-1-(4-bromophenyl 1)-1-(3-pyridyl) propene (H 102/09), a new selective inhibitor of the neuronal 5-hydroxytryptamine uptake, *Acta pharmacol. toxicol.* **39**, 152–166.

Ross, S. B. and Renyi, A. L. (1977). Inhibition of the neuronal uptake of 5-hydroxytryptamine and noradrenaline in rat brain by (Z)- and (E)-3-(4-bromophenyl)-*N*, *N*-dimethyl-3-(3-pyridyl) allylamines and their secondary analogues, *Neuropharmacology,* **16**, 57–63.

Saletu, B., Schjerve, M., Grunberger, J., Schanda, H. and Arnold, O. H. (1977). Fluvoxamine—a new serotonin re-uptake inhibitor: first clinical and psychometric experiences in depressed patients, *J. Neural Transmission,* **41**, 17–36.

Samanin, R., Bendotti, C., Candelaresi, G., and Garattini, S. (1977). Specificity of serotoninergic involvement in the decrease of food intake induced by quipazine in the rat, *Life Sci.* **21**, 1259–1266.

Samanin, R., Caccia, S., Bendotti, C., Borsini, F., Borroni, E., Invernizzi, R.,

Paraccini, R., and Mennini, T. (1980). Further studies on the mechanism of serotonin-dependent anorexia in rats, *Psychopharmacology*, **68**, 99–104.

Samanin, R., Mennini, T., Ferraris, A., Bendotti, C., Borsini, F., and Garattini, S. (1979). *m*-Chlorophenylpiperazine: a central serotonin agonist causing powerful anorexia in rats, *Naunyn-Schmiedeberg's Arc. Pharmacol.* **308**, 159–163.

Sanders-Bush, E., Bushing, J. A., and Sulser, F. (1972). Long-term effects of *p*-chloroamphetamine on tryptophan hydroxylase activity and on the levels of 5-hydroxytryptamine and 5-hydroxyindoleacetic acid in brain, *Eur. J. Pharmacol.* **20**, 385–388.

Savage, D. D., Frazer, A., and Mendels, J. (1979). Differential effects of monoamine oxidase inhibitors and serotonin reuptake inhibitors on ^{3}H-serotonin receptor binding in rat brain, *Eur. J. Pharmacol.* **58**, 87–88.

Schachter, M. and Parkes, J. D. (1980). Fluvoxamine and clomipramine in the treatment of cataplexy, *J. Neurol. Neurosurg. Psychiat.* **43**, 171–174.

Segawa, T., Mizuta, T., and Nomura, Y. (1979). Modifications of central 5-hydroxytryptamine binding sites in synaptic membranes from rat brain after long-term administration of tricyclic antidepressants, *Eur. J. Pharmacol.* **58**, 75–83.

Shopsin, B., Friedman, E., and Gershon, S. (1976). Parachlorophenylalanine reversal of tranylcypromine effects in depressed patients, *Arch. Gen. Psychiat.* **33**, 811–819.

Shopsin, B., Gershon, S., Goldstein, M., Friedman, E., and Wilk, S. (1975). Use of synthesis inhibitors in defining a role for biogenic amines during imipramine treatment in depressed patients, *Psychopharmacol. Commun.* **1**, 239–249.

Smith, S. A., Marston, F. A. O., Dickson, A. J., and Pogson, C. I. (1979). Control of enzyme activities in rat liver by tryptophan and its metabolites, *Biochem Pharmacol.* **28**, 1645–1651.

Sugrue, M. F., Goodlet, I., and Mireylees, S. E. (1976). On the selective inhibition of serotonin uptake *in vivo* by Org 6582, *Eur. J. Pharmacol.* **40**, 121–130.

Sved, A. F., Fernstrom, J. D., and Wurtman, R. J. (1979). Tyrosine administration reduces blood pressure and enhances brain norepinephrine release in spontaneously hypertensive rats, *Proc. Nat. Acad. Sci.* **76**, 3511–3514.

Trulson, M. E. and Jacobs, B. L. (1976). Behavioral evidence for the rapid release of CNS serotonin by PCA and fenfluramine, *Eur. J. Pharmacol.* **36**, 149–154.

Van Praag, H. M. and Korf, J. (1973). 4-Chloroamphetamines. Chance and trend in the development of new antidepressants, *J. Clin. Pharmacol.* **13**, 3–14.

Van Woert, M. H., Magnussen, I., Rosenbaum, D., and Hwang, E. C. (1980). Effect of fluoxetine on intention muoclonus, *Neurology*, **30**, 384.

Van Woert, M. H., Rosenbaum, D., Howieson, H., and Bowers, Jr., M. B. (1977). Long-term therapy of myoclonus and other neurologic disorders with L-5-hydroxytryptophan and carbidopa, *New England J. Med.* **296**, 70–75.

Vargaftig, B. B., Coignet, J. L., De Vos, C. J., Grijsen, H., and Bonta, J. L. (1971). Mianserin hydrochloride: peripheral and central effects in relation to antagonism against 5-hydroxytryptamine and tryptamine, *Eur. J. Pharmacol.* **16**, 336–346.

Waldmeier, P. C., Baumann, P. A., and Maitre, L. (1979). CGP 6085 A, a new, specific, inhibitor of serotonin uptake: neurochemical characterization and comparison with other serotonin uptake blockers, *J. Pharmacol. Exp. Ther.* **211**, 42–49.

Whitaker, P. M. and Seeman, P. (1978). High-affinity ^{3}H-serotonin binding to caudate: inhibition by hallucinogens and serotoninergic drugs, *Psychopharmacology*, **59**, 1–5.

Wirz-Justice, A., Krauchi, K., Lichtsteiner, M., and Feer, H. (1978). Is it possible to modify serotonin receptor sensitivity, *Life Sci.* **23**, 1249–1254.

Wong, D. T., Horng, J. S., Bymaster, F. P., Hauser, K. L., and Molloy, B. B. (1974). A selective inhibitor of serotonin uptake: Lilly 110140, 3-(*p*-trifluoromethyl-phenoxy)-*N*-methyl-3-phenylpropylamine, *Life Sci.* **15**, 471–479.

Wurtman, J. J. and Wurtman, R. J. (1979). Fenfluramine and other serotonergic drugs depress food intake and carbohydrate consumption while sparing protein consumption, *Curr. Med. Res. Opin.* Suppl. 1, **6**, 28–33.

Yaryura-Tobias, J. A. (1977). Obsessive-compulsive disorders: a serotonergic hypothesis, *J. Orthomol. Psychiat.* **6**, 317–326.

Zabik, J. E., Liao, S. -S., Jeffreys, M., and Maickel, R. P. (1978). The effects of DL-5-hydroxytryptophan on ethanol consumption by rats, *Res. Commun. Chem. Pathol. Pharmacol.* **20**, 69–78.

Biology of Serotonergic Transmission
Edited by N. N. Osborne
© 1982, John Wiley & Sons Ltd.

Chapter 10

Serotonin Neurotoxins

H. G. BAUMGARTEN and S. JENNER
Department of Neuroanatomy and Electron Microscopy,
Free University of Berlin, Berlin-West (GFR)

A. BJÖRKLUND
Department of Histology, University of Lund, Lund, Sweden

H. P. KLEMM
Department of Neurology, University of Hamburg,
Universitatskrankenhaus Eppendorf, Hamburg, GFR

H. G. SCHLOSSBERGER
Max-Planck-Institute for Biochemistry, Martinsried, GFR

INTRODUCTION

The widespread application of serotonin neurotoxins in experimental neurobiology and the progress in our understanding of the mode and mechanism of action of 5,6- and 5,7-dihydroxytryptamine (5.6- and 5.7-DHT) justify an up-to-date review on the properties and actions and optimum mode of administration of these compounds. *p*-Chloroamphetamine—which in addition to its serotonin-releasing property has also been considered to act as a toxin in some CNS 5-HT neurones (for review, see Jonsson, 1980)—will not be discussed in this review since it causes only region-specific and poor long-term 5-HT reductions when compared to 5,7-DHT.

CHEMICAL PROPERTIES OF 5,6- AND 5,7-DHT

A basic prerequisite for 5,6- and 5,7-DHT to act as toxins is their tendency to become oxidized at neutral or alkaline pH. At biological pH (7.2), both compounds react with oxygen when incubated in buffer at $37\,^{\circ}$C; during this (aut)-oxidation process, coloured quinoidal intermediates are produced that are soluble in case of 5,7-DHT and precipitate in case of 5,6-DHT. The black, insoluble product formed during 5,6-DHT-oxidation resembles melanin formed by interaction of dopamine with O_2 under comparable conditions.

There are remarkable differences in the mode of interaction of 5,6- and 5,7-DHT with O_2. First, the speed of oxidation of 5,7-DHT during the initial, near-linear phase of the reaction is about 10-fold that of 5,6-DHT (3.2 nmol O_2/min and 34.5 nmol O_2/min, respectively); secondly, the (kinetic) characteristics of the reaction with O_2 differ significantly: while 5,7-DHT oxidation obeys a reaction of second-order type that of 5,6-DHT exhibits autocatalytic promotion following an initial, near-linear, slow phase of the reaction (Fig. 1). During the phase of progressive increase in the oxidation velocity of 5,6-DHT, significant amounts of H_2O_2 are generated which account for approximately 40 per cent of the O_2 released from the incubation medium following addition of catalase. The reaction sequence can be formulated as follows:

$$5,6\text{-DHT} + O_2 \longrightarrow 5,6,\text{-DHT-quinone} + H_2O_2$$

$$5,6,\text{-DHT} + H_2O_2 \longrightarrow 5,6\text{-DHT-quinone} + 2H_2O$$

H_2O_2 does not appear to be formed during 5,7-DHT oxidation. Since the oxidation products that are generated by the oxidation of 5,7-DHT are not yet characterized, no comparable equation for the oxidation of 5,7-DHT can be formulated. Based on theoretical reasoning one would suppose that the *p*-quinoneimine of 5,7-DHT could constitute an important intermediate in the oxidation of 5,7-DHT since its formation is favoured by the negative redox potential of 5,7-DHT. Model studies with a 5,7-DHT analogue indicate that oxygen can be incorporated into the indole molecule during oxidation (Schlossberger, unpublished observations); whether this occurs *in vivo* is unknown.

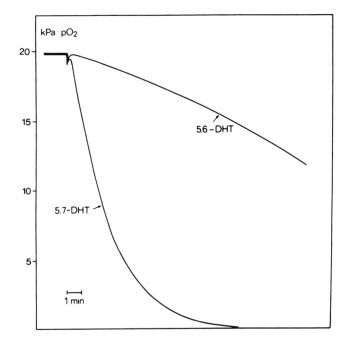

Fig. 1   Oxygen consumption curves of 5,6-DHT, 1 mM, and 5,7-DHT, 1 mM, in air-saturated 0.1 M phosphate buffer pH 7.2, at 37 °C, measured with the oxygen electrode. From Klemm *et al.* (1980)

Presumably, radicals (e.g. superoxide anion, hydroxyl radical and singlet oxygen) are generated during the oxidation of 5,7-DHT since the *in vivo* toxicity of the latter on peripheral noradrenergic axons can be abolished by pretreatment with radical scavengers (Cohen and Heikkila, 1978).

The overall importance of oxidation in the events that render the DHT's (neuro) toxins is immediately apparent when one compares the autoxidation rates of 5,6- and 5,7-DHT with that of a substituted, 5,7-DHT analogue, 5-hydroxy-7-methoxytryptamine. This compound which has about the same affinity to the 5-HT-transport sites as 5,7-DHT, does not consume measurable amounts of oxygen under comparable *in vitro* conditions (see Table 1) and is hardly toxic to serotonergic axons when injected into the CSF in high doses. If, on the other hand, autoxidation proceeds at too high a rate, as in case of 5,6,7-trihydroxytryptamine, the compound exhibits no toxicity in serotonergic fibres because it does not exist as a hydroxy-tryptamine in solution but rather as a quinone which is not accepted as a substrate by the high-affinity 5-HT transport sites. The poor toxicity of this compound has also been demonstrated *in vivo* (Baumgarten *et al.*, 1975).

In fact, the oxidation rate of 5,7-DHT at pH 7.2 as measured *in vitro* would also probably be too high to permit substantial amounts of this drug to be taken up and accumulated inside the 5-HT neurones. However, as clearly evidenced by

Table 1. Protective effects of various antioxidants and other compounds on the initial oxygen consumption rate by 5,6-DHT and 5,7-DHT (expressed as percentage decrease in oxygen consumption)

Protective agent	5,7-DHT (1 mM)	5,6-DHT (1 mM)
none	$(33.4\pm2.1$ nmol O_2/min)	$(2.7\pm0.5$ nmol O_2/min)
dithiothreitol (1 nM)	46.7	50.0
ascorbate (1 mM)	97.5	65.6
ascorbate (0.5 mM)	90.1	n.m.
ascorbate (0.1 mM)	68.8	n.m.
cysteine (1 mM)	73.0	54.0 above control
GSH (1 mM)	67.0	23.0
superoxide dismutase (0.5 mg)	80.7	33.5
albumin (0.5 mg)	63.6	39.0
catalase (0.5 mg)	10.0 above control	47.4

n.m. = not measured.
From Klemm et al., 1980.

Table 2. Comparison of the initial oxygen consumption rate (in nmoles O_2/min) of various hydroxylated tryptamines with their uptake affinity (IC50: concentration of inhibitor required to produce a 50 per cent inhibition of the uptake of 1 X 10^{-7} [^{3}H] 5-HT into synaptosome-rich homogenates of rat hypothalamus; cf. Horn *et al.*, 1973) (*n* = 3)

Compound (1 mM)	Oxygen consumption rate	(*n*) IC50	Forebrain 5-HT content of control* (%)
5-OH-7-CH$_3$O-T	insignificant	(3) 7.9 X 10^{-7} M	90**
5,6,-DHT	2.7±0.5	(3) 6.0 X 10^{-7} M	56
α-methyl-5,7-DHT	20.9±1.8	(3) 4.4 X 10^{-6} M	62
5,7,-DHT	33.4±2.1	(4) 4.0 X 10^{-6} M	48
5,6,7-THT	986.0	(1)***	97

* Measured at 12 days after intraventricular injection of 50 µg (free base) of either compound (data from Baumgarten *et al.*, 1975)
** Value from Baumgarten *et al.*, 1980
*** No inhibition of uptake [^{3}H] 5-HT at 10^{-6} or 10^{-5} M (cf. Björklund *et al.*, 1975)
n = Number of determinations
From Klemm *et al.*, 1980.

the data in Table 2 (and by evidence to be discussed below), the *in vitro* measurements do not reflect the true autoxidation rates *in vivo* since the addition of various types of peptides, proteins, boiled mitochondria, SH-reagents or reducing agents such as ascorbate result in significant retardation of the oxidation of 5,7-DHT suggesting that brain tissue provides a protective milieu that helps to maintain 5,7-DHT reduced. Similar conditions are valid for 5,6,-DHT but the nature of the reaction products formed (the *o*-quinone and H_2O_2) render this drug generally cytotoxic and thus seriously limit its applicability as a tool to induce serotonin fibre degeneration in brain.

INTERACTION OF 5,6- AND 5,7-DHT WITH MONOAMINE OXIDASE

In vitro studies by Klemm and Baumgarten (1978) and Klemm *et al.* (1979) have shown that both DHTs are substrates for MAO-A, and to a much lesser degree, for MAO-B as judged from inhibition studies with clorgyline and deprenyl. *In vitro*, 5,6,-DHT irreversibly inhibited MAO, a process paralleled by extensive binding of reactive intermediates to proteins and by precipitation of a melanin-like polymer. It was thus argued that both cross-linkage of proteins and specifically of MAO by the 5,6-DHT quinone might be involved in the *in vivo* toxicity of 5,6-DHT. No such correlations were established for 5,7-DHT since it did not inactivate MAO, did not undergo significant binding to proteins and since there was no measurable alteration in the time-dependent weak binding of radioactivity to proteins following MAO inhibition. Therefore, the role of MAO in the toxicity of 5,7-DHT remained obscure. By contrast, a significant fraction of radioactivity was irreversibly bound to proteins of rat brain (following *in vivo* administration of [^{14}C] 5,7,-DHT (Baumgarten *et al.*, 1978). This result is not compatible with *in vivo* findings by Creveling *et al.* (1975) and Breese and Cooper (1975) indicating that pretreatment with pargyline protected peripheral noradrenergic fibres in mice from damage by systemically administered 5,7-DHT and that pretreatment of rats with various MAO-inhibitors abolished the toxicity of intracisternally administered 5,7-DHT in central noradrenergic fibres. Both these findings suggested a key role of MAO in the toxicity of 5,7-DHT. However, as shown by Baumgarten *et al.* (1975, 1978), the MAO-resistant analogue of 5,7-DHT, α-methyl-5,7-DHT, exerted toxic actions on central serotonergic, noradrenergic and dopaminergic fibre systems but its toxic potency in injuring 5-HT and NA fibres was approximately 15–20 per cent and 30–40 per cent less than that of comparable doses of 5,7-DHT suggesting that metabolism of 5,7-DHT by MAO enhanced the neurotoxicity of 5,7-DHT. In harmony with this, Björklund *et al.* (1975) have shown that MAO inhibition (by nialamide) counteracts the acute damage of the serotonin uptake mechanism caused by 5.6- and 5.7-DHT *in vitro* ('uptake impairment'). These findings imply that the MAO-metabolism of 5,7-DHT accounts for only a (minor) portion of the overall neurotoxicity of 5,7-DHT, and that this effect of MAO is quantitatively more important in NA than in 5-HT axons.

THE METABOLIC FATE OF 5,6- AND 5,7-DHT *IN VIVO*

The potential role of MAO in the toxicity of 5,7-DHT was further explored by studies on the *in vivo* metabolism of [^{14}C] 5,7-DHT and the binding of radioactivity to rat brain proteins (Baumgarten *et al.* 1980). This study revealed that protein-binding of radioactivity from [^{14}C] 5,7-DHT depended on, (a) initial binding of [^{14}C] 5,7-DHT to MAO itself, (b) a time-dependent binding of an intermediate of 5,7-DHT formed by MAO (most probably the aldehyde) to the protein fraction, and (c) of an unknown, reactive product; the latter component in the protein binding was independent of MAO and progressed slowly with time.

This MAO-independent component of the total binding of radioactivity to proteins may be mediated by quinoidal oxidation products of 5,7-DHT that are formed *in vivo* through interaction of 5,7-DHT with cytochrome-*c* oxidase as recently described in detail by Klemm *et al.* (1980) and originally proposed by Cohen and Heikkila (1978). This process has been termed mitochondria-promoted oxidation of 5,7-DHT and occurs primarily in those structures that actively accumulate 5,7-DHT, i.e. the serotonergic axons and, to a much lesser degree, the noradrenergic and dopaminergic fibres. This was revealed in experiments in which uptake into monaminergic structures was prevented by pretreatment of animals

Table 3. MAO-dependent metabolism and protein binding of [^{14}C]5,6-DHT and [^{14}C]5,7-DHT in rat brain, 30 or 60 min following intraventricular injection of the DHTs. The type and dose of MAO- and of aldehyde inhibitor are indicated in the table. All drugs were administered ip. The frozen brains were homogenized in perchloric acid (containing ascorbic acid) and the deaminated products measured in toluene-isoamylalcohol extracts of the supernatant. Radioactivity in the acid-insoluble pellet represents the protein-bound fraction. All values are mean values of 5–6 separate determinations and expressed as per cent of control values

Treatment		Deaminated fraction		Protein-bound fraction	
		30 min (%)	60 min (%)	30 min (%)	60 min (%)
Clorgyline, 2 mg/kg, 6 h before [^{14}C]5,6-	5,6-DHT	14.4*	23.6*	61.9*	75.0*
or [^{14}C]5,7-DHT	5,7-DHT	8.9*	19.3*	34.9*	37.4*
Pargyline 10 mg/kg, 6 h before [^{14}C]5,6-	5,6-DHT	83.5*	85.4	66.5*	96.2
or [^{14}C]5,7-DHT	5,7-DHT	105.5	109.7	80.9*	87.4
Disulfiram, 200 mg/kg, Barbital 200 mg/kg, 2 and 0.5 h before [^{14}C]5,7-DHT	5,7-DHT	161.9*	181.9*	101.3	130.5*

*significantly different from control values.
Unpublished data by Klemm, Blaumgarten, and Schlossberger.

with nomifensine and citalopram (NA/DA and 5-HT uptake blocker, respectively); under these conditions, MAO-metabolism of 5,7-DHT was hardly attenuated but the time-dependent increase in the protein binding of radioactivity was reduced.

Blockade of MAO-A with clorgyline also significantly reduced the extent of protein binding (Table 3); in this situation, metabolism of 5,7-DHT was completely prevented and persisting high levels of the unchanged amine were recovered from brain even at 120 min after injection of [^{14}C]5,7-DHT indicating (what has been stated above) that brain matter provides 5,7-DHT with a highly effective reductive milieu which prevents its auto-oxidative destruction. This finding also implies that metabolism by MAO-A represents the only degradative pathway for 5,7-DHT in brain and that unmetabolized 5,7-DHT cannot be cleared from brain since there exists no outward transport mechanism. This contrasts with both 5-HT and 5,6-DHT which continue to be cleared from brain despite inhibition of MAO-A or -A and -B (by Table 3 pretreatment with clorgyline and pargyline). One of the enzymes involved in inactivation of 5,6-DHT has been characterized in our study by TLC analysis of the [^{14}C]labelled metabolites following [^{14}C]5,6-DHT injection into the CSF and found to be COMT. In rat brain, about 50 per cent of the acid metabolites formed from [^{14}C] 5,6-DHT at 120 min after injection is accounted for by O-methylated, deaminated products with 5-methoxy-6-hydroxyindoleacetic acid being the major MAO/COMT product.

This finding suggests that 5-methoxy-6-hydroxytryptamine is formed in brain by interaction of 5,6-DHT with COMT; this monomethoxylated derivative may be a substrate for both MAO-A and -B.

DRUG TREATMENT FOR IMPROVING THE SPECIFICITY OF 5,7-DHT

Already in their first report on the effects of intraventricularly administered 5,7-DHT, Baumgarten *et al* (1973) described scattered lesions of some catecholamine-containing axons in the mesolimbic basal forebrain projection areas; however, due to the failure of the conventional Falck–Hillarp method to distinguish between dopaminergic and noradrenergic axons, this finding was thought to reflect damage to by-passing NA rather than DA axons. With the advent of the glyoxylic acid technique and the magnesium and aluminum formaldehyde procedures developed by Björklund and associates (Björklund *et al.*, 1972; Lorén *et al.*, 1980; Björklund *et al.*, 1980). It became evident that in addition to the well-known lesions that occurred in many of the NA projection systems high intraventricular doses of 5,7-DHT (100 μg free base or more) injured a minor proportion of the ventricle-close DA axon terminals in the septum, the head of the caudate and nucleus accumbens and surface-near DA axons in the olfactory tubercles and adjacent structures (cf. Wuttke *et al.*, 1978; Baumgarten *et al.*, 1978). While desmethylimin-pramine (DMI) was able to prevent the damage to most of the NA projection systems as outlined by different authors (cf. Björklund *et al*, 1975b; Gerson and Baldessarini, 1975; Breese and Cooper, 1975) it naturally failed to protect the DA systems from damage by 5,7-DHT. In animals that received 5,7-DHT via the IVth ventricle (so called intracisternal injection), similar discrete DA fibre lesions occurred in the periventricular hypothalamus and the basal forebrain limbic structures and DMI pretreatment failed to provide complete protection for all of the NA terminal systems (cf. Baumgarten *et al.*, 1978). These findings made Baumgarten *et al.* use nomifensine (hydrogenmaleate; 25 mg/kg, ip), a blocker of the NA and DA uptake mechanisms, rather than DMI (see Baumgarten *et al.*, 1979) to protect all central catecholamine fibre systems from undesired damage by 5,7-DHT (150 μg free base, injected either intraventricularly or intracisternally). The same study also disclosed that the proposal by Breese and Cooper (1975) to employ MAO-inhibitors for specificity improvement rather than DMI (particularly in newborn rats) could not be substantiated since pargyline attenuated the long-term 5-HT reductions in some CNS-regions, did not completely protect the NA projection systems and, actually, enhanced the toxicity in ventricle- and surface-near DA axons. Taken together these observations indicate that measurements of brain amine levels (as performed by most investigators) are poor indicators of the specificity of neuro-toxins and that detailed analysis of regional changes in uptake capacity for tritiated amines, regional amine and metabolite content and morphology and fluorescence intensity of CA axons are required for specificity evaluation.

For optimum-extensive and long-lasting-selective 5-HT reductions in various regions of the CNS, a number of factors have to be considered; these include: the site of injection, the speed of injection, the type of anaesthetic, the dose of catecholamine uptake blocker in relation to the amount of neurotoxin employed. Furthermore, due to the fact that many 5-HT terminals do not degenerate acutely but by slow anterograde mechanisms that follow a primary lesion to the corresponding main, non-terminal axons (cf. Nobin *et al.*, 1973 for details), and that regeneration occurs in certain 5-HT systems with a region-specific time-course, the time of maximum 5-HT reductions has to be carefully assessed by regional analysis if physiological experiments are to be carried out in lesioned animals. If pharmacological testing with postsynaptic 5-HT agonists is to be performed, another variable has to be carefully controlled by regional measurements, i.e. the development and recovery of the post-synaptic 5-HT receptor supersensitivity which profoundly alters drug response and also affects recovery of physiological effects mediated by endogenous 5-HT released from sprouting 5-HT fibres which exhibit increased 5-HT turnover (cf. Wuttke *et al.*, 1977, 1978; Björklund and Wiklund, 1980; Wiklund and Björklund, 1980; McCall and Aghajanian, 1979). The importance of these parameters for selective 5-HT depletion by 5,7-DHT will now be discussed.

The site of introduction of 5,7-DHT into the CSF (generally the lateral ventricle or the IVth ventricle) has consequences for the extent of regional 5-HT reductions and the time-course pattern of recovery. As shown by the analysis of radioactivity retention in various CNS regions following either left lateral or IVth ventricle injection of [^{14}C] 5,7-DHT, there is a cranio-caudal activity gradient in the CNS after intraventricular injection and a reverse pattern after intracisternal drug administration (see Table 4). A comparison of these divergent radioactivity patterns with the time-course of depletion and recovery of 5-HT levels in the forebrain and

Table 4. Effect of injection route on regional retention of radioactivity, derived from 5,7-DHT

	Radioactivity in acid-insoluble fraction (dpm/mg tissue)	
	Intraventricular	Intracisternal
Left striatum	42.1±1.2	1.9±0.4
Right striatum	19.8±5.5	1.4±0.2
Hypothalamus	117.9±13.6	93.3±16.7
Pons-medulla oblongata	65.2±5.4	58.5±7.2
Cervical spinal cord	41.5±3.2	54.0±6.1
Thoracic spinal cord	24.3±3.7	38.0±7.8
Lumbar spinal cord	8.4±1.3	15.7±5.8

brainstem, respectively, reveals that minimum levels are obtained by 5 days in ivt injected animals and only at 12 days in ic injected rats, that the percentage reductions are larger and long-lasting in the forebrain of ivt treated rats and that the 5-HT levels in all CNS regions (except the spinal cord) show time-dependent recovery in the ic injected animals. Thus, there is some correlation between the amount of neurotoxin bound to protein in various brain regions and the time-course and extent of 5-HT depletion and recovery in the forebrain, and these correlations depend in part on the site of injection of 5,7-DHT into the CSF. Whether or not such relationships do exist also depends on the amount of neurotoxin injected. Consequently, the findings obtained with 100 μg (free base) 5,7-DHT must not necessarily be representative for lower or higher doses of this serotonin neurotoxin.

Recently, Gershanik *et al.* (1979) suggested that the speed of injection of 6-hydroxydopamine (6-OH-DA) and of 5,7-DHT may be of considerable importance for the symmetry (i.e. extent) of toxic action on regional aminergic systems though they did not analyze the consequences of this parameter for the resulting specificity of action of the toxins. They compared two different velocities of injection of either drug into the left lateral ventricle of the rat, namely 1 and 3μl/min, and measured ^{3}H-dopamine uptake for 6-OH-DA and ^{3}H-5-HT uptake after 5,7-DHT administration. From the results obtained they concluded that the lower speed of injection of 6-OH-DA or 5,7-DHT resulted in highly asymmetric, localized destruction of DA and 5-HT terminals in the ipsilateral caudate and that this technique may be useful for restricted, unilateral deafferentation of the aminergic input into areas bordering on the site of introduction of the neurotoxin into the CSF.

In a detailed study on the time-course of reduction of 5-HT, NA and DA in various CNS regions following either a slow infusion (2μl/min) or rapid, pulse injection (20μl/sec) of 5,7-DHT in sodium pentobarbitone or ether or sodium pentobarbitone plus ketamine anaesthetized rats (cf. Baumgarten *et al.* 1980; Baumgarten, Jenner, and Schlossberger, unpublished findings, c.f. Tables 5-7), we have not been able to confirm their results and conclusion; instead, we have found that the asymmetric effects that are obtained with slow infusion of 5,7-DHT are significant only at early time-points after drug administration (cf. Fig. 2 and Table 5) and tend to become insignificant after 7-8 days. This is surprising in view of the fact that the differences in the speed of injection that were compared by Gershanik *et al.* (1979) are negligible when contrasted with the injection velocities chosen in our study; thus, the side differences postulated by Gershanik *et al.* (1979) to result from the small differences in the injection speed employed should have been much more dramatic in our approach. The reasons for these discrepancies rest mainly on the fact that Gershanik *et al.* (1979) pooled data from animals at 2-7 days after neurotoxin administration; this should be avoided because of the well-established fact that the maximum long-term destructive effect of any intrathecally administered chemical neurotoxin is the sum of, (a) initial direct destruction of terminals residing within the zone of diffusion of the neurotoxin (a process evident within 24-48 h, and (b) retarded anterograde loss of aminergic terminals (outside

Table 5. Hippocampus: 5-hydroxytryptamine (pg/mg fresh tissue weight ± S.D.)

Days post injection	Uptake blocker	Anaesthetic		Slow infusion (20 μl/10 min)					Fast injection (20 μl/sec)				
				Left	n	Right	n	$\frac{1}{r}$	Left	n	Right	n	$\frac{1}{r}$
1	Nom.	Pentob.	Control	420.1±39.4	4	381.8±43.7	4	1.10	473.4±39.2	6	438.0±32.9	5	1.08
			5,7-DHT	37.2±12.8	4	99.3±16.3	4	0.37	30.4±7.6	5	3.7±16.8	6	0.90
7	Nom.	Pentob.	Control	433.9±70.1	3	381.8±35.8	4	1.14	412.6±56.1	5	470.8±27.6	5	0.88
			5,7-DHT	36.3±13.3	6	38.9±14.9	7	0.93	56.5±23.2	8	44.2±14.0	6	1.28
1	Nom.	Pentob.	Control	364.0±40.0	4	444.1±69.8	4	0.82	395.3±80.3	5	331.0±3.8	3	1.19
		Ket.	5,7-DHT	61.2±22.9	5	134.2±80.0	5	0.46	25.8±1.37	4	37.4	1	0.69
7	Nom.	Pentob.	Control	366.0±38.9	5	389.0±47.3	4	0.94	33.4±30.5	5	419.0±66.9	5	0.80
		Ket.	5,7-DHT	36.3±12.0	6	38.5±10.5	6	0.94	26.8±9.55	5	23.2±5.93	5	1.16
1	Nom.	Pentob.	Contol	365.0±46.7	5	470.0±30.9	5	0.89	386.1±25.8	5	450.0±42.8	5	0.86
	DMI	Ket.	5,7-DHT	50.8±14.7	4	116.0±13.9	4	0.44	45.7±9.95	5	33.5±6.0	3	1.36
7	Nom.	Pentob.	Control	351.0±42	5	442.0±25	5	0.79	388.3±67.4	5	411.0±39.0	5	0.94
	DMI	Ket.	5,7-DHT	45.9±12.8	5	49.9±13.6	5	0.92	39.8±10.4	5	28.0±6.1	5	1.42
1	Nom	Ether	Control						403.0±61.3	7	419.0±45.4	6	0.96
			5,7-DHT						37.3±11.3	5	35.5±8.7	5	1.05
7	Nom.	Ether	Control						384.0±38.8	4	402.0±45.2	4	0.96
			5,7-DHT						30.4±15.9	4	27.6±2.7	4	1.10

Nom.: Nomifensine (hydrogenmaleate)
Pentob.: Pentobarbital (sodium)
DMI: Desmethylimipramine (HCl)
Ket.: Ketamine (HCl)

Table 6. Hippocampus: noradrenaline (pg/mg fresh tissue weight ±S.D.), 1 or 7 days after 150µg 5,7-DHT

Days post injection	Uptake blocker	Anaesthetic drug		Slow infusion (20 µl/10 min)					Fast injection (20 µl/sec)				
				Left	n	Right	n	$\frac{1}{r}$	Left	n	Right	n	$\frac{1}{r}$
1	Nom.	Pentob.	Control	309.5±60.1	4	359.5±64.8	3	0.86	310.3±62.5	4	285.3±44.0	3	1.09
		Ket.	5,7-DHT	315.5±15.9	4	312.8±20.0	5	1.01	275.3±24.6	3	277.0	2	0.99
7	Nom.	Pentob.	Control	336.2±41.7	5	336.7±61.8	4	1.00	345.2±60.7	3	340.0±58.0	5	1.02
		Ket.	5,7-DHT	296.4±37.3	4	335.0±56.0	5	0.88	287.6±72.3	5	283.8±66.1	4	1.01
1	Nom.	Pentob.	Control	281.5±50.2	4	313.9±65.2	5	0.90	302.5±35.3	4	312.7±70.1	3	0.97
	DMI	Ket.	5,7-DHT	279.6±41.5	4	338.6±90.0	4	0.8	335.2±47.2	4	353.3±29.4	4	0.95
7	Nom.	Pentob.	Control	388.5±26.3	4	431.7±37.4	4	0.90	379.7	2	383.1±22.7	3	0.99
	DMI	Ket.	5,7-DHT	338.6±64.4	5	265.6±47.3	5	0.93	309.5±69.8	5	333.8±40.1	4	0.93
1	Nom	Ether	Control						282.3		299.0	2	0.94
		Ether	5,7-DHT						285.6±39.4	3	318.2±16.3	3	0.90
7	Nom.	Ether	Control						336.5±55.4	4	337.0±109.0	4	1.00
		Ether	5,7-DHT						357.0	2	279.3	2	1.26

Nom.: Nomifensine (hydrogenmaleate)
Pentob.: Pentobarbital (sodium)
DMI: Desmethylimipramine (HCl)
Ket.: Ketamine

Table 7a. Hippocampus: 5-Hydroxytryptamine in per cent of control (±S.D.) after 150 µg 5,7-DHT f.b. (controls: 20 µl of saline) and different forms of pretreatment and anaesthesia

Days post-injection	Uptake inhibition	Anaesthetic	Slow infusion (20 µl/10 min)						Fast injection	
			Left	n	Right	n	Left	n	Right	n
1	Nom.	Pentob.	8.86±3.0	4	26.0±4.3	4	6.4±1.6	5	7.69±3.8	6
7	Nom.	Pentob.	8.37±3.1	6	10.2±3.9	6	13.7±3.2	8	9.39±3.0	6
1	Nom.	Pentob. Ket.	16.8±6.3	5	30.2±18.0	5	6.53±0.3	4	11.3	1
7	Nom.	Pentob. Ket.	9.92±3.3	6	9.90±2.7	6	8.02±2.9	5	5.54±1.4	5
1	Nom. DMI	Pentob. Ket.	13.9±4.0	4	28.3±3.4	4	11.8±2.6	5	7.44±1.3	2
7	Nom.	Pentob. Ket.	13.1±3.6	5	11.3±3.1	5	10.3±2.7	5	6.81±1.5	5
1	Nom.	Ether					9.26±2.8	5	8.47±2.1	5
7	Nom.	Ether					7.92±4.1	4	6.87±0.67	4

Nom.: Nomifensine (hydrogenmaleate)
Pentob.: Pentobarbital (sodium)
DMI: Desmethylimipramine (HCl)
Ket.: Ketamine (HCl)

f.b. = free base

Table 7b. Hippocampus: Noradrenaline in per cent of control (±S.D.) after 150 µg 5,7-DHT f.b. (controls: 20 µl of saline) and different forms of pretreatment and anaesthesia

Days post-injection	Uptake inhibition	Pretreatment Anaesthetic	Slow infusion (20 µl/10 min) Left	n	Right	n	Fast injection Left	n	Right	n
1	Nom.	Pentob.					63.5	2	78.0	2
7	Nom.	Pentob.					79.2	2	74.8	2
1	Nom.	Pentob. Ket.	101.9±5.6	4	87.0±5.60	5	88.7±7.90	3	97.0	2
7	Nom.	Pentob. Ket.	88.2±11.1	4	99.4±16.6	5	83.3	5	83.5±19.4	4
1	Nom.	Pentob. Ket.	99.3±14.7	4	107.9±28.1	4	11.8±15.6	4	113.0±0.4	4
7	Nom.	Pentob. Ket.	87.2±16.6	5	84.7±11.0	5	81.5±11.0	5	87.1±10.5	4
1	Nom.	Ether					101.2±14.0	3	106.4±5.5	3
7	Nom.	Ether					104.6	2	82.9	2

Nom.: Nomifensine (hydrogenmaleate)
Pentob.: Pentobarbital (sodium)
DMI: Desmethylimipramine (HCl)
Ket.: Ketamine (HCl)
f.b. = free base

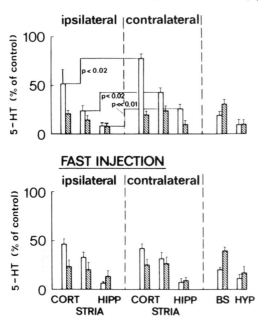

Fig. 2 Serotonin reductions in CNS regions following slow infusion or fast injection of 5,7-DHT into the left lateral ventricle of nomifensine pretreated, pentobarbitone anaesthetized rats. (Control levels of 5-HT in pg/mg fresh tissue weight: cortex, 384.4 ± 28.2; striatum, 547.7 ± 50.5; hippocampus, 426.6 ± 43.1; brainstem, 994 ± 105.5; hypothalamus, 1155.2 ± 99.1; the values represent means ± SEM of four to seven rats). From Baumgarten *et al.* (1981)

the diffusion area of the neurotoxin), the parent non-terminal axons of which have been initially lesioned by the drug, a process not evident before 2-4 days and not complete until 8-12 days after treatment. The data presented in Fig. 3 indicate that the radioactivity distribution patterns at 24 h after [^{14}C] 5,7-DHT infusion and pulse injection, respectively, roughly correlate with the percentage 5-HT reductions in most of the corresponding CNS regions but that the 1 h distribution patterns of radioactivity do not correlate with the depletion patterns at 24 h.

An important point not considered in the study by Gershanik *et al.* (1979) concerns the selectivity of action of 5,7-DHT in relation to the catecholamine uptake blocker and the type of anaesthetic used. Their approach is based on the assumption that a dose of 25 mg/kg ip DMI protects the CNS NA systems and that 5,7-DHT does not affect the DA systems. While DMI protects most of the NA axons from damage by 5,7-DHT provided the latter is given as a rapid injection

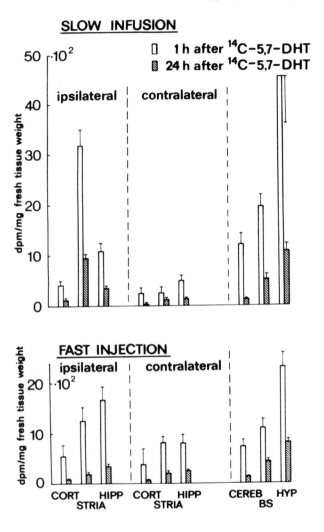

Fig. 3   Radioactivity retention in CNS regions following slow or fast injection of [14C] 5,7-DHT into the left lateral ventricle of nomifensine treated, pentobarbitone anaesthetized rats. The data represent total radioactivity in the brain regions (means ± SEM of five determinations). Slow infusion of [14C] 5,7-DHT results in highly asymmetric radioactivity retention at 1 h (open columns in upper graph) in the striatum and hippocampus; slight left/right differences are also apparent in the pulse injected animals (lower graph) at 1 h. These differences tend to disappear by 24 h in pulse injected rats but persist in the slowly infused animals (hatched columns). In the remaining CNS regions (cerebellum), brainstem (BS), and hypothalamus (HYP) and in the contralateral forebrain regions, the retention patterns at 24 h are similar following slow infusion or fast injection of [14C] 5,7-DHT into the left lateral ventricle. From Baumgarten *et al.* (1981)

into the CSF of animals anaesthetized with ether or a short-acting barbiturate this is not necessarily so if it is infused slowly into the lateral ventricle since the latter procedure results in highly asymmetric retention of radioactivity in the ipsilateral striatum by 1 h after infusion as demonstrated in our study (see Fig. 3). As a consequence, there is an increased chance for damage to the NA and DA terminal systems in the ipsilateral, cannula-close structures that border on the left lateral ventricle and also on the periventricular hypothalamus which retains significantly more radioactivity in the slow infusion experiments. In harmony with this assumption, we have found that a dose of nomifensine (25 mg/kg ip) that is largely protective to the NA and DA systems in animals that receive a pulse injection of 150 μg 5,7-DHT under ether anaesthesia (cf. Baumgarten *et al.*, 1979) is unable to provide complete protection of NA and DA axons in slow infusion and rapid injection studies carried out on rats anaesthetized with sodium pentobarbitone (Baumgarten *et al.*, 1981). This is more evident at 24 h after 5,7-DHT injection than at 7–8 days when the NA levels are still slightly reduced but no longer significantly different from those in saline injected controls (cf. Table 7).

The failure of nomifensine to give complete protection of the central CA systems is due to two factors: the unfavourable effects of high local concentrations of 5,7-DHT which override the NA/DA uptake blockade by nomifensine and the unfavourable effects of the long-acting barbiturate on the protecting capacity of nomifensine. Since the protection of the NA systems by nomifensine was less effective than that of the DA ones, a combination of DMI and nomifensine (15 mg/kg of either drug, given ip) was considered more suitable for optimally protecting the central catecholamine fibres under the experimental conditions described (sodium pentobarbitone plus ketamine/slow 5,7-DHT infusions) and found more satisfactory than nomifensine in preliminary experiments (cf. Table 6). The effects of intraventricularly administered 5,7-DHT on hippocampal amine concentrations may not be representative for other regions of the rat CNS.

THE USE OF 5,7-DHT FOR LOCALIZED INTERRUPTION OF SEROTONERGIC PROJECTIONS

The ideal approach for studying the role of individual serotonergic fibre projections or of serotonergic terminals within a given target area would necessitate localized microinjections into fibre bundles or brain nuclei via very thin cannulae that cause as little unspecific traumatic lesions as possible. This goal has hardly been successfully accomplished up to now, mainly because of the use of too large cannulae, because of the choice of unfavourable injection sites and because of the use of too high amounts of 5,7-DHT and insufficient specificity control. The literature thus contains only a few reports which satisfy the rigid criteria that have to be fulfilled for localized, specific 5-HT axon lesions (Björklund *et al.*, 1973; Azmitia *et al.*, 1978). Selective removal of 5-HT terminals within a small target area does not appear to have been achieved so far.

Due to intermingling of 5-HT and catecholamine fibres in many regions of the CNS, specificity is one of the problems the investigator is confronted with. Catecholamine axons are not only sensitive to damage by 5,7-DHT itself but are also sensitive to injury by ascorbic acid, the most commonly used reducing agent. Dopaminergic axons are particularly vulnerable to ascorbic acid, which when present in the solution enhances the potential toxicity of 5,7-DHT in this neurone system. The concentration of ascorbic acid in the vehicle should therefore either be reduced (from 1 mg/ml to 0.1 mg/ml) or replaced by another biologically inert reducing agent such as GSH. When dissolving 5,7-DHT, the vehicle should be kept at about 4 °C and deoxygenated with N_2. The latter precautions protect 5,7-DHT from rapid autoxidation.

Sites in the rat CNS where serotonergic axons are largely separated from ascending to descending catecholamine axons include the paramedian and ventromedian regions of the midbrain tegmentum dorsal to the interpenduncular nucleus and rostral to the median raphe nucleus; the paramedian tegmentum also has some 5-HT fibres ventral to the median longitudinal fascicles. A major portion of the ascending 5-HT axons emanating in both the dorsal and median raphe nucleus penetrate into this anterior, mediobasal region of the midbrain. Björklund *et al.* (1973) obtained rather selective and extensive serotonergic deafferentation of the forebrain by injecting small amounts of 5,7-DHT (4µg) into this area without pretreating the rats with a catecholamine uptake blocker; the extent of the traumatic central, the surrounding unspecific, and of the adjacent, peripheral area of selective 5-HT axon damage by 5,7-DHT were investigated in detail (cf. also the review by Baumgarten *et al.*, 1977). A careful analysis of the topography of the different aminergic components of the median forebrain bundle reveals that, in the caudal diencephalon, part of the 5-HT fibre bundles are separated from the catecholamine fibre tracts; however, when cannulae of the conventional size are used, there is a high risk for mechanical damage of the more dorsally located dopaminergic tracts. As shown by Azmitia *et al.* (1978), selective degeneration of the serotonergic input to parts of the hippocampus can be achieved by local injections of 5,7-DHT into the fimbria fornicis. Concomitant with increasing proximity to the forebrain target areas, serotonergic fibres become increasingly intermingled with catecholaminergic axons, particularly with the noradrenergic ones with which they show considerable overlap what concerns their projection areas. In preliminary studies we have found that selective lesions of these target-close fibre bundles requires simultaneous systemic and central protection of the CA fibres by DMI and nomifensine.

Another example of separation of CA and 5-HT axon pathways concerns the bulbospinal projection systems. At the junction of the medulla oblongata and spinal cord, most of the descending 5-HT fibres that originate in the nuclei B_1-B_3 approach the ventral and ventrolateral subpial surface of the anterolateral funiculus and can selectively be lesioned by small amounts of 5,7-DHT injected into the IVth ventricle of DMI pretreated animals.

At present there are only few target areas from which 5-HT terminals can be removed with some degree of local precision and specificity. These include the 5-HT terminals in the suprachiasmatic nuclei, the intra-ventricular endings of the subependymal 5-HT fibre plexuses and in the subcommissural organ (SCO). Only very small doses of 5,6- or 5,7-DHT are required for destruction of the intra-ventricular serotonergic endings (5–10 μg, free base), whereas selective degeneration of the 5-HT terminals in the suprachiasmatic nuclei and the SCO is possible by infusion of very small quantities of 5,7-DHT into the preoptic recess of the third ventricle or the exit of the third ventricle where it merges with the aqueduct, respectively.

More experiments are necessary to elaborate optimum conditions for micro-injection of 5,7-DHT into ventricle-distant target areas containing both CA and 5-HT terminals.

THE ROLE OF SEROTONIN NEUROTOXINS TO UNRAVEL THE FUNCTIONAL IMPORTANCE OF SEROTONIN IN BRAIN

Serotonin has been implicated to have a modulatory—though not indispensable—role in many CNS functions; 5,6-DHT and 5,7-DHT have been widely used to study the importance of 5-HT in all those functions that can be analyzed in animal experiments. The intention of the present review is to give but a few examples of such experiments and to critically analyze the problems that arise in the interpretation of results obtained by neurotoxic lesions when compared to those gained by pharmacological manipulation of central serotonergic transmission (cf. Fuller, 1980).

The role of serotonin in the release of LH, FSH, prolactin, ACTH and corticosterone

The controversy concerning the role of brain serotonin in the control of hypothalamic hypophysiotropic hormones and thus—indirectly—of the hypophysial hormones prompted us to study the effect of chemically induced lesions of brain serotonergic fibres on the serum concentration of LH, FSH, prolactin, ACTH and corticosterone (Wuttke *et al.*, 1977, 1978; Lachenmayer *et al.*, 1980; Baumgarten *et al.*, 1978).

Mildly stressed male rats exhibit decreased mean serum prolactin levels on day 5 (significantly different from vehicle treated controls) and on day 12 (non-significant difference) following a single rapid injection of 100 μg (free base) 5,7-DHT into the left lateral ventricle (30 min after pretreatment with 25 mg/kg ip DMI). However, if rats with an even more extensive destruction of their central serotonergic projection (150 μg 5,7-DHT) are subjected to ether stress before death, they are able to secrete as much prolactin as vehicle-treated controls indicating unimpaired stress responsiveness of the hypothalamic mechanisms for prolactin release. Hypothalamic 5-HT content, although reduced in the 5,7-DHT treated rats by 12 days after injection, does not change in response to stress and the situation is similar in the vehicle treated animals. By contrast, hypothalamic NA levels are significantly reduced suggesting that this amine is involved in mediating the stress-increased release of a PRF or inhibition of PIF. This concept is supported by the results of Fenske and Wuttke (1975) who observed a blockade of the stress-induced prolactin release following lesions of the central NA fibre systems and Hancke *et al.* (1977), who destroyed the ascending ventral tegmental NA projections by localized injections of 6-OH-DA. It appears, therefore, that brain 5-HT neurones have only a permissive facilitatory importance in the regulation of stress-induced prolactin secretion.

This concept does not fully agree with findings obtained by pharmacological manipulation of central serotonergic mechanisms, such as enhanced transmission by overflow of 5-HT from nerve endings in animals treated with a re-uptake blocker (fluoxetine) and/or 5-hydroxytryptophan (Krulich, 1975; Clemens *et al.*, 1978) or direct stimulation of 5-HT receptors by quipazine for example (Meltzer *et al.*, 1976), procedures which result in highly significant increases of serum prolactin levels. However, both types of treatment cannot be considered to mimick a physio-logical situation because serotonin is prevented from being inactivated by re-uptake following release in the fluoxetine treated animals and thus causes overstimulation of post-synaptic receptors and stimulation of receptors not normally reached by synaptically released 5-HT in the 5-HTP and, particularly, in the 5-HTP-fluoxetine treated animals.

In contrast to this 'pharmacological' situation, we have been able to support the concept of the facilitatory role of 5-HT on prolactin secretion mechanisms (most likely by stimulation of a PRF) by administering very small doses of 5-HTP to con-trol and 5,7-DHT-lesioned rats; invariably, the small dose of 5-HTP (10 mg/kg ip) caused a much higher elevation of the mean serum prolactin levels in the lesioned animals despite smaller increases in hypothalamic 5-HT content indicating a super-sensitivity of 5-HT receptors. This exaggerated response of the prolactin releasing mechanisms occurred also in haloperidol treated rats (which already have elevated mean prolactin levels due to blockade of the PIF activity at the pituitary). This result strongly argues in favour of the possibility that there exist at least some serotonergic nerve terminals in the hypothalamus that facilitate prolactin release; but it is uncertain under which physiological circumstances these mechanisms are activated (in the adult male rat) and their relative importance in the overall pro-lactin stimulating mechanisms appears to be small.

A similar situation is noted when the results of neurotoxic lesions of the 5-HT system—showing only slight impairment of the spontaneous and stress-induced ACTH release in 5-HT deficient male rats (Vogt, Lang, Baumgarten, and Jenner, unpublished observations)—are compared with those obtained in animals sub-jected to an overstimulation of the postsynaptic 5-HT receptors by fluoxetine and fluoxetine plus 5-HTP (Fuller *et al.*, 1975, 1976). As for prolactin, one might arrive at the conclusion that serotonin has but a permissive facilitatory role in ACTH/CRF release if one considers the results of the lesion studies. However, there are a number of reasons that could explain why the effects of the neurotoxic lesions were only transitory and at the borderline of significance. First, there might exist serotonergic projection systems which have opposing effects on prolactin and ACTH release; the outcome of a generalized neurotoxic lesion to 5-HT systems by the intraventricular pulse injection technique might cause an unpredictable imbalance in the opposing systems. Consequently, more localized interruption of defined pathways and terminal systems would provide new insights into region-specific modulatory actions of potentially antagonistic 5-HT projections. Secondly, the hypothalamic PRF/PIF and CRF modulating 5-HT systems might have a peculiar insensitivity against the neurotoxic action of 5,7-DHT, e.g. the postulated intra-hypothalamic short-projection 5-HT neurones (Sladek and Garver, 1976; Chan-Palay, 1978). This possibility cannot be ruled out until, (a) an unequivocal demon-stration of the existence and termination of these 5-HT neurones has been achieved, and (b) their modulatory role in prolactin and ACTH has been proven.

That localized destructions of 5-HT fibre input to hypothalamic regions known to be involved in neuroendrocrine control may provide results different from those obtained by diffuse neurotoxic lesions or pharmacological interference with 5-HT synthesis (by *p*-chlorophenylalanine. *p*-CPA) is evidenced by the findings of Szaf-arczyk *et al.* (1980) who injected 5,7-DHT into the suprachiasmatic nuclei (SCN) and found a moderate decrease in the amplitude and mean level of the ACTH rhythms without any change in the period and phase of the rhythms; in *p*-chloro-phenylalamine treated rats, the circadian ACTH rhythms were obliterated. Their preliminary data do not allow us to comment on the specificity of the local 5,7-DHT injection into the SCN which contains defined sets of NA terminals and perhaps also scattered endings of the intrahypothalamic DA projection systems.

By contrast, clear-cut effects on the regulation of serum LH levels are obtained by a toxic lesion in the brain 5-HT fibre systems by intraventricular 5,7-DHT (Wuttke *et al.*, 1977, 1978). First, the degree of the reduction in the mean serum LH levels depends on the dose of 5,7-DHT used (50, 100 and 150 μg, respectively); secondly, the time-course pattern of recovery of the impaired LH secretion (i.e. the recovery of the 5-HT actions on the LRF neurones) depends on the dose of 5,7-DHT administered; thirdly, there is some correlation between the restoration of the LH secretion and the time-dependent recovery of hypothalamic [^{3}H] 5-HT uptake and 5-HT synthesis capacity and metabolism, i.e. indicators of the anatomical and functional regeneration of the 5-HT systems. However, if the time-course pattern of recovery of the LH secretion is compared with that of the [^{3}H] 5-HT uptake capacity in the anterior hypothalamus, it becomes evident that both para-meters do not recover in parallel but that the LH secretion is restored to control levels at a time when the uptake capacity is still reduced by about 40–50 per cent below control. However, this discrepancy does not contradict the concept that both parameters are somehow interrelated for the following reasons: first, it has been documented that the destruction of serotonergic terminals by 5,7-DHT results in the development of a postsynaptic type supersensitivity of [^{3}H] 5-HT binding sites (i.e. the receptors), and, secondly, it has been shown that the newly developed serotonergic axons have an increased synthesis and release capacity for serotonin which compensates for the reduction in the number of serotonergic fibres in a denervated target (see the papers by Wiklund and Björklund, 1980; Björklund and Wiklund, 1980). Both these mechanisms can be presumed to restore an apparently 'normal' serotonergic stimulatory effect on the LRF neurones at a time when the regeneration of the 5-HT terminal systems is still incomplete. These findings and considerations imply that both the functional activity of the serotonergic axons and the serotonin-binding characteristics of the receptors should be monitored at various times after a neurotoxic lesion to provide the basis for understanding the dynamics of functional recovery of serotonin-modulated functions in the CNS.

Some further insight into the mechanism by which serotonin might facilitate release of LRF and thus of LH was gained in studies in which 4 h profiles of LH level fluctuations were measured in ovariectomized rats treated with a low dose of 5,7-DHT (50 μg) following DMI pretreatment. On day 5 after 5,7-DHT (and less so on day 12 and 28), the frequency of the pulsatile LH secretion was reduced to one-half that in vehicle injected controls, but the regularity of the LH peaking was maintained as was the mean amplitude of each individual LH burst indicating that 5-HT acted by increasing the frequency of a mechanism responsible for the oscillation (i.e. temporal spacing) of the secretion of LRF. The same dose of 5,7-DHT given to cycling rats did not interfere with the preovulatory surges in LH (following a postoperative period of prolonged dioestrus in both vehicle-injected or 5,7-DHT treated rats). This result does not exclude the possibility that a more extensive destruction of the hypothalamic serotonergic axons involved in the control of LRF release might prolong the interpeak interval of the pulsatile LRF secretion to such an extent that the critical surge of LRF on the afternoon of proestrous fails to occur at the critical time thus causing either temporal anomalities in the cycle or even failure of ovulation; the latter was indeed claimed to occur in female rats treated with an enormous dose (intrathecal) of 5,7-DHT (200 μg free base; see Meyer, 1978) or local infusion of *p*-CPA (10 μg) into the anterior preoptic recess of the third ventricle (Walker *et al.*, 1980) which, admittedly, reduced NA significantly besides 5-HT in the hypothalamus. According to our own experience, a dose of 200 μg 5,7-DHT causes NA reduction in ventricle-near forebrain areas despite DMI pretreatment. Therefore, the failure of ovulation observed in the latter studies might well be explained by simultaneous (structural and/or functional) damage to NA and 5-HT axons involved in LRF control. Further studies in which localized and selective destruction of hypothalamic 5-HT terminals are achieved will be necessary to decide whether serotonin fibres do have an indispensable excitatory action on LRF neurones as postulated in the latter two studies or whether they have but a permissive facilitatory action on the pulsatile release of LRF provided NA transmission is unimpaired as suggested by our own studies.

The role of serotonin in abnormal motor neurone excitability (Myoclonus)

Since the clinical observation that certain forms of myoclonus react favourably to the administration of 1-5-hydroxytryptophan (1-5-HTP), the search for an animal model of this disabling syndrome has continued. Four models have been proposed.

1. Administration of high doses of 5-HTP to young (about 4-weeks-old) guinea-pigs (Chadwick *et al.*, 1978).
2. Administration of tryptophan or of 5-HTP to adult rats or guinea-pigs following pretreatment with MAO-inhibitors (Grahame-Smith, 1971; Modigh, 1972);
3. Administration of 5-HTP to adult rats subjected to destruction of their central serotonergic systems by high doses of 5,7-DHT (200 µg) following protection of the central NA systems by DMI (Stewart *et al.*, 1979).
4. Hypoxia and subsequent reventilation during which there occurs a transitory period of myoclonic jerking (Sharma *et al.*, 1979).

A pharmacological analysis of the 5-HTP and the 5-HTP plus MAO inhibition models indicates that the resulting motor disturbances cannot solely be attributed to serotonin since 5-HTP will be taken up by neurones and cells other than serotonin-containing ones and transformed into serotonin provided they contain *l*-aromatic acid decarboxylase (Melamed *et al.*, 1980), and released from these neurones or cells when oxidative deamination is blocked or when these neurones accept and handle 5-HT as a false transmitter; given the latter situation, 5-HT may also act indirectly by promoting the release of the natural transmitter. One of the central neurone systems which is particularly rich in decarboxylase activity is the dopaminergic system which, in addition, has been shown to actively accumulate $[^3H]$ serotonin and related hydroxytryptamines such as $[^{14}C]$ 5,7-DHT, $[^{14}C]$ 5,6-DHT and their methoxylated analogues (Baumgarten, unpublished observation; Yamamoto *et al.*, 1980).

Thus, there is a dopaminergic component in the action of 5-HT synthesized from 5-HTP in brain which can partly be unravelled by pretreating the animals with high doses of a dopamine receptor blocker such as haloperidol (cf. Fozard and Palfreyman, 1979). The fraction of 5-HT formed from 5-HTP in dopaminergic neurones is even greater in animals with most of their serotonergic projections destroyed by a neurotoxic lesion (the model proposed by Stewart *et al.*, 1979). The latter model has the drawback of direct-toxic damage to ventricle- and surface-near dopaminergic axons which are not protected by DMI pretreatment as evidenced in the fluorescence microscopical studies reported and discussed by Baumgarten *et al.* (1978).

Thus it appears that most of the models proposed so far do have a variable dopaminergic component and that the neurochemical mechanisms involved in the different models and, particularly, of myoclonus in humans, are not homogeneous. This is also evident from the fact that only those forms of myoclonus in man that are associated with decreased lumbar CSF 5-hydroxyindoleacetic acid (C5-HIAA) concentrations respond favourably to L-5-HTP therapy whereas other forms do not respond or become worsened by this treatment (see Chadwick *et al.*, 1978; van Woert *et al.*, 1977; Magnussen *et al.*, 1978; van Woert and Hwang, 1978; van Woert and Rosenbaum, 1979, Lhermite *et al.*, 1972; Guilleminault *et al.*, 1973; van Woert and Sethy, 1975). This observation also suggests that the site of origin of the increased motor neurone excitability in the CNS may be different. A brainstem/ spinal cord origin seems most likely in those forms that have reduced 5-HIAA levels in the lumbar CSF (which is only indicative of a reduced function in the bulbo-spinal 5-HT projection system; cf. Aizenstein and Korf, 1979) and the fact that they respond positively to 5-HT replacement therapy by L-5-HTP administration points to a serotonin deficiency in the pathophysiology of this form of myoclonus (so called reticular reflex myoclonus, cf. Hallett *et al.*, 1977).

By contrast, 5-HTP insensitive forms of myoclonus that have normal lumbar CSF 5-HIAA levels may have a cortical origin (Hallett *et al.*, 1979) but their neuro-chemical background is unknown. We have attempted to find models of myoclonus by using 5,7-DHT as a tool (Lachenmayer, Baumgarten and Janzen, unpublished observations). One of our models consists of adult rats injected with a high dose of 5,7-DHT (150 μg) into the left lateral ventricle under ether or a short-acting barbitu-rate, pretreated with DMI or nomifensine (25 mg/kg ip, 30 min before 5,7-DHT) or with saline. Myoclonic jerking is apparent by 10-15 min after 5,7-DHT injection (high frequency, disseminated myoclonus in the limbs and, later on, in the trunk). These hyperkinetic phenomena then become generalized and are accompanied by vegetative stimulation, such as piloerection and hypersalivation. In the recovery period (after 30–40 min), voluntary movements are accompanied and interrupted by intention-myoclonus. Animals treated with 5,7-DHT alone or in combination with DMI show more intense and generalized myoclonic jerking than those pre-treated with nomifensine (a combined NA/DA uptake blocker) suggesting that 5,7-DHT-mediated release of DA in addition to 5-HT is involved in this myoclonic syndrome. Our model of serotonin/dopamine stimulation myoclonus thus resembles the 5-HTP induced myoclonus.

Since human myoclonus is—in most cases—caused by hypoxia, we have attempted to evaluate the potential anticonvulsive action of serotonin in the rat by subjecting them to prolonged hypoxia (97 per cent N_2 and 3 per cent O_2), seven days follow-ing lesioning of the central serotonergic projections by a high intraventricular dose of 5,7-DHT or saline following protection of central catecholaminergic systems by DMI or nomifensine (25 mg/kg), 30 min before 5,7-DHT or saline injection. The motor behaviour during hypoxia and subsequent reventilation with air was con-tinuously monitored by video recording. Myoclonic jerking starts within 60 s after reventilation in both 5,7-DHT or vehicle injected rats; initially, myoclonic phen-omena occur as random, disseminated and irregular jerks; these are followed by strong flexion—and extension myoclonus; finally, intention myoclonus is observed regularly in the 5,7-DHT treated rats for about 2 min. While generalized intense myoclonic jerking lasts for about 5 min in the 5,7-DHT treated rats, disseminated myoclonus is mild and short-lasting (1 min duration) in the vehicle treated rats with little tendency towards generalization and occasional, transitory subsequent inten-tion myoclonus. These findings clearly indicate that serotonin deficiency may be associated with increased tendency to motor neurone excitability if the animals are subjected to hypoxia. However, posthypoxic myoclonus is only transitory in the experimental animals whereas it occurs for long periods of time in the human despite normal oxygenation. Therefore one has to conclude that the syndrome of persistent posthypoxic myoclonus in the human (Lance-Adams syndrome) does not only depend on distributed serotonergic transmission but also on deficiences in other types of neurones that inhibit motor neurone excitability.

The evaluation of the results and observations presented here suggests that serotonin has both facilitatory and inhibitory actions on motor neurone excitability which are probably exerted at different levels and circuits of the CNS.

1. Facilitatory effects in the brainstem motor nuclei (by enhancing the effect of excitatory inputs; McCall and Aghajanian, 1979).
2. Tremorogenic facilitatory effects in the inferior olive (de Montigny and Lamarre, 1973; Wiklund *et al.*, 1977; Sjölund *et al.*, 1977; Wiklund *et al.*, 1980).
3. Inhibitory effects in sensory relay nuclei of the spinal cord and brainstem (via excitation of inhibitory interneurones; cf. the discussion in the paper by Ruda and Gobel 1980).
4. Inhibitory effects in limbic centres that provide excitatory projections to neocortical sensory-motor reflex arcs.

Much work needs to be done to unravel the neurochemical background of the various models of experimental myoclonus in animals and, particularly, in the human; however, as shown here, 5.7-DHT has proved particularly powerful to explore the role of serotonin in various hyperkinetic syndromes.

ACKNOWLEDGEMENTS

H. G. Baumgarten was supported by grants from the Deutsche Forschungsgemeinschaft.

REFERENCES

Aizenstein, M. L. and Korf, J. (1979). On the elimination of centrally formed 5-hydroxyindoleacetic acid by cerebrospinal fluid and urine, *J. Neurochem.* **32**, 1227-1233.

Azmitia, E. C., Buchan, A. M., and Williams, J. H. (1978). Structural and functional restoration by collateral sprouting of hippocampal 5-HT axons, *Nature*, **274**, 374.

Baumgarten, H. G., Björklund, A., Lachenmayer, L., and Nobin, A. (1973). Evaluation of the effects of 5,7-dihydroxytryptamine on serotonin and catecholamine neurons in the rat CNS, *Acta physiol. scand.* Suppl. 391.

Baumgarten, H. G., Björklund, A., Nobin A., Rosengren, E., and Schlossberger, H. G. (1975). Neurotoxicity of hydroxylated tryptamines: structure-activity relationship, *Acta physiol. scand.* Suppl. 429, 1-27.

Baumgarten, H. G., Lachenmayer, L., and Björklund, A. (1977). Chemical lesioning of indoleamine pathways, in *Methods in Psychobiology*, Vol. 3, pp. 47-98, Academic Press, New York, San Francisco, London.

Baumgarten, H. G., Klemm, H. P., Lachenmayer, L., Björklund, A., Lovenberg, W., and Schlossberger, H. G. (1978). Mode and mechanism of action of neurotoxic indoleamines: A review and a progress report, *Ann. N.Y. Acad. Sci.* **305**, 3–24.

Baumgarten, H. G. Björklund, A., and Wuttke, W. (1978). Neural control of pituitary LH, FSH and prolactin secretion: the role of serotonin, in *Brain-endocrine Interaction III. Neural Hormones and Reproduction*, pp. 327–343, (Eds D. E. Scott, G. P. Kozlowski, and G. P. Weindel. S. Karger, Basel.

Baumgarten, H. G., Jenner, S., and Schlossberger, H. G. (1979). Serotonin neurotoxins: effects of drugs on the destruction of brain serotonergic, noradrenergic and dopaminergic axons in the adult rat by intraventricularly, intracisternally or intracerebrally administered 5,7-dihydroxytryptamine and related compounds, in *Neurotoxins. Fundamental and Clinical Advances.* (Eds I. W. Chubb, and L. B. Geffen) pp. 221–226, Adelaide University Union Press.

Baumgarten, H. G., Klemm, H. P. and Schlossberger, H. G. (1980). Monoamine oxidase-dependent metabolism of 5,6- and 5,7-dihydroxytryptamine and of serotonin in rat brain, in *Biochemical and Medical Aspects of Tryptophan Metabolism*, p. 334 (Eds. O. Hayaishi and R. Kido), Elsevier/North Holland, Amsterdam.

Baumgarten, H. G., Jenner, S., and Klemm, H. P. (1981). Serotonin Neurotoxins: Recent advances in the mode of administration and molecular mechanism of action, *J. de Physiologie (Paris).* **77**, 309–314.

Björklund, A. and Wiklund, L. (1980). Mechanisms of regrowth of the bulbospinal serotonin system following 5,6-dihydroxytryptamine induced axotomy. I. Biochemical correlates, *Brain Res.* **191**, 109–127.

Björklund, A., Lindvall, O., and Svensson, L. A. (1972). Mechanisms of fluorophore formation in the histochemical glyoxylic acid method for monoamines, *Histochemie*, **32**, 113–131.

Björklund, A., Nobin A., and Stenevi, U. (1973). The use of neurotoxic dihydroxytryptamines as tools for morphological studies and localized lesioning of central indolamine neurons, *Z. Zellforsch.* **145**, 479–501.

Björklund, A., Baumgarten, H. G., and Rensch, A. (1975b). 5,7-Dihydroxytryptamine: improvement of its selectivity for serotonin neurons in the CNS by pretreatment with desipramine, *J. Neurochem.* **24**, 833–835.

Björklund, A., Baumgarten, H. G., Horn, A. S., Nobin, A., and Schlossberger, H. G. (1975a). Neurotoxicity of hydroxylated tryptamines: structure-activity relationships. 2. *In vitro* studies on monoamine uptake inhibition and uptake impairment, *Acta physiol. scand.* Suppl. 429, 31–61.

Björklund, A., Falck, B., Lindvall, and Loren, I. (1980). The aluminium-formaldehyde. (ALFA) histofluoresence method for improved visualization of catecholamines and indoleamines. 2. Model experiments, *J. Neuroscience Methods,* **2**, 301–318.

Breese, G. R. and Cooper, B. R. (1975). Behavioral and biochemical interactions of 5,7-dihydroxytryptamine with various drugs when administered intracisternally to adult and developing rats, *Brain Res.* **98**, 517–527.

Chadwick, D., Hallett, M., Jenner, S., and Marsden, C. D. (1978). 5-Hydroxytryptophan-induced myoclonus in guinea pigs, *J. Neurol. Sci.* **35**, 157–165.

Chan-Palay, V. (1978). Indoleamine neurons and their processes in the normal rat brain and in chronic diet-induced thiamine deficiency demonstrated by uptake of ^{3}H-serotonin, *J. comp. Neurol.* **176**, 467–494.

Clemens, J. A., Roush, M. E., and Fuller, R. W. (1978). Evidence that serotonin neurons stimulate secretion of prolactin releasing factor, *Life Sci.* **22**, 2209–14.

Cohen, G. and Heikkila, R. E. (1978). Mechanisms of action of hydroxylated phenylethylamine and indoleamine neurotoxins, *Ann. N.Y. Acad. Sci.,* **305,** 74–78.

Creveling, C. R., Lundström, J., McNeal, E. T., Tice, L., and Daly, J. W. (1975). Dihydroxytryptamines: effects on nordrenergic functions in mouse heart *in vivo. Mol. Pharmacol.* **11,** 211–222.

Fenske, M. and Wuttke, W. (1975). Effects of intraventricular 6-hydroxydopamine injection on serum prolactin and LH levels: absence of stress induced pituitary prolactin release, *Brain Res.* **104,** 63–70.

Fozard, J. R. and Palfreyman, M. G. (1979). Metoclopramide antagonism of 5-hydroxytryptophan-induced 'wet-dog' shake behaviour in the rat, *Naunyn-Schmiedeberg's Arch. Pharmacol.* **307,** 135–142.

Fuller, R. W. (1980). Pharmacology of central serotonin neurons, *Ann. Rev. Pharmacol. Toxicol.* **20,** 111–27.

Fuller, R. W., Snoddy, H. D., and Molloy, B. B. (1975). Potentiation of the L-5-hydroxytryptophan-induced elevation of plasma corticosterone levels in rats by a specific inhibitor of serotonin uptake, *Res. Commun. Chem. Pathol. Pharmacol.* **10,** 193–96.

Fuller, R. W., Snoddy, H. D., and Molloy, B. B. (1976). Pharmacologic evidence for a serotonin neural pathway involved in hypothalamus-pituitary-adrenal function in rats, *Life Sci.* **19,** 337–46.

Gershanik, O. S., Heikkila, R. E., and Duvoisin, R. C. (1979). Asymmetric action of intraventricular monoamine neurotoxins, *Brain Res.* **174,** 345–350.

Gerson, S. and Baldessarini, R. J. (1975). Selective destruction of serotonin terminals in rat forebrain by high doses of 5,7-dihydroxytryptamine, *Brain Res.* **85,** 140–145.

Grahame-Smith, D. G. (1971). Studies *in vivo* on the relationship between brain tryptophan, brain 5-HT synthesis and hyperactivity in rats treated with a monoamine oxidase inhibitor and L-tryptophan, *J. Neurochem.* **18,** 1053–1066;

Guilleminault, C., Tharp, B. R., and Cousin, D. (1973) HVA and 5-HIAA CSF measurements and 5-HTP trials in some patients with involuntary movements, *J. Neurol. Sci.* **18,** 435–441.

Hallett, M., Chadwick, D., Adam, J., and Marsden, C. D. (1977). Reticular reflex myoclonus: a physiological type of human post-hypoxic myoclonus. *J. Neurol. Neurosurg., Psychiat.* **40,** 253–264.

Hallett, M., Chadwick, D., and Marsden, C. D. (1979). Cortical reflex myoclonus. Neuroclonus. *Neurology,* **29,** 1107–1125.

Hancke, J. L., Beck, W., Baumgarten, H. G. Höhn, K., and Wuttke, W. (1977). Modulatory effect of noradrenaline and serotonin on circhoral LH release in adult ovariectomized rat, *Acta endocr., Copenh.* Supp. 208, 22–23.

Horn, A. S., Baumgarten, H. G., and Schlossberger, H. G. (1973). Inhibition of the uptake of 5-hydroxytryptamine, noradrenaline and dopamine into rat brain homogenates by various hydroxylated tryptamines, *J. Neurochem.,* **21,** 233–236.

Jonsson, G. (1980). Chemical neurotoxins as denervation tools in neurobiology. *Ann. Rev. Neurosci.* **3,** 169–87.

Klemm, H. P. and Baumgarten, H. G. (1978). Interaction of 5,6- and 5,7-dihydroxytryptamine with tissue monoamine oxidase, *Ann. N.Y. Acad. Sci.* **305,** 36–56.

Klemm, H. P., Baumgarten, H. G., and Schlossberger, H. G. (1979). *In vitro* studies on the interaction of brain monoamine oxidase with 5,6- and 5,7-dihydroxytryptamine, *J. Neurochem.* **32,** 111–120.

Klemm, H. P., Baumgarten, H. G., and Schlossberger, H. G. (1980). Polarographic measurements of spontaneous and mitochondria-promoted oxidation of 5,6- and 5,7-dihydroxytryptamine, *J. Neurochem.* **35,** 1400–1408.

Krulich, L. (1975). The effect of a serotonin uptake inhibitor (Lilly 110140) on the secretion of prolactin in the rat, *Life Sci.* 17, 1141–44.

Lachenmayer, L., Baumgarten, H. G., and Wuttke, W. (1980). The central serotonin system: anatomy, physiology and clinical relevance, in *Pathologische Erregbarkeit des Nervensystems und ihre Behandlung* (Eds H. G. Mertens and H. Przuntek), pp. 480–486, Springer-Verlag, Berlin, Heidelberg, New York.

Lhermitte, F., Marteau, R., and Degos, C. F. (1972). Analyse pharmacologique d'un nouveau cas de myoclonies d'intention et d'action post-anoxique, *Revue Neurologique*, 126, 107–114.

Lorén, I., Björklund, A., Falck, B., and Lindvall, O. (1980). The luminum- ormaldehyde (Alfa) histofluorescence method for improved visualization of catecholamines and indoleamines. 1. A detailed account of the methodology for central nervous tissue using paraffin, cryostat or vibratome sections, *J. Neuroscience Methods*, 2, 277–300.

Magnussen, I., Dupont, E., Engbaek, F., and de Fine Olivarious, B. (1978). Posthypoxic intention myoclonus treated with 5-hydroxytryptophan and an extracerebral decarboxylase inhibitor, *Acta neurol scand.* 57, 289–294.

McCall, R. B. and Aghajanian, G. K. (1979). Serotonergic facilitation of facial motoneurone excitation, *Brain Research*, 169, 11–27.

Melamed, E., Jefti, F, and Wurtman, R. J. (1980). L-3,4-Dihydroxyphenylalanine and L-5-hydroxytryptophan decarboxylase activities in rat striatum: Effect of selective destruction of dopaminergic or serotonergic input, *J. Neurochem*, 34, 1753–1756.

Meltzer, H. Y., Fang, V. S., Paul, S. M., and Kaluskar, R. (1976). Effect of quipazine on rat plasma prolactin levels, *Life Sci.* 19, 1073–78.

Meyer, D. C. (1978). Hypothalamic and raphe serotonergic systems in ovulation control. *Endocrinology*, 103, 1067–1074.

Modigh, K. (1972). Central and peripheral effects of 5-hydroxytryptophan on motor activity of mice, *Psychopharmacologia (Berl.)*, 23, 48–54.

de Montigny, C. and Lamarre, Y. (1973). Rhythmic activity induced by harmaline in the olivo-cerebello-bulbar system of the cat, *Brain Res.*, 53, 81–95.

Nobin, A., Baumgarten, H. G., Björklund, A., Lachenmayer, L., and Stenevi, U. (1973). Axonal degeneration and regeneration of the bulbospinal indolamine neurons after 5,6-dihydroxytryptamine treatment, *Brain Res.*, 56, 1–24.

Ruda, M. A. and Gobel, St. (1980). Ultrastructural characterization of axonal endings in the substantia gelatinosa which take up [^{3}H] serotonin, *Brain Res.*, 184, 57–83.

Sharma, J. N., Snider, S. R., Fahn, S., and Hiesinger, E. (1979). Anoxic myoclonus in the rat, in *Advances in Neurology*, Vol. 26, pp. 181–189, (Ed S. Fahn), Raven Press, New York.

Sjölund, B., Björklund, A., and Wiklund, L. (1977). The indolaminergic innervation of the inferior olive. 2. Relation to harmaline induced tremor. *Brain Res.*, 131, 23–37.

Sladek, J. and Garver, D. L. (1976). Serotonin containing perikarya and pathways in the stump-tailed macaque (Macaca arctoides), *Neurosci. Abst.*, 2, 475.

Stewart, R. M., Campbell, S., Sperk, G., and Baldessarini, R. J. (1979). Electronic monitoring of myoclonus induced by 5-hydroxytryptophan after intracranial 5,7-dihydroxytryptamine. *Psychopharmacologia (Berl.)*, 60, 281–289.

Szafarczyk, A., Alonso, G., Ixart, G., Malaval, F., Nougier-Soule, J., and Assenmacher, I. (1980). The serotonergic system and the circadian rhythms of ACTH and corticosterone in rats. *Amer. J. Physiol.* (In press.)

Walker, R. F., Cooper, R. L., and Timiras, P. S. (1980). Constant oestrus: Role of rostral hypothalamic monoamines in development of reproductive dysfunction in aging rats, *Endocrinology*, **107**, 249–255.

Wiklund, L. and Björklund, A. (1980). Mechanisms of regrowth in the bulbospinal serotonin system following 5,6-dihydroxytryptamine induced axotomy. II. Fluorescence histochemical observations, *Brain Res.*, **191**, 129–160.

Wiklund, L., Björklund, A., and Sjölund, B. (1977). The indolaminergic innervation of the inferior olive. 1. Convergence with the direct spinal afferents in the areas projecting to the cerebellar anterior lobe, *Brain Res.*, **131**, 1–21.

van Woert, M. H. and Rosenbaum, D. (1979). L-5-Hydroxytryptophan therapy in myoclonus, in *Advances in Neurology*, Vol. 26, pp. 107–115, (Ed S. Fahn), Raven Press, New York.

van Woert, M. H., Rosenbaum, D., Howieson, J., and Bowers, M. B. (1977). Long-term therapy of myoclonus and other neurologic disorders with L-5-hydroxytryptophan and carbidopa, *New Engl. J. Med.* **296**, 70–75.

van Woert, M. H., and Sethy, V. H. (1975). Therapy of intention myoclonus with L-5-hydroxytryptophan and a peripheral decarboxylase inhibitor, MK 486. *Neurology*, **25**, 135–140.

van Woert, M. H. and Hwang, E. C. (1978). Biochemistry and pharmacology of myoclonus, in *Clinical Neuropharmacology*, Vol. 3, pp. 167–184, (Ed H. L. Klawans), Raven Press, New York.

Wuttke, W., Björklund, A., Baumgarten, H. G., Lachenmayer, L., Fenske, M. and Klemme, H. P. (1977). De- and regeneration of brain serotonin neurons following 5,7-dihydroxytryptamine treatment: Effects on serum LH, FSH and prolactin levels in male rats, *Brain Res.* **134**, 317–331.

Wuttke, W., Hancke, L., Höhn, K. G., and Baumgarten, H. G. (1978). Effect of intraventricular injection of 5,7-dihydroxytryptamine on serum gonadotropins and prolactin, *Ann. N.Y. Acad. Sci.* **305**, 423–436.

Yamamoto, Miyuki, Chan-Palay, V., and Palay, S. L. (1980). Autoradiographic experiments to examine uptake, anterograde and retrograde transport of tritiated serotonin in the mammalian brain, *Anat. Embryol.* **159**, 137–149.

Biology of Serotonergic Transmission
Edited by N. N. Osborne
© 1982, John Wiley & Sons Ltd.

Chapter 11

Radioactive Ligand Binding Studies: Identification of Multiple Serotonin Receptors

S. J. Peroutka and S. H. Snyder

*Johns Hopkins Medical School, Department of Pharmacology,
725 North Wolfe Street, Baltimore, Maryland 21205, USA*

INTRODUCTION

The development of radioreceptor binding assays during the past 10 years represents a major breakthrough in pharmacological methodology (Snyder, 1978). Early pharmacological studies relied on *in vivo* biological assays to measure drug potencies. Inherent problems with such assays included the determination of drug concentration at receptor sites, specificity of drug action and correlation with other biological systems.

Radioreceptor binding assays developed from a series of technological advances in the early 1970s. Affinity chromatography (Cuatrecasas, 1972), centrifugation separation (El-Allaway and Glieman, 1972) and especially rapid filtration methods (Cuatrecasas, 1971) led to the design of radioreceptor binding techniques. When applied to neurotransmitter and drug receptors in the brain, these methods allow for the direct determination of parameters such as receptor number, kinetic constants and drug affinity. Previously, these values could only be indirectly derived from classical pharmacological studies. Perhaps most importantly, the relative simplicity of an *in vitro* radioreceptor assay allows a large series of drugs to be tested simultaneously under the same experimental conditions.

Beginning with the identification of specific brain receptor sites for opiates, radiolabelled ligand techniques have been applied successfully to most known neurotransmitter receptor candidate. A number of general principles applies to all radioreceptor assay studies. First, the radioligand should be an agonist or antagonist drug with high (i.e. nanomolar) affinity for the receptor in question. Ideally, the ligand should also be selective in that it should not interact with other receptor sites at low concentrations. The binding should be saturable and reversible. In addition, the binding of the ligand must be 'specific' in terms of its displacement. That is, the binding should be inhibited only by drugs which are known to interact with the receptor site. In drug competition studies, the ligand should be displaced by a series of drugs in a rank order that reflects their known pharmacological potencies. Finally, the radioligand binding should display regional variations that tend to coincide with known physiological actions of the compound under study.

During the 1970s, an extensive literature arose concerning the application of these techniques to neurotransmitter and drug receptors in the brain (Yamamura *et al.*, 1978). An important outgrowth of this research concerned the direct application of radioreceptor studies to clinical pharmacological problems. The affinity of a drug for an *in vitro* receptor site, of course, cannot be assumed to measure the *in vivo* effect of the drug. Absorption, tissue distribution, protein binding and metabolism are all important factors that influence *in vivo* but not *in vitro* studies. Even more importantly, the demonstration of a 'specific' binding site does not necessarily prove a functional role for the receptor site under study. That is, a particular binding site can only be considered relevant after a correlation has been made between its pharmacological profile and some physiological function.

RADIOLIGAND ANALYSIS OF BRAIN SEROTONIN RECEPTORS

The first attempts to label serotonin (5-hydroxytryptamine, 5-HT) receptors in the central nervous system occurred before the development of rapid filtration techniques (Marchbanks, 1966, 1967; Fiszer and De Robertis, 1969). Measurements were made of [^{3}H] 5-HT interactions with synaptosomal fractions from rat brain.

After varying lengths of incubation, the tissue suspension was centrifuged and the supernatant discarded. The [³H] 5-HT which remained bound in the tissue pellet had an equilibrium constant (K_D) in the micromolar range and could be displaced by D-LSD. However, the bound [³H] 5-HT could also be displaced with similar concentrations of L-LSD, the non-hallucinogenic isomer of D-LSD. In addition, reserpine, a drug which disrupts the vesicular storage of 5-HT, could also displace [³H] 5-HT at relatively low concentrations. The relevance of this specific binding site was therefore questionable since its drug specificity did not correlate with any known effects of the 5-HT related agents studied.

Because of the known interaction of 5-HT and D-LSD, Farrow and Van Vanukis (1972) attempted to label 5-HT receptors with [³H] LSD. Using an equilibrium dialysis technique to detect specific binding, they observed that [³H] LSD bound with high affinity (K_D = 9 nM) to subcellular fractions of rat cerebral cortex. The binding was also stereospecific since L-LSD was an ineffective displacing agent. Among known neurotransmitters, 5-HT was the most potent displacer. As a result, the binding site labelled by [³H]LSD during equilibrium dialysis appeared to represent a central 5-HT receptor.

Shortly thereafter, rapid filtration techniques were applied to D-LSD binding (Bennett and Aghajanian, 1974). The binding of [³H] LSD was saturable, reversible and displayed high affinity (K_D = 7.5 nM). The binding was also stereospecific since L-LSD was over three orders of magnitude weaker than D-LSD. Moreover, the binding displayed regional variations with the highest levels of binding observed in brain regions known to receive a large 5-HT innervation. Lesions of the raphe nuclei had minimal effect on [³H] LSD binding which suggested that the receptor site was associated with postsynaptic membranes. These findings were soon confirmed and extended by others (Bennett and Snyder, 1975; Lovell and Freedman, 1976).

A second major development in the study of central 5-HT receptors was the use of [³H]5-HT in radioreceptor assays (Bennett and Snyder, 1976). In marked contrast to previous studies of [³H] 5-HT receptor labelling (Marchbanks, 1966, 1967; Fiszer and De Robertis, 1969), the radiolabel bound with very high affinity to rat brain membranes (K_D = 7.0 nM). The binding was saturable, stereospecific and displayed appropriate regional variations. As was observed with [³H]LSD binding, raphe lesions which result in a degeneration of 5-HT neurones did not lower [³H] 5-HT binding and therefore indicated a postsynaptic localization of receptor sites.

However, although the pharmacological profiles of both [³H] 5-HT and [³H] LSD binding suggested the labelling of 5-HT receptors, important differences were noted. For instance, D-LSD was one of the few drugs tested that had similar affinity for both [³H] 5-HT and [³H] LSD sites. The agonist 5-HT, on the other hand, had almost 100-fold greater affinity for [³H] 5-HT than [³H] LSD binding sites. By contrast, the classical peripheral antagonists such as cyproheptadine and cinanserin had markedly higher affinity for [³H] LSD than [³H] 5-HT sites. Moreover, [³H] LSD was invariably found to bind to a larger number of receptor sites than [³H] 5-HT.

At the time, Bennett and Snyder (1976) suggested that [³H]5-HT and [³H]LSD bound to two different 'states' of the 5-HT receptor in brain membranes. This 'flip-flop' model of neurotransmitter receptor function was based on the assumption that 5-HT bound to the agonist 'state' of the receptor while D-LSD, a mixed agonist-antagonist in physiological experiments, bound to both the agonist and antagonist 'state' of the 5-HT receptor. If restricted interconversion of receptor sites was assumed, then the experimental observations could be explained. That is, pure agonists would be more potent in displacing another agonist then in displacing a mixed agonist-antagonist. This explanation could also account for the increased number of [³H]LSD binding sites since, unlike [³H]5-HT, [³H]LSD could be expected to label both states of the 5-HT receptor.

A third [³H]ligand became available to study 5-HT receptors when Leysen *et al.* (1978b) reported that [³H]spiroperidol also labelled apparent 5-HT receptors in rat brain. Previously, [³H]spiroperidol had been used exclusively as a label for dopamine receptors. Like [³H]haloperidol, also a butyrophenone neuroleptic, [³H]spiroperidol binding in striatal membranes displayed properties which were characteristic of brain dopamine receptors (Leysen *et al.*, 1978a). But unlike [³H]haloperidol binding, [³H]spiroperidol also bound to a significant number of receptors in the rat frontal cerebral cortex where dopamine projections are minimal. When the pharmacological profile of the cortical [³H]spiroperidol binding was compared to cortical [³H]LSD binding, a significant correlation was found (Leysen *et al.*, 1978b). In addition, some of the most potent displacers of [³H] spiroperidol binding in the cerebral cortex were classical 5-HT antagonists. Of known neurotransmitters, 5-HT was the most potent displacing agent. These findings led Leysen *et al.* (1978b) to conclude that the [³H]spiroperidol binding sites in rat cerebral membranes are 'virtually identical to those labelled by [³H]LSD'. The ability of [³H]spiroperidol to label apparent 5-HT receptors in certain brain areas was confirmed and extended by other laboratories (Creese and Snyder, 1978; Quik *et al.*, 1978).

DIFFERENTIATION OF MULTIPLE SEROTONIN RECEPTORS

Thus, three distinct [³H]ligands had been reported to specifically label serotonin receptors in the central nervous system. However, subtle differences were noted between the binding characteristics of each [³H]ligand. With these differences in mind, we began a detailed investigation of the binding properties of [³H]5-HT, [³H]LSD and [³H]spiroperidol in the brain. As suggested by the evidence described below, we conclude that two distinct, non-interconverting serotonin receptors are present in the central nervous system.

Displacement of [³H]ligands associated with 5-HT receptors by 5-HT, spiroperidol and LSD

Unlabelled 5-HT, spiroperidol and LSD all mutually compete for their respective [³H]ligands (Fig. 1). If the three [³H]ligands label the same population of receptor

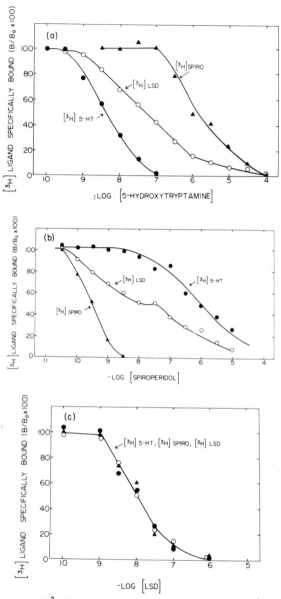

Fig. 1 Inhibition of [³H] ligand binding to 5-HT receptors. [³H] Ligand binding assays were performed under standard assay conditions in the presence of increasing concentrations of 5-hydroxytryptamine (Peroutka and Snyder, 1979). Data shown are the means of triplicate assays performed in a single experiment. The experiment was replicated four times. The [³H] ligands studies are [³H] 5-HT (●), [³H] spiroperidol (▲) and [³H] LSD (○). (a) Inhibition of [³H] ligand binding by 5-hydroxytryptamine. (b) Inhibition of [³H] ligand binding by spiroperidol. (c) Inhibition of [³H] ligand binding by LSD

sites, then 5-HT, spiroperidol and LSD should have the same potencies and displacement slopes in competing for all three agents (Peroutka and Snyder, 1979). However, there are marked differences among the three agents. For instance, 5-HT is almost a thousand times more potent in reducing [^{3}H] 5-HT binding than in competing for [^{3}H] spiroperidol binding (Fig. 1a). Hill coefficients for 5-HT displacement of these two agents are close to 1.0. In contrast, 5-HT displacement of [^{3}H]LSD binding is intermediate in potency when compared to the effects on [^{3}H] 5-HT and [^{3}H] spiroperidol binding. Moreover, the displacement curve of [^{3}H] LSD is quite shallow and has a Hill coefficient of 0.5, suggesting negative cooperativity or multisite transitions.

An inverse type of pattern occurs when spiroperidol competes for each of the three [^{3}H] ligands (Fig. 1b). Thus, spiroperidol is a thousand-fold less potent in reducing [^{3}H] 5-HT binding than in competing for [^{3}H] spiroperidol binding, while [^{3}H] LSD is displaced with an intermediate potency. Spiroperidol displacement of [^{3}H] spiroperidol has a Hill coefficient of 1.1 while displacement of [^{3}H] 5-HT binding by spiroperidol has a more shallow slope and a Hill coefficient of 0.64. The most shallow displacement occurs in the competition of spiroperidol for [^{3}H] LSD binding with a Hill coefficient of 0.30. Moreover, in repeated experiments, a biphasic displacement curve is obtained with a plateau for inhibition of [^{3}H] LSD binding between 10 and 30 nM spiroperidol. The difference in apparent half-maximal displacement values for spiroperidol displacement of [^{3}H] 5-HT and [^{3}H] spiroperidol is 1500-fold. A plateau is, therefore, consistent with biphasic displacement of [^{3}H] LSD from two distinct sites with more than three orders of magnitude difference in affinity. No plateau was consistently noted in 5-HT displacement of [^{3}H] LSD, where only a 700-fold difference exists between 5-HT affinity for [^{3}H] 5-HT and [^{3}H] spiroperidol binding.

In marked contrast to the behaviour of 5-HT and spiroperidol, LSD itself has identical potencies in reducing [^{3}H] 5-HT, [^{3}H] spiroperidol or [^{3}H] LSD binding (Fig. 1c). The slopes of the three displacement curves are essentially the same with Hill coefficients of approximately 1.0. This type of displacement behaviour suggests that [^{3}H] LSD has similar affinity for the receptors labelled by the three [^{3}H] ligands.

Saturation of [^{3}H] 5-HT, [^{3}H] spiroperidol and [^{3}H] LSD binding as influenced by 5-HT and spiroperidol

All three [^{3}H] ligands display monophasic saturation with a single binding component in evidence on Scatchard analysis (Peroutka and Snyder, 1979). The maximal number of [^{3}H] 5-HT and [^{3}H] spiroperidol binding sites are both approximately equal to 10 picomoles/g wet weight tissue. The amount of [^{3}H] LSD bound, however, is slightly greater than 20 picomoles/g wet weight tissue. This finding is in agreement with earlier observations that the maximal number of [^{3}H] LSD binding sites in rat cerebral cortex is approximately twice the number of [^{3}H] 5-HT sites in the same region (Bennett and Snyder, 1976).

The binding of low concentrations (i.e. nM) of [³H] 5-HT is completely displaced by 300 nM 5-HT. Similarly, low concentrations of [³H] spiroperidol are displaced by 30 nM spiroperidol. However, 300 nM 5-HT does not affect [³H] spiroperidol binding while 30 nM spiroperidol has a minimal effect on [³H] 5-HT binding. Interestingly, both 300 nM 5-HT and 30 nM spiroperidol inhibit [³H] LSD binding by about 50 per cent. The decrease in [³H] LSD binding represents a decrease in the number of binding sites with no change in the affinity of [³H] LSD for the remaining sites.

Influence of drugs on [³H] 5-HT, [³H] spiroperidol and [³H] LSD binding

To examine further the apparent labelling of different sites by the three [³H] ligands, the effects of various drugs on the binding of [³H] 5-HT, [³H] spiroperidol and [³H] LSD were evaluated (Table 1). Tryptamines are substantially more potent in reducing [³H] 5-HT binding than [³H] spiroperidol binding while their effects on [³H] LSD binding are intermediate in potency. By contrast, classical 5-HT antagonists and neuroleptics are much more potent in displacing [³H] spiroperidol than [³H] 5-HT binding with, again, intermediate effects on [³H] LSD binding. D-LSD is unique in that it has similar potency in competing for all three [³H] ligands while bromo-LSD and methysergide are more potent on [³H] spiroperidol binding than [³H] 5-HT binding. Importantly, no correlation exists between the *Ki* values of each of these drugs against [³H] 5-HT and [³H] spiroperidol binding (Fig. 2).

For all drugs tested, the *Ki* values versus [³H] LSD are intermediate between those for [³H] 5-HT and [³H] spiroperidol. If [³H] LSD labels both the [³H] 5-HT and [³H] spiroperidol binding sites, then a 'predicted *Ki*' value against [³H] LSD can be computed as the logarithmic mean of the *Ki* values for the inhibition of [³H] 5-HT and [³H] spiroperidol binding. The 'predicted *Ki*' values agree well with those observed experimentally (Table 1; Columns 3 and 4), indicating that [³H] LSD binds to a similar extent to the sites labelled by [³H] 5-HT and [³H] spiroperidol.

Additionally, this conclusion is supported by drug influences on [³H] LSD binding in the presence of 30 nM spiroperidol or 300 nM 5-HT. Since 30 nM spiroperidol completely blocks [³H] spiroperidol binding and reduces [³H] LSD binding by 50 per cent, then the residual [³H] LSD binding should involve the same sites labelled by [³H] 5-HT. When [³H] LSD binding is assayed in the absence of spiroperidol, no significant correlation is found with *Ki* values against [³H] 5-HT and [³H] LSD ($r = 0.43; p > 0.05$) (Fig. 3a). However, when [³H] LSD binding is assayed in the presence of 30 nM spiroperidol, the potencies of all drugs closely approximate their effect on [³H] 5-HT binding with a correlation coefficient of 0.98 ($p > 0.001$) (Fig. 3c).

Conversely, the binding of [³H] LSD in the presence of 300 nM 5-HT should represent labelling of the same sites as [³H] spiroperidol. While a correlation does exist between drug effects on [³H] spiroperidol and [³H] LSD binding ($r = 0.71$;

Table 1. Drug effects on $[^3H]$5-HT, $[^3H]$spiroperidol and $[^3H]$LSD binding to rat frontal cerebral cortex membranes. Rat frontal cortex membranes were incubated under standard assay conditions with 2.0 nM $[^3H]$5-HT, 0.26 nM $[^3H]$spiroperidol and 4.2 nM $[^3H]$LSD together with four concentrations of unlabelled drugs. $[^3H]$LSD assays were also performed in the presence of 30 nM spiroperidol or 300 nM 5-HT. IC_{50} values were determined by log-probit analysis and apparent K_i values were calculated from the equation $K_i = IC_{50}/(1 + [^3H]\text{ligand}/K_D)$. The K_D values were obtained from Peroutka and Snyder (1979). Values given are the means of three or more experiments, each performed in triplicate.

	Apparent K_i (nM)					
Drugs	$[^3H]$5-HT	$[^3H]$LSD + spiroperidol	$[^3H]$LSD	'Predicted' $[^3H]$LSD	$[^3H]$LSD + 5-HT	$[^3H]$spiroperidol
Tryptamines						
5-HT	3.8	17	110	100	1500	2700
5-methoxytryptamine	11	21	210	170	1600	2700
Bufotenine	37	83	220	180	2600	840
LSD analogues						
D-LSD	10	13	7.3	11	13	13
Bromo-LSD	89	75	13	15	1.9	2.5
Methysergide	88	110	8.4	15	2.8	2.6
5-HT antagonists						
Cyproheptadine	1500	480	90	56	1.7	2.0
Mianserin	860	1500	33	64	5.7	4.8
Cinanserin	1800	2000	130	180	13	18
Neuroleptics						
Spiroperidol	730	630	18	19	0.76	0.51
Haloperidol	16000	13000	900	820	36	42
Promethazine	10000	13000	2000	1300	190	170

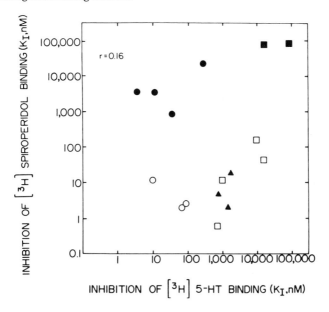

INHIBITION OF $\left[{}^{3}\text{H}\right]$ 5-HT BINDING (K_I.nM)

Fig. 2   Correlation between drug affinities for $[{}^{3}\text{H}]$ 5-HT and $[{}^{3}\text{H}]$ spiroperidol binding sites in rat frontal cerebral cortex. *Ki* values are taken from Table 1 and from Peroutka and Snyder (1979). The correlation coefficient ($r = 0.16$) is not statistically significant. The drug classes studies are tryptamines (●), LSD analogues (○), 5-HT antagonists (▲), neuroleptics (□) and others (■)

$p < 0.05$) (Fig. 3b), the addition of 300 nM 5-HT to the $[{}^{3}\text{H}]$ LSD assay markedly improves the correlation ($r = 0.99; p < 0.001$) (Fig. 3d). In addition, the slope of the linear regression line is 1.01 which suggests that the sites labelled by $[{}^{3}\text{H}]$ spiroperidol are identical to those labelled by $[{}^{3}\text{H}]$ LSD in the presence of 300 nM 5-HT.

Regional variations in serotonin receptor binding sites

The above data suggest that $[{}^{3}\text{H}]$ 5-HT binding sites are distinct from $[{}^{3}\text{H}]$ spiroperidol labelled sites. However, the possibility exists that these binding sites may be located on different portions of the same molecule since a similar number of both sites exist in the rat frontal cerebral cortex. To examine this possibility more carefully, we analysed regional variations in both $[{}^{3}\text{H}]$ 5-HT, $[{}^{3}\text{H}]$ LSD and $[{}^{3}\text{H}]$ spiroperidol binding in multiple regions of rat, guinea pig and bovine brain (Peroutka and Snyder, 1980b). In addition, $[{}^{3}\text{H}]$ mianserin, a tetracyclic antidepressant with anti-serotonergic and anti-histaminic properties, was utilized (Peroutka and Snyder, 1980b).

Marked regional variations in both $[{}^{3}\text{H}]$ 5-HT labelled receptors (designated 5-HT$_1$ sites) and $[{}^{3}\text{H}]$ spiroperidol or $[{}^{3}\text{H}]$ mianserin labelled receptors (designated 5-HT$_2$ sites) are noted. In general, 5-HT$_1$ receptors predominate in all brain regions

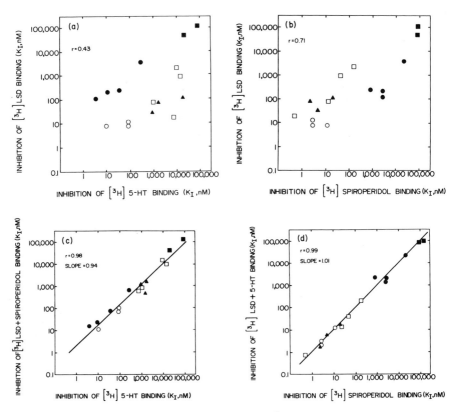

Fig. 3   Correlations between drug affinities for [³H]ligands in rat frontal cerebral cortex. *Ki* values are taken from Table 1 and from Peroutka and Snyder (1979). The drug classes include tryptamines (●), LSD analogues (○), 5-HT antagonists (▲), neuroleptics (□) and others (■). (a) Correlation between affinities for [³H] 5-HT and [³H] LSD. (b) Correlation between affinities for [³H] spiroperidol and [³H] LSD. (c) Correlation between affinities for [³H] 5-HT and [³H] LSD in the presence of 30 nM spiroperidol. (d) Correlation between affinities for [³H]spiroperidol and [³H] LSD in the presence of 300 nM 5-HT

with the exception of the cerebral cortex where similar numbers of 5-HT$_1$ and 5-HT$_2$ receptors are found. Highest levels of 5-HT$_1$ sites are in the hippocampus while the cerebellum contains the lowest amount of 5-HT$_1$ binding. In contrast, the highest levels of 5-HT$_2$ receptors are obtained in the cerebral cortex and caudate, with all other areas having substantially fewer receptor sites. This differential regional pattern strongly suggests that 5-HT$_1$ receptors labelled by [³H] 5-HT and 5-HT$_2$ receptors labelled by [³H] spiroperidol or [³H] mianserin are distinct molecular entities. The cerebral cortex, with essentially identical numbers of 5-HT$_1$ and 5-HT$_2$ receptors, is thus the exception rather than the rule in terms of the relative numbers of 5-HT$_1$ and 5-HT$_2$ receptors in any given brain region.

Conclusions

Analysis of radioactive ligand binding studies clearly suggests that two distinct serotonin receptors are present in the central nervous system. The receptor populations labelled by $[^3H]$ 5-HT are designated as 5-HT$_1$ receptors while $[^3H]$ spiroperidol and $[^3H]$ mianserin label 5-HT$_2$ receptors. $[^3H]$ LSD appears to label both receptor populations to a similar extent although pharmacological manipulations may limit specific $[^3H]$ LSD binding to either 5-HT$_1$ or 5-HT$_2$ sites.

The proposal that multiple 5-HT receptors exist in the brain differs from earlier conclusions from this laboratory which suggested that $[^3H]$ 5-HT and $[^3H]$ LSD label interconvertible states of a single 5-HT receptor (Bennett and Snyder, 1976). Early studies of the 5-HT receptor in brain membranes (Bennett and Snyder, 1976; Lovell and Freedman, 1976) had also noted the subtle differences between $[^3H]$ 5-HT and $[^3H]$ LSD binding. By analogy to similar differences in the opiate and dopamine systems, an interconvertible, two-state receptor was proposed as a means of explaining the binding data. However, the recent availability of $[^3H]$ spiroperidol and $[^3H]$ mianserin as ligands for 5-HT receptors has generated a large amount of new data which is not consistent with an interconverting 'flip-flop' model. Rather, all experimental data suggest that two distinct, non-interconverting populations of 5-HT receptors exist in the central nervous system.

FUNCTIONAL ROLE OF MULTIPLE SEROTONIN RECEPTORS

Of perhaps more interest is the demonstration of the possible functional roles of multiple 5-HT receptors in the central nervous system. If 5-HT$_1$ and 5-HT$_2$ receptors represent distinct molecular entities which both mediate the action of 5-HT, it follows that different effects of 5-HT may be associated with 5-HT$_1$ and/or 5-HT$_2$ receptors. For example, each of the two 5-HT receptors appears to correspond to different physiological processes.

Serotonin-sensitive adenylate cyclase

A 5-HT sensitive adenylate cyclase has been described in brain regions innervated by 5-HT nerve terminals (von Hungen *et al.*, 1975; Enjalbert *et al.*, 1978a). A number of observations indirectly suggest a possible linkage of the 5-HT adenylate cyclase with 5-HT$_1$ receptors labelled by $[^3H]$ 5-HT. For instance, 5-HT was initially reported to stimulate the cyclase with an apparent affinity constant (K_A) in the micromolar range (von Hungen *et al.*, 1975). Since only a group of relatively equipotent neuroleptics were tested as antagonists of this system (Enjalbert *et al.*, 1978b) a correlation with binding data cannot be made. On the other hand, Fillion *et al.* (1979a) have more recently obtained K_A values in the nanomolar range for 5-HT stimulation of adenylate cyclase. In addition, the potencies of some classical 5-HT antagonists correlate with drug affinities for 5-HT$_1$ receptors labeled by

[^{3}H] 5-HT (Peroutka *et al.*, 1980). However, another group of drugs fails to display such correlations (Nelson *et al.*, 1980). Moreover, Fillion *et al.* (1979b) have reported that the decrease in 5-HT cyclase stimulation after kainic acid lesions of the rat caudate corresponds to a concurrent loss of [^{3}H] 5-HT binding sites.

Further evidence supporting a linkage between an adenylate cyclase and 5-HT$_1$ receptors concerns the effects of nucleotides. Regulation of neurotransmitter receptor binding by guanine nucleotides often reflects an association of the receptor site with an adenylate cyclase (Maguire *et al.*, 1977). Guanine nucleotides decrease the binding of [^{3}H] 5-HT to 5-HT$_1$ but do not alter the binding of [^{3}H] spiroperidol to 5-HT$_2$ receptors, while [^{3}H] LSD binding is affected in an intermediate manner (Fig. 4) (Peroutka *et al.*, 1979). Furthermore, both Enjalbert (1978a) and Fillion (1979a) report a profound synergistic effect of GTP on 5-HT stimulation of adenylate cyclase. Although a direct linkage has not been documented, these findings do suggest that 5-HT$_1$ receptors labelled by [^{3}H] 5-HT mediate the activity of the 5-HT sensitive adenylate cyclase.

Serotonin behavioural syndrome

Behavioural hyperactivity results after central 5-HT stimulation with drugs such as 5-hydroxytryptophan (5-HTP) (Jacobs, 1976). The syndrome includes resting tremor, hindlimb abduction, splayed hindlimbs, snake tail, side-to-side weaving and head twitching. It follows that some, if not all, of these behavioural manifestations of central 5-HT stimulation may be mediated by 5-HT$_1$ and/or 5-HT$_2$ receptors. Of these behaviours, the head twitch is an easily monitored and reliable measure of the presence of the syndrome that, importantly, can be quantified. Accordingly, the potencies of a wide range of drugs in inhibiting 5-HTP induced head twitches were evaluated.

A number of drugs including classical 5-HT antagonists such as cyproheptadine and metergoline, neuroleptics such as spiroperidol and pipamperone, and antidepressants such as mianserin and amitriptyline display considerable potency in preventing 5-HTP induced head twitches (Peroutka *et al.*, 1981). The most potent inhibitor of 5-HTP induced head twitches is the butyrophenone neuroleptic spiroperidol. Interestingly, spiroperidol displays the same potency in blocking head twitches (ID$_{50}$ = 0.18 μmoles/kg) as in inhibiting apomorphine induced stereotypy (Creese *et al.*, 1976), a dopamine linked behavioural syndrome. Pipamperone (1.78 μmoles/kg) and chlorpromazine (2.41 μmoles/kg), on the other hand, are over 100 and 8 times, respectively, more potent inhibitors of the 5-HT rather than the dopamine behavioural syndrome. Conversely, haloperidol (4.36 μmoles/kg) is 10 times more potent in blocking the dopamine than the 5-HT syndrome. Thus, while neuroleptic drugs are potent antagonists of the 5-HT behavioural syndrome, their effects do not appear to be mediated through the dopamine system.

No correlation exists between drug affinities for 5-HT$_1$ receptors labelled by [^{3}H] 5-HT and head twitch blockade (Fig. 5A). By contrast, potencies of drugs in

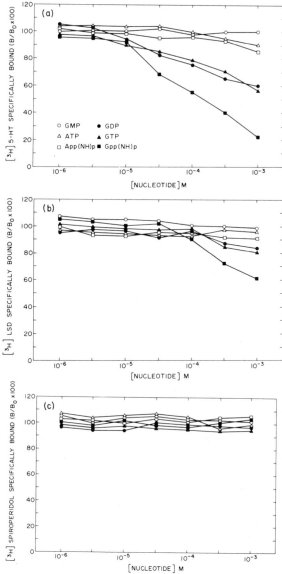

Fig. 4   Effects of increasing concentrations of nucleotides on the specific binding of [³H] ligands to rat cerebral cortex membranes. Increasing concentrations of GTP (▲), GDP (●), Gpp (NH) p (■), GMP (○), ATP (△) and App (NH) p (□) were added to the standard binding assay. Values are expressed as the percentage of specific binding as determined in the absence of nucleotides. Each point is the mean of triplicate determinations performed in a single experiment. The experiment was repeated three times with values which varied less than 20 per cent (A) [³H] 5-HT; (B) [³H] LSD; (C) [³H] spiroperidol

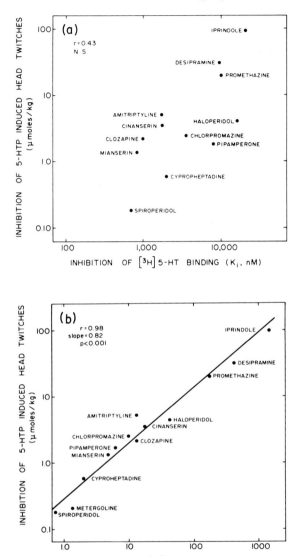

Fig. 5   Comparison of drug affinities at 5-HT receptors with *in vivo* blockade of 5-HTP induced head twitches. *Ki* values were determined as described in Table 1. (a) Inhibition of [³H] 5-HT binding to 5-HT₁ receptors *vs*. inhibition of 5-HTP induced head twitches. (b) Inhibition of [³H] spiroperidol binding to 5-HT₂ receptors *vs*. inhibition of 5-HTP induced head twitches

inhibiting 5-HTP induced head twitches correlate closely with affinities for 5-HT_2 receptors labelled by $[^3\text{H}]$ spiroperidol ($r = 0.98; p < 0.001$) (Fig. 5B). It is particularly striking that *in vivo* blockade of head twitches correlates with *in vitro* receptor affinity suggesting that the drugs tested differ little in bioavailability at target sites in the brain. Thus, at least one distinct component of the 5-HT hyperactivity behavioural syndrome, the head twitch, appears to be mediated via 5-HT_2 receptors.

Antidepressants and serotonin receptors

Chronic antidepressant treatment also differentiates 5-HT_1 and 5-HT_2 receptors (Peroutka and Snyder, 1980a). A significant reduction in the number of 5-HT_2 receptors, with no change in the number or affinity of 5-HT_1 receptors, occurs after long-term treatment with a variety of antidepressant agents. Tricyclic and atypical antidepressants as well as monoamine oxidase inhibitors are specific in this effect since the 5-HT antagonist methysergide, the neuroleptics chlorpromazine and haloperidol and the 5-HT uptake inhibitor fluoxitene have no effect on the binding parameters of 5-HT_1 or 5-HT_2 receptors. Detailed analysis of the effect of chronic amitriptyline treatment on 5-HT_2 receptors reveals that a significant decrease in the number of 5-HT_2 sites is observed after 1 week of treatment (10 mg/kg/day) and is maximal after 3-6 weeks of chronic treatment (Peroutka and Snyder, 1980b). Since a decrease in 5-HT_2 receptors is observed at clinically effective doses, alterations in 5-HT_2 receptors may have important implications for the role of 5-HT in human affective disorders.

PHYSIOLOGICAL CORRELATES OF SEROTONIN RECEPTORS

Significant pharmacological differences in the microiontophoretic effects of 5-HT have led neurophysiologists to suggest that at least two types of 5-HT receptor exist in the central nervous system (Roberts and Straughan, 1967; Banna and Anderson, 1968; Haigler and Aghajanian, 1974a; McCall and Aghajanian, 1979). The excitatory effects of 5-HT are invariably blocked by classical 5-HT antagonists such as D-LSD, bromo-LSD, methysergide and cyproheptadine. In marked contrast, the inhibitory actions of 5-HT are not affected by the classical antagonists and, in fact, are mimicked by D-LSD and lisuride. Only a single drug, metergoline, has been reported to antagonize specifically the inhibitory effect of 5-HT in microiontophoretic studies (Sastry and Phillis, 1977).

A comparison of drug potencies at 5-HT receptors with drug effects in microiontophoretic studies is summarized in Table 2. Only D-LSD, lisuride and metergoline have been reported to influence 5-HT induced inhibition. Importantly, each of these drugs has nanomolar affinity for 5-HT_1 receptors. The peripheral antagonists are without effect on the inhibitory actions of 5-HT and are one to three orders of magnitude less potent at 5-HT_1 sites. Conversely, each of these 5-HT related agents has been shown to specifically antagonize the excitatory effects of 5-HT. At 5-HT_2 receptors, these agents all display nanomolar affinity.

Table 2. Comparison of drug potencies at 5-HT receptors with neurophysiological actions. The apparent affinity constants (K_i) of various drugs for 5-HT receptors were determined as described in the legend to Table 1. Values given are the mean of three to six experiments, each performed in triplicate. The influence of drugs on the neurophysiological action of 5-HT was derived from the literature

5-HT related agents	K_i vs. $[^3H]$ 5-HT (nM)	K_i vs. $[^3H]$ spiroperidol (nM)	Effect on 5-HT synaptic inhibition	Effect on 5-HT synaptic excitation
5-HT	2.7	2700	Agonist[1-9]	Agonist[1-6]
D-LSD	9.8	8.9	Agonist[3,4,7,9]	Antagonist[1-4]
Lisuride	6.2	11	Agonist[9,10]	–
Metergoline	9.9	2.1	Antagonist[8]	Antagonist[5]
Cyproheptadine	1100	2.1	No effect[5]	Antagonist[5]
Bromo-LSD	100	2.5	No effect[1,7]	Antagonist[1]
Methysergide	150	3.1	No effect[1,5]	Antagonist[1,2,5,6]
Methiothepin	310	4.1	No effect[5]	Antagonist[5]
Cinanserin	1800	16	No effect[1,5]	Antagonist[1,5]

The references listed are as follows:
1. Roberts and Straughan, 1967
2. Boakes *et al.*, 1970
3. Bramwell and Gonye, 1973
4. Haigler and Aghajanian, 1974a
5. Haigler and Aghajanian, 1974b
6. Belcher *et al.*, 1978
7. Aghajanian *et al.*, 1972
8. Sastry and Phillis, 1977
9. Laurent and Pieri, 1978
10. Rogawski and Aghajanian, 1979

If both the inhibitory and excitatory effects of 5-HT are accepted, then an interesting pattern emerges. The classical 5-HT antagonists are capable of blocking excitatory but not inhibitory effects of 5-HT. Moreover, this finding corresponds to the selectivity of the peripheral antagonists for $5\text{-}HT_2$ over $5\text{-}HT_1$ receptors. Cyproheptadine, for example, is over 500 times more potent at $5\text{-}HT_2$ than $5\text{-}HT_1$ sites. On the other hand, D-LSD and metergoline have high affinity for both receptor binding sites and interact with both the excitatory and inhibitory effects of 5-HT.

Microiontophoretic studies are limited in that only a few drugs can be tested on a single neurone. The pattern which emerges from a review of the literature suggests an association between $5\text{-}HT_1$ receptors and the inhibitory effects of 5-HT and between $5\text{-}HT_2$ receptors and excitatory effects. Further studies aimed at increasing the knowledge of the pharmacological effects of microiontophoretically applied 5-HT would have important implications. As the actions and behaviours mediated by the two 5-HT receptors become more apparent, drugs could be specifically designed to stimulate or antagonize only one population of receptors. At the present time, specific high affinity antagonists exist for $5\text{-}HT_2$ receptors, the 5-HT behavioural syndrome and the excitatory effects of 5-HT. No such antagonist has yet been described for $5\text{-}HT_1$ receptors, the 5-HT sensitive adenylate cyclase or 5-HT-induced inhibition. Development of such drugs appears essential for a more complete understanding of 5-HT in the central nervous system.

SUMMARY AND CONCLUSIONS

Two distinct 5-HT receptors have been differentiated in mammalian brain on the basis of radioactive ligand binding studies (Table 3). The binding to $5\text{-}HT_1$ receptors, which are labelled by $[^3H]$ 5-HT and $[^3H]$ LSD, is regulated by guanine

Table 3. Characteristics of two distinct 5-HT receptors

	$5\text{-}HT_1$	$5\text{-}HT_2$
5-HT affinity	nanomolar	micromolar
Labelled by	$[^3H]$ 5-HT	$[^3H]$ spiroperidol
	$[^3H]$ LSD	$[^3H]$ LSD
		$[^3H]$ mianserin
Effect of LSD	agonist	antagonist
High affinity antagonists	metergoline	classical 5-HT antagonists; neuroleptics; certain anti-depressants
Drug potencies correlate with	5-HT sensitive adenylate cyclase (in part)	blockade of head twitches
Probable neurophysiological action	inhibition	excitation

nucleotides and may be associated with adenylate cyclase. In contrast, 5-HT$_2$ receptors, labelled by [^{3}H] spiroperidol and [^{3}H] LSD, are not affected by guanine nucleotides but are sensitive to chronic treatment with antidepressant medications. Moreover, 5-HT$_2$ receptors mediate the behavioural syndrome which follows central 5-HT stimulation.

The implications of these findings are extensive. At the molecular pharmacological level, the data clarify previously noted anomalies in the binding data. The additional number of [^{3}H] LSD binding sites as well as the differences in drug potencies between [^{3}H] 5-HT and [^{3}H] LSD can now be easily explained on the basis of two distinct 5-HT receptors. The sensitivity of 5-HT$_2$ sites to chronic antidepressant treatment represents a simple *in vitro* method for the determination of possible antidepressant efficacy. Also of interest is the potency of certain neuroleptics against 5-HT$_2$ receptors. This finding clarifies the controversy concerning the specificity of the 5-HT behavioural syndrome. Spiroperidol, in fact, is the most potent inhibitor of both 5-HT$_2$ receptors and the head twitch component of the 5-HT behavioural syndrome. Finally, there are neurophysiologic implications. Although the small number of drugs reported in any single microiontophoretic study precludes a detailed comparison, it should be noted that 5-HT induced inhibition of neuronal cells cannot be blocked by classical antagonists, a characteristic shared by 5-HT$_1$ receptors. On the other hand, both 5-HT induced excitation and 5-HT$_2$ receptors can be blocked by classical 5-HT antagonists.

REFERENCES

Aghajanian, G. K., Haigler, H. J., and Bloom, F. E. (1972). Lysergic acid diethylamide and serotonin: direct actions on serotonin-containing neurons in rat brain, *Life Science*, 11, 615–622.
Banna, N. R. and Anderson, E. G. (1968). The effects of 5-hydroxytryptamine antagonists on spinal neuronal activity, *J. Pharmacol. Exp. Ther.* 162, 319–325.
Belcher, G., Ryall, R. W., and Schaffner, R. (1978). The differential effects of 5-hydroxytryptamine, noradrenaline and raphe stimulation on nocioceptive and non-nocioceptive dorsal horn interneurons in the cat, *Brain Res.* 151, 307–321.
Bennett, J. L. and Aghajanian, G. K. (1974). D-LSD binding to brain homogenates: possible relationship to serotonin receptors, *Life Science*, 15, 1935–1944.
Bennett, J. P., Jr. and Snyder, S. H. (1975). Stereospecific binding of D-lysergic acid diethylamide (LSD) to brain membranes; relationship to serotonin receptors, *Brain Res.*, 94, 523–544.
Bennett, J. P., Jr. and Snyder, S. H. (1976). Serotonin and lysergic acid diethylamide binding to rat brain membranes: relationship to postsynaptic serotonin receptors, *Mol. Pharmacol.* 12, 373–389.
Boakes, R. J., Bradley, P. B., Briggs, I., and Dray, A. (1970). Antagonism of 5-hydroxytryptamine by LSD-25 in the central nervous system: a possible neuronal basis for the actions of LSD-25, *Br. J. Pharmacol.* 40, 202–218.
Bramwell, G. J. and Gonye, T. (1973). Responses of midbrain neurons to iontophoretically applied 5-hydroxytryptamine, *Br. J. Pharmacol.* 48, 357P–358P.
Creese, I., Burt, D. R., and Snyder, S. H. (1976). Dopamine receptor binding

predicts clinical and pharmacological potencies of antischizophrenic drugs, *Science*, **192**, 481–483.

Creese, I. and Snyder, S. H. (1978). ^{3}H-Spiroperidol labels serotonin receptors in rat cerebral cortex and hippocampus, *Eur. J. Pharmacol.* **49**, 201–202.

Cuatrecasas, P. (1971). Insulin-receptor interactions in adipose tissue cells; direct measurement and properties, *Proc. Natl. Acad. Sci.* **68**, 1264–1268.

Cuatrecasas, P. (1972). Affinity chromatography and purification of the insulin receptor of liver cell membranes, *Proc. Natl. Acad. Sci.* **69**, 1277–1281.

El-Allaway, R. M. M. and Gliemann, J. (1972). Trypsin treatment of adipocytes: effect on sensitivity to insulin, *Biochem. Biophys. Acta.* **273**, 97–109.

Enjalbert, A., Bourgoin, S., Hamon, M., Adrien, J., and Bockaert, J. (1978a). Postsynaptic serotonin-sensitive adenylate cyclase in the central nervous system, *Mol. Pharmacol.* **14**, 2–10.

Enjalbert, A., Hamon, M., Bourgoin, S., and Bockaert, J. (1978b). Postsynaptic serotonin-sensitive adenylate cyclase in rat brain, *Mol. Pharmacol.* **14**, 11–23.

Farrow, J. T. and Van Vanukis, H. (1972). Binding of D-lysergic acid diethylamide to subcellular fractions from rat brain, *Nature (Lond.)*, **237**, 164–166.

Fillion, G., Rousselle, J. C., Beaudoin, D., Pradelles, P., Goiny, M., Dray, F., and Jacob, J. (1979a). Serotonin sensitive adenylate cyclase in horse brain membranes, *Life Sci.* **24**, 1813–1822.

Fillion, G., Beaudoin, D., Rousselle, J. C., Deniau, J. M., Fillion, M. P., Dray, F., and Jacob, J. (1979b). Decrease of ^{3}H-5-HT high affinity binding and 5-HT adenylate cyclase activation after kainic acid lesion in rat brain striatum, *J. Neurochem.* **33**, 567–570.

Fiszer, S. and De Robertis, E. (1969). Subcellular distribution and chemical nature of the receptor site for 5-hydroxytryptamine in the central nervous system, *J. Neurochem.* **16**, 1201–1209.

Haigler, H. J. and Aghajanian, G. K. (1974a). Lysergic acid diethylamide and serotonin: a comparison of effects on serotonergic neurons and neurons receiving a serotonergic input, *J. Pharmacol. Exp. Ther.* **188**, 688–699.

Haigler, H. J. and Aghajanian, G. K. (1974b). Peripheral serotonin antagonists: failure to antagonize serotonin in brain areas receiving a prominent serotonergic input, *J. Neural Trans.* **35**, 257–273.

Jacobs, B. L. (1976). An animal behaviour model for studying central serotonergic synapses, *Life Sci.* **19**, 777–786.

Laurent, J. P. and Pieri, L. (1978). Lisuride and LSD decrease spontaneous firing of neurons in the raphe, *Experientia*, **34**, 926.

Leysen, J. E., Gommeren, W., and Laduron, P. M. (1978a). Spiperone: a ligand of choice for neuroleptic receptors, *Biochem. Pharmacol.* **27**, 307–316.

Leysen, J. E., Niemegers, C. J. E., Tollenaere, J. P., and Laduron, P. M. (1978b). Serotonergic component of neuroleptic receptors, *Nature (Lond.)*, **272**, 163–166.

Lovell, R. A. and Freedman, D. X. (1976). Stereospecific receptor sites for D-lysergic diethylamide in rat brain: effects of neurotransmitters, amine antagonists, and other psychotropic drugs, *Mo. Pharmacol.* **12**, 620–630.

McCall, R. B. and Aghajanian, G. K. (1979). Serotonergic facilitation of facial motoneuron excitation. *Brain Res.* **169**, 11–27.

Maguire, M. E., Ross, E. M., and Gilman, A. G. (1977). Beta-adrenergic receptor: ligand binding properties and the interaction with adenyl cyclase, *Adv. Cyclic Nucleotide Res.* **8**, 1–83.

Marchbanks, R. M. (1966). Serotonin binding to nerve ending particles and other preparations from rat brain, *J. Neurochem.* **13**, 1481–1493.

Marchbanks, R. M. (1967). Inhibitory effects of lysergic acid derivatives on 5-HT binding to nerve ending particles, *Biochem. Pharmacol.* **16**, 1971–1979.

Nelson, D. L., Herbet, A., Enjalbert, A., Bockaert, J., and Hamon, M. (1980). Serotonin-sensitive adenylate cyclase and [^{3}H] serotonin binding sites in the central nervous system of the rat -I. *Biochem. Pharmacol.* **29**, 2445-2453.

Peroutka, S. J., Lebovitz, R. M. and Snyder, S. H. (1979). Serotonin receptors affected differentially by guanine nucleotides, *Mol. Pharmacol.* **16**, 700-708.

Peroutka, S. J., Lebovitz, R. M., and Snyder, S. H. (1981). Two distinct central serotonin receptors with different physiological functions, *Science*, **212**, 827-829.

Peroutka, S. J. and Snyder, S. H. (1979). Multiple serotonin receptors: differential binding of ^{3}H-serotonin, ^{3}H-lysergic acid diethylamide and ^{3}H-spiroperidol, *Mol. Pharmacol.* **16**, 687-699.

Peroutka, S. J. and Snyder, S. H. (1980a). Long-term antidepressant treatment decreases spiroperidol-labeled serotonin receptor binding, *Science*, **210**, *88-90*.

Peroutka, S. J. and Snyder, S. H. (1980b). Regulation of serotonin$_2$ (5-HT$_2$) receptor labeled with ^{3}H-spiroperidol by chronic treatment with the antidepressant amitriptyline, *J. Pharmacol. Exp. Ther.* **215**, 582-587.

Peroutka, S. J. and Snyder, S. H. (1981a). ^{3}H-Mianserin: differential labeling of serotonin$_2$ and histamine$_1$ receptors in rat brain, *J. Pharmacol. Exp. Ther.* **216**, 142-148.

Peroutka, S. J. and Snyder, S. H. (1981b). Two distinct serotonin receptors: regional variations in receptor binding in mammalian brain, *Brain Res.* **208**, 339-347.

Quik, M., Iversen, L. L., Lardner, A., and Mackay, A. V. P. (1978). Use of ADTN to define specific ^{3}H-spiperone binding to receptors in brain, *Nature (Lond.)*, **274**, 513-514.

Roberts, M. H. T. and Straughan, D. W. (1967). Excitation and depression of cortical neurones by 5-hydroxytryptamine, *J. Physiol.* **193**, 269-294.

Rogawski, M. A. and Aghajanian, G. K. (1979). Response of central monoaminergic neurons to lisuride: comparison with LSD, *Life Science*, **24**, 1289-1298.

Sastry, B. S. R. and Phillis, J. W. (1977). Metergoline as a selective 5-hydroxytryptamine antagonist in the cerebral cortex, *Can. J. Physiol. Pharmacol.* **55**, 130-133.

Snyder, S. H. (1978). Overview of neurotransmitter receptor binding, in *Neurotransmitter Receptor Binding* (eds H. I. Yamamura, S. J. Enna, and M. J. Kuhar), pp. 1-12, Raven Press, New York.

von Hungen, K., Roberts, S., and Hill, D. F. (1975). Serotonin-sensitive adenylate cyclase activity in immature rat brain, *Brain Res.* **84**, 257-267.

Yamamura, H. I., Enna, S. J., and Kuhar, M. J. (1978). *Neurotransmitter Receptor Binding*, Raven Press, New York.

Biology of Serotonergic Transmission
Edited by N. N. Osborne
© 1982, John Wiley & Sons Ltd.

Chapter 12

Some Functional Aspects of Central Serotonergic Neurones

MARTHE VOGT

A.R.C. Institute of Animal Physiology,
Babraham, Cambridge CB2 4AT, UK

DISTRIBUTION OF SEROTONERGIC NEURONES IN THE CNS

As discussed in detail in other chapters, most serotonergic neurones originate from the so-called raphe nuclei found in a restricted midline region of medulla and midbrain. These nuclei contain a varying proportion of 5-HT-containing cells and send their axons into ascending and descending pathways. The site of origin of these neurones is further restricted by the fact that in, for example, the dorsal raphe nucleus, only one third of the cells are serotonergic. The physiological importance of this fact was demonstrated (Aghajanian *et al.*, 1978) by the observation that electrical stimulation of the central end of the sciatic nerves suppressed firing in the 5-HT-laden cells, and activated the remaining (non-fluorescent) cells. Since dorsal raphe lesions are very often carried out to determine the function of serotonergic neurones, much caution is needed in the interpretation of work done by electrolytic rather than the more selective chemical lesions.

The serotonergic axons were originally only traced by fluorescence microscopy after condensation of the amine with formaldehyde. However, this method is far less sensitive for 5-HT than for catecholamines and the fluorophore fades rapidly. A much wider distribution of 5-HT-containing axons than originally suspected has been established by methods using autoradiography: either $[^3H]$ 5-HT was visualized in the serotonergic terminals of the cortex thanks to the potent uptake process for the amine shown by all serotonergic neurones (Descarries, Beaudet, and Watkins, 1975), or tritiated amino acids were injected into the raphe nuclei and their anterograde transport studied (Bobillier *et al.*, 1976).

Recently, immunohistochemical methods have become feasible. They are carried out by coupling 5-HT with a protein and immunizing a rabbit or guinea-pig with the protein complex. Antibodies are formed against the complex as well as against the protein moiety; these latter antibodies are removed by affinity chromatography, leaving a fraction which, when applied to tissue slices, is deposited only where 5-HT is present in the tissue (Steinbusch *et al.*, 1978). The antibody can be visualized by fluorescence or a peroxidase–antiperoxidase reaction.

In contrast to former views that 5-HT terminals were only present in some cortical layers, these methods have shown (Lidov, *et al.*, 1980) that the terminals impinge on every or nearly every cortical cell, and certainly form plexus in all cortical layers; two regions are exceptional, in that only some layers display serotonergic terminals, the posterior cingulate cortex and the hippocampus.

Another assessment of the density of 5-HT terminals may be made by the serotonin concentration of the tissue; this is feasible because the cells of origin, apart from a few perikarya in the hypothalamus, are congregated in the raphe nuclei; outside this region, there are essentially only axons and terminals. The axons contain much less 5-HT than the terminals, so that the tissue concentration of the amine reflects mainly the density of terminals. The majority of subcortical grey structures have a high concentration of serotonin. Examples are: medulla oblongata, substantia nigra, superior colliculi, central grey, tuberculum olfactorium, hypothalamus and thalamus. This contrasts with low figures for hippocampus and olfactory bulb, and minute concentrations in cerebellum. Descending axons innervate the spinal cord, concentrations in the dorsal exceeding those in the ventral parts.

This intense arborization which allows terminals from a very restricted number of cells in the raphe nuclei to innervate the majority of cells in the brain, probably justifies the search for a general 'modulatory' activity of the serotonergic system on brain activity. Any specificity of the action exerted would then lie in the neurones impinged upon and not in a unique functional property of the discharging raphe neurones. There are, however, other factors which may also lead to specific effects of raphe activity. First, the electrophysiological response to 5-HT (usually applied by iontophoresis) is not always the same. A diversity of the receptors (and, therefore, of responses) is known to exist in the invertebrate, and a heterogeneity of terminals has been described in the rat striatum (Arluison and de la Manche, 1980). Aghajanian and Haigler (1974) have shown that mammalian cells with a heavy serotonergic input respond to injected 5-HT with inhibition, whereas cells with little such input are frequently excited (see also Boakes *et al.*, 1970). The general rule, therefore, that serotonin release will always cause inhibition, has its exceptions.

Secondly, specific effects may be elicited by 5-HT if a serotonergic terminal enters a neuronal circuit with a circumscript immutable function. Several such instances are known in the field of endocrinology. Taleisnik *et al.* (1973/1974) have established that the release of melanocyte-stimulating hormone (MSH) from the pituitary which follows physiological stimuli such as suckling, is mediated by serotonin and can be imitated by its intraventricular injection. By the judicious use of antagonists, the authors showed that the serotonergic neurone achieves its purpose by a very circuitous route. It impinges on a GABAergic neurone which, in turn, inhibits a catecholamine-containing (probably dopaminergic) third neurone. This exerts a tonic stimulating influence on a fourth neurone which contains and secretes MSH-R-IF (melanophore stimulating hormone release-inhibitory factor). The arrest of secretion of the inhibitory factor finally leads to the rise in secretion of MSH by pituitary cells.

Prolactin secretion is controlled by a similar circuit (Caligaris and Taleisnik, 1974), and one may wonder how many equally involved multineuronal circuits of which one link is serotonergic exist in the brain. These considerations appear to justify the search both for generalized and for specific effects of the serotonergic system on normal brain function.

It will have been noted that, in the circuits studied by Taleisnik and co-workers, 5-HT has a depolarizing (activating) action on a GABAergic neurone; this raises the question whether other inhibitory actions of serotonergic neurones are mediated by the release of GABA. Such a possibility has been suggested by Giambalvo and Snodgrass (1978) for the inhibition by 5-HT of striatal dopamine terminals, but they considered a cholinergic interneurone as an alternative. Wang and Aghajanian (1977), on the other hand, obtained evidence in the amygdala for a direct inhibitory effect of serotonergic neurones originating in the dorsal raphe nucleus.

DENSITY OF TERMINALS AND OF SYNAPSES

When compared with some other transmitters, the serotonergic terminals are fairly sparse. In the striatum, they are estimated to amount to 5 per cent of the total number of endings. This is certainly no under-estimate, as suggested by the findings of Melamed *et al.* (1980). These authors showed that the decarboxylase which is necessary for the formation of dopamine as well as of serotonin, is hardly diminished by destruction of the raphe nuclei, but is reduced to 20 per cent after lesion of the dopaminergic pathways. Cholinergic terminals are, of course, abundant in the striatum.

Another feature of serotonergic neurones which is bound to have functional consequences is the paucity of synapses; only about 5 per cent of boutons along an axon make synaptic contacts (Descarries *et al.*, 1975); this contrasts with about 50 per cent found on varicosities of non-aminergic nerves (Dismukes, 1977). The interpretation given to this finding is that release may take place from all varicosities, and that the serotonin liberated from non-synaptic boutons diffuses in the neuropil and affects a whole region of brain tissue. This phenomenon might be the basis for generalized effects of activity of raphe nuclei. Similar anatomical features are shown by the central noradrenergic system: it originates from a few thousand cells in the lower brain stem, undergoes very extensive arborization in many parts of the brain, and is characterized by a paucity of varicosities which form synapses.

OPPOSITE EFFECTS OF NEURONES CONTAINING 5-HT AND NORADRENALINE

In spite of the structural similarities between the serotonergic and the adrenergic system, the two types of neurones tend to antagonize each other. Thus, noradrenaline inhibits electrically induced serotonin release from brain slices (Göthert and Huth, 1980). Motor effects mediated by dopaminergic terminals in the striatum and nucleus accumbens are reduced by serotonergic activity (Carter and Pycock, 1979). This also applies to dopamine-dependent motor effects of amphetamine.

Since there are serotonergic terminals in the locus coeruleus and adrenergic terminals in the raphe nuclei, there is much scope for a physiological interaction of the two aminergic systems in this region. The infrequent spontaneous firing of serotonergic dorsal raphe cells is activated by noradrenaline. Antagonists of α-adrenoreceptors abolish this effect which appears to be produced directly, not by inhibiting GABA interneurones (Baraban and Aghajanian, 1980). Whereas some modification of serotonergic discharge always results from local application of noradrenaline, the response need not always be in the same direction. Baraban and Aghajanian (1980) found that only low doses of noradrenaline applied iontophoretically activated cells in the dorsal raphe, large doses inhibited them. Key, Boakes and Candy (1980) obtained both inhibitory and stimulatory effects, depending on the part of the dorsal raphe nucleus involved. Thus the results seem to depend on experimental conditions.

An inhibitory, not stimulatory, noradrenergic input in the nucleus raphe magnus is thought to be the basis for the observation that α-adrenoreceptor antagonists applied locally cause hypoalgesia. The prevention of this effect by intrathecal methysergide suggested that it is caused by stimulation of serotonergic neurones (Hammond *et al.*, 1980). Very clear antagonism is shown in the lateral geniculate of the rat. Norepinephrine accelerated spontaneous activity and facilitated evoked responses, whereas serotonin slowed discharge rate and depressed evoked responses (Rogawski and Aghajanian, 1980).

NEURONES CONTAINING 5-HT AND SUBSTANCE P

Before dealing with individual effects of lack or excess of serotonin, it is necessary to discuss a recent development in transmitter distribution which may have important physiological consequences. In the last decade, more and more substances, mostly peptides, have been discovered within neurones of the central nervous system and, on the strength of this localization, accepted into the family of 'putative transmitters'. Their identification was based on immunofluorescence shown in brain slices incubated in antisera to identified peptides isolated from nervous or gastro-intestinal tissue. The technique has led to the discovery of many neurones which contain, and may transmit by, a single peptide such as an enkephalin, β-endorphin or cholecystokinin. However, immunofluorescence has also demonstrated such peptides in neurones already characterized by their content of a 'classical' transmitter, among others of serotonin. The compound found in some serotonergic neurones is the undecapeptide discovered in 1931 by von Euler and Gaddum, called 'substance P' and identified by Chang and Leeman (1970). This peptide usually occurs in neurones devoid of serotonin; an example is its presence in a group of primary afferent pain fibres, the corresponding sensory ganglion cells and their spinal terminals. At present it seems that the occurrence of 5-HT and substance P in the same neurone is the exception rather than the rule; it has been established (Chan-Palay *et al.*, 1978; Hökfelt *et al.*, 1978) in the three caudal raphe nuclei, some neighbouring nuclei and in the spinal cord; even here not all serotonergic neurones contain the peptide.

When the fairly specific toxins 5,6- or 5,7-dihydroxytryptamine (which preferentially damage serotonergic axons and terminals) are used to destroy serotonergic neurones, the axons containing substance P disappear in the same way as axons lacking substance P (Hökfelt *et al.*, 1978). There is no recovery of these descending neurones for at least many months, and both 5-HT and substance P in the cord remain quite low. There is no information on the role played by substance P in the serotonergic spinal neurones. A possibility is that the polypeptide acts as presynaptic 'modulator' of transmission. The action of substance P in other parts of the central nervous system is known to be excitatory, and one assumes that the same may hold for the fibres containing two transmitters, but there is as yet nothing known about release of the peptide from these unusual terminals. Reserpine causes 5-HT to disappear from these neurones, but substance P is retained.

p-CHLOROPHENYLALANINE

The most generalized reversible inactivation of serotonergic neurones is probably achieved by the administration of *p*-chlorophenylalanine (PCPA). It acts by inhibiting tryptophanhydroxylase, and its effects can, therefore, be antagonized by giving the precursor of serotonin, 5-hydroxytryptophan. Some authors doubt the specificity of the action of PCPA because of a slight additional effect exerted on catecholamine synthesis. Not only is this effect small, it has also a completely different time course from the changes in serotonin. It starts on the first day, and remains unchanged thereafter. The depleting effect on the 5-HT in the brain has its maximum 2-3 days after a single high dose of PCPA. Any clinical signs shown when the 5-HT content of the tissue is at its lowest can with confidence be attributed to lack in serotonergic activity.

Rats or cats given doses of PCPA which reduce cerebral 5-HT to about 12 per cent of normal are overactive, sleep little, and react excessively to environmental stimuli. Males mount other males and non-oestrous females, thus exhibiting indiscriminate sexual activity which has lost its biological purpose (Hoyland *et al.*, 1970). There is no immediate threat to life, although the condition would lead to loss of weight and deterioration of health if it could be maintained for a long period.

These observations alone direct attention to the possibility that serotonergic activity protects the organism from damaging responses to environmental stimuli. The following sections will illustrate this view further.

SLEEP

Sleep is one of the most important protective mechanisms permitting optimal performance of the central nervous system during the waking state. Insomnia was the first serious deficiency shown by Jouvet (1962) and his co-worker Renault (1967) to be caused by either PCPA or by surgical lesion of raphe nuclei. Administration of 5-hydroxytryptophan temporarily restores the capacity to sleep in animals in which PCPA has depleted raphe neurones of their transmitter, but not in animals in which the insomnia has been caused by lesions.

Raphe discharges greatly diminish during slow-wave sleep, become frequent in arousal, and are fastest at the onset of paradoxical sleep (Shen *et al.*, 1974; McGinty, 1976). Application of serotonin to the region of the area postrema causes EEG synchronization (Bronzino *et al.*, 1972). All these findings show that raphe nuclei play an essential part in the phenomenon of sleep, but so do dorsal noradrenergic systems: damage to the locus coeruleus or to the dorsal noradrenaline bundle increase REM sleep, produce several days of hypersomnia and a rise in telencephalic 5-hydroxyindoleacetic acid (Petitjean *et al.*, 1975). Furthermore, α-methyltyrosine injected intraperitoneally can antagonize the insomnia produced by raphe lesions (Jouvet, 1971). A great deal of work has tried to establish the exact contribution which serotonergic and adrenergic pathways play in the causation or inhibition of the different phases of sleep. Yet, the picture remains confused. It is also well known that cholinergic mechanisms are involved in the production of sleep, and that a number of humoral factors produced by sleep deprivation appear also to play a role.

Perhaps it is not surprising that REM sleep, which is supposed to be essential for memory consolidation, and slow-wave sleep, which has recuperative effects on brain activity, are very difficult to inhibit permanently. Their role may be so vital that compensatory phenomena take over if single links in the chain of the normal complex neuronal interactions are damaged. If PCPA is given daily to cats for up to 37 days, the 5-HT content of the brain remains low, but sleep returns spontaneously between days 6 and 8. During the insomniac phase, small doses of chlorpromazine (1-5 mg/kg) temporarily restore REM and slow-wave sleep (Cohen *et al.*, 1973). This finding suggested to the authors that loss of brain serotonin and schizophrenia had certain features in common.

In contrast to the effect of PCPA, insomnia following surgical destruction of the raphe is not reversible; this may be caused by the destruction not only of serotonergic neurones, but of many other neighbouring structures. Such an explanation would agree well with the observation (Froment *et al.*, 1974) that intracerebroventricular injection of 5,6-dihydroxytryptamine, which selectively destroys serotonergic neurones, seriously disturbed sleep for a period of 3–4 days only. However, some diminution of REM sleep persisted for the whole observation period of 15 days.

Attempts at combating human insomnia by giving 5-hydroxytryptophan or tryptophan have met with little success. The reasons may be that, in man, insomnia is rarely or never due to simple lack of serotonin. On the other hand, when L-DOPA treatment of patients has caused a decrease in REM sleep time, 5-hydroxytryptophan administration restores the balance of amines and brings REM sleep back.

TEMPERATURE REGULATION

This is another vital autonomic function which is known to be influenced by cholinergic, adrenergic and serotonergic neurones. Serotonin injection into the hypothalamus (Feldberg and Myers, 1964) raises the body temperature of the cat. In the conscious monkey, Myers *et al.* (1969) showed that, in response to a blast of cold air, serotonin was released from the hypothalamus into a permanently implanted collecting cannula. PCPA, on the other hand, reduced the febrile response of the cat to injection of bacterial pyrogen (Harvey and Milton, 1974). There are as yet unexplained species differences in the effect of 5-HT on temperature regulation. Rabbit, rat and sheep respond to serotonin injected into the lateral ventricle with a fall, not with a rise in body temperature. These differences could well be adaptations to different environment or 'life styles' of the animals, which may have caused development of some and neglect of other pathways which impinge on temperature regulation. In some instances, it might be a matter of differences in the sites affected by the administation of 5-HT. Thus Komiskey and Rudy (1977) pointed out that injection of 5-HT into the anterior hypothalamus caused shivering and *hyper*thermia in the cat, whereas similar injections into pre-optic sites elicited vasodilatation and *hypo*thermia.

HORMONE SECRETION

Since both lobes of the pituitary gland are under hypothalamic control, and serotonergic neurones abound in the hypothalamus, including the median eminence, it is not surprising that serotonin exerts an influence on the secretion of most hormones. The influence is always indirect, and usually mediated through inhibition of the liberation of either hypothalamic releasing or of release inhibiting factors. It is thus generally true to say that serotonin reduces secretion of hormones which are normally activated by a releasing factor, and enhances secretion of those kept in check by a release inhibiting factor. Examples were given above (p. 301) which showed the complex circuitry involved in the inhibition, by serotonin, of the liberation of two release inhibiting factors, thus enhancing blood concentrations of prolactin and melanophore stimulating hormone. Growth hormone secretion is also enhanced by 5-HT, probably as a result of a reduction in the production of somatostatin, which acts as a release inhibiting factor (Arnold and Fernstrom, 1980). On the other hand, secretion of pituitary hormones which are, like gonadotrophins, mainly controlled by releasing factors, is decreased by serotonin. Yet even this statement does not go unchallenged. Van de Kar *et al.* (1980) showed a fall in serum LH after 5,7-dihydroxytryptamine lesions of the dorsal raphe nucleus. The disagreement with the results of other workers might be due to the different techniques of inactivating serotonergic neurones, or to the precise site of the anatomical lesions. Yet another explanation is suggested by Héry *et al.* (1976) who showed restoration, by 5-hydroxytryptophan given to PCPA-treated rats, of the surge in luteinizing hormone which determines ovulation. They think that 5-HT has a permissive effect on the surge, and that the surge takes place independently of stimulation by releasing factor.

Inhibitory effects on secretion have also been reported for corticotrophin and thyrotrophin although the findings do not all agree. Reflex release of oxytocin during suckling is inhibited by serotonin or its precursor 5-hydroxytryptophan (Mizuno *et al.*, 1967). Nothing is known about the neurones involved.

While destruction of the raphe nuclei undoubtedly impairs lactation, ovulation, and the compensatory effects of hemicastration, the role of serotonin in another feedback control, that of adrenocortical hormones, is suggested by the following observation: hydrocortisone implants into the rat hypothalamus increased the 5-HT content of the basomedial hypothalamus, whereas peripheral hormone injections lowered it (Ulrich *et al.*, 1975).

OTHER AUTONOMIC FUNCTIONS

There is evidence that peripheral sympathetic activity and adrenomedullary secretion are attenuated by serotonin (Ito and Schanberg, 1972). Vasomotor reflexes are damped by serotonergic activity. Accumulation of 5-HT slows respiratory rate: this is probably a consequence of diminished response to CO_2 in the inspired air.

5-HT containing terminals have been found between the ependymal cells of cerebral ventricles and in the chorioid plexus (Lorez and Richards, 1973). Their origin is from cells in raphe nuclei (Moskowitz *et al.*, 1979). These findings might suggest a role of 5-HT terminals on formation of cerebrospinal fluid, and possibly on blood flow in the plexus. It is not clear whether these terminals innervate the ependyma or secrete into the cerebral ventricles and in this way affect distant cells. Another possibility is a local influence on ciliary movements.

MOTOR ACTIVITIES

Monosynaptic reflexes as well as motoneurone activity are *enhanced* by 5-HT, but polysynaptic reflexes are diminished. So are convulsions, whether elicited by external stimuli, by electric shock or by pentetrazol. Thus administration of 5-HTP to mice raises the threshold for audiogenic seizures or the hypersensitivity to noise caused by reserpine (Boggan and Seiden, 1973). The threshold for convulsive shock or pentetrazol is lowered by a reduction, and elevated by an increase in cerebral 5-HT (Kilian and Frey, 1973).

Elevation of brain concentration of 5-HT to non-physiological levels causes motor hyperactivity (Brodie and Shore, 1957; Grahame-Smith, 1974). Production of this syndrome by simultaneous administration of 5-hydroxytryptophan and an inhibitor of monoamine oxidase depends on the activation of dopaminergic neurones. This illustrates how one neuronal system may depend on the integrity of another for the manifestation of its effects. Another example is provided by the observation that catalepsy, produced by depleting the brain of dopamine or by antagonizing dopamine receptors, is masked if there is simultaneous loss of cerebral serotonin (Kostowski *et al.*, 1972; Fuenmayor and Vogt, 1979).

RESPONSE TO SENSORY STIMULI

As indicated earlier, the PCPA-treated animal shows exaggerated responses to sensory stimuli. Cats scratch themselves excessively and rub their coat against all available surfaces. Males indiscriminately mount any other cat. Rats have often been described as 'aggressive' after the administration of PCPA; their tendency to kill mice, present in a small percentage of 'normal' rats, increases after treatment. In mice, isolation-induced fighting is worsened by lowering, and attenuated by raising, brain serotonin (Hodge and Butcher, 1974).

The startle response produced in rats by acoustic stimuli is influenced by activity of serotonergic neurones. It is briefly inhibited by *p*-chloroamphetamine while the drug releases 5-HT, and greatly enhanced a few hours later, when the tissue is depleted of the transmitter (Davis and Sheard, 1976). Rats lesioned in the raphe show enhanced speed of acquisition of avoidance responses, which is equivalent to little tolerance towards unusual environmental stimuli. An attempt was made by Srebro and Lorens (1975) to attribute enhanced reactivity towards sensory stimuli to a specific part of the raphe system. They found that destruction of the dorsal raphe nucleus, in spite of reducing forebrain 5-HT to 35 per cent of normal, did hardly change behaviour at all, whereas lesion of the median nucleus, while leaving 74 per cent of forebrain 5-HT intact, increased open field activity and responses to novel experience. Even self-stimulation of rats following the implantation of electrodes into the medial forebrain bundle was greatly accelerated by lowering the serotonin content of the brain (Poschel and Ninteman, 1971).

The damping-down of the impact of environmental stimuli is beneficial since, provided the stimuli are not danger signals, interference with the normal activities of the animal is reduced.

The indifference to, or 'tuning out of', irrelevant stimuli requires an intact serotonergic system (Solomon *et al.,* 1978). Normal rats pre-exposed 30 times to a tone later used as a conditioned stimulus in an avoidance task show delayed learning of avoidance. PCPA erased the effect of pre-exposure by preventing the development of tuning out of irrelevant acoustic stimuli.

There are other instances of serotonin shielding the animal from overreacting to the environment. Brain 5-HT has an antiappetite action shown by overeating after PCPA and anorexia after 5-hydroxytryptophan, particularly when this is combined with an inhibitor of 5-HT uptake. Serotonin neurones damped taste-aversion, which is found greatly enhanced after lesions of the raphe nuclei (Lorden and Margules, 1977; Lorden, 1978).

Most of the clues to the functions of serotonin described in this section were obtained by increasing or decreasing neuronal activity or supply of transmitter, and noting the change produced in the organism. The next examples make use of an increased 5-HT turnover to indicate the circumstances under which the system is brought into action.

STRESS

In view of the close link between serotonin and environmental stimuli it is not surprising to find an increased turnover of 5-HT in stress. In the rat, it is a fairly short-lived response, for example, it does not outlast immobilization stress for more than an hour (Morgan and Rudeen, 1976). It is, however, a very sensitive index of stress: after mild disturbances, such as the changing of cages, Knott *et al.* (1977) found a significant rise in brain indole turnover within a few minutes.

The fact that any stress increases turnover is important in the interpretation of experiments in which the drug or procedure under investigation may cause stress.

VISUAL STIMULI

Serotonin is present in certain cells of the retina, and its role in the eye will be discussed in another chapter. There is a high concentration of 5-HT in the superior colliculi, and this raised the question whether visual stimuli modified the activity in the serotonergic terminals of that region. Experiments were carried out on rabbits which have almost completely crossed optic pathways. This permitted to affect only one colliculus by exposing one eye to light flashes, and to use the other colliculus as a control (Fukui and Vogt, 1975). The result was a small, but significant increase in 5-HT turnover in the colliculus contralateral to the stimulated eye, whereas no such differences were found between other symmetrical parts taken from the two hemispheres. Exogenous serotonin has been invariably found to inhibit collicular cells. Thus the activation of collicular serotonergic terminals by light seems to be another example of their counteracting excessive stimulation by environmental stimuli.

ANTINOCICEPTION

It has been known since 1968 through the work of Tenen that administration of PCPA reduces the analgesic action of morphine; the dose-response curve is shifted to the right, and the same effect is seen after 5-HT antagonists or after lesions of raphe nuclei. That one site of this effect is the cord, was shown (Vogt, 1974) by localized destruction of the descending 5-HT neurones which originate mainly in the nucleus raphe magnus. This procedure reduced the efficacy of morphine, but the effect was less than after using PCPA; this suggested supraspinal sites of interaction of morphine and 5-HT neurones. A measure of this interaction is the acceleration by morphine of 5-HT turnover. Such an acceleration occurs in the brain. The reacting sites were mapped in some detail in the brain of the rat (Snelgar and Vogt, 1981).

The following conclusions were reached:

1. There was no correlation between local density of serotonergic neurones and susceptibility of the region to turnover acceleration by morphine.
2. There was equally no correlation between basal 5-HT turnover of a site in untreated rats and its responsiveness to morphine.

The absence of such correlations show that certain serotonergic neurones preferentially interact with morphine. Whether the reactivity is determined by the density of binding sites or by some more subtle factor remains to be discovered.

Recent work, using local microinjections of morphine, has shown that in the caudate nucleus, the interaction between the drug and the serotonergic neurones takes place at the nerve terminals. In other regions, the perikarya may also, or even exclusively, be involved (Vasko and Vogt, 1981).

Although many of the regions in which morphine accelerates turnover of serotonin are connected with pain pathways or with emotional reactions to pain, others, such as tuberculum olfactorium, superior colliculi and lateral geniculate bodies, while apparently lacking this connection, are also highly responsive. The situation in the cord is interesting, because we know that structures in the dorsal cord are responsible for pain perception, whereas no such function is known in the ventral cord. In spite of this, both divisions of the cord show accelerated turnover of 5-HT after morphine; however, the response in the dorsal part is the larger of the two. Morphine, of course, has many actions unconnected with pain, and one of them is on spinal reflexes; serotonergic neurones appear to take part in this effect, since it has been shown that motor root discharges are facilitated by iontophoretic application of serotonin.

For the physiologist, it is important to remember that morphine has endogenous analogues in the enkephalins and endorphins. Both types of compound have been shown to accelerate 5-HT turnover (Van Loon *et al.*, 1978; Algeri *et al.*, 1978), so that the findings on morphine have their natural equivalents whenever endogenous opioids are active in the organism.

It is not possible to complete this list of actions of serotonin without referring to its effect on mood, although the subject will be discussed in its clinical aspect in two other chapters of this book. The low 5-HIAA content of the CSF in depressed patients has suggested a lack of serotonergic activity as long ago as 1966 (Ashcroft *et al.*). Infusions of an ester of 5-HTP (given with an inhibitor of DOPA-decarboxylase) into normal volunteers is indeed reported to cause temporary elation, but depression has been very rarely improved by the treatment (Pühringer *et al.*, 1976). There are many possible reasons for the failure. One is that in these patients there is malfunction of other systems, for example of the catecholamine neurones. Another possibility, suggested by the administration of antidepressive drugs to animals, is that the low 5-HT content of the CSF is not a sign of lack of serotonergic activity. Antidepressive drugs take several weeks of administration before a therapeutic effect is detectable in patients. In animals treated chronically for a similar time, certain noradrenaline receptors develop a reduced sensitivity to noradrenaline (Sulser *et al.*, 1978). The authors suggest that receptor hypersensitivity to noradrenaline, not reduced activity of noradrenergic neurones, may underlie some depressions. In contrast, *hypo*sensitivity of serotonergic receptors may possibly exist in depressive patients. Prolonged, not brief, treatment with tricyclic antidepressants increases the inhibition produced in several forebrain regions by 5-HT applied iontophoretically (De Montigny and Aghajanian, 1978). Thus receptor hypersensitivity develops as a long-term effect of administering these drugs. Their therapeutic effect might therefore be attributed to correction of a reduced sensitivity to 5-HT in the depressive patient. Presynaptic changes (in the serotonergic neurone itself) have not been found (Bier and de Montigny, 1980). The low concentration of 5-hydroxyindoleacetic acid in the CSF is not explained by these findings.

Related to 'mood elevation' is the observation that stimulation of the rat's dorsal raphe can suppress, and PCPA administration enhance, the expression of a form of fear shown by rats on stimulation of the periaqueductal grey (Kiser and Levovits, 1975; Kiser *et al.*, 1980). However, in other circumstances, in which 'anxiety' was measured by inhibition of social interaction between male rats, such interaction was increased by PCPA, suggesting that 5-HT was necessary for the manifestation of anxiety (File and Hyde, 1977). It is obvious that there is a contribution of serotonergic activity in the complex phenomena of anxiety or fear, but that different circumstances leading to what appears to be a similar emotional response involve the 5-HT system in a different way. The effects of serotonergic neurones on mood is clearly no exception to many other actions discussed in this article, namely the support they give to cope with the demands of the outside world.

CONCLUSION

The nearly ubiquitous distribution of serotonergic terminals in the brain is paralleled by the observation that practically every brain function investigated is somewhat modified by their activity. Their action is often, but not always, inhibitory, either directly or by stimulation of GABAergic and, probably, other inhibitory neurones. Serotonergic neurones are active in homeostasis, in damping down the effects of excessive tactile, auditory, visual and nociceptive impulses, and in counteracting the sensations of depression and fear. The sytem seems to create a shield against environmental stimuli which endanger an appropriate amd self-preserving performance of the organism.

REFERENCES

Aghajanian, G. K. and Haigler, H. J. (1974). Mode of action of LSD on serotonergic neurons. *Adv. biochem. Psychopharmacology*, **10**, 167–177.

Aghajanian, G. K., Wang, R. Y., and Baraban, J. (1978). Serotonergic and non-serotonergic neurons of the dorsal raphe: reciprocal changes in firing induced by peripheral nerve stimulation, *Brain Res.* **153**, 169–175.

Algeri, A., Consolazione, A., Calderini, G., Achilli, G., Puche Canas, E., and Garattini, S. (1978). Effect of the administration of $(d$-Ala$)^2$ methionine-enkephalin on the serotonin metabolism in rat brain. *Experientia*, **34**, 1488–1489.

Arluison, M. and de la Manche, I. S. (1980). High resolution radioautographic study of the serotonin innervation of the rat corpus striatum after intraventricular administration of $[^3H]$ 5-hydroxytryptamine, *Neuroscience*, **5**, 229–240.

Arnold, M. A. and Fernstrom, J. D. (1980). Administration of antisomatostatin serum to rats reverses the inhibition of pulsatile growth hormone secretion produced by injection of metergoline but not yohimbine, *Neuroendocrinology*, **31**, 194–199.

Ashcroft, G. W., Crawford, T. B. B., Eccleston, D., Sharman, D. F., MacDougall, E. J., Stanton, J. B. and Binns, J. K. (1966). 5-Hydroxyindole compounds in the CSF of patients with psychiatric and neurological disease, *Lancet*, **2**, 1049–1052.

Baraban, J. M. and Aghajanian, G. K. (1980). Suppression of serotonergic neuronal firing by α-adrenoceptor antagonists: evidence against GABA mediation, *Eur. J. Pharmac.* **66**, 287–294.

Bier, P. and de Montigny, C. (1980). Effect of chronic tricyclic antidepressant treatment on the serotonergic autoreceptor, *Naunyn-Schmiedeberg's Arch. Pharmacol.* **314**, 123–128.

Boakes, R. J., Bradley, P. B., Briggs, P. B., and Dray, A. (1970). Antagonism of 5-hydroxytryptamine by LSD25 in the central nervous system: a possible neuronal basis for the actions of LSD25, *Br. J. Pharmac.* **40**, 202–218.

Bobillier, P., Seguin, S., Petitjean, F., Salvert, D., Touret, M., and Jouvet, M. (1976). The raphe nuclei of the cat brain stem: a topographical atlas of their efferent projections as revealed by autoradiogaphy, *Brain Res.* **113**, 449–486.

Boggan, W. O. and Seiden, L. S. (1973). 5-Hydroxytryptophan reversal of reserpine enhancement of audiogenic seizure susceptibility in mice, *Physiol. & Behav.* **10**, 9–12.

Brodie, B. B. and Shore, P. A. (1957). A concept for a role of serotonin and norepinephrine as chemical mediators in the brain, *Ann. NY. Acad. Sci.* **66**, 631–642.

Bronzino, J. D., Morgane, P. J. and Stern, W. C. (1972). EEG synchronization following application of serotonin to area postrema, *Am. J. Physiol.* **223**, 376–383.

Caligaris, L. and Taleisnik, S. (1974). Involvement of neurones containing 5-hydroxytryptamine in the mechanism of prolactin release induced by oestrogen, *J. Endocr.* **62**, 25–33.

Carter, C. J. and Pycock, C. J. (1979). The effects of 5,7-dihydroxytryptamine lesions of extrapyramidal and mesolimbic sites on spontaneous motor behaviour and amphetamine-induced stereotypy, *Naunyn-Schmiedeberg's Arch. Pharmacol.* **308**, 51–54.

Chan-Palay, V., Jonsson, G., and Palay, S. L. (1978). On the co-existence of serotonin and substance P in neurons of the rat's central nervous system, *Proc. Natn. Acad. Sci.* **75**, 1582–1586.

Chang, M. M. and Leeman, S. E. (1970). Isolation of a sialogogic peptide from bovine hypothalamic tissue and its characterization as substance P, *J. biol. Chem.* **245**, 4784–4790.

Cohen, H. B., Dement, W. C., and Barchas, J. D. (1973). Effects of chlorpromazine on sleep in cats pretreated with *para*-chlorophenylalanine, *Brain Res.* **53**, 363–371.

Davis, M. and Sheard, M. H. (1976). *p*-Chloroamphetamine (PCA): acute and chronic effects on habituation and sensitization of the acoustic startle response in rats, *Eur. J. Pharmac.* **35**, 261–273.

De Montigny, C. and Aghajanian, G. K. (1978). Tricyclic antidepressants: long-term treatment increases responsivity of rat forebrain neurons to serotonin, *Science,* **202**, 1303–1306.

Descarries, L., Beaudet, A. and Watkins, K. C. (1975). Serotonin nerve terminals in adult rat neocortex, *Brain Res.* **100**, 563–588.

Dismukes, K. (1977). New Look at the aminergic nervous system. *Nature* (Lond.), **269**, 557–558.

Euler, U. S. von and Gaddum, J. H. (1931). An unidentified depressor substance in certain tissue extracts, *J. Physiol.* **72**, 74–87.

Feldberg, W. and Myers, R. D. (1964). Effects on temperature of 5-hydroxytryptamine, adrenaline and noradrenaline injected into the cerebral ventricles or the hypothalamus of cats, *J. Physiol.* **173**, 25P.

File, S. E. and Hyde, J. R. G. (1977). The effects of *p*-chlorophenylalanine and ethanolamine-*O*-sulphate in an animal test of anxiety. *J. Pharm. Pharmac.* **29**, 735–738.

Froment, J. L., Petitjean, F., Bertrand, N., Cointy, C., and Jouvet, M. (1974). Effects de l'injection intracérébrale de 5,6-hydroxytryptamine sur les monoamines cérébrales et les états de sommeil du chat, *Brain Res.* **67**, 405–417.

Fuenmayor, L. D. and Vogt, M. (1979). The influence of cerebral 5-hydroxytryptamine on catalepsy induced by brain-amine depleting neuroleptics or by cholinomimetics. *Br. J. Pharmac.* **67**, 309–318.

Fukui, K. and Vogt, M. (1975). The effect of visual stimuli on the turnover of 5-hydroxytryptamine (5-HT) in the superior colliculi, *J. Physiol.* **248**, 39–40P.

Giambalvo, C. T. and Snodgrass, S. R. (1978). Biochemical and behavioural effects of 5-hydroxytryptamine neurotoxins on the nigrostriatal dopamine system: comparison of injection sites, *Brain Res.* **152**, 555–566.

Göthert, M. and Huth, H. (1980). Alpha-adrenoceptor mediated modulation of 5-hydroxytryptamine release from rat brain cortex slices, *Naunyn-Schmiedeberg's Arch. Pharmacol.* **313**, 21-26.

Grahame-Smith, D. G. (1974). How important is the synthesis of brain 5-hydroxytryptamine in the physiological control of its central function? *Adv. biochem. Psychopharmacology,* **10**, 83-91.

Hammond, D. L., Levy, R. A., and Proudfit, H. K. (1980). Hypoalgesia induced by microinjection of a norepinephrine antagonist in the raphe magnus: reversal by intrathecal administration of a serotonin antagonist, *Brain Res.* **201**, 475-479.

Harvey, C. A. and Milton, A. S. (1974). The effect of parachlorophenylalanine on the response of the conscious cat to intravenous and intraventricular bacterial pyrogen and to intraventricular prostaglandin E_1, *J. Physiol.* **236**, 14-15P.

Héry, M., Laplante, E., and Kordon, C. (1976). Participation of serotonin in the phasic release of L.H. I. Evidence from pharmacological experiments, *Endocrinology,* **99**, 496-503.

Hodge, G. K. and Butcher, L. L. (1974). 5-Hydroxytryptamine correlates of isolation-induced aggression in mice, *Eur. J. Pharmac.* **28**, 326-337.

Hökfelt, T., Ljungdahl, A., Steinbusch, H., Verhofstad, A., Nilsson, G., Brodin, E., Pernow, B. and Goldstein, M. (1978). Immunohistochemical evidence of substance P-like immunoreactivity in some 5-hydroxytryptamine-containing neurons in the rat central nervous system, *Neuroscience,* **3**, 517-538.

Hoyland, V. J., Shillito, E. E., and Vogt, M. (1970). The effect of parachlorophenylalanine on the behaviour of cats, *Br. J. Pharmac.* **40**, 659-667.

Ito, A. and Schanberg, S. M. (1972). Central nervous system mechanisms responsible for blood pressure elevation induced by *p*-chlorophenylalanine, *J. Pharmac. exp. Ther.* **181**, 65-71.

Jouvet, M. (1962). Recherches sur les structures nerveuses et les mécanismes responsables des différentes phases du sommeil physiologique, *Arch. ital. Biol.* **100**, 125-206.

Jouvet, M. (1971). L'α-CH₃ tyrosine supprime l'insomnie consécutive à la lésion du système du raphé, *C.R. Soc. Biol.* **165**, 2128-2131.

Key, B. J., Boakes, R. J. and Candy, J. M. (1980). Responses of cat dorsal raphe neurones to iontophoretically applied noradrenaline. *Neuropharmacology,* **19**, 139-142.

Kilian, M. and Frey, H.-H. (1973). Central monoamines and convulsive thresholds in mice and rats, *Neuropharmacology,* **12**, 681-692.

Kiser, R. S., Brown, C. A., Sanghera, M. K., and German, D. C. (1980). Dorsal raphe nucleus stimulation reduces centrally-elicited fearlike behaviour, *Brain Res.* **191**, 265-272.

Kiser, R. S. and Lebovitz, R. M. (1975). Monoaminergic mechanisms in aversive brain stimulation, *Physiol. & Behav.* **15**, 47-53.

Knott, P. J., Hutson, P. H., and Curzon, G. (1977). Fatty acid and tryptophan changes on disturbing groups of rats and caging them singly, *Pharmacol. Biochem. Behav.* **7**, 245-252.

Komiskey, H. L. and Rudy, T. A. (1977). Serotonergic influences on brain stem thermoregulatory mechanisms in the cat. *Brain Res.* **134**, 297-315.

Kostowski, W., Gumulka, W., and Czlonkowski, A. (1972). Reduced cataleptogenic effects of some neuroleptics in rats with lesioned midbrain raphe and treated with *p*-chlorophenylalanine, *Brain Res.* **48**, 443-446.

Lidov, H. G. W., Grzanna, R., and Molliver, M. E. (1980). The serotonin innervation of the cerebral cortex in the rat—an immunohistochemical analysis, *Neuroscience,* **5**, 207-227.

Lorden, J. F. (1978). Alteration of the characteristics of learned taste aversion by manipulation of serotonin levels in the rat, *Phaarmacol. Biochem. Behav.* **8**, 13-18.

Lorden, J. F. and Margules, D. L. (1977). Enhancement of conditioned taste aversions by lesions of the midbrain raphe nuclei that deplete serotonin, *Physiol. Psychol.* **5**, 273-279.

Lorez, H. P. and Richards, J. G. (1973). Distribution of indolealkylamine nerve terminals in the ventricles of the rat brain, *Z. Zellforsch.* **144**, 511-522.

McGinty, D. J. (1976). Dorsal raphe neurons: depression of firing during sleep in cats, *Brain Res.* **101**, 569-575.

Melamed, E., Hefti, F., and Wurtman, R. J. (1980). L-3,4-dihydroxyphenylalanine and L-5-hydroxytryptophan decarboxylase activities in rat striatum: effect of selective destruction of dopaminergic or serotonergic input, *J. Neurochem.* **34**, 1753-1756.

Mizuno, H., Talwalker, P. K., and Meites, J. (1967). Central inhibition by serotonin of reflex release of oxytocin in response to suckling stimulus in the rat, *Neuroendocrinology*, **2**, 222-231;

Morgan, W. W. and Rudeen, P. K. (1976). Temporal study of 5-hydroxyindoleacetic acid normalization during recovery from immobilization stress, *Exp. Neurol.* **51**, 259-262.

Moskowitz, M. A., Liebmann, J. E., Reinhard, J. F. Jr., and Schlosberg, A. (1979). Raphe origin of serotonin containing neurons within choroid plexus of the rat, *Brain Res.* **169**, 590-594.

Myers, R. D., Kawa, A., and Beleslin, D. B. (1969). Evoked release of 5-HT and NEFA from the hypothalamus of the conscious monkey during thermoregulation, *Experientia*, **25**, 705-706.

Petitjean, F., Sakai, K., Blondaux, C., and Jouvet, M. (1975). Hypersomnie par lésion isthmique chez le chat. II. Étude neurophysiologique et pharmacologique, *Brain. Res.* **88**, 439-453.

Poschel, B. P. H. and Ninteman, F. W. (1971). Intracranial reward and the forebrain's serotonergic mechanism: studies employing *para*-chlorophenylalanine and *para*-chloroamphetamine, *Physiol. Behav.* **7**, 39-46.

Pühringer, W., Wirz-Justive, A., Graw, P., Lacoste, V. and Gastpar, M. (1976). Intravenous L-5-hydroxytryptophan in normal subjects: an interdisciplinary precursor loading study. Part I: Implications of reproducible mood elevation, *Pharmakopsychiatrie Neuropsychopharmakologie*, **9**, 259-268.

Renault, J. (1967). *Monoamines et sommeils III'* (Thèse), Tixier et fils, Lyon.

Rogawski, M. A. and Aghajanian, G. K. (1980). Norepinephrine and serotonin: opposite effects on the activity of lateral geniculate neurons evoked by optic pathway stimulation, *Exp. Neurol.* **69**, 678-694.

Shen, Y.-S., Nelson, J. P., and Bloom, F. E. (1974). Discharge patterns of cat raphe neurons during sleep and waking, *Brain Res.* **73**, 263-276.

Snelgar, R. S. and Vogt, M. (1981). Mapping, in the rat central nervous system, of morphine-induced changes in turnover of 5-hydroxytryptamine, *J. Physiol.* **314**, 395-410.

Solomon, P. R., Kiney, C. A., and Scott, D. R. (1978). Disruption of latent inhibition following systemic administration of *para*chlorophenylalanine (PCPA). *Physiol. and Behav.* **20**, 265-271.

Srebro, B., and Lorens, S. A. (1975). Behavioural effects of selective midbrain raphe lesions in the rat, *Brain Res.* **89**, 303-325.

Steinbusch, H. W. M., Verhofstad, A. A. J., and Joosten, H. W. J. (1978). Localization of serotonin in the central nervous system by immunohistochemistry: description of a specific and sensitive technique and some applications, *Neuroscience,* **3**, 811–819.

Sulser, F., Vetulani, J., and Mobley, P. L. (1978). Mode of action of antidepressant drugs, *Biochem. Pharmac.* **27**, 257–261.

Taleisnik, S., Celis, M. E. and Tomatis, M. E. (1973/74). Release of melanocyte-stimulatory hormone by several stimuli through the activation of a 5-HT-mediated inhibitory neuronal mechanism, *Neuroendocrinology,* **13**, 327–338.

Tenen, S. S. (1968). Antagonism of the analgesic effect of morphine and other drugs by *p*-chlorophenylalanine, a serotonin depletor. *Psychopharmacologia,* **12**, 278–285.

Ulrich, R. Yuwiler, A. and Geller, E. (1975). Effects of hydrocortisone on biogenic amine levels in the hypothalamus, *Neuroendocrinology,* **19**, 259–268.

Van de Kar, L. D., Lorens, S. A., Vodraska, A., Allers, G., Green, M., Van Orden, D. E., and Van Orden, L. S. III (1980). Effect of selective midbrain and diencephalic 5,7-dihydroxytryptamine lesions on serotonin content in individual preopticohypothalamic nuclei and on serum luteinizing hormone level, *Neuroendocrinology,* **31**, 309–315.

Van Loon, G. R. and De Souza, E. B. (1978). Effect of β-endorphin on brain serotonin metabolism, *Life Sci.* **23**, 971–978.

Vasko, M. R. and Vogt, M. (1981). Site of interaction between morphine and 5-hydroxytryptamine-containing neurones in the rat brain, *Br. J. Pharmac.* **73**, 245P–246P.

Vogt, M. (1974). The effect of lowering the 5-hydroxytryptamine content of the rat spinal cord on analgesia produced by morphine, *J. Physiol.* **236**, 483–498.

Wang, R. Y. and Aghajanian, G. K. (1977). Inhibition of neurons in the amygdala by dorsal raphe stimulation: mediation through a direct serotonergic pathway, *Brain Res.* **120**, 85–102.

Biology of Serotonergic Transmission
Edited by N. N. Osborne
© 1982, John Wiley & Sons Ltd.

Chapter 13

Central Serotonin Neurones and Learning in the Rat

S. O. ÖGREN

Research Laboratories, Astra Läkemedel AB,
S-151 85 Södertälje, Sweden

INTRODUCTION

The role of the catecholamine noradrenaline (NA) and dopamine (DA) neurones in learning and memory processes has been a subject of intensive study during the last decade. Somewhat surprisingly, investigations on the role of serotonin (5-HT) in these functions have received far less attention. Despite the relative scarcity of experimental data, there is accumulated evidence that central 5-HT pathways in the rat may play a role in both learning and memory (Hunter *et al.*, 1977; Lorens, 1978; Ögren *et al.*, 1981). It has been reported that increases or decreases of brain and/or spinal 5-HT concentrations by pharmacological or physiological manipulations can markedly interfere with the acquisition, retention and extinction of a variety of learning tasks. A putative involvement for 5-HT has been suggested in several different learning procedures including active (Brody, 1970; Lorens and

Yunger, 1974; Ögren *et al.,* 1976; Tenen, 1967) and passive avoidance learning (Stevens *et al.,* 1969; Tenen, 1967), punishment (Graeff and Schoenfeld, 1970; Tye *et al.,* 1977), retrograde amnesia (Essman, 1970; Rake, 1973), conditioned suppression (Stein *et al.,* 1975) extinction (Beninger and Phillips, 1979), habituation and sensitization processes (Carlton and Advokat, 1973; Davis and Sheard, 1976), latent inhibition (Solomon, 1978) and intracranial self-stimulation reward (Poschel and Ninteman, 1971).

HYPOTHESIS ON THE ROLE OF SEROTONIN IN LEARNING

Studies on the role of 5-HT in learning have mainly been focussed on its function in aversively motivated behaviours such as active and passive avoidance learning. The choice of these procedures has been guided by the belief that 5-HT has a complementary or reciprocal role to that of NA (Brodie and Shore, 1957). Catecholamine (CA) neurones and in particular the ascending NA neurones from the locus coeruleus have been identified in positive reinforcement or reward processes (for review see Stein *et al.,* 1977; Crow and Deakin, 1979). The activation of CA neurones is generally thought to facilitate behaviours which are maintained by favourable consequences, e.g. reward. Ascending serotonin neurones originating from the dorsal and medial raphe nuclei, on the other hand, have been implicated in non-reward (negative reinforcement) and extinction processes (Stein *et al.,* 1977; Crow and Deakin, 1979). Thus, the physiological consequences of non-reward or witholding of positive reinforcement which results in reduced behavioural responding, are suggested to involve 5-HT mechanisms. In addition, neurochemical theories of behaviour have suggested that 5-HT neurones mediate the suppression of behaviour induced by aversive stimuli (punishment) (Stein *et al.,* 1977). Much of the support for this hypothesis is derived from studies showing that depletion of brain 5-HT attenuates the suppressive effect of punishment in passive avoidance situations or 'conflict tests'. In this situation, the rat performs a lever press response (an operant response) at a high rate to obtain a reward (pellet or sweet milk). When signalled shock (punishment) is made contingent upon the operant response, response rate is drastically reduced. In a series of work Stein, Wise and coworkers have shown that procedures which reduce central 5-HT activity (electrolytic midbrain raphe lesions, receptor blockers, synthesis inhibitors) reduce the rate-depressant effect of punishment (Stein *et al.,* 1977). This and related work has led to the conceptualization that 5-HT reduction will result in response or behavioural disinhibition (Brody, 1970; Stein *et al.,* 1977). This hypothesis predicts that manipulations that reduce central 5-HT activity interfere with passive avoidance learning situations in which the witholding of a specified response is required for optimal performance (Bignami, 1976). In addition, 5-HT depletion is predicted to result in facilitation of two-way active avoidance acquisition by attenuation of response suppression (Steranka and Barrett, 1974), as well as attenuation of behavioural inhibition of unconditioned behaviours, i.e. locomotor activity and exploration (Carlton and Advokat, 1973).

Table 1 Experimental approaches used to study the role of serotonin in learning

---

A. Manipulations that are inferred to enhance central 5-HT activity
1. Putative 5-HT agonists: tryptophan, 5-HT, D-LSD, quipazine etc.
2. Inhibition of 5-HT metabolism or neuronal re-uptake: MAO-inhibitors, 5-HT uptake inhibitors
3. 5-HT releasing compounds: *p*-chloroamphetamine (PCA).

B. Manipulations that are inferred to decrease central 5-HT activity
1. Electrolytic midbrain raphe lesions.
2. Systemic administration of the tryptophan hydroxylase inhibitor *p*-chloro phenylalanine (PCPA).
3. Use of indolemine neurotoxins: Systemic administration of *p*-chloroamphetamine (PCA) or intracerebral injections of 5,7-dihydroxytryptamine (5.7-DHT).

RESEARCH METHODS

Several different approaches (see Table 1) including the use of intracranial injections of 5-HT, 5-HT releasing agents, the use of synthesis inhibitors, electrolytic lesions of the midbrain raphe nuclei as well as serotonin neurotoxins have provided evidence that modulations of 5-HT activity in the brain are associated with changes in acquisitional and/or memory-storage processes. It is not known, however, whether the effects caused by these 5-HT manipulations reflect changes in performance rather than an alteration of associative processes.

In the present paper, the putative role of 5-HT in learning will mainly be analyzed in terms of its rôle in active avoidance learning where most of the data regarding the involvement of 5-HT are available at present. In the normal active avoidance situation the rat has to perform a discrete response to avoid shock. Typically, the animal has to jump to a platform or run to the safe side of a shuttle-box to avoid shock. Shock is in most cases signalled and there is a delay of 5–10 sec. between the onset of the signal and the application of shock. In the one-way situation the animal has to perform a unidirectional response to avoid shock while in the two-way situation the animal is required to shuttle back and forth between two compartments to avoid shock. The normal rat acquires learning crition ($\sim$ 80% avoidance performance) in about 50–80 trials in the two-way situation and after about 15 trials in the one-way situation. Thus, the one-way situation is an easier learning task for the rat than the two-way situation.

EFFECTS OF THE SEROTONIN PRECURSOR 5-HYDROXYTRYPTOPHAN

It was originally suggested by Wooley and Van der Hoeven (1963) that increases in brain 5-HT may result in an inferior learning which may be independent of changes in CA concentrations. When 5-HT in the brain of adult mice was elevated by administration of the 5-HT precursor 5-hydroxytryptophan (5-HTP) in combination with 1-benzyl-2-methyl-5-methoxy-tryptamine (BAS), the acquisition of a maze learning task was clearly impaired. Administration of dihydroxyphenylalanine (D L-dopa), on the other hand, which elevated both NA and DA concentrations, did not change acquisition. Administration of 5-HTP has also been found to impair acquisition of other tasks. Thus, administration of 5-HTP to rats was found to impair both active avoidance acquisition and performance (Joyce and Hurwitz, 1964; Roffman and Lal, 1971). In addition, Essman reported that 5-HTP administration given 1 h before acquisition of a one-trial passive avoidance task caused a marked deficit in retention when tested 24 h later (Essman, 1973a). This finding implicates 5-HT both in encoding and memory storage processes.

An interpretation of the results using systemic 5-HTP administration is difficult for several reasons. Systemic injections of 5-HTP primarily elevates peripheral 5-HT stores and can produce a dose-related suppression of ongoing behaviour in particular locomotor activity (Aprison and Hingten, 1970). It should be noted, however, that there is no observational evidence that changes in locomotion could explain the results with 5-HTP. Moreover, there is no evidence that 5-HTP (37.5–100 mg/kg ip) causes any apparent change in pain sensitivity in rats (Harvey and Lints, 1971). Another complication is that 5-HTP is partly taken up by CA neurones, and can at least in high doses cause a depletion of central CA stores (Fuxe *et al.*, 1971) and possibly alter CA neuronal activity. There is at present no information whether the effects of 5-HTP on learning may involve disruption in CA activity.

INTRACRANIAL INJECTION OF SEROTONIN

In subsequent studies intracranial microinjections of 5-HT have been applied mainly in tasks believed to reflect memory storage processes i.e. one-trial passive avoidance situations (Essman, 1973b). Injections of 5-HT but not NA into the hippocampus immediately following the acquisition of the passive avoidance response caused a marked retention deficit when the animals were tested 24 h later (Essman, 1973b). Of particular interest is the observation that the effect of 5-HT followed a temporal gradient. Injection times later than 16 min resulted in no retention impairment indicating that 5-HT may be involved in memory consolidation. The methodological problems associated with intracranial 5-HT administration are in part similar to the use of 5-HTP. It cannot be denied that the effects of 5-HT reflect some interference with CA and in particular NA transmission. However, the observation that intrahippocampal NA injection did not produce any retention deficit argues against this interpretation.

SEROTONIN RELEASING COMPOUNDS

An alternative approach in studying the role of 5-HT has been through the use of compounds which release 5-HT from presynaptic nerve terminals. If endogenous 5-HT exerts a role upon learning processes, compounds which cause a potent 5-HT release should theoretically produce the same effect as intracranial administration of 5-HT or systemic administration of 5-HTP. A drug which has received considerable attention in this context is the halogenated amphetamine *p*-chloroamphetamine (PCA). There is good evidence indicating that PCA administration in low doses produces a rapid release of 5-HT from presynaptic nerve endings (Sanders-Bush and Steranka, 1978). Part of the acute behavioural effects of PCA in mice and rats appears to be related to the rapid release of 5-HT and a subsequent increase in 5-HT activity. Thus, pretreatment with the 5-HT synthesis inhibitor *p*-chlorophenylalanine (PCPA) completely blocks the 5-HT dependent behaviours produced by PCA in rats (Trulson and Jacobs, 1976) or mice (Ögren and Ross, 1977). In addition, drugs thought to block postsynaptic 5-HT receptors such as methergyline inhibit the 5-HT dependent behavioural effects caused by PCA (Ögren and Ross, 1977). Acute administration of PCA effects also CA neurones. Thus, PCA is a potent NA uptake inhibitor and also releases NA and DA in high doses (Sanders-Bush and Steranka, 1978). Some of the behavioural effects of PCA, e.g. the increase in locomotor activity appears to be mediated via CA neurones (Sanders-Bush and Steranka, 1978). The acute effect of PCA-induced release of 5-HT *in vivo* is blocked by inhibition of the 5-HT reuptake mechanisms without affecting the acute effect of PCA monoaminergic neurones. Thus, zimelidine and fluoxetine (see Ross, chap. 7, this volume), two selective 5-HT uptake blockers, have been shown to block the PCA-induced 5-HT depletion and the 5-HT dependent behaviours produced by PCA. In contrast, the specific NA uptake inhibitor desipramine has consistently failed to block the acute action of PCA upon 5-HT neurones (Ross, 1976; Sanders-Bush and Steranka, 1978).

In a series of experiments administration of PCA prior to training was shown to severely retard the acquisition of a one-way avoidance response (see Fig. 1). As noted previously, this type of response, which is unidirectional is easily acquired by the normal rat. PCA caused a significant impairment of avoidance acquisition at a dose level of 0.63 mg/kg and doses of 1.25 and 2.5 mg/kg ip caused a very marked impairment (see Fig. 1). The impaired avoidance acquisition could not be attributed to any loss of motor initiation or any apparent change in foot-shock sensitivity, since the impairment was independent of foot-shock intensity. In subsequent studies PCA was also found to cause a dose-related deficiency of the retention of an already acquired response (Ögren *et al.*, 1977; Ögren, 1981). Some interference of memorial processes seems possible. If a serotoninergic system in the brain is involved and the action of PCA is dependent upon 5-HT release, then serotonin depletion should block the learning deficiency caused by PCA. To test this hypothesis rats were pretreated with the tryptophan hydroxylase inhibitor PCPA (300 mg/kg ip, 24 h prior to PCA). Lowering of 5-HT concentration by PCPA pretreatment completely prevented the avoidance impairment caused by PCA (Fig. 2). In a similar study, pretreatment with the tyrosine hydroxylase inhibitor α-methyl-*p*-tyrosine failed to block the actions of PCA (Ögren, 1981).

There is also evidence that PCA might utilize the 5-HT uptake mechanism at the

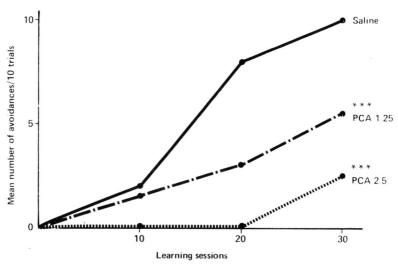

Fig. 1   The acute effect of *p*-chloroamphetamine (PCA) on the acquisition of an one-way active avoidance response in the rat. Groups of eight animals were injected with saline or PCA (2.5 and 1.25 mg/kg). Acquisition trials started 30 min after injection. The results are expressed in mean number of avoidance responses per ten trials. ***$p < 0.001$, ANOVA (Data taken from Ögren, 1981)

neuronal membrane to enter the 5-HT presynaptic terminals (Fuller *et al.*, 1975). If this hypothesis is correct, PCA has to be taken up into the presynaptic neurones and release endogenously stored 5-HT to produce an effect on avoidance learning. To assess this hypothesis rats were pretreated with zimelidine (10 mg/kg), fluoxetine (10 mg/kg) or desipramine (10 mg/kg) prior to injection of PCA (5 mg/kg). Zimelidine and fluoxetine (but not desipramine) pretreatment completely blocked the impaired avoidance caused by PCA (see Fig. 2). Neither zimelidine nor fluoxetine caused by themselves any significant depression of avoidance retention. In addition, methergoline, which blocks at least some postsynaptic 5-HT receptors in the brain (Fuxe *et al.*, 1978a), was found to block the action of PCA on avoidance (Ögren, 1981). Since PCA causes a rapid release of 5-HT which subsequently develops in a marked 5-HT depletion (Sanders-Bush and Steranka, 1978), the learning impairment could be due to the 5-HT depleting action. However, a study on the time-course effect showed that the effect of PCA on avoidance was maximal within 30-60 min and had terminated with 4 h in low doses (Fig. 3). Thus, the learning impairment was related to the acute phase of 5-HT release as established in previous behavioural and in electrophysiological studies (Marsden *et al.*, 1979; Ögren and Ross, 1977, 1979; Trulson and Jacobs, 1976).

Measurements of monoamine concentrations in several brain regions showed that PCA treatment (2.5 mg/kg) preferentially affected forebrain 5-HT terminal systems (cerebral cortex, striatum and hippocampus) while having slight effects on the hypothalamus, midbrain and spinal cord (Fig. 4). PCA caused marginal changes

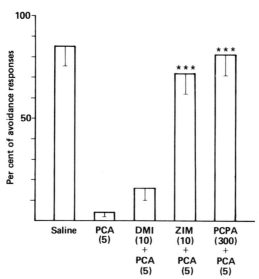

Fig. 2   The acute effect of *p*-chloroamphetamine (PCA) administration on one-way avoidance retention in the rat. Groups of five to six rats were trained to a learning criterion of nine out of ten consecutive avoidances. Zimelidine (ZIM) and desipramine (DES) were given at a dose of 10 mg/kg i.p. 30 min prior to PCA (5 mg/kg ip) *p*-Chlorophenylalanine (PCPA) was injected at a dose of 300 mg/kg 24 h prior to the PCA administration. Retention was tested 60 min after injection of PCA or saline. The results are expressed in per cent of the mean number of avoidance responses (the bar indicates SEM) ***$p < 0.001$ *vs.* PCA (Mann–Whitney U-test). Data taken from Ögren and co-workers (1977, 1981)

in NA and DA levels. Interestingly, NA concentration in the cerebral cortex was slightly increased. Thus, the acute PCA action seems to involve mainly those regions innervated by the ascending 5-HT pathways arising from both the dorsal (B7) and median (B8) raphe nculei with less effects on the hypothalamic projection and on the descending pathways arising from the caudally located raphe nuclei in the pons-medulla region (Azmitia and Segal, 1978; Fuxe, 1965).

Taken together, there is evidence that the impaired avoidance following PCA injection is due to the release of 5-HT from presynaptic nerve endings and the subsequent increase in postsynaptic 5-HT activity. The available data do not support a direct role for NA and DA neurones in the PCA-induced learning deficit. In this respect PCA differs from drugs such as amphetamine which release NA and/or DA from presynaptic nerve endings and which facilitate active avoidance acquisition in the rat (Evangelista *et al.*, 1970).

SEROTONIN AGONISTS

A number of drugs are thought to exert their pharmacological effects through activation of postsynaptic 5-HT receptors. There is both behavioural, biochemical

Dose of PCA (mg/kg)

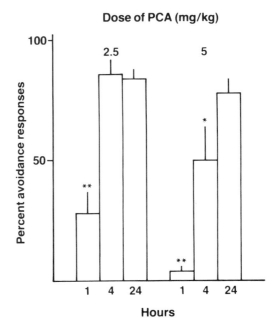

Hours

Fig. 3   Time course effect of *p*-chloroamphetamine (PCA) on avoidance retention in the rat. Groups of rats (*n* = 5–6) were injected with PCA (2.5 or 5 mg/kg) and tested at various intervals after injection. For details of treatment, see Fig. 2. $**p < 0.01$, $*p < 0.05$ *vs.* saline control (Mann–Whitney U-test). Data taken from Ögren (1981)

and neurophysiological evidence that hallucinogenic drugs such as 5-methoxy-*N*, *N*-dimethyltryptamine (5-MeO-DMT) and D-lysergic acid diethylamide (D-LSD) activate receptor sites in the brain associated with the 5-HT receptor complex (see Aghajanian and Wang, 1978). Interestingly, D-LSD which is more potent to activate the presynaptic receptors located at the cell membrane of the dorsal raphe ('the autoreceptors'), than the postsynaptic receptors, behaves differently from 5-MeO-DMT. Hallucinogenic drugs such as D-LSD and mescaline have been reported to cause both enhanced and retarded acquisition of a two-way avoidance response depending on dose. In a low dose range D-LSD enhanced avoidance acquisition while in a high dose range D-LSD retarded acquisition (Bignami, 1972). In a recent study D-LSD was shown to also produce a biphasic effect on the acquisition of a classically conditioned nictitating response in the rabbit (Gimpl *et al.*, 1979). In this procedure, D-LSD (1–100 μmol/kg) significantly enhanced the rate of acquisition while higher doses 300 μmol/kg tended to retard acquisition. The effect of D-LSD seemed to exclusively be related to an effect on the acquisitive stage but not to retention processing. It is not known whether the effect of D-LSD on acquisition is exclusively related to an action on 5-HT neurones. It could be speculated that the biphasic action of D-LSD on acquisition is associated with the

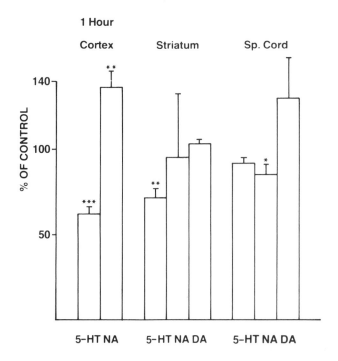

Fig. 4   Effects of *p*-chloroamphetamine (PCA) on monoamine concentrations in some rat-brain regions. PCA (2.5 mg/kg ip) treated rats were sacrificed 60 min after injection and cerebral cortex, striatum and spinal cord were rapidly dissected. The results are the mean values of six determinations. *$p < 0.05$, **$p < 0.01$, ***$p < 0.001$ (Mann–Whitney U-test). Data taken from Ögren (1981)

preferential pre- versus postsynaptic depression of raphe unit firing reflecting the relative potency at pre- versus postsynaptic 5-HT receptors.

Injection of 5-MeO-DMT and related indole hallucinogens to rats cause a clearly dose related acquisition learning impairment in low doses (Stoff *et al.*, 1978). The relative potency of the indole hallucinogens to retard acquisition was shown to parallel their potency as 5-HT receptor agonist as well as their preferential pre- versus postsynaptic depression of raphe firing (de Montigny and Aghajanian, 1977).

SEROTONIN SYNTHESIS INHIBITORS

In quite a few of the early experiments on the role of 5-HT in behavioural studies, 5-HT depletion was achieved by *p*-chlorophenylalanine (PCPA). PCPA is an irreversible tryptophanhydroxylase inhibitor which produces a marked 5-HT depletion throughout the brain and spinal cord as well as in the periphery. PCPA is not specific for 5-HT and when given at a high dose reduces 5-HT concentrations by approximately 90 per cent and NA and DA concentrations by about 10–15 per cent of control values (Koe and Weissman, 1966). In addition, PCPA increased phenylalanine levels in the brain which, however, appears not to influence the behavioural effect of PCPA in acute experiments (see Brody, 1970).

Tenen (1967) originally reported that PCPA injections to rats enhanced the acquisition of a signalled one-way avoidance response (a jump response to a platform) when a low shock intensity was used. PCPA was also found to decrease jump-threshold and Tenen concluded that the enhanced acquisition probably was due to the higher reactivity to shock of the animals. Also Schlesinger and Schreiber (1968) and Brody (1970) report facilitated avoidance acquisition following PCPA administration in similar one-way avoidance situations (pole jump). These investigators found that both shock (US) level and strain were important. Thus, PCPA injection caused enhancement of acquisition in the Fisher strain at low or moderate shock-intensities (0.25–0.5 mA) while the Buffalo strain were facilitated at high US intensities (Schlesinger and Schreiber, 1968). PCPA have also been reported to produce facilitation in other tasks. Stevens *et al.* (1967) reported facilitation of a successive discrimination task but not of a position habit in the rat. In this study thirst instead of shock was the incentive employed. Köhler and Lorens (1978) failed to observe facilitated two-way avoidance acquisition following PCPA treatment. In a recent study however, PCPA was found to produce facilitated two-way avoidance acquisition during the first learning sessions (Ögren *et al.*, 1981). In this particular study the avoidance facilitation was apparently related to the increase in locomotion in PCPA-treated rats (see Fig. 5). Thus, the effects of PCPA on avoidance learning appear to critically depend upon the test situation. Facilitation of avoidance learning has mainly been observed in one-way situations, in which quite rigorous avoidance responses have been employed such as jump-up responses. PCPA treatment increases pain sensitivity to shock, which hypothetically could be favourable in certain one-way avoidance situations. However, changes in pain sensitivity cannot explain the facilitation of discrimination learning nor the retardation of passive avoidance learning reported by some authors (Tenen, 1967; Stevens *et al.*, 1969). Taken together, PCPA seems, under certain conditions, to facilitate the acquisition of active avoidance acquisition and to retard passive avoidance learning. The results with PCPA in general do not support the hypothesis that depletion of brain serotonin increases learning ability (Wooley and Van der Hoeven, 1963).

MIDBRAIN RAPHE LESIONS

The effect of electrolytic midbrain raphe lesions has recently been summarized by Lorens (1978). In a series of experiments, Lorens, Köhler and Srebro have shown that the effects of raphe lesions are highly test-dependent. Thus, combined electrolytic lesions of the dorsal (B7) and median raphe nuclei (B8) have been shown to facilitate two-way avoidance acquisition in the rat. In contrast to PCPA, electrolytic midbrain raphe lesions impair one-way avoidance acquisition (Lorens and

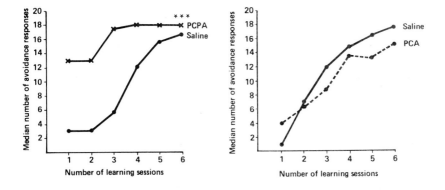

Fig. 5   The effects of *p*-chlorophenylalanine (PCPA) and the long-term effect of *p*-chloroamphetamine (PCA) on the acquisition of a two-way active avoidance response. PCPA was injected in a dose of 200 mg/kg ip 72 h before testing and in a dose of 100 mg/kg 48 h and 24 h before the start of acquisition trials. PCA (10 mg/kg ip) or saline (0.9 per cent NaCl) was given on two consecutive days, the last injection 7 days before start of the first learning session. ***$p < 0.001$, ANOVA. Data from Ögren and co-workers (1981)

Yunger, 1974, Köhler and Lorens, 1978; Srebro and Lorens, 1975). The combined dorsal and median raphe lesions caused a profound depletion of 5-HT in the forebrain but did not effect spinal 5-HT concentrations (Srebro and Lorens, 1975). Selective lesions of either the median or the dorsal raphe nuclei did not significantly change two-way avoidance acquisition learning although they caused a marked reduction of 5-HT concentrations in their respective terminal regions. However, lesions restricted to the median raphe nuclei retarded the acquisition of a one-way avoidance response (Srebro and Lorens, 1975). On the other hand, Steranka and Barrett (1974) reported, following lesion of the median raphe nuclei, a facilitated acquisition in a discriminated Y-maze situation which essentially is a one-way avoidance task.

It is not clear at present whether the effects of avoidance learning caused by dorsal and/or median raphe lesions are related to the depletion of forebrain 5-HT. For instance, lesions of both the dorsal and ventral tegmental nuclei of Gudden are reported to produce similar changes in avoidance learning as the combined dorsal and median raphe lesions (Lorens, 1978; Köhler and Lorens, 1978). The existence of 5-HT cells in these nuclei remains in doubt (Fuxe, 1965) and lesions of the dorsal tegmental nuclei do not lower forebrain 5-HT concentrations (Lorens, 1978). As a result of these findings Lorens and his co-workers have suggested that the behavioural effects of electrolytic midbrain raphe lesions are not associated with lesioning of the ascending 5-HT pathways but may be due to damage to tegmental nuclei and/or their projection which are located in the vicinity of the midbrain raphe nuclei (Lorens *et al.*, 1976).

SEROTONIN NEUROTOXINS

The seemingly contradictory results obtained with PCPA and electrolytic mid-brain raphe lesions led to the employment of more selective tools. The discovery of the neurotoxic properties of 5,6 and 5,7-dihydroxytryptamine (5,6-DHT; 5,7-DHT) on central 5-HT neurones (Baumgarten and Lachenmayer, 1972) opened up the possibility to produce a rather selective degeneration of central 5-HT neurones and their terminal systems. In functional studies the use of 5,7-DHT seems to be preferable to that of 5,6-DHT since injections of 5,6-DHT cause a quite substantial damage at the injection site in cerebral tissue. 5,7-DHT causes a much smaller unspecific lesion than produced by electrolysis (Lorens *et al.*, 1976). Intracerebral injections of 5,7-DHT produce a quite substantial neurotoxic action also on NA neurones which partially but not totally is counteracted by pretreatment with NA uptake blocking agents such as protriptyline and desipramine (Björklund *et al.*, 1975; Hole *et al.*, 1976). In contrast to what is found for PCA (see below), potent and specific 5-HT uptake inhibitors such as zimelidine can only partially block the neurotoxic action of 5,7-DHT on 5-HT pathways (Fuxe *et al.*, 1978b). Thus, when using 5,7-DHT the effects of this compound on both NA and 5-HT neurones as well as its non-specific neurotoxic action must be taken into account.

In addition to 5.7-DHT, the halogenated amphetamine derivative *p*-chloroamphetamine (PCA) may be a useful tool in the analysis of the role of 5-HT. Systemic injections of high doses of PCA produces a long-lasting reduction of brain 5-HT, 5-hydroxyindoleacetic acid (5-HIAA) and synaptosomal high-affinity uptake of 5-HT (Sanders-Bush and Steranka, 1978). This long-term effect of PCA develops within 3 days and is clearly distinguished from the acute 5-HT releasing action (see above). The degenerative effects of PCA seem mainly to be restricted to the ascending 5-HT projections originating from the B7-B9 groups with marginal effect on the descending 5-HT terminal systems (Köhler *et al.*, 1978). In contrast to 5,7-DHT, the neurotoxic action of PCA involves exclusively 5-HT neurones and its neurotoxic action on 5-HT is completely blocked by 5-HT reuptake inhibitors such as zimelidine (Ross, 1976; Köhler *et al.*, 1978).

In contrast to lesions of the midbrain raphe nuclei, degeneration of the ascending 5-HT pathways by 5,7-DHT failed to produce any impairment of one-way avoidance acquisition and facilitation of two-way avoidance acquisition. Thus, injections of 5,7-DHT into the dorsal tegmentum just in front of the raphe nuclei did not produce any significant change in one-way (Hole *et al.*, 1976; Fuxe *et al.*, 1978b), and two-way avoidance acquisition (Fig. 6; Ögren *et al.*, 1981). Similar effects on one-way active avoidance learning were observed when the 5,7-DHT injection was placed in the vicinity of the dorsal and median raphe nuclei (Lorens *et al.*, 1976). PCA treatment (2 X 10 mg/kg, 7 days before testing) caused similar effects as 5,7-DHT, i.e. no significant effects on one-way and two-way avoidance acquisition (Fig. 5) PCA-treatment, 5,7-DHT microinjections into the ascending 5-HT pathways or into the raphe nuclei produced a marked reduction of 5-HT concentration in forebrain regions (cerebral cortex, striatum, hippocampus). These data show that different

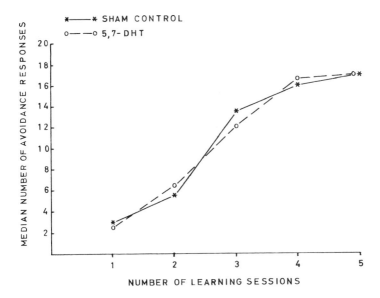

Fig. 6   The effect of a 5,7-DHT (4 μg/3 μl) induced lesion of the ascending 5-HT pathways in the midbrain on the acquisition of a two-way active avoidance response. Training started 7 days following 5,7-DHT injection. No overall significant difference was found between the sham and 5,7-DHT treated groups, ANOVA. Data from Ögren and co-workers (1980)

experimental manipulations resulting in 5-HT depletion can differentially affect one-way and two-way avoidance learning. In view of the effects of PCA and 5,7-DHT on avoidance acquisition, a role of the ascending 5-HT pathways in the avoidance deficit caused by electrolytic midbrain raphe lesions seems less likely.

CONCLUSIONS

Treatments that result in an enhanced 5-HT activity in the brain or that increase the availability of 5-HT at the synaptic cleft generally cause an impairment of both the acquisition and retention of various avoidance tasks. Thus, 5-HTP injections, micro-injections of 5-HT into the hippocampus, injection of the 5-HT releasing agent PCA or 5-HT agonists such as 5-MeO-DMT or D-LSD in high doses can cause acquisition or retention deficits. In addition, electrical stimulation of the dorsal raphe causes an avoidance learning impairment which seems to be related to increased 5-HT release (Fibiger *et al.*, 1978). In view of the preferential action of PCA on the ascending 5-HT pathways, a 5-HT system in the forebrain region (cerebral cortex, striatum or hippocampus) is implicated in avoidance learning. In summary, these findings suggest that treatments which enhance 5-HT activity have more pronounced effects on the acquisitional phase of avoidance conditioning than on the already established responses. In the case of PCA, this avoidance impairment seems to be clearly related to a release of endogenous 5-HT. The observed learning deficiency seem in general not to be due to a general suppressive effect on behaviour, since many of these treatments increase locomotion or exploration. In addition the learning deficit following 5-HT activation appears not to be related to loss of foot-shock sensitivity or response failure. Thus, these findings are consistent with the view that 5-HT may have a specific role in learning and memory processes.

The effects of 5-HT manipulations on learning are not bidirectional. Thus, treatments which reduce 5-HT activity do not generally produce facilitated avoid-ance acquisition. Compounds such as D-LSD which at very low doses reduce the firing rate of the raphe neurones facilitate avoidance acquisition. On the other hand, both facilitated and impaired avoidance learning can be observed following 5-HT depletion (Ögren *et al.*, 1980). Both response topology, shock-levels and the regional effect on various 5-HT terminal system appear to determine the avoidance perform-ance following 5-HT depletion. There is suggestive evidence that different 5-HT systems may subserve different roles in this respect. The different effects on avoid-ance learning caused by PCPA on one hand and 5,7-DHT/PCA on the other may be due to their differential action on brain 5-HT neurones. PCA and 5,7-DHT when injected into the ascending 5-HT neurones primarily affect the ascending 5-HT pathways while PCPA affects both the ascending and descending systems. This dif-ference suggests that forebrain 5-HT terminals may not be involved in the facilitory action of PCPA on avoidance acquisition as observed in certain situations.

These findings do not in general support the response disinhibition hypothesis (see Introduction). Both facilitation and impairment of avoidance learning are observed following 5-HT depletion depending on the test situation (Lorens, 1978; Ögren *et al.*, 1981). In addition, quite a number of studies have failed to demon-strate an attenuation of response suppression following 5-HT depletion with the synthesis inhibitor PCPA (Brody, 1970; Blakely and Parker, 1973). In addition, PCPA did not change passive-avoidance learning in situations where punishment is applied for performing a well-defined response, e.g. a step-down response (Fibiger *et al.*, 1978). Serotonergic depletion in the rat can result in both increases and decreases of locomotion depending on the test situation (Köhler *et al.*, 1978). For instance, rats treated with PCPA displayed decreased locomotion in a quiet novel environment (Ellison and Bresler, 1974). Although such findings are inconsistent with the global hypothesis that 5-HT neurones are specifically involved in behavioural suppression, there is suggestive evidence for a role of 5-HT in various learning pro-cesses in the rat. The available data suggest that 5-HT neurones in some way may modify the encoding and/or consolidation of recent information. There is at present little evidence that the putative role of 5-HT in learning may directly involve associative processes. There is both behavioural and neurophysiological evidence that 5-HT neurones may be involved in sensory integrative and arousal processes (Davis and Sheard, 1976). Changes in 5-HT neurotransmission could possibly modify sensory input and thereby modify information-storage.

Further studies are required to define the specific role(s) for serotonin in learning and memory. As already pointed out there are several critical methodological problems with regard to the presently used tools. For future research, selective lesioning of 5-HT pathways to specified terminal regions by, for example, 5,7-DHT seems to be preferable to the gross insult caused by PCPA treatment. Moreover, there is a neeed for parametric studies in relationship to various 5-HT manipulations and in relationship to different tasks. It is clear from the present review that every manipulation must be assessed in terms of the behavioural context. In addition, there is also a need for more information on the possible role of 5-HT in more cognitive oriented tasks such as compound conditioning as well as spatially oriented tasks such as maze learning. So far the role of 5-HT has mainly been analysed in terms of punishment or aversively motivated learning such as avoidance Some data indicate, however, that 5-HT neurones may also play an important part in positive reinforcement processes such as intracranial self-stimulation reward (Poschel and Ninteman, 1971) and ethanol and morphine intake (Rockman *et al.*, 1980).

REFERENCES

Aghajanian, G. K. and Wang, R. Y. (1978). Physiology and pharmacology of central serotonergic neurons, in *Psychopharmacology: A Generation of Progress* (Eds M. A. Lipton, A. DiMascio, and K. F. Killman), pp. 171–183, Raven Press, New York.

Aprison, M. H. and Hingten, J. M. (1970). Neurochemical correlates of behavior. *Int. Rev. Neurobiol.* **13**, 325–342.

Azmitia, E. C. and Segal, M. (1978). An autoradiographic analysis of the differential ascending projections of the dorsal and median raphe nuclei in the rat, *J. Comp. Neurol.* **179**, 641–668.

Baumgarten, H. G. and Lachenmayer, L. (1972). 5,7-dihydroxytryptamine: improvement in chemical lesioning of indoleamine neurons in the mammalian brain, *Z. Zellforsch.* **135**, 399–414.

Beninger, R. J. and Phillips, A. G. (1979). Possible involvement of serotonin in extinction. *Pharmacol. Biochem. Behav.* **10**, 37–41.

Bignami, G. (1972). Facilitation of avoidance acquisition by LSD-25. Possible effects on drive modulating systems. *Psychopharmacologia*, **25**, 146–151.

Bignami, G. (1976). Nonassociative explanations of behavioral changes induced by central cholinergic drug, *Acta Neurobiol. Exp.* **36**, 5–90.

Björklund, A., Baumgarten, H. G., and Rensch, A. (1975). 5,7-Dihydroxytryptamine: improvement of its selectivity for serotonin neurons in the CNS by pretreatment with desipramine, *J. Neurochem.* **24**, 833–835.

Blakely, T. A. and Parker, L. F. (1973). The effects of *para*chlorphenylalanine on experimentally induced conflict behavior, *Pharmacol. Biochem. Behav.* **1**, 609–613.

Brodie, B. B. and Shore, P. A. (1957). A concept for a role of serotonin and norepinephrine as chemical mediators in the brain, *Ann. N. Y. Acad. Sci.* **66**, 631–642.

Brody, J. F. Jr. (1970). Behavioural effects of serotonin depletion and of p-chlorophenylalanine (a serotonin depletor) in rats, *Psychopharmacologia*, **17**, 14–33.

Carlton, P. L. and Advokat, C. (1973). Attenuated habituation due to parachlorophenylalanine, *Pharmacol. Biochem. Behav.* **1**, 657–663.

Crow, T. J. and Deakin, J. F. W. (1979). Monoamines and the psychoses, in *Chem-*

ical Influences on Behaviour (Eds K. Brown and S. J. Cooper), pp. 503–532, Academic Press, New York.

Davis, M. and Sheard, M. H. (1976). *p*-Chloroamphetamine (PCA): Acute and chronic effects on habituation and sensitization of the acoustic startle response in rats, *Eur. J. Pharmacol.* **35**, 261–273.

de Montigny, C. and Aghajanian, G. K. (1977). Preferential action of 5-methoxy-tryptamine and 5-methoxydimethyltryptamine on presynaptic serotonin recep-tors: A comparative iontophoretic study with LSD and serotonin, *Neuro-pharmacol.* **16**, 811–818.

Ellison, G. D. and Bresler, D. E. (1974). Tests of emotional behavior in rats follow-ing depletion of norepinephrine, of serotonin, or of both, *Psychopharmacologia*, **34**, 275–288.

Essman, W. B. (1970). Some neurochemical correlates of altered memory con-solidation, *Trans. N.Y. Acad. Sci.* **32**, 948–973.

Essman, W. B. (1973a). Neuromolecular modulation of experimentally-induced retrograde amnesia, *Confin. Neurol.* **35**, 1–22.

Essman, W. B. (1973b). Age-dependent effects of 5-hydroxytryptamine upon memory consolidation and cerebral protein synthesis, *Pharmacol. Biochem. Behav.* **1**, 7–14.

Evangelista, A. M., Gattioni, R. C., and Izquierdo, I. (1970). Effect of amphetamine, nicotine and hexamethonium on performance of a conditioned response during acquisition and retention trials, *Pharmacology*, **3**, 91–96.

Fibiger, H. C., Lepiane, F. G., and Phillips, A. G. (1978). Disruption of memory produced by stimulation of the dorsal raphe nucleus: mediation by serotonin, *Brain Res.* **155**, 380–386.

Fuller, R. W., Perry, K. W., and Molloy, B. B. (1975). Effect of 3-(*p*-trifluoro-methylphenoxy)-*N*-methyl-3-phenylpropylamine on the depletion of brain serotonin by 4-chloroamphetamine, *J. Pharmacol. Exp. Ther.* **193**, 796–803.

Fuxe, K. (1965). Evidence for the existence of monoamine neurons in the central nervous system. IV. Distribution of monoamine nerve terminals in the central nervous system, *Acta Physiol. Scand.* **64**, Suppl. 247, 37–85.

Fuxe, K., Butcher, L. L., and Engel, J. (1971). DL-5-hydroxytryptophan-induced changes in central monoamine neurons after peripheral decarboxylase inhibition, *J. Pharm. Pharmacol.* **23**, 420–424.

Fuxe, K., Ögren, S. O., Agnati, L. F., and Jonsson, G. (1978a). Further evidence that methergoline is a central 5-hydroxytryptamine receptor blocking agent, *Neurosci. Lett.* **9**, 195–200.

Fuxe, K., Ögren, S. O., Agnati, L. F., Jonsson, G., and Gustafsson, J. Å. (1978b). 5.7-Dihydroxytryptamine as a tool to study the functional role of central 5-hydroxytryptamine neurons, *Ann. N.Y. Acad. Sci.* **305**, 346–369.

Gimpl, M. P., Gormezano, I., and Harvey, J. A. (1979). Effects of LSD on learning as measured by classical conditioning of the rabbit nictitating membrane res-ponse, *J. Pharmacol. Exp. Ther.* **208**, 330–334.

Graeff, F. G. and Schoenfeld, R. I. (1970). Tryptaminergic mechanisms in punished and nonpunished behavior, *J. Pharmacol. Exp. Ther.* **173**, 277–283.

Harvey, J. A. and Lints, C. E. (1971). Lesions in the medial forebrain bundle: relationship between pain sensitivity and telencephalic content of serotonin, *J. Comp. Physiol. Psychol.* **74**, 28–36.

Hole, K., Fuxe, K., and Jonsson, G. (1976). Behavioural effects of 5,7-dihydroxy-tryptamine lesions of ascending 5-hydroxytryptamine pathways, *Brain Res.* **107**, 385–399.

Hunter, B., Zornetzer, S. F., Jarvik, M. E., and McGaugh, J. L. (1977). Modulation of learning and memory: Effects of drugs influencing neurotransmitters, in *Handbook of Psychopharmacology* Vol. 8 (Eds L. L. Iversen, S. D. Iversen and S. H. Snyder), pp. 531–577, Plenum Press, New York.

Joyce, D. and Hurwitz, H. M. B. (1964). Avoidance behaviour in the rat after 5-hydroxytryptophan (5-HTP) administration, *Psychopharmacologia*, 5, 424–430.

Koe, B. K. and Weissman, A. (1966). *p*-Chlorophenylalanine: A specific depletor of brain serotonin, *J. Pharmacol. Exp. Ther.* **154**, 499–516.

Köhler, C. and Lorens, S. A. (1978). Open field activity and avoidance behaviour following serotonin depletion: A comparison of the effects of *para*chlorophenylalanine and electrolytic midbrain raphe lesions, *Pharmacol. Biochem. Behav.* **8**, 223–233.

Köhler, C., Ross, S. B., Srebro, B. and Ögren, S. O. (1978). Long-term biochemical and behavioural effects of *p*-chloroamphetamine in the rat, *Ann. N.Y. Acad. Sci.* **305**, 645–663.

Lorens, S. A. and Yunger, L. M. (1974). Morphine analgesia, two-way avoidance, and consummatory behaviour following lesions in the midbrain raphe nuclei of the rat, *Pharmacol. Biochem. Behav.* **2**, 215–221.

Lorens, S. A., Guldberg, H. C., Hole, K., Köhler, C. and Srebro, B. (1976). Activity, avoidance learning and regional 5-hydroxytryptamine following intrabrain stem 5,7-dihydroxytryptamine and electrolytic midbrain raphe lesions in the rat, *Brain Res.* **108**, 97–113.

Lorens, S. A. (1978). Some behavioural effects of serotonin depletion depend on method: A comparison of 5,7-dihydroxytryptamine, *p*-chlorophenylalanine, *p*-chloroamphetamine and electrolytic raphe lesions, *Ann. N.Y. Acad. Sci.* **305**, 532–555.

Marsden, C. A., Conti, J., Strope, E., Curzon, G., and Adams, R. N. (1979). Monitoring 5-hydroxytryptamine release in the brain of the freely moving unanaesthetized rat using *in vivo* voltammetry, *Brain Res.* **171**, 85–99.

Ögren, S. O., Köhler, C., Ross, S. B., and Srebro, B. (1976). 5-hydroxytryptamine depletion and avoidance acquisition in the rat. Antagonism of the long-term effects of *p*-chloroamphetamine with a selective inhibitor of 5-hydroxytryptamine uptake, *Neurosci. Lett.* **3**, 341–347.

Ögren, S. O., Ross, S. B., Holm, A. C., and Baumann, L. (1977). 5-hydroxytryptamine and avoidance performance in the rat. Antagonism of the acute effect of *p*-chloroamphetamine by zimelidine, an inhibitor of 5-hydroxytryptamine uptake, *Neurosci. Lett.* **3**, 331–336.

Ögren, S. O. and Ross, S. B. (1977). Substituted amphetamine derivatives. II Behavioural effects in mice related to monoaminergic neurons, *Acta Pharmacol. Toxicol.* **41**, 353–368.

Ögren, S. O., Fuxe, K., Archer, T., Hall, H., Holm, A. C., and Köhler, C. (1981). Studies on the role of central 5-HT neurons in avoidance learning: A behavioral and biochemical analysis in *Serotonin: Current Aspects of Neurochemistry and Function. Advances in Experimental Medicine and Biology* Vol. 133 (Eds B. Haber, S. Gabay, M. R. Issidorides and S. G. A. Alivisatos), pp. 681–705, Plenum Press, New York.

Ögren, S. O. (1981) Forebrain serotonin and avoidance learning. Behavioural and biochemical studies on the acute effect of *p*-chloroamphetamine on one-way active avoidance learning in the male rat, *Pharmacol. Biochem. Behav.* (accepted for publication).

Poschel, B. P. H. and Ninteman, F. W. (1971). Intracranial reward and the forebrain's serotonergic mechanism: studies employing *para*-chlorophenylalanine and *para*-chloroamphetamine. *Physiol. Behav.* **7**, 39–46.

Rake, A. V. (1973). Involvement of biogenic amines in memory formation: The central nervous system indole amine involvement, *Psychopharmcologia*, **29**, 91–100.

Rockman, G. E., Amit, Z., Bourque, C., Brown, Z. W., and Ögren, S. O. (1980). Reduction of voluntary morphine consumption following treatment with zimelidine. *Arch. Int. Pharmacodyn.* **244**, 123–129.

Roffman, M. and Lal, H. (1971). Facilitatory effect of amphetamine on learning and recall of an avoidance response in rats, *Arch. Int. Pharmacodyn.* **193**, 87–91.

Ross, S. B. (1976). Antagonism of the acute and long-term biochemical effects of 4-chloroamphetamine on the 5-HT neurones in the rat brain by inhibitors of the 5-hydroxytryptamine uptake, *Acta Pharmacol. Toxicol.* **39**, 456–476.

Sanders-Bush, E. and Steranka, L. R. (1978). Immediate and long-term effects of *p*-chloroamphetamine on brain amines. *Ann. N.Y. Acad. Sci.* **305**, 208–221.

Schlesinger, K. and Schreiber, R. A. (1968). Effects of *p*-chlorphenylalanine on conditioned avoidance learning, *Psychon. Sci.* **11**, 225–226.

Solomon, P. R., Colleen, A. K., and Scott, D. R. (1978). Disruption of latent inhibition following systemic administration of *para*chlorophenylalanine (PCPA), *Physiol. Behav.* **20**, 265–271.

Srebro, B. and Lorens, S. A. (1975). Behavioral effects of selective midbrain raphe lesions in the rat, *Brain Res.* **89**, 303–325.

Stein, L., Wise, C. D., and Belluzzi, J. D. (1975). Effects of benzodiazepines on central serotonergic mechanisms, in *Mechanism of Action of Benzodiazepines* (Eds E. Costa and P. Greengard), pp. 29–44, Raven Press, New York.

Stein, L., Wise, C. D., and Belluzzi, J. D. (1977). Neuropharmacology of reward and punishment, in *Handbook of Psychopharmacology*. Vol. 8 (Eds L. L. Iversen, S. D. Iversen and S. H. Snyder), pp. 25–53, Plenum Press, New York.

Steranka, L. R. and Barrett, R. J. (1974). Facilitation of avoidance acquisition by lesion of the median raphe nucleus: evidence for serotonin as a mediator of shock induced suppression, *Behav. Biol.* **11**, 205–213.

Stevens, D. A., Resnick, O. and Krus, D. M. (1967). The effects of *p*-chlorophenylalanine, a depletor of brain serotonin on behavior: I. Facilitation of discrimination learning, *Life Sci.* **6**, 2215–2220.

Stevens, D. A., Fechter, L. D. and Resnick, O. (1969). The effects of *p*-chlorophenylalanine, a depletor of brain serotonin on behavior: II. Retardation of passive avoidance learning, *Life Sci.* **8**, 379–385.

Stoff, D. M., Gorelick, D. A., Bozewicz, T., Bridger, W. H., Gillin, J. C., and Wyatt, R. J. (1978). The indole hallucinogens *N,N*-dimethyltryptamine (DMT) and 5-methoxy *N,N*-dimethyltryptamine (5-MeO-DMT) have different effects from mescaline on rat shuttlebox avoidance, *Neuropharmacol.* **17**, 1035–1040.

Tenen, S. S. (1967). The effects of *p*-chlorophenylalanine, a serotonin depletor, on avoidance acquisition, pain sensitivity and related behaviour in the rat, *Psychopharmacologia*, **10**, 204–219.

Trulson, M. E. and Jacobs, B. L. (1976). Behavioural evidence for the rapid release of CNS serotonin by PCA and fenfluramine, *Eur. J. Pharmacol.* **36**, 149–154.

Tye, N. C., Everitt, B. J., and Iversen, S. D. (1977). 5-hydroxytryptamine and punishment, *Nature (Lond.)*, **268**, 741–743.

Woolley, D. W. and Van der Hoeven, T. (1963). Alteration in learning ability caused by changes in cerebral serotonin and catecholamines, *Science*, **139**, 610–611.

Biology of Serotonergic Transmission
Edited by N. N. Osborne
© 1982, John Wiley & Sons Ltd.

Chapter 14

Serotonin in the Basal Ganglia

A. DRAY

MRC Neurochemical Pharmacology Unit, Medical Research Council Centre, Medical School, Hills Road, Cambridge CB2 2QH, UK

INTRODUCTION: THE SIGNIFICANCE OF SEROTONIN IN NEUROLOGICAL DISORDERS INVOLVING THE BASAL GANGLIA

The serotonergic innervation of the basal ganglia (caudate nucleus, putamen, globus pallidus, substantia nigra and subthalamic nucleus) has received relatively little attention compared with that given to other neurotransmitters (see Dray, 1980). Nevertheless, there is compelling evidence from clinical and experimental observations for an important role for serotonin in these brain regions. Thus neuro-

335

chemical findings of disturbed serotonin metabolism and changes in serotonin receptors in a number of neurological diseases have highlighted the potential importance of serotonin in the etiology of the disease though the significance of these findings is still debatable. For example, Parkinson's disease, in which pathological lesions of dopamine-containing cells in the substantia nigra and loss of dopamine from the basal ganglia are common findings, is also associated with significant losses of serotonin and its major metabolite 5-hydroxyindoleacetic acid (5-HIAA) from several brain regions including the striatum, globus pallidus, thalamus and substantia nigra (Fahn *et al.*, 1971; Lloyd and Hornykiewicz, 1974) (see Table 1). These losses may result from impaired synthesis of serotonin in the midbrain raphe where a reduction in decarboxylase activity has been found in Parkinson's disease (Lloyd, 1977). Disturbances of serotonin metabolism are additionally reflected by a reduction of 5-HIAA, measured in samples of CSF (see Guldberg *et al.*, 1967; Curzon, 1972, 1977, 1978). Moreover, Parkinson's disease patients with low CSF, 5-HIAA (compared with the major dopamine metabolite, homovanillic acid) respond less readily to dopamine replacement therapy with L-DOPA (Gumpert *et al.*, 1973) suggesting that impaired serotonin neurotransmission may be more significant than that of dopamine in some patients. Administration of the serotonin precursor 5-hydroxytryptophan (5-HT) may in fact reduce some symptoms, especially tremor (Sano and Taniguchi, 1972) but in other patients may exacerbate them (Chase *et al.*, 1972). However, the significance of serotonin disturbances in Parkinson's disease can be questioned by the observation that its concentration was similar in both caudate nuclei whereas that of dopamine was reduced on the contralateral side in a patient with unilateral Parkinsonism (see Curzon, 1972).

Receptor-binding studies in post-mortem Parkinson's disease brain show that there are significant decreases in 5-HT receptors in the putamen but not in globus pallidus (Reisine *et al.*, 1977). The associated losses of cholinacetyltransferase activity, an index of the integrity of cholinergic neurones, led these authors to postulate that 5-HT receptors in the striatum, but not the pallidum, were localized to cholinergic neurones.

Serotonergic nerve terminals in the basal ganglia were also considered to be important for the therapeutic effects of L-DOPA in Parkinson's disease. Thus *in vitro* studies provided evidence that L-DOPA may be taken up, decarboxylated (Butcher *et al.*, 1970; Ng *et al.*, 1970; 1972b) and released as dopamine (Ng *et al.*, 1970, 1972a) by serotonergic terminals. On the other hand, recent experimental evidence, following combined destruction of dopaminergic and serotonergic neurones, suggests that serotonin neurones are not involved in the actions of L-DOPA in Parkinson's disease to any significant extent (Melamed *et al.*, 1980).

Reduced levels of serotonin have also been measured in post-mortem Huntington's disease brain particularly from the globus pallidus though there appeared to be an increase or unchanged striatal serotonin content (Lloyd and Hornykiewicz, 1974; Bernheimer and Hornykiewicz, 1973; Table 1). Interestingly CSF, 5-HIAA

Table 1. Serotonin in the basal ganglia: changes in concentration and receptor binding in neurological disease

Region	Concentration/g wet weight	In disease % control		Receptor binding % control		Source of transmitter
		PD	HC	PD	HC	
Striatum	0.35 μg[4]	40[1]	377[4] or NC	63[6]	38[2]	DRN/MRN (mainly DRN)
Globus pallidus	301 ng[3]	57[4]	60.5[4]	NC[6]	25[2]	DRN
Subthalamic nucleus	11.2 ng/mg[5] protein	NT	NT	NT	NT	DRN/MRN
Substantia nigra	1.02 μg[4]	40[1]	NT	NT	NT	DRN/MRN (mainly DRN)

1. Curzon, 1977;
2. Enna *et al.*, 1976;
3. Fahn *et al.*, 1971;
4. Lloyd and Hornykiewicz, 1974;
5. Palkovits *et al.*, 1974;
6. Reisine *et al.*, 1977.

PD = Parkinsons disease; HC = Huntingtons chorea; NC = No change; NT = Not estimated.

concentrations of an akineto-rigid Huntington's patient were also significantly reduced (Curzon, 1972). The significance of these neurochemical changes is not presently apparent. However, receptor-binding data (see Yamamura *et al.*, 1978) shows a significant reduction of serotonin receptors in the globus pallidus and striatum (Enna *et al.*, 1976; Reisine *et al.*, 1977). Since there are marked losses of cholinergic and GABA-containing neurones from these regions, serotonin receptors have been postulated to exist on these chemically defined neurones. Additionally, the loss of serotonin-receptors suggest that serotonin-replacement therapy would be inappropriate for the symptomatic treatment of Huntington's disease.

Low CSF, 5-HIAA levels have been reported in a number of other dyskinesias (see Johansson and Roos, 1974; Curzon, 1978). For example, in the Parkinson-like dementia associated with amyotropic lateral sclerosis (Brody *et al.*, 1970), in torticollis (Johansson and Roos, 1974; Lal *et al.*, 1979) and in myoclonus (Chadwick *et al.*, 1975). 5-Hydroxytryptophan therapy appeared to improve spasmodic torticollis (Mori *et al.*, 1975) and the hypotonia observed in Down's syndrome (Dubowitz and Rogers, 1969; Bazelon *et al.*, 1967, 1968). Also serotonin replacement therapy was efficacious in the treatment of certain types of myoclonus (Chadwick *et al.*, 1975; Van Woert and Sethy, 1975). It is uncertain, however, how these neurological syndromes relate to disturbances of serotonin systems in the basal ganglia.

Overall it would appear from the clinical observations that dysfunction of the serotonergic systems of the basal ganglia are not entirely consistent findings in neurological disorders. Neither does serotonin replacement (with 5-HTP) constitute a predictably efficacious approach for their management. Perhaps closer monitoring of relative transmitter deficiences (e.g. in CSF samples) might select patients who would receive greatest benefit from such replacement therapy. Of course, the relevance of serotonin in basal ganglia disease may become more apparent as the functions of these nuclei are better understood. In this respect this review serves to co-ordinate the evidence for serotonergic neurotransmission in the basal ganglia and discuss its functional role in concert with other neurotransmitters within this region of the brain.

Some basic requirements seem necessary to appreciate the relationships between neural subsystems. These include the identification of the source and distribution of projections linking different components of the system. Moreover, the functional signs of these links should be defined in terms of synaptic activation of the target elements. The synaptic activation should correlate with the actions of the transmitter in the projection and both synaptic activity and that produced by the direct post-synaptic actions of the transmitter should exhibit the same susceptibility to pharmacological antagonists (see Werman, 1966).

THE SEROTONIN INNERVATION OF THE STRIATUM: ANATOMICAL NEUROCHEMICAL CONSIDERATIONS

Several studies have described the presumed serotonergic projections to the basal ganglia. These originate principally from the serotonin-containing cell bodies of the raphe nuclei, in particular the dorsal (B7) and medial (B8) raphe (Dahlstrom and Fuxe, 1964). These projections are summarized in Fig. 1 and will be discussed systematically.

The serotonergic innervation of the striatum derives exclusively from the serotonin-containing neurones of the midbrain raphe (Fuxe, 1965; Ungerstedt, 1971a; Olsen and Seiger, 1972). Both retrograde (HRP or fluorescent dye studies) and antero-grade (radio-labelled amino acids) fibre-tracing experiments show a prominent projection mainly from the dorsal raphe nucleus (DRN) with a minimal projection from the median raphe nucleus (MRN) (Conrad *et al.*, 1974; Miller *et al.*, 1975; Bobillier *et al.*, 1976; Pasquier *et al.*, 1977; Jacobs *et al.*, 1978; Azmitia and Segal, 1978; Royce, 1978; Van der Kooy and Kuypers, 1979; Van der Kooy, 1979; Steinbusch *et al.*, 1980. 1981; Szabo, 1980; Veening *et al.*, 1980). However, Pierce *et al.* (1976) could not confirm the DRN projection possibly due to a more caudal placement of the labelled tracer within the DRN.

The raphe-striatal pathway takes a course, in the dorsal raphe forebrain and dorsal raphe cortical tracts (Azmitia and Segal, 1978), along the ventrolateral aspect of the medial forebrain bundle. Some of the raphe cells send collaterals to the ipsilateral substantia nigra (Van der Kooy and Hattori, 1980a) and their terminals

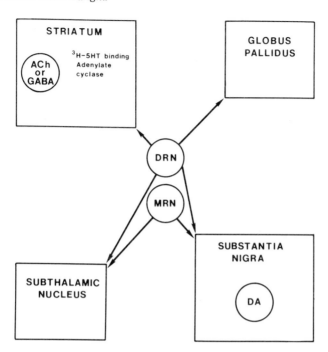

Fig. 1 The serotonergic innervation of the basal ganglia, arising from the dorsal (DRN) and medial (MRN) raphe nuclei. Projections are mainly unilateral and it should be noted that many other forebrain areas are innervated by the DRN (e.g. amygdala, nucleus accumbens, septum, hypothalamus) and MRN (e.g. cortex, habenula, thalamus, hippocampus).

Serotonergic terminals in the striatum probably contact cholinergic or GABA-containing neurones. These cells posses binding sites for $[^3H]$ 5-HT (displaced by neuroleptics), and contain serotonin stimulated adenylate cyclase. Striatal glial cells also possess serotonin binding sites and adenylate cyclase activity. Serotonin terminals in the substantia nigra are associated with dopamine-containing neurones and dendrites

end diffusely in the striatum with the densest innervation to the caudal parts of the caudate nucleus. The DRN projection is mainly unilateral though some fibres from both the DRN and MRN cross to the contralateral side in the supraoptic commissure, supramammilary commissure or in the decussation of the brachium conjunctivum (Björklund *et al.*, 1973; Conrad *et al.*, 1974; Bobillier *et al.*, 1976; Jacobs *et al.*, 1978; Moore *et al.*, 1978). Studies using retrograde transport of fluorescent markers injected into the striata show that cell bodies of the DRN project ipsilaterally or contralaterally to the striatum, but not bilaterally (Van der Kooy and Hattori, 1980b). Also, the striatal-projecting neurones are found in all parts of the DRN whereas those that also send collateral projections to the substantia nigra are localized mainly to the dorsal parts of the nucleus (Van der Kooy and Hattori, 1980a).

Histochemical observations suggest that the major proportion (80 per cent) of the raphe-striatal pathway may be serotonergic (Fuxe, 1965; Ungerstedt, 1971a; Olsen and Seiger, 1972; Seiger and Olsen, 1973; Steinbusch et al, 1981). Indeed, the striatum contains serotonin and mechanisms for its synthesis, degradation and reuptake into fine varicose nerve fibre terminals (Bogdanski et al, 1957; Aghajanian and Bloom, 1967; Broch and Marsden, 1972; Palkovits et al, 1974, 1977; Calas et al, 1976). Raphe stimulation produces an increase in the release of serotonin from the striatum (Ashkenazi et al, 1972; Holman and Vogt, 1972), whereas lesions or cooling of the raphe significantly reduce striatal serotonin release, serotonin content, serotonin uptake, and produce a loss of serotonin fluorescence (Kuhar et al, 1972; Marsden and Guldberg, 1973; Costall et al, 1975; Moore and Halaris, 1975; Halaris et al, 1976; Lorens et al, 1976; Bourgoin et al, 1977; Palkovits et al, 1977; Azmitia and Segal, 1978; Van de Kar and Lorens 1979; Hery et al, 1979). In particular discrete lesions show that the major serotonergic contribution to the striatum arises from the DRN (Jacobs et al, 1974; Lorens and Guldberg, 1974; Moore and Halaris, 1975; Geyer et al, 1976; Lorens et al, 1976; Jacobs et al, 1978; Dray et al, 1978). The distribution of serotonin in the striatum is ventro-caudal and apparently similar to that of glutamic acid decarboxylase-activity (a marker enzyme fro GABA neurones and terminals) but differs from that of dopamine (rostral striatum) (Bockaert et al, 1976; Tassin et al, 1976; Fonnum et al, 1978).

ELECTROPHYSIOLOGY AND PHARMACOLOGY OF STRIATAL SEROTONIN INNERVATION

In electrophysiological experiments, stimulation of the DRN produced long-lasting inhibition of firing in the majority of responsive striatal neurones recorded extra-cellularly (Miller et al, 1975; Olpe and Koella, 1977; Davies and Tongroach, 1978) (Fig. 2). In these experiments the activity of a few cells was facilitated by the stimulation. On the other hand, stimulation of the MRN produced few evoked effects (Olpe and Koella, 1977) and this data supports the differential projections of the raphe-striatal fibres postulated from the anatomical tracing and neuro-chemical studies.

Raphe-striatal fibres were found to be slowly conducting (0.3-0.7 m/sec) (Davies and Tongroach, 1978), compatible with their being poorly myelinated. The constant latency responses of striatal neurones following raphe stimulation also suggested the possible monosynaptic nature of this pathway. Recent evidence from intra-cellular studies indicates, however, that DRN stimulation evokes monosynaptic EPSPs and EPSP-IPSP sequences in striatal neurones (Van der Maelen et al, 1979) (Fig. 3). Thus it is possible that the extracellular recordings were unable to detect the primary subthreshold increased excitability changes.

The likelihood that raphe evoked EPSPs were due to synaptically released serotonin was concluded from the observations that fewer or lower-amplitude

(a)

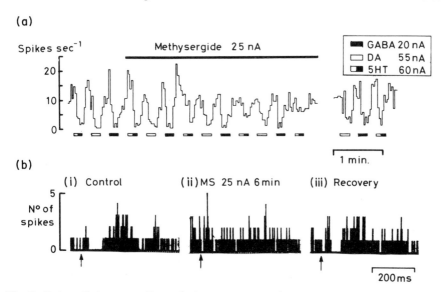

(b)

Fig. 2 Extracellular recording of the responses of a rat neostriatal neurone to dopamine (DA), serotonin (5-HT), GABA and to stimulation of the dorsal raphe nucleus (DRN). Activity was reduced by the putative transmitters (upper ratemeter trace) and by DRN stimulation (lower PSTH). The serotonin antagonist methysergide reduced the responses to serotonin and also raphe-evoked inhibition. (From Davies and Tongroach 1978; reproduced with permission)

EPSPs could be recorded following systemic *p*-chlorophenylalanine treatment. The subsequent administration of 5-HTP restored the EPSP amplitude (Park *et al.*, 1980). Interestingly, some 20 per cent of striatal neurones showed shorter latency EPSPs which remain unchanged following serotonin depletion (Van der Maelen *et al.*, 1979; Park *et al.*, 1980). This accords with the anatomical observations of Steinbusch *et al.* (1980, 1981) showing the existence of non-serotonergic raphe striatal neurones with faster conducting fibres. Indeed non-serotonergic neurones in the DRN have been previously described (Aghajanian *et al.*, 1978; Jacobs *et al.*, 1978; Descarries *et al.*, 1979) some of which may contain dopamine (Ochi and Shimizu, 1978; Nagatsu *et al.*, 1979), noradrenaline (Grzanna and Molliver, 1980; Steinbusch *et al.*, 1980, 1981), GABA (Nanopoulos *et al.*, 1980; Gamrani *et al.*, 1979), enkephalin (Uhl *et al.*, 1979; Hökfelt *et al.*, 1979) or VIP (Sims *et al.*, 1980). However, it is uncertain whether any of these chemically defined cells give rise to the non-serotonergic raphe striatal pathway neither is it entirely certain whether the shorter latency striatal EPSPs were evoked from DRN stimulation (Park *et al.*, 1980).

On the other hand pharmacological studies support the probable inhibitory serotonergic nature of the DRN projection since the electrophoretic adminis-tration of serotonin inhibited the firing of most striatal neurones tested (Herz and

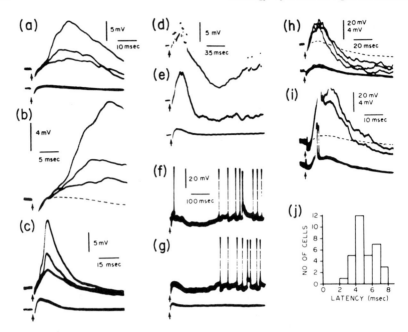

Fig. 3 Intracellular recordings from rat striatal neurones during dorsal raphe (DRN) stimulation. (a) and (b) are the EPSP records evoked by stimulation of DRN at three different intensities showing increased amplitude but constant latency. (c) The passage of hyperpolarizing current through the recording electrode increased the EPSP amplitude whereas depolarizing current reduced it. The middle of the three superimposed traces shows the EPSP with no intracellular current injection. (d) and (e) EPSP–IPSP sequences in response to DRN stimulation. The lower gain records of (f) and (g) show a single spike (f) or no spike (g) triggered by the initial EPSP, followed by supression of action potentials during the IPSP. Traces (h) and (i) show convergence of inputs onto the same neurone from DRN stimulation (h) and cortical stimulation (i). (j) shows the frequency histogram of EPSP latencies of striatal neurones following DRN stimulation. Extracellular control recordings are shown in bottom traces a, c, e, g and by dashed lines in b, h and i. (From Van Der Maelen *et al.*, 1979, reproduced with permission)

Zieglgansberger, 1968; Davies and Tongroach, 1978) while the 5-HT-receptor antagonist drug, methysergide reduced the inhibitory effects of raphe stimulation as well as those of 5-HT on the same striatal neurones (Olpe and Koella, 1977, Davies and Tongroach, 1978; Fig. 2). However, Bevan *et al.* (1975) reported that most striatal cells (64 per cent) (halothane anaesthetized rats) were excited by 5-HT with a smaller number showing inhibition (24 per cent) or biphasic responses (12 per cent). The reasons for these different findings is not clear though differences in neuronal sampling or anesthetic conditions may be contributory factors. Interestingly similar differences have been reported regarding the effects of dopamine administered by electrophoresis to striatal neurones (see e.g. Siggins, 1978; Dray,

1979, 1980). It is also possible that raphe evoked synaptic inhibition may have resulted from the indirect release of other inhibitory transmitters such as GABA or glycine. However, neither bicuculline nor tetanus toxin administered locally in the striatum affected DRN-evoked inhibition (Davies and Tongroach, 1978, 1979).

It is not certain what type of striatal neurone receives raphe projections though many cells receive a converging input from the substantia nigra (Davies and Tongroach, 1978) and cerebral cortex (Van der Maelen *et al.*, 1979; Park *et al.*, 1980). However, the intracellular injection of HRP into striatal cells receiving DRN or substantia nigra projections suggests that they are medium sized spiny neurones (Kocsis *et al.*, 1977; Van der Maelen *et al.*, 1979) which might be chemically defined as cholinergic or GABA-ergic (McGeer and McGeer, 1975; Hattori *et al.*, 1976). Interestingly serotonin-receptor binding data from neurological disease also suggests the presence of serotonin receptors on similar striatal neurones (see Introduction). It is likely that many of the raphe-striatal target neurones are output cells (Somogyi and Smith, 1979; Bolam *et al.*, 1980; Preston *et al.*, 1980).

It would seem reasonable to conclude that the raphe-striatal pathway is serotonergic, at least in part, and serves to inhibit the activity of striatal target cells. The biophysical mechanisms of this inhibition are presently unclear and on the contrary the preliminary intracellular data suggests that the raphe-striatal transmitter produces membrane depolarization. Possibly depolarization might be accompanied by an increase in membrane ionic permeability and this would shunt the synaptic currents generated by tonic facilitatory inputs from other brain regions in a manner similar to that proposed to explain the actions of synaptically released dopamine on striatal neurones (see discussion in Dray, 1980). Clearly the membrane permeability changes produced by synaptically-released or electrophoretically-administered serotonin need to be compared in order to resolve these discrepancies.

Finally, it is of interest that the caudate nucleus itself sends efferents to the raphe (the raphe magnus exclusively) (Usunoff *et al.*, 1974). The significance of this postulated projection has not been determined but its presence might serve as another example of the reciprocal innervation found within the basal ganglia (see Webster, 1975; Nieuwenhuys, 1977).

THE SEROTONERGIC INNERVATION OF THE SUBSTANTIA NIGRA

Anatomical studies clearly show the existence of projections (mainly ipsilateral) to the substantia nigra from the DRN (the major projection) and MRN (Conrad *et al.*, 1974; Bak *et al.*, 1975; Bunney and Aghajanian, 1976; Kanazawa *et al.*, 1976; Pierce *et al.*, 1976; Bobillier *et al.*, 1976; Fibiger and Miller, 1977; Pasquier *et al.*, 1977; Van der Kooy and Hattori, 1980a). Many of the raphe-nigral fibres (from the DRN) appear to be collaterals of raphe-striatal fibres (Poirier *et al.*, 1969; Van der Kooy and Hattori, 1980a,b). These fibres take their origin from cells located mainly in the dorsal part of the DRN (Van der Kooy and Hattori, 1980a). There is abundant evidence indicating that the raphe-nigral projections are serotonergic. Thus,

[3 H] serotonin may be transported by these fibres from substantia nigra terminals to their cell bodies in the raphe (Streit *et al.*, 1979). In addition, both serotonin and its synthesizing enzyme tryptophan hydroxylase are found in the nigra mainly within the reticulata region. The serotonin fibre terminals appear to make contact with nigral dopamine neurones and dendrites (Fuxe, 1965; Fahn *et al.*, 1971; Parizek *et al.*, 1971; Hajdu *et al.*, 1973; Palkovits *et al.*, 1974; Brownstein *et al.*, 1975; Pickel *et al.*, 1975; Reubi and Emson, 1978) and *in vitro* studies have suggested that nigral dopamine release may stimulate serotonin release from raphe nigral terminals (Reubi *et al.*, 1978).

A K^+-evoked, Ca^{2+} dependent release of serotonin has been demonstrated from substantia nigra slice preparations and this was reduced following chronic raphe (DRN and MRN) lesions (Reubi and Emson, 1978). Such lesions also lowered nigral tryptophan hydroxylase activity and the serotonin content of the substantia nigra (Kuhar *et al.*, 1972; Dray *et al.*, 1976b, 1978; Fibiger and Miller, 1977). Additionally *in vivo* experiments showed that cooling the DRN reversibily reduced nigral release of [3 H] serotonin newly synthesized from superfused [3 H] tryptophan. In similar studies superfusion of the DRN with serotonin decreased the release of [3 H] serotonin from the substantia nigra and caudate nuclei (Bourgoin *et al.*, 1981) possibly by an inhibition of raphe neuronal activity (Haigler and Aghajanian, 1974). Curiously, superfusion of the DRN with the serotonin antagonist, methergoline, did not change [3 H] serotonin release from the substantia nigra but produced the expected increase in release from the caudate nuclei (Bourgoin *et al.*, 1981).

ELECTROPHYSIOLOGY AND PHARMACOLOGY OF NIGRAL SEROTONERGIC INNERVATION

Raphe-nigral fibres appear to be slowly conducting (0.2–1.0 m/sec) and either DRN or MRN stimulation inhibited the activity of the majority of nigral neurones recorded extracellularly (Dray *et al.*, 1976b, 1978; Fibiger and Miller, 1977; Fig. 4). In these studies some neurones were also excited by raphe stimulation. However, the correlation of raphe-evoked inhibition and the predominantly inhibitory effect of serotonin in the substantia nigra supports the concept of an inhibitory serotonergic projection (Crossman *et al.*, 1974; Aghajanian and Bunney, 1975; Dray *et al.*, 1976b; Fibiger and Miller, 1977; Collingridge and Davies, 1980). On the other hand, microelectrophoretic serotonin also excites some nigral neurones though these effects were not readily correlated with synaptic activation (Dray *et al.*, 1976b, 1978). However, these data imply that the serotonin receptors in the substantia are heterogenous. It is also possible that the raphe-nigral pathways comprise both inhibitory and facilitatory fibres. Indeed cell bodies of the raphe, as mentioned earlier, contain putative transmitters other than 5-HT (Björklund *et al.*, 1971; Moore and Halaris, 1975; Jacobs *et al.*, 1978; Hökfelt *et al.*, 1978). Further studies are obviously required since preliminary intracellular data suggest that most orthodromic responses in substantia nigra neurones following DRN stimulation consist of EPSPs rather than the expected IPSPs (Karabelas and Purpura, 1979).

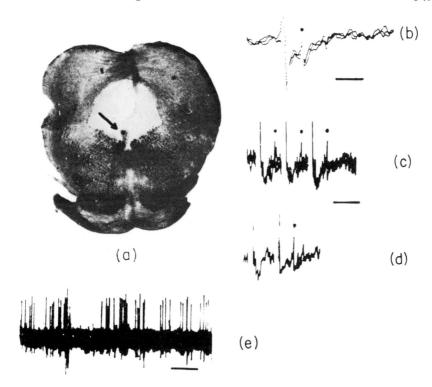

(b)

(c)

(d)

(a)

(e)

Fig. 4 Extracellular recording in the raphe and substantia nigra. Traces (b) and (c) show constant latency antidromic spikes from a medial (b) and dorsal (c) raphe neurone evoked by stimulation of the substantia nigra. The histological section in (a) shows the position of a dye mark (arrow) locating the tip of the micropipette in the DRN. Trace (d) shows the extinction of an antidromic spike of a DRN neurone due to collision with a spontaneous spike. (e) shows inhibition in the substantia nigra evoked by DRN stimulation. Calibrations are 5 msec. in (b); 10 msec. in (c) and (d), and 20 msec. in (e). (From Dray *et al.*, 1978; reproduced with permission)

Pharmacological observations support the existence of an inhibitory raphe-nigral pathway that uses serotonin rather than GABA. Thus raphe evoked inhibition may be blocked by the serotonin antagonist methiothepin (Dray and Oakely, 1977) and also by *p*-chlorophenylalanine pretreatment (Fibiger and Miller, 1977) but not by bicuculline or tetanus toxin (Davies and Tongroach, unpublished observations).

There is also evidence for reciprocal connections from the substantia nigra (compacta and reticulata) to the entire DRN and the parts of the MRN (Sakai *et al.*, 1978; Pasquier *et al.*, 1977; Beckstead *et al.*, 1979). Though there is little evidence confirming these projections, electrophysiological data suggests that substantia nigra stimulation produces inhibition or excitation of the raphe neurones located in the posterior midbrain and anterior pons (Stern *et al.*, 1979).

THE SEROTONERGIC INNERVATION TO OTHER PARTS OF THE BASAL GANGLIA

A number of autoradiographic studies have demonstrated a diffuse projection with moderately dense terminal fields from the DRN to the globus pallidus and ento-pedunuclar nucleus (the internal pallidal segment in primates) (Bobillier *et al.,* 1976; Azmitia and Segal, 1978; DeVito *et al.,* 1980). Axoplasmic transport in these fibres is abolished by 5,6-dihydroxytryptamine pretreatment, suggesting their serotonergic identity (Halaris *et al.,* 1976). Since the globus pallidus has a significant serotonin content (Palkovits *et al.,* 1974; Table 1) and terminals in this region fluoresce for serotonin (Fuxe, 1965) it is tempting to speculate that this serotonin may originate from the DRN. Preliminary electrophysiological studies suggest that DRN stimulation may evoke inhibition in pallidal neurones when background firing is maintained by the administration of an excitant amino acid. The activity of such cells was depressed by serotonin. On the other hand spontaneously firing neurones were seldom and weakly affected by raphe stimulation or by serotonin (Perkins and Stone, unpublished observations). Clearly further electrophysiological and pharmacological confirmation of this pathway and its transmitter identity are required. Also, the functional role of serotonin in this region of the basal ganglia requires exploration.

Finally orthograde axoplasmic transport of labelled amino acids suggest a sparse projection from the MRN to the subthalamic nucleus (STN) (Bobillier *et al.,* 1976) and HRP studies suggest a projection from the DRN (Rinvik *et al.,* 1979). Since the STN contains significant amounts of serotonin (Palkovits *et al.,* 1974) it is tempting to speculate that this may derive from nerve terminals of the raphe projection. Unfortunately no further evidence concerning this pathway is presently available.

THE FUNCTIONS OF SEROTONIN IN BASAL GANGLIA

Striatum

The role of serotonin in the striatum is unclear, though as mentioned earlier its content in this area is reduced in some neurological diseases and its actions are implicated in certain aspects of behaviour (discussed below).

Microinjections of serotonin into the cat striatum produced little effect besides increased licking and cleaning activity (Cools, 1973) though lesions of the raphe suppressed the actions (head turning and steriotypies) of dopaminergic agents in the striatum (Cools and Janssen, 1974). Similarly in rats, serotonin microinjections into the striatum produced only weak contralateral asymmetric or turning behaviour which could be potentiated by nialamide or the anticholinergic hyoscine (Costall and Naylor, 1974; James and Starr, 1980). Injections of quipazine (a serotonin agonist) produced marked contralateral asymmetry whereas methysergide caused weak ipsiversive turning (Costall and Naylor, 1975; James and Starr, 1980). The behaviour elicited by serotonin was qualitatively similar to that produced by dopaminergic agonists and could be partially diminished by haloperidol and eserine (James and Starr, 1980).

It has been hypothesized that the functions of serotonin and dopamine are closely related (Cools and Janssen, 1974; Costall and Naylor, 1975; Samamin and Garattini, 1976) and that they may act in a cooperative manner on striatal mechanisms (Waddington and Crow, 1979). Indeed reductions in forebrain serotonin concentration enhance the behavioural effects of dopaminergic agents while increased serotonin has the opposite effect (Baldessarini *et al.*, 1975; Milson and Pycock, 1976). Additionally serotonin can activate the release of dopamine from the striatum while serotonin-uptake inhibitors potentiate the haloperidol-induced increase in striatal dopamine metabolites (Besson *et al.*, 1969; Waldmeier and Delini-Stula, 1979; Glowinski *et al.*, 1981).

Dopamine-receptor stimulation may increase cerebral serotonin turnover (Grabowska *et al.*, 1973) but the local administration of dopamine unilaterally into the substantia nigra reduced [^{3}H] serotonin release both in the striatum and substantia nigra (Héry *et al.*, 1980). The exact mechanisms of these interactions are not clear. Neurochemically, however, activation of striatal serotonin receptors (e.g. with quipazine) appears to have similar effects to activation of dopamine receptors: there is an inhibition of acetylcholine turnover which may be blocked by serotonin antagonists (Euvrard *et al.*, 1977; Samanin *et al.*, 1978). Thus, the serotonin innervation in the striatum may serve to inhibit cholinergic neurones (Samanin *et al.*, 1978). On the other hand, lesions of the raphe result in a decrease in acetylcholine synthesis suggesting that the synaptically released transmitter (assumed to be serotonin) would have a net facilitatory effect on cholinergic neurones (Butcher *et al.*, 1976). This is contrary to the expectation from the bulk of electrophysiological data suggesting a predominantly inhibitory role for both serotonin and the raphe-striatal pathway (but see intracellular data discussed earlier). This might also imply that the effect of raphe lesions on striatal acetylcholine synthesis were indirectly mediated.

Lesions of the raphe (DRN) produced relatively few behavioural signs though a transient increase in exploratory behaviour has been noted (Jacobs *et al.*, 1974; Srebro and Lorens, 1975; Geyer *et al.*, 1976; Dray *et al.*, 1978). By contrast MRN lesions commonly produced hyperactivity (Kostowski *et al.*, 1968; Jacobs *et al.*, 1974, 1975; Srebro and Lorens, 1975; Costall *et al.*, 1976a; Geyer *et al.*, 1976; Lorens *et al.*, 1976; Dray *et al.*, 1978) which has been attributed to impaired serotonergic function in mesolimbic areas, especially the hippocampus (Jacobs *et al.*, 1974, 1975; Geyer *et al.*, 1976; but see Lorens *et al.*, 1976). The role of striatal serotonin mechanisms in these behaviours is uncertain as is the involvement of striatal dopaminergic function since neither DRN or MRN lesions modify striatal dopamine turnover or content (Samanin and Garattini, 1976; Costall *et al.*, 1976a, b; Dray *et al.*, 1978). The data suggest however that the DRN and MRN have different functions in behaviour (Srebro and Lorens, 1975). Indeed, in so far as basal ganglia mechanisms are involved, experiments where unilateral raphe lesions have been made may be more informative (see below). However, it has been argued that selective raphe lesions (especially electrolytic) are difficult to make since the DRN and MRN overlap to some extent, project to each other and damage to fibres of passage cannot always be avoided (Srebro and Lorens, 1975; Bobillier *et al.*, 1976; Aghajanian and Wang, 1977; Mosko *et al.*, 1977). In addition selective raphe lesions would be unlikely to clarify the role of serotonin on striatal neurotransmitter systems since the raphe probably contain non-serotonergic projecting neurones and these nuclei may modify striatal activity indirectly through other basal ganglia (substantia nigra) or limbic circuits.

Serotonin receptors in the striatum

It is possible that some of the neurochemical discrepancies discussed earlier may be explained by the fact that striatal serotonin receptors activated by pharmacological agents or by synaptically released serotonin may produce different effects. In fact several types of serotonin receptor have been recognized in striatal tissue. For example, Fillion *et al.* (1979) describe a loss of high affinity [³H] 5-HT binding sites (probably neuronal) with an increase in low affinity sites following kainic acid lesions of the striatum. The glial proliferation observed following these lesions suggested that the low affinity sites were located to glial cells. These cells appear to possess a low-sensitivity 5-HT-stimulated adenylate cyclase which was distinguishable from dopamine-stimulated adenylate cyclase (see Fillion *et al.*, 1980; Nelson *et al.*, 1981). The parallel decrease in high affinity [³H] serotonin binding and loss of high-sensitivity serotonin-stimulated adenylate cyclase activity following striatal kainic acid injections suggested that these serotonin receptors were closely linked (Fillion *et al.*, 1979).

In addition, it is noteworthy that [³H] serotonin binding (high-affinity neuronal sites) in the striatum may be displaced by a number of neuroleptic drugs (Nelson *et al.*, 1981) while these agents also inhibit striatal serotonin stimulated adenylate cyclase activity (Enjalbert *et al.*, 1978). Indeed neuroleptic agents, presumed to be selective antagonists for dopamine receptors have been shown to attenuate the electrophoretic effects of serotonin both in striatum (York, 1972; Davies and Tongroach, 1978) and in the substantia nigra (Dray *et al.*, 1976a). Such interactions may be involved in neuroleptic-induced catalepsy and contribute to the disorders of movement produced in humans during chronic neuroleptic therapy. Thus neuroleptic-induced catalepsy may be reduced by the administration of serotonin-antagonists or by lesions of the raphe nuclei and substantia nigra (Kostowski *et al.*, 1972; Gumulka *et al.*, 1973; Costall *et al.*, 1975; Carter and Pycock, 1978). Moreover, the chronic administration of neuroleptics produces a significant increase in serotonin binding sites in the striatum (Müller and Seeman, 1977).

The substantia nigra

As discussed earlier the raphe-nigral pathway appears to inhibit tonically the activity of dopamine neurones, since lesions of the raphe not only lower the serotonin-content and turnover but increase nigral dopamine turnover (Dray *et al.*, 1978; Giambalvo and Snodgrass, 1978; Nicolaou *et al.*, 1979; Blackburn *et al.*, 1980). In keeping with this hypothesis, unilateral microinjections of serotonin into the rat substantia nigra produced ipsiversive circling which, according to the model pro-posed by Ungerstedt (1971b), was indicative of a unilateral reduction in nigro-striated dopaminergic activity and indeed was accompanied by a decrease in striatal homovanillic acid content. Conversely intra-nigral microinjections of methysergide, or reducing nigral 5-HT content with *para*-chlorophenylalanine or the neurotoxin 5,7-dihydroxytryptamine, produced the opposite behavioural and neurochemical effects (Straughan and James, 1978; Tanner, 1978; Giambalvo and Snodgrass, 1978; James and Starr, 1980). Similarly unilateral lesions placed in the median forebrain bundle or in the raphe caused either no behavioural bias or spontaneous contraversive circling. Contraversive circling could be induced or intensified by the systemic administration of 5-hydroxytryptophan, serotonin agonists, or dopamin-ergic agonists and could be blocked by serotonergic or dopaminergic antagonists (Costall and Naylor, 1974; Jacobs *et al.*, 1977; Costall *et al.*, 1976b; Giambalvo and Snodgrass, 1978; Hodge and Butcher, 1979; Waddington and Crow, 1979; Blackburn *et al.*, 1980). The indirectly acting serotonin agonist, *p*-chloroamphetamine, produced ipsilateral turning behaviour in animals with unilateral neurotoxic lesions of the DRN (Blackburn *et al.*, 1980).

On the other hand, Azmitia and Segal (1978) suggested that similar neurotoxin-induced lesions of the raphe produced spontaneous ipsiversive turning which was only reversed to contraversive turning following 5-hydroxytryptophan adminis-tration. Finally, Nicolaou *et al.* (1979) have reported that differential turning may be produced by selective unilateral lesions of either the DRN or the MRN. Thus DRN-lesioned animals turned contraversively following the administration of a serotonergic agonist while MRN lesioned animals turned ipsiversively. Opposite turning behaviour was produced by a dopaminergic agonist and both dopamine and serotonin stimulated behaviours were blocked by haloperidol. Only DRN-lesions decreased nigral serotonin turnover and increased nigral dopamine turnover whereas MRN-lesions decreased striatal serotonin but increased striatal dopamine turnover. Clearly such findings are at variance with the differential projections of the DRN and MRN outlined earlier.

While changes in dopaminergic neurotransmission appear to be implicated in the behaviours produced by manipulating basal ganglia serotonin systems it is not clear to what extent altered nigral or striated mechanisms are involved. In some respects a nigral mechanism might be favoured by the suggestion (Blackburn *et al.*, 1980) that unilateral lesions of the DRN may produce nigral serotonin-receptor super-sensitivity and that additional ipsilateral neurotoxic lesions of the nigro striatal pathway reverses the contralateral rotation induced by a serotonin agonist into ipsilateral rotation. However, it is still apparent that the discrepancies regarding the neurochemical sequelae of raphe lesions and their correlation with behaviour need to be resolved.

CONCLUSIONS

The serotonergic innervation of some basal ganglia structures has now been confirmed by a number of techniques, though it is only' the projections to the striatum and substantia nigra that have received the greatest attention. These inputs (and serotonin) appear to be inhibitory in both regions although data from electrophysiological studies suggest the possibility of facilitatory actions as well. Interestingly, preliminary intracellular studies show that the raphe inputs to both striatum and nigra produce monosynaptic EPSPs rather than the expected IPSP. Further studies in this area are still required while additional information on the serotonergic innervation of the globus pallidus and subthalamic nucleus is essential.

Evidence from behavioural studies shows that serotonin, like dopamine, produces contraversive asymmetries when administered into the striatum which might be mediated through altered cholinergic or GABA neuronal activity. Though the cooperative nature of striatal serotoninergic and dopaminergic mechanisms is indicated from pharmacological studies, lesions of the raphe whether total or unilateral, suggest that the DRN and MRN serotoninergic projections serve different functional roles. While these roles appear to be linked to changes in dopaminergic neurotransmission (other systems having received little attention) the significance of the different loci involved is unclear. It is probable that the behaviour which is ultimately expressed following raphe lesions would involve complex changes in several basal ganglia regions. Note should be taken of the non-serotonergic projections of the raphe and of the branching projections from the raphe to the striatum and substantia nigra. Limbic areas (e.g. hippocampus, nucleus accumbens, habenula) also receive prominant serotonergic innervation from the same raphe nuclei, and a number of anatomical connections have been described linking limbic structures with the basal ganglia (see Dray, 1980).

Defective serotoninergic neurotransmission has been implicated in a number of neurological disorders, e.g. Parkinson's disease, though a rational regimen of replacement therapy has been difficult to establish. This has tended to undermine the significance of serotonin in the etiology of the diseases. Possibly closer monitoring of relative transmitter deficiencies might select patients who would receive greatest benefit from replacement therapy. The recognition of different types of serotonin receptors in basal ganglia also allow some speculation concerning the etiology of drug-induced motor disorders. Thus, significantly altered serotonin transmission would be involved in the catalepsy or dyskinesias produced by neuroleptic therapy.

SUMMARY

The serotonergic innervation of the striatum and substantia nigra has been confirmed by a number of techniques and originates from the serotonin cell groups of the dorsal and medial raphe nuclei. Electrophysiologically these inputs appear to be inhibitory, as do the actions of serotonin. However, the possibility that serotonin receptors are heterogeneous and that serotonergic transmission may have an important facilitatory component requires further elucidation. Also the non-serotonergic component of the raphe-basal ganglia pathways should be explored. No conclusion can be reached concerning the function of serotonin in the globus pallidus or subthalamic nucleus as electrophysiological and pharmacological evidence is lacking.

Asymmetrical behaviour may be produced by the unilateral manipulation of serotonergic transmission in the striatum or substantia nigra by pharmacological agents. These behaviours may be mediated through altered dopaminergic, cholinergic or GABA-ergic neuronal activity. Unilateral or bilateral lesions of the DRN or MRN suggest that the serotonergic projections from these nuclei may serve different functional roles in the basal ganglia but these are not clearly understood.

Defective serotonergic transmission has been implicated in a number of neurological disorders though the significance of this has been undermined by the difficulty in establishing rational replacement therapy. Also it is possible that altered serotonin transmission is involved in the catalepsy and dyskinesias produced by neuroleptic therapy.

ACKNOWLEDGEMENTS

I would like to thank Drs John Davies, Steven Kitai and Elsevier/North-Holland Biomedical Press for permission to reproduce published figures. I am also grateful to Irene M. Anderson and Mary Wynn for typing this manuscript.

REFERENCES

Aghajanian, G. K. and Bunney, B. S. (1975). Dopaminergic and non-dopaminergic neurons of the substantia nigra: differential responses to putative transmitters, *Exerpta Medica*, **359**, 444–452.

Aghajanian, G. K. and Bloom, F. E. (1967). Localization of tritiated serotonin in rat brain by electron microscopic autoradiography, *J. Pharmacol. Exp. Ther.* **156**, 23–30.

Aghajanian, G. K. and Wang, R. Y. (1977). Habenular and other midbrain raphe afferents demonstrated by a modified retrograde tracing technique, *Brain Res.* **122**, 229–242.

Aghajanian, G. E., Wang, R. Y., and Baraban, J. (1978). Serotonergic and non-serotonergic neurons of the dorsal raphe: reciprocal changes in firing induced by peripheral nerve stimulation, *Brain Res.* **153**, 169–175.

Ashkenazi, R., Holman, R. B., and Vogt, M. (1972). Release of transmitters on stimulation of the nucleus linearis raphe in the cat, *J. Physiol. (Lond.)*, **233**, 255–259.

Azmitia, E. C. and Segal, M. (1978). An autoradiographic analysis of the differential ascending projections of the dorsal and median raphe nuclei in the rat, *J. Comp. Neurol.* **179**, 641–668.

Bak, I. J., Choi, W. B., Hassler, R., Usunoff, K. G., and Wagner, A. (1975). Fine structural synaptic organization of the corpus striatum and substantia nigra in the rat and cat, in *Advances in Neurology*, Vol. 9. (Eds D. B. Calne, T. N. Chase, and A Barbeau), pp. 25–41, Raven Press, New York.

Baldessarini, R. J., Amatruda, T. T., Griffith, F. F., and Gerson, S. (1975). Differential effects of serotonin on turning and stereotypy induced by apomorphine, *Brain Res.* **93**, 158–163.

Bazelon, M., Barnet, A., Lodge, A., and Shelburne, S. S. (1968). The effect of high doses of 5-Hydroxytryptophan on a patient with Trosomy 21. *Brain*, **11**, 397–411.

Bazelon, M., Paine, R. S., Cowie, V. A., Hunt, P., Houck, J. C. and Mahanand, D. (1967). Reversal of hypotonia in infants with Down's Syndrome by administration of 5-Hydroxytryptophan. *Lancet*, **1**, 1130–1133.

Beckstead, R. M., Domesick, V. B., and Nauta, W. J. H. (1979). Efferent connections of the substantia nigra and ventral tegmental area in the rat, *Brain Res.* **175**, 191–217.

Berhnheimer, H. and Hornykiewicz, O. (1973). Brain amines in Huntington's chorea, *Adv. Neurol.* **1**, 525–531.

Besson, M. J., Cheramy, A., Feltz, P., and Glowinski, J. (1969). Release of newly synthesized dopamine from dopamine-containing terminals in the striatum of the rat, *Proc. Nat. Acad. Sci. USA*, **62**, 741–748.

Bevan, P., Bradshaw, C. M., and Szabadi, E. (1975). Effects of desipramine on neuronal responses to dopamine, noradrenaline, 5-hydroxytryptamine and acetylcholine in the caudate nucleus of the rat. *Br. J. Pharmacol.* **54**, 285–293.

Björklund, A., Falck, B., and Stenevi, U. (1971). Classification of monoamine neurones in the rat mesencephalon: distribution of a new monoamine neurone system, *Brain Res.* **32**, 269–285.

Björklund, A., Nobin, A., and Stenevi, U. (1973). The use of neurotoxic dihydroxy-tryptamines as tools for morphological studies and localized lesioning of central indolamine neurons, *Z. Zellforsch.* **145**, 479–501.

Blackburn, T. P., Foster, G. A., Heapy, C. G., and Kemp, J. D. (1980). Unilateral 5,7-dihydroxytryptamine lesions of the dorsal raphe nucleus (DRN) and rat rotational behaviour, *Eur. J. Pharmacol.* **67**, 427–438.

Bobillier, P., Seguin, S., Petitjean, F., Salvert, D., Touret, M., and Jouvet, M. (1976). The raphe nuclei of the cat brain stem: a topographical atlas of their efferent projections as revealed by auto-radiography, *Brain Res.* **113**, 449–486.

Bockaert, J., Premont, J., Glowinski, J., Thierry, A. M., and Tassin, J. P. (1976). Topographical distribution of dopaminergic innervation and of dopaminergic receptors in the rat striatum. II. Distribution and characteristics of dopamine adenylate cyclase-interaction of D-LSD with dopaminergic receptors, *Brain Res.* **107**, 303–315.

Bogdanski, D. F., Weissbach, H., and Udenfriend, S. (1957). The distribution of serotonin, 5-hydroxytryptophan decarboxylase, and monoamine oxidase in the brain, *J. Neurochem.* **1**, 272–278.

Bolam, J. P., Somogyi, P., and Smith, A. D. (1980). Input and output of identified striatonigral medium-sized spiny neurons in the rat, *Neuroscience Lett.* Suppl. **5**, S39.

Bourgoin, S., Enjalbert, A., Adrien, J., Jéry, F., and Hamon, M. (1977). Midbrain raphe lesions in the newborn rat: II. Biochemical alterations in serotoninergic innervation, *Brain Res.* **127**, 111-126.

Bourgoin, S., Soubrie, P., Artaud, F., Reisine, T. D., and Glowinski, J. (1981). Control of 5-HT release in the caudate nucleus and the substantia nigra of the cat. *J. de Physiologie.* (In press.)

Broch, O. J. and Marsden, C. A. (1972). Regional distribution of monoamines in the corpus striatum of the rat, *Brain Res.* **38**, 425-428.

Brody, J. A., Chase, T. N., and Gordon, E. K. (1970). Depressed monoamine catabolite levels in cerebrospinal fluid of patients with Parkinsonism Dementia of Gaum. *New Eng. J. Med.* **282**, 947-950.

Brownstein, M. J., Palkovits, M., Saavedra, J. M., and Kizer, J. S. (1975). Tryptophan hydroxylase in the rat brain, *Brain Res.* **97**, 163-166.

Bunney, B. S. and Aghajanian, G. K. (1976). The precise localization of nigral afferents in the rat as determined by a retrograde tracing technique, *Brain Res.* **117**, 423-435.

Butcher, L. L., Engel, J., and Fuxe, J. (1970). L-Dopa induced changes in central monoamine neurons after peripheral decarboxylase inhibition, *J. Pharm. Pharmacol.* **22**, 313-316.

Butcher, S. G., Butcher, L. L., and Cho, A. K. (1976). Modulation of neostriatal acetylcholine in the rat by dopamine and 5-hydroxytryptamine afferents, *Life Sci.* **18**, 733-744.

Calas, A., Besson, M. J., Gaughy, C., Alonso, G., Glowinski, J., and Cheramy, A. (1976). Radioautographic study of *in vivo* incorporation of [³H] monoamines in the cat caudate nucleus: identification of serotoninergic fibres, *Brain Res.* **118**, 1-13.

Carter, C. J. and Pycock, C. J. (1978). A study of the sites of interaction between dopamine and 5-hydroxytryptamine for the production of fluphenazine-induced catalepsy, *Naunyn-Schmeideberg's Arch. Pharmacol.* **304**, 135-139.

Chadwick, D., Harris, R., Jenner, P., Reynolds, E. H., and Marsden, C. D. (1975). Manipulation of brain serotonin in the treatment of myoclonus, *Lancet*, **2**, 434-435.

Chase, T. N., Ng, L. K. Y., and Watanabe, A. M. (1972). Parkinson's disease. Modification by 5-hydroxytryptophan, *Neurology (Minneap.)*, **22**, 479-484.

Collingridge, G. L. and Davies, J. (1980). Responses of nigral compacta and reticulata neurones to putative inhibitory neurotransmitters and striatal stimulation, *J. Physiol. (Lond.)*, **301**, 41-42.

Conrad, L. C. A., Leonard, C. M., and Pfaff, D. W. (1974). Connections of the median and dorsal raphe nuclei in the rat: an autoradiographic and degeneration study, *J. Comp. Neurol.* **156**, 179-205.

Cools, A. R. (1973). Chemical and electrical stimulation of the caudate nucleus in freely moving cats: the role of dopamine. *Brain Res.* **58**, 437-451.

Cools, A. R. and Janssen, H. -J. (1974). The nucleus linearis intermedius raphe and behaviour evoked by direct and indirect stimulation of dopamine-sensitive sites within the caudate nucleus of cats, *Eur. J. Pharmacol.* **28**, 266-275.

Costall, B., Fortune, D. H., Naylor, R. J., Marsden, C. D., and Pycock, C. (1975). Serotonergic involvement with neuroleptic catalepsy, *Neuropharmacology*, **14**, 859-868.

Costall, B. and Naylor, R. J. (1974). Specific asymmetric behaviour induced by the direct chemical stimulation of neostriatal dopaminergic mechanisms, *Naunyn-Schmeideberg's Arch. Pharmacol.* **285**, 83-98.

Costall, B. and Naylor, R. J. (1975). The role of raphe and extra-pyramidal nuclei in the stereotyped and circling responses to quipazine, *J. Pharm. Pharmacol.* **27**, 368–371.

Costall, B., Naylor, R. J., Marsden, C. D., and Pycock, C. J. (1976a). Serotoninergic modulation of the dopamine response from the nucelus accumbens, *J. Pharm. Pharmacol.* **28**, 523–526.

Costall, B., Naylor, R. J., Marsden, C. D., and Pycock, C. J. (1976b). Circling behaviour produced by asymmetric medial raphe nuclei lesions in rats, *J. Pharm. Pharmacol.* **28**, 248–249.

Crossman, A. R., Walker, R. J., and Woodruff, G. N. (1974). Pharmacological studies on single neurones in the substantia nigra of the rat, *Brit. J. Pharmacol.* **51**, 137–138P.

Curzon, G. (1972). Brain Amine Metabolism in some neurological and psychiatric disorders, in *Biochemical Aspects of Nervous Diseases* (Ed J. N. Cumings), pp. 151–212, Plenum Press, London.

Curzon, G. (1977). The biochemistry of basal ganglia and Parkinson's disease, *Postgraduate Med. J.* **53**, 719–725.

Curzon, G. (1978). Serotonin and neurological disease, in *Serotonin in Health and Disease*, Vol. III. *The Central Nervous System* (Ed W. B. Essman), pp. 403–443, Spectrum, New York.

Dahlström, A. and Fuxe, K. (1964). Evidence for the existence of monoamine-containing neurons in the central nervous system. I. Demonstration of mono-amines in the cell bodies of brain stem neurons, *Acta Physiol. Scand.* Supp. 232, **62**, 1–55.

Davies, J. and Tongroach, P. (1978). Neuropharmacological studies on the nigro-striatal and raphe-striatal system in the rat, *Eur. J. Pharmacol.* **51**, 91–100.

Davies, J. and Tongroach, P. (1979). Tetanus toxin and synaptic inhibition in the substantia nigra and striatum of the rat, *J. Physiol. (Lond.)*, **290**, 23–36.

Descarriers, L., Beaudet, A., Watkins, K. C., and Garcia, S. (1979). The serotonin neurons in nucleus raphe dorsalis of adult rat, *Anat. Rec.* **193**, 520.

DeVito, J. L., Anderson, M. E., and Walsh, K. E. (1980). A horseradish peroxidase study of afferent connections of the globus pallidus in macaca mulatta. *Exp. Brain Res.* **38**, 65–73.

Dray, A. (1979). The striatum and substantia nigra: a commentary on their relation-ships, *Neuroscience*, **4**, 1407–1439.

Dray, A. (1980). The physiology and pharmacology of mammalian basal ganglia, *Prog. Neurobiol.* **14**, 221–335.

Dray, A., Davies, J., Oakley, N. R., Tongroach, P., and Vellucci, S. (1978). The dorsal and medial raphe projections to the substantia nigra in the rat. Electro-physiological, biochemical and behavioural observations, *Brain Res.* **151**, 431–442.

Dray, A., and Oakley, N. R. (1977). Methiothepin and a 5-HT pathway to rat substantia nigra, *Experientia*, **33**, 1198–1199.

Dray. A., Gonye, T. J., and Oakley, N. R. (1976a). Effects of α-flupenthixol on dopamine and 5-hydroxytryptamine responses of substantia nigra neurones, *Neuropharmacology*, **15**, 793–796.

Dray, A., Gonye, T. J., Oakley, N. R., and Tanner, T. (1976b). Evidence for the existence of a raphe projection to the substantia nigra in rat, *Brain Res.* **113**, 45–57.

Dubowitz, V. and Rogers, K. J. (1969). 5-Hydroxyindoles in the cerebrospinal fluid of infants with Down's Syndrome and muscle hypotonia, *Dev. Med. Child. Neurol.* **11**, 730–734.

Enjalbert, A., Hamon, M., Bourgoin, S., and Bockaert, J. (1978). Postsynaptic serotonin-sensitive adenylate cyclase in the central nervous system. II. Comparison with dopamine- and isoproteronol-sensitive adenylate cyclase in rat brain, *Mol. Pharmacol.* **14**, 11–23.

Enna, S. J., Bennett, J. P., Bylund, D. B., Snyder, S. H., Bird, E. D., and Iversen, L. L. (1976). Alterations of brain neurotransmitter receptor binding in Huntington's chorea. *Brain Res.* **116**, 531–537.

Euvrard, C., Javoy, F., Herbet, A., and Glowinski, J. (1977). Effects of quipazine, a serotonin-like drug, on striatal cholinergic interneurons, *Eur. J. Pharmacol.* **41**, 281–289.

Fahn, S., Libsch, L. R., and Cutler, R. W. (1971). Monoamines in the human neo-striatum: topographic distribution in normal and in Parkinson's disease and their role in akinesia, rigidity, chorea and tremor, *J. Neurol. Sci.* **14**, 427–455.

Fibiger, H. C. and Miller, J. J. (1977). An anatomical and electrophysiological investigation of the serotonergic projection from the dorsal raphe nucleus to the substantia nigra in the rat, *Neuroscience*, **2**, 975–987.

Fillion, G., Beaudoin, D., Rousselle, J. C., Deniau, J. M., Fillion, M. P., Dray, F., and Jacob, J. (1979). Decrease of [^{3}H]5-HT high affinity binding and 5-HT adenylate cyclase activation after kainic acid lesion in rat brain striatum, *J. Neurochem.* **33**, 567–570.

Fillion, G. Beaudoin, D., Rousselle, J. C., and Jacob, J. (1980). [^{3}H]5-HT binding sites and 5HT-sensitive adenylate cyclase in glial cell membrane fraction, *Brain Res.* **198**, 361–374.

Fonnum, F., Gottesfeld, Z., and Grofova, I. (1978). Distribution of glutamate decarboxylase, choline acetyl-transferase and aromatic amino acid decarboxylase in the basal ganglia of normal and operated rats. Evidence for striatopallidal, striatoentopeduncular and striatonigral GABAergic fibres, *Brain Res.* **143**, 125–138.

Fuxe, K. (1965). Evidence for the existence of monoamines in the central nervous system. IV. Distribution of monoamine nerve terminals in the central nervous system, *Acta Physiol. Scand. Suppl.* **247**, 37–85.

Gamrani, H., Calas, A., Belin, M. F., Aguera, M., and Pujol, J. F. (1979). High resolution radiographic identification of [^{3}H]GABA labelled neurons in rat nucleus raphe dorsalis, *Neuroscience Lett.* **15**, 43–48.

Geyer, M. A., Puerto, A., Dawsey, W. J., Knapp, S., Bullard, W. P., and Mandell, A. J. (1976). Histologic and enzymatic studies of the mesolimbic and meso-striatal serotoninergic pathways, *Brain Res.* **106**, 241–256.

Giambalvo, C. T. and Snodgrass, S. R. (1978). Effect of *p*-chloroamphetamine and 5,7,-dihydroxytryptamine on rotation and dopamine turnover, *Brain Res.* **149**, 453–467.

Glowinski, J., Cheramy, A., and Giorguieff-Chesselet, M. -F. (1981). Presynaptic control of dopamine release from terminals and dendrites of nigrostriatal dopaminergic neurones, in *Psychopharmacology and Biochemistry of Neurotransmitter Receptors* (Ed H. I. Yamamura, R. W. Olsen and E. Usdin). *Developments in Neuroscience* Vol II, 483–498.

Grabowska, M., Antikiewicz, L., Maj, J., and Michaluk, J. (1973). Apomorphine and central serotonin neurones, *Pol. J. Pharmac. Pharm.* **25**, 29–39.

Grzanna, R. and Molliver, M. E. (1980). The locus coeruleus in the rat: an immuno-histochemical delineation, *Neuroscience*, **5**, 21–41.

Guldberg, H. C., Turner, J. W., Hanieh, A., Ashcroft, G. W., Crawford, T. B. B., Perry, W. L. M., and Gillingham, F. J. (1967). On the occurrence of homovanillic acid and 5-hydroxyindole-3-acetic acid in the ventricular C.S.F. of patients suffering from Parkinsonism, *Confina Neurol.* **29**, 73–77.

Gumpert, E. J. W., Sharpe, D. M., and Curzon, G. (1973). Amine metabolites in cerebrospinal fluid in Parkinson's disease and the response to L-dopa, *J. Neurol. Sci.* **19**, 1-12.

Gumulka, W., Kostowski, W. and Czlonkowski, A. (1973). Role of 5-HT in the action of some drugs affecting extrapyramidal system, *Pharmacology*, **10**, 363-372.

Haigler, H. J. and Aghajanian, G. K. (1974). Lysergic acid diethylamide and serotonin: a comparison of effects on serotoninergic neurons and neurons receiving a serotonergic input. *J. Pharmacol. exp. Ther.* **186**, 688-699.

Hajdu, F., Hassler, R., and Bak, I. J. (1973). Electron microscopic study of the substantia nigra and the strio-nigral projection in the cat, *Z. Zellforsch.* **146**, 207-221.

Halaris, A. E., Jones, B. E., and Moore, R. Y. (1976). Axonal transport in serotonin neurons of the midbrain raphe, *Brain Res.* **107**, 555-574.

Hattori, T., Singh, V. K., McGeer, E. G., and McGeer, P. L. (1976). Immunohistochemical localization of choline acetyltransferase containing neostriatal neurones and their relationship with dopaminergic synapses, *Brain Res.* **102**, 164-173.

Héry, F., Simonnet, G., Bourgoin, S., Soubrie, P., Artaud, F., Hamon, M., and Glowinski, J. (1979). Effect of nerve activity on the *in vivo* release of ^{3}H-serotonin continuously formed from L-^{3}H-tryptophan in the caudate nucleus of the cat, *Brain Res.* **169**, 317-334.

Héry, F., Soubrie, P., Bourgoin, S., Montasruc, J. L., Artaud, F., and Glowinski, J. (1980). Dopamine released from dendrites in the substantia nigra controls the nigral and striatal release of serotonin, *Brain Res.* **193**, 143-151.

Herz, A. and Zieglgansberger, W. (1968). The influence of microelectrophoretically applied biogenic amines, cholinomimetics and procaine on synaptic excitation in the corpus striatum, *Int. J. Neuropharmacol.* **7**, 221-230.

Hodge, G. K. and Butcher. L. L. (1979). Role of pars compacta of the substantia nigra in circling behaviour, *Pharmacol. Biochem. Behav.* **10**, 695-709.

Hökfelt, T., Ljungdahl, A., Steinbusch, H., Verhofstad, A., Nilsson, G., Brodin, E., Pernow, B., and Goldstein, M. (1978). Immumohistochemical evidence of substance P-like immunoreactivity in some 5-hydroxytryptamine-containing neurons in the rat central nervous system, *Neuroscience*, **3**, 517-538.

Hökfelt, T., Terenius, L., Kuypers, H. G. J., and Dann, O. (1979). Evidence for enkephalin-immumoreactive neurons in the medulla oblongata projecting to the spinal cord, *Neuroscience Lett.* **14**, 55-60.

Holman, R. B. and Vogt, M. (1972). Release of 5-hydroxytryptamine from caudate nucleus and septum, *J. Physiol. (Lond.),* **223**, 243-254.

Jacobs, B. L., Foote, S. L., and Bloom, F. E. (1978). Differential projections of neurons within the dorsal raphe nucleus of the rat: a horseradish peroxidase (HRP) study. *Brain Res.* **147**, 149-153.

Jacobs, B. L., Simon, S. M., Ruimy, D. D., and Trulson, M. E. (1977). A quantitative rotational model for studying serotoninergic function in the rat, *Brain Res.* **124**, 271-281.

Jacobs, B. L., Trimbach, C., Eubanks, E. E., and Trulson, M. (1975). Hippocampal mediation of raphe lesion- and PCPA-induced hyperactivity in the rat, *Brain Res.* **94**, 253-261.

Jacobs, B. L., Wise, W. D., and Taylor, K. M. (1974). Differential behavioural and neurochemical effects following lesions of the dorsal or medial raphe nuclei in rats, *Brain Res.* **79**, 353-361.

James, T. A. and Starr, M. S. (1980). Rotational behaviour elicited by 5-HT in the rat: evidence for an inhibitory role of 5-HT in the substantia nigra and corpus striatum, *J. Pharm. Pharmac.* **32**, 196-200.

Johansson, B. and Roos, B. H. (1974). 5-Hydroxindoleacetic acid and homovanillic acid in cerebrospinal fluid of patients with neurological disease, *Eur. Neurol.* **11**, 37–45.

Kanazawa, I., Marshall, G. R., and Kelly, J. S. (1976). Afferents to the rat substantia nigra studies with horseradish peroxidase, with special reference to fibres from the subthalamic nucleus, *Brain Res.* **115**, 485–491.

Karabelas, A. B. and Purpura, D. P. (1979). Functional properties of dorsal raphe-substantia nigra projections in the cat, *Neurosci. Abst.* **5**, 73.

Kocsis, J. D., Sugimori, M., and Kitai, S. T. (1977). Convergence of excitatory synaptic inputs to caudate spiny neurons, *Brain Res.* **124**, 403–413.

Kostowski, W., Giacolone, E., Garattini, S., and Valzelli, L. (1968). Studies on behavioural and biochemical changes in rats after lesions of midbrain raphe, *Eur. J. Pharmacol.* **4**, 371–376.

Kostowski, W., Gumulaka, W., and Czlonkowski, A. (1972). Reduced catalepto-genic effects of some neuroleptics in rats with lesioned midbrain raphe and treated with *p*-chlorophenylalanine, *Brain Res.* **48**, 443–446.

Kuhar, M. J., Aghajanian, G. K., and Roth, R. H. (1972). Tryptophan hydroxy-lase activity and synaptosomal uptake of serotonin in discrete brain regions after midbrain raphe lesions: correlations with serotonin levels and histochemical fluorescence, *Brain Res.* **44**, 165–176.

Lal, S., Hoyte, K., Kiely, M. E., Sourkes, T. L., Baxter, D. W., Missala, K., and Andermann, F. (1979). Neuropharmacological investigation and treatment of spasmodic torticollis, in *Advances in Neurology,,* Vol. 24 (Ed L. J. Poirier, T. L. Sourkes and P. J. Bedard), pp. 335–351, Raven Press, New York.

Lloyd, K. G. (1977). Neurochemistry of the substantia nigra (SN) in Parkinson's disease, *Int. Soc. for Neurochemistry, Copenhagen*, pp. 252.

Lloyd, K. G. and Hornykiewicz, O. (1974). Dopamine and other monoamines in the basal ganglia: relation to brain dysfunctions, in *Frontiers in Neurology and Neuroscience Research* (Ed P. Seeman and G. M. Brown), pp. 26–35, Neuro-science Institute, University of Toronto, Toronto.

Lorens, S. A. and Guldberg, H. C. (1974). Regional 5-hydroxytryptamine following selective midbrain raphe lesions in the rat, *Brain Res.* **78**, 45–56.

Lorens, S. A., Guldberg, H. C., Hole, K., Koehler, C., and Srebro, B. (1976). Activity, avoidance learning and regional 5-hydroxytryptamine following intra-brain stem 5,7-dihydroxytryptamine and electrolytic midbrain raphe lesions in the rat, *Brain Res.* **108**, 97–113.

Marsden, C. D. and Guldberg, H. C. (1973). The role of monoamines in rotation induced or potentiated by amphetamine after nigral, raphe and mesencephalic reticular lesions in the rat brain, *Neuropharmacology,* **12**, 195–211.

McGeer, P. L. and McGeer, E. G. (1975). Evidence for glutamic acid decarboxylase-containing interneurons in the neostriatum, *Brain Res.* **91**, 331–335.

Melamed, E., Hefti, F., Liebman, J., Schlosberg, A. J., and Wurtman, R. J. (1980). Serotonergic neurones are not involved in action of L-dopa in Parkinson's disease, *Nature (Lond.),* **283**, 772–774.

Miller, J. J., Richardson, T. L., Fibiger, H. C., and McLennan, H. (1975). Anatom-ical and electrophysiological identification of a projection from the mesencephalic raphe to the caudate putamen in the rat, *Brain Res.* **97**, 133–138.

Milson, J. A. and Pycock, C. J. (1976). Effects of drugs acting on cerebral 5-dihy-droxytryptamine mechanisms on dopamine-dependent turning behaviour in mice, *Br. J. Pharmacol.* **56**, 77–85.

Moore, R. Y. and Halaris, A. E. (1975). Hippocampal innervation by serotonin neurons of the midbrain raphe in the rat, *J. Comp. Neurol.* **164**, 171–184.

Moore, R. Y., Halaris, A. E., and Jones, B. E. (1978). Serotonin neurons of the midbrain raphe: ascending projections, *J. Comp. Neurol.* **180**, 417–438.

Mori, K., Fugita, Y., Shimabukuro, H., Ito, M., and Handa, H. (1975). Some considerations for the treatment of spasmodic torticollis. Clinical and experimental studies, *Confin. Neurol.* **37**, 265–269.

Mosko, S. S., Haubrich, D., and Jacobs, B. L. (1977). Serotonergic afferents to the dorsal raphe nucleus: evidence from HRP and synaptosomal uptake studies, *Brain Res.* **119**, 269–290.

Müller, P. and Seeman, P. (1977). Brain neurotransmitter receptors after long-term haloperidol: dopamine, acetylcholine, serotonin, α-noradrenergic and naloxone receptors, *Life Sci.* **21**, 1751–1758.

Nagatsu, I., Inagaki, S., Kondo, Y., Karasawa, N., and Nagatsu, T. (1979). Immuno-fluorescent studies on the localization of tyrosine hydroxylase and dopamine-β-hydroxylase in the mes-, di-, and telencephalon of the rat using unperfused fresh frozen sections, *Acta Histochem. Cytochem,* **12**, 20–37.

Nanopoulos, D., Belin, M. -F., Maitre, M., and Pujol, J. F. (1980). Immunocyto-chimie de la glutamate décarboxylase mise en évidence d'éléments neuronanaux GABA-ergiques dans le noyau. Raphe Dorsalis du Rat. *C.R. Acad. Sci. Paris.* **290**, 1153–1156.

Nelson, D. L., Pedigo, N. W., and Yamamura, H. I. (1981). Multiple types of serotonin receptors, in *Psychopharmacology and Biochemistry of Neurotransmitter Receptors* (Eds H. I. Yamamura, R. W. Olsen, and E. Usdin), Elsevier-North Holland, New York. (In press.)

Ng. L. K. Y., Chase, T. N., Colburn, R. W., and Kopin, I. J. (1970). L-dopa induced release of cerebral monoamines. *Science,* **170**, 76–77.

Ng. L. K. Y., Chase, T. N., Colburn, R. W., and Kopin, I. J. (1972a). L-dopa in Parkinsonism, *Neurology,* **22**, 688–696.

Ng. L. K. Y., Colburn, R. W., and Kopin, I. J. (1972b). Effects of L-dopa on accumulation and efflux of monoamines in particles of rat brain homogenates, *J. Pharmacol. Exp. Ther.* **183**, 316–325.

Nicolaou, N. M., Garcia-Munoz, M., Arbuthnott, G. W., and Eccleston, D. (1979). Interactions between serotonergic and dopaminergic systems in rat brain demonstrated by small unilateral lesions of the raphe nuclei, *Eur. J. Pharmacol.* **57**, 295–305.

Nieuwenhuys, R. (1977). Aspects of the morphology of the striatum, in *Psychobiology of the Striatum* (Eds A. R. Cools, A. H. M. Lohman, and J. H. L. Vanden Bercken), pp. 1–19, Elsevier-North Holland Biomedical Press, Amsterdam.

Ochi, J. and Shimizu, K. (1978). Occurrence of dopamine-containing neurons in the midbrain raphe nuclei of the rat, *Neuroscience Lett.* **8**, 317–320.

Olpe, H. R. and Koella, W. P. (1977). The response of striatal cells upon stimulation of the dorsal and median raphe nuclei, *Brain Res.* **122**, 357–360.

Olsen, L. and Seiger, A. (1972). Early prenatal ontogeny of central monoamine neurons in the rat: fluorescence histochemical observations, *Z. Anat. Entwickl. Gesch.* **137**, 301–316.

Palkovits, M., Brownstein, M., and Saavedra, J. M. (1974). Serotonin content of the brain stem nuclei in the rat, *Brain Res.* **80**, 237–249.

Palkovits, M., Saavedra, J. M., Jacobwitz, D. M., Kizer, J. S., Zaborszky, L., and Brownstein, M. J. (1977). Serotonergic innervation of the forebrain: effect of lesions of serotonin and tryptophan hydroxylase levels, *Brain Res.* **130**, 121–134.

Parizek, J., Hassler, R., and Bak, I. J. (1971). Light and electron microscopic auto-radiography of rat after intraventricular administration of tritium labelled norepinephrine, dopamine, serotonin and the precursors, *Z. Zellforsch.* **115**, 137–148.

Park, M. R., Gonzales-Vegas, J. A., Lighthall, J. W., and Kitai, S. T. (1980). Excitatory postsynaptic potentials recorded in neurons of the rat caudate-putamen from stimulation of the dorsal raphe nucleus are serotonergic, *Society for Neuroscience Abstracts.* **6**, 187.

Pasquier, D. A., Kemper, T. L., Forbes, W. B., and Morgane, R. J. (1977). Dorsal raphe, substantia nigra and locus coeruleus interconnections with each other and the neostriatum, *Brain Res. Bull.* **2**, 323-339.

Pickel, V. M., Joh, T. H., and Reis, D. J. (1975). Immunocytochemical demonstration of a serotonergic innervation of catecholamine neurons in locus coeruleus and substantia nigra, *Society for Neuroscience Abs.* **1**, 320.

Pierce, E. T., Foote, W. E., and Hobson, J. A. (1976). The efferent connection of the nucleus raphe dorsalis, *Brain Res.* **107**, 137-144.

Poirier, L. J., Bedard, P., Boucher, R., Bouvier, G., Larochelle, L., Olivier, A., and Singh, P. (1969). The origin of different striato and thalamopetal neurochemical pathways and their relationship to motor activity, in *Third Symposium on Parkinson's Disease* (Eds F. J. Gillingham and I. M. L. Donaldson), pp. 60-66. Livingstone, London.

Preston, R. J., Bishop, G. A., and Kitai, S. T. (1980). Medium spiny neuron projection from the rat striatum: an intracellular horseradish peroxidase study, *Brain Res.* **183**, 253-263.

Reisine, T. D., Fields, J. Z., Yamamura, H. I., Bird, E. D., Spokes, E., Schreiner, P. S., and Enna, S. J. (1977). Neurotransmitter receptor alterations in Parkinson's disease. *Life Sci.* **21**, 335-344.

Reubi, J. -C. and Emson, P. C. (1978). Release and distribution of endogenous 5-HT in rat substantia nigra, *Brain Res.* **139**, 164-168.

Reubi, J. C., Emson, P. C., Jessell, T. M., and Iversen, L. L. (1978). Effects of GABA, dopamine, and substance P on the release of newly synthesized ^{3}H-5-hydroxytryptamine from rat substantia nigra *in vitro*, *Naunyn-Schmiedeberg's Arch. Pharmacol.* **304**, 271-275.

Rinvik, E., Grofova, I., Hammond, C., Feger, J., and Deniau, J. M. (1979). A study of the afferent connections to the subthalamic nucleus in the monkey and the cat using the HRP technique, in *Advances in Neurology*. Vol. 24 (Eds L. J. Poirier, T. L. Sourkes, and P. J. Bedard). pp. 53-70, Raven Press, New York.

Royce, G. J. (1978). Cells of origin of subcortical afferents to the caudate nucleus: a horseradish peroxidase study in the cat, *Brain Res.* **153**, 465-475.

Sakai, K., Salvert, D., Touret, M., and Jouvet, M. (1978). Afferent connections of the nucleus raphe dorsalis in the cat as visualized by the horseradish peroxidase technique, *Brain Res.* **137**, 11-35.

Samanin, R. and Garattini, S. (1976). The serotonergic system in the brain and its possible functional connections with other aminergic systems, *Life Sci.* **17**, 1201-1210.

Samanin, R., Quattrone, A., Peri, G., Ladinsky, H., and Consolo, S. (1978). Evidence of an interaction between serotoninergic and cholinergic neurons in the corpus striatum and hippocampus of the rat brain, *Brain Res.* **151**, 73-82.

Sano, I. and Taniguchi, K. (1972). L-5-Hydroxytryptophan (L-5-HTP)-Therapie des Morbus Parkinson. *Munch. Med. Wochenschr.* **114**, 1717-1719.

Seiger, A. and Olsen, L. (1973). Late prenatal otogeny of central monoamine neurons in the rat: fluorescence histochemical observations, *Z. Anat. Entwickl. Gesch.* **40**, 281-318.

Siggins, G. R. (1978). Electrophysiological role of dopamine in striatum: excitatory or inhibitory, in *Psychopharmacology: A Generation of Progress* (Eds M. A. Lipton, A. DiMascio and K. F. Killman), pp. 143-157, Raven Press, New York.

Sims, K. B., Hoffman, D. L., Said, S. I. and Zimmerman, E. A. (1980). Vasoactive intestinal polypeptide (VIP) in mouse and rat brain: an immunocytochemical study. *Brain Res.* **186**, 165–183.

Somogyi, P. and Smith, A. D. (1979). Projections of neostriatal spiny neurons to the substantia nigra. Application of a combined Golgi-staining and horseradish peroxidase transport procedure at both light and electron microscopic levels, *Brain Res.* **178**, 3–15.

Srebro, B. and Lorens, S. A. (1975). Behavioural effects of selective midbrain raphe lesions in the rat, *Brain Res.* **89**, 303–325.

Steinbusch, H. W. M., Nieuwenhuys, R., Verhofstad, A. A. F., and Van Der Kooy, D. (1981). The nucleus raphe dorsalis of the rat and its projection upon the caudatoputamen. *J. de Physiologie.* **77**, 157–174.

Steinbusch, H. W. M., Van Der Kooy, D., Verhofstad, A. A. F. and Pellegrino, A. (1980). Serotonergic and non-serotonergic projections from the nucleus raphe dorsalis to the caudate-putamen complex in the rat, studied by a combined immunofluorescence and fluorescent retrograde axonal labelling technique, *Neuroscience Letts.* **19**, 137–142.

Stern, W. C., Johnson, A., Bronziono, J. D., and Morgane, P. J. (1979). Influence of electrical stimulation of the substantia nigra on spontaneous activity of raphe neurons in the anesthetized rat, *Brain Res. Bull.* **4**, 561–565.

Straughan, D. W. and James, T. A. (1978). Microphysiological and pharmacological studies on transmitters in the substantia nigra, in *Advances in Pharmacology and Therapeutics*, Vol. 2, *Neurotransmitters* (Ed P. Simon), pp. 87–96, Pergamon Press, Oxford.

Streit, P., Knecht, E., and Cuenod, M. (1979). Transmitter-specific retrograde labelling in the striato-nigral and raphe-nigral pathways, *Science*, 205, 306–308.

Szabo, J. (1980). Organisation of the ascending striatal afferents in monkey, *J. Comp. Neurol.* **189**, 307–321.

Tanner. T. (1978). Circling behavior in the rat following unilateral injections of *p*-chlorophenylalanine and ethanolamine-*O* sulphate into the substantia nigra, *J. Pharm. Pharmacol.* **30**, 158–161.

Tassin, J. P., Cheramy, A., Blanc, G., Thierry, A. M., and Glowinski, J. (1976). Topographical distribution of dopaminergic innervation and of dopaminergic receptors in the rat striatum. 1. Microestimation of [^{3}H] dopamine uptake and dopamine content in microdiscs, *Brain Res.* **107**, 291–301.

Uhl, G. R., Goodman, R. R., Kuhar, M. J., Childers, S. R., and Snyder, S. H. (1979). Immunohistochemical mapping of enkephalin containing cell bodies, fibres and nerve terminals in the brain stem of the rat, *Brain Res.* **166**, 75–94.

Ungerstedt, U. (1971a). Stereotaxic mapping of the monoamine pathways in the rat brain, *Acta Physiol. Scand. Suppl.* **367**, 1–48.

Ungerstedt, U. (1971b). Postsynaptic supersensitivity after 6-hydroxydopamine induced degeneration of the nigro-striatal dopamine system, *Acta Physiol. Scand Suppl.* **367**, 69–93.

Usunoff, K. G., Hassler, R., Wagner, A., and Bak, I. J. (1974). The efferent connections of the head of the caudate nucleus of the cat: an experimental morphological study with special reference to a projection to the raphe nuclei, *Brain Res.* **74**, 143–148.

Van de Kar, L. D. and Lorens, S. A. (1979). Differential serotonergic innervation of individual hypothalamic nuclei and other forebrain regions by the dorsal and median midbrain raphe nuclei, *Brain Res.* **162**, 45–54.

Van der Kooy, D. (1979). The organization of the thalamic, nigral and raphe cells projecting to the medial versus lateral caudate-putamen in rat. A fluorescent retrograde double labeling study, *Brain Res.* **169**, 381–387.

Van der Kooy, D. and Hattori, T. (1980a). Dorsal raphe cells with collateral projections to the caudate-putamen and substantia nigra: a fluorescence retrograde double labeling study in the rat, *Brain Res.* **186**, 1–7.

Van der Kooy, D. and Hattori, T. (1980b). Bilaterally situated dorsal raphe cell bodies have only unilateral forebrain projections in rat, *Brain Res.* **192**, 550–554.

Van der Kooy, D. and Kuypers, H. G. J. M. (1979). Retrograde fluorescent double labeling: axon collaterals in the ascending raphe and nigral projections, *Science*, **204**, 873–875.

Van der Maelen, C. P., Bonduki, A. C., and Kitai, S. T. (1979). Excitation of caudate-putamen neurons following stimulation of the dorsal raphe nucleus in the rat, *Brain Res.* **175**, 356–361.

Van Woert, M. H. and Sethy, V. H. (1975). Therapy of intention myoclonus with L-5-hydroxytryptophan and a peripheral decarboxylase inhibitor, MK 486. *Neurology (Minneapolis)*, **25**, 135–140.

Veening, J. G., Cornelissen, F. M., and Lieven, P. A. J. M. (1980). The topical organization of the afferents to the caudatoputamen of the rat. A horseradish peroxidase study, *Neuroscience*, **5**, 1253–1268.

Waddington, J. L. and Crow, T. J. (1979). Rotational responses to serotonergic and dopaminergic agonists after unilateral dihydroxytryptamine lesions of the medial forebrain bundle, co-operative interactions of serotonin and dopamine in neostriatum, *Life Sci.* **25**, 1307–1314.

Waldmeier, P. C. and Delini-Stula, A. A. (1979). Serotonin-dopamine interactions in the nigrostriatal system, *Eur. J. Pharmacol.* **55**, 363–373.

Webster, K. E. (1975). Structure and function of the basal ganglia—a non-clinical view, *Proc. Roy. Soc. Med.* **68**, 202–310.

Werman, R. (1966). Criteria for identification of a central nervous system transmitter, *Comp. Biochem. Physiol.* **18**, 745–766.

Yamamura, H. I., Enna, S. J., and Kuhar, M. J. (1978). *Neurotransmitter Receptor Binding* (Eds H. I. Yamamura, S. J. Enna, and M. J. Kuhar), Raven Press, New York.

York, D. H. (1972). Dopamine receptor blockade—a central action of chlorpromazine on striatal neurones, *Brain Res.* **37**, 91–99.

Biology of Serotonergic Transmission
Edited by N. N. Osborne
© 1982, John Wiley & Sons Ltd.

Chapter 15

Enteric Serotonergic Neurones

M. D. GERSHON

Department of Anatomy and Cell Biology, Columbia University,
College of Physicians and Surgeons, New York, NY 10032, USA

INTRODUCTION

It has become apparent that serotonin (5-HT) acts as a neurotransmitter both in vertebrates and invertebrates (see Gershon, 1977, for references). With respect to mammals, most of the attention given to this function of the amine has focused on the serotonergic neurones of the central nervous system. This attention is understandable, as it relates both to the historic fact that knowledge of central serotonergic neurones preceded that of peripheral serotonergic neurones, and the natural primacy investigators tend to place on understanding the brain. Peripheral serotonergic neurones, however, are more accessible than their central counterparts and they do resemble central serotonergic neurones in many ways; consequently, the peripheral serotonergic neurones possess certain advantages that facilitate study of the cellular biology of serotonergic neurones as a class and thus they are, or ought to be, of interest as objects of investigation.

In addition to their value as models of the serotonergic neuronal set, peripheral serotonergic neurones are intriguing because of their presence in the enteric nervous system (ENS; see Gershon, 1979a). The ENS is a complex, autonomous region of the autonomic nervous system (ANS) that contains a number of neurones that approximates the number contained in the spinal cord, and that can function independently of CNS control (Furness and Costa, 1980; Gershon, 1981a). The ENS also contains, as intrinsic elements within its ganglia, many different types of neurone, including many that also are found in the CNS (Furness and Costa, 1980; Gershon, 1981a). These neurones form local circuits within the gut and relatively little is known about how the activity of individual enteric neurones relates to the behavioural output of the gastrointestinal tract. In fact, more is known about the interconnections made between neurones in the CNS than of the interconnections of neurones in the seemingly simpler ENS. Peripheral serotonergic neurones are valuable to study therefore, not simply because of what they can reveal about serotonergic neurones as a group, but also for what they can reveal about the ill-understood function of the massive intrinsic nervous system of the gut.

SEROTONIN AND THE GUT

Although serotonin was discovered as a vasoconstrictor material present in blood serum (Rapport *et al.,* 1948), it was not long before an association was made between serotonin and the gastrointestinal tract. In fact, a vast amount of serotonin was found to be present in the enteroendocrine (enterochromaffin; EC) cells of the epithelium of the gastrointestinal mucosal epithelium (Erspamer, 1966). The function of this large store of serotonin in EC cells was not immediately apparent; nor is it clear what that function might be, even today. Subsequently, however, careful studies of the distribution of serotonin in the wall of the gut revealed that serotonin was also present in layers below the mucosa (Feldberg and Toh, 1953); moreover, serotonin was soon shown to affect both enteric smooth muscle and enteric ganglia (Gaddum and Picarelli, 1957; Robertson, 1953; Roche e Silva and DoValle, 1953). While these observations were being made, a great deal of investigation was being done on the distribution and role of serotonin in invertebrates (Welsh, 1957) and in the mammalian CNS (Amin *et al.,* 1954). This work suggested that serotonin might be a neurotransmitter molecule. It was natural, therefore, for investigators to try to determine if serotonin was an enteric neurotransmitter as well.

MUCOSAL SEROTONIN

The earliest studies to attempt to relate gastrointestinal serotonin to the neural activity of the gut concerned themselves with the larger mucosal store of the amine, rather than with the smaller quantity of serotonin found in the other layers of the bowel. Serotonin, added to the mucosal side of a loop of intestine maintained *in vitro*, was found to stimulate the peristaltic reflex, while the reflex was blocked by adding serotonin to the serosal side of such a preparation (Bulbring and Lin, 1958). The peristaltic reflex, discovered by Bayliss and Starling (1899) at the close of the nineteenth century, is a neurally mediated, descending wave of oral contraction and anal relaxation that is elicited by increasing intraluminal pressure. This reflex can be provoked and manifested by segments of the gut *in vitro* (Trendelenburg, 1917); therefore, all of the components, sensory receptors, primary afferent neurones, inter-neurones, and excitatory and inhibitory motor neurones that are necessary for the manifestation of the reflex are intrinsic to the gut itself. Since serotonin was also found to be released from the intestinal mucosa by pressure, the postulate was framed that the EC cells are pressure receptors, that they release serotonin to the lamina propria when stimulated appropriately, and that this serotonin stimulates afferent nerve terminals within the mucosa that in turn initiate the peristaltic reflex (Bulbring and Lin, 1958; Bulbring and Crema, 1958, 1959). The blockade of the reflex by serosal application of serotonin was not linked to any specific function of the amine, but rather was explained as a non-specific effect due to flooding 5-HT-sensitive ganglia with that substance. The difference in effects elicited by serotonin, depending upon which surface of the gut it was applied to, was taken as evidence that the mucosal application mimicked the physiological release of the amine (Bulbring and Crema, 1959). This may not have logically followed, but the observation clearly demonstrated that 5-HT is not free to diffuse throughout the wall of the gut, and that serotonin added to the mucosal surface reaches different receptors from those reached by serotonin applied to the serosa. This clear suggestion of an intra-enteric barrier to the movement of serotonin from mucosa to myenteric ganglia was not remarked upon at the time, but the existence of such a barrier has since been confirmed (Gershon and Tamir, 1981; see below). In any case, when it was subsequently found that animals whose mucosal serotonin was depleted by maintenance on a tryptophan-free diet still manifested a normal peristaltic reflex, it became apparent that the serotonin in EC cells was not *necessary* for initiation of the reflex (Boullin, 1964). Modulation of the peristaltic reflex, however, by mucosal serotonin was still a possibility and no evidence against such modulation has yet been produced. As scepticism increased concerning a role for mucosal serotonin in the regulation of gastrointestinal motility, attention turned to the smaller concentration of serotonin in the remainder of the bowel wall.

ENTERIC UPTAKE OF SEROTONIN

The first suggestion that there might be serotonergic neurones in the gut came from radioautographic analyses of the localization of [3H]serotonin ([3H]5-HT) synthesized *in vivo* from administered [³H] 5-hydroxytryptophan ([³H] 5-HTP; Gershon *et al.,* 1965; Gerson and Ross, 1966a,b). Intensive labelling was found in the myenteric plexus throughout the gut but, under the conditions of the experiments, no labelling was seen in perivascular noradrenergic nerves that might have been expected to become labelled non-specifically by [³H] 5-HTP (Geffen and Jarrott, 1977). These experiments demonstrated, therefore, that the gut contained either a unique population of peripheral nerves characterized by their content of aromatic amino acid decarboxylase and the ability to store serotonin (Furness and Costa, 1978), or a unique subset of noradrenergic axons with a particular affinity for serotonin.

It soon became evident that the enteric nerves that were radioautographically labelled when [³H] 5-HTP was injected into animals also took up [³H] 5-HT itself (Robinson and Gershon, 1971). Radioautographic labelling of enteric neural elements by [³H] 5-HT is prominent (see Figs. 1-3). The uptake of serotonin by enteric nerves was soon characterized. The uptake of serotonin, by the myenteric plexus of the guinea pig, was found to be a saturable, temperature-dependent process with a Km of about 0.7 μM and a Q_{10} (27-37 °C) of 3.6 (Gershon and Altman, 1971). The enteric uptake of serotonin is antagonized by inhibitors of glycolysis, such as fluoride, iodoacetate, 2-deoxy-D-glucose, or glucose deprivation; however, serotonin uptake is not reduced by inhibitors of aerobic metabolism, such as 2,4-dinitrophenol, cyanide, or oxygen deprivation (Gershon and Altman, 1971). When glycolysis is inhibited, pyruvate can substitute for glucose and overcome the antagonism of serotonin uptake (personal observation). It thus seems likely that metabolic energy is required for serotonin uptake; this energy is derived from the breakdown of glucose, and can be provided by glycolysis. Aerobic metabolism is probably not necessary for uptake of serotonin, but can provide the energy if an appropriate substrate is made available. Since the uptake of serotonin is also inhibited by oaubain, transport adenosine triphosphatase appears to be at least indirectly involved in the process as well.

The myeneteric serotonin uptake mechanism has additional distinguishing characteristics. Ionic requirements are fairly precise; the uptake of serotonin is Na^+-dependent (Gershon and Altman, 1971) and it is antagonized by raising the external K^+ concentration (Gershon *et al.,* 1976). Accumulation of serotonin by the myenteric plexus is inversely related to the logarithm of the concentration of K^+ and is also decreased in the presence of 0 mM Ca^{2+} or 12 mM Ca^{2+}. Molecular structural requirements for affinity for the enteric serotonin uptake site are also quite restrictive (Gershon *et al.,* 1976). Affinity is much reduced in compounds that have no alkyl amino side chain, in which the amino group is methylated, or that have no 5-hydroxyl group on the indole ring. Most indoleamine analogues of serotonin competitively inhibit the uptake of serotonin itself and do, like serotonin, enter axons in the myenteric plexus. This can be demonstrated by histofluorescence with 6-hydroxytryptamine (Gershon *et al.,* 1976), which has a better fluorescent yield than 5-HT and thus is very useful for visualizing serotonergic neurones (Jonsson *et*

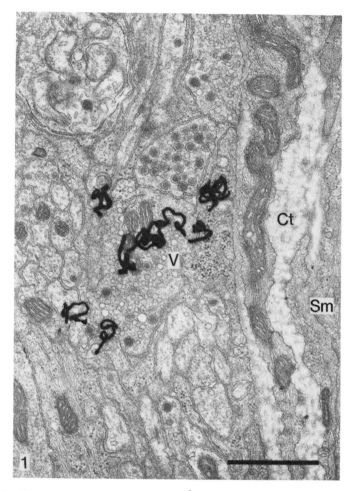

Fig. 1   Radioautographic localization of [³H] 5-HT in the myenteric plexus of a mouse. The animal was pre-treated with 6-hydroxydopamine (100 mg/kg) to prevent uptake of [³H] 5-HT by noradrenergic axons. The labelled axonal varicosity (V) contains a mixture of small lucent and large dense-cored synaptic vesicles. No pre- or postsynaptic membrane modifications are visible. Ct = connective tissue. Sm = smooth muscle. Marker = 1.0 μm

al., 1969). Entry of analogues of serotonin into enteric axons can also be demonstrated with 5,6- or 5,7-dihydroxytryptamine (5,7-DHT). These indolic neurotoxins induce the formation of lesions (Figs. 4 and 5) within those enteric axons that take up the compounds (Gershon *et al.,* 1980c). These lesions develop gradually and appear soon after administration of the neurotoxins. The first effect of 5,7-DHT is a filling of synaptic vesicles with the neurotoxin. This filling is not obvious in

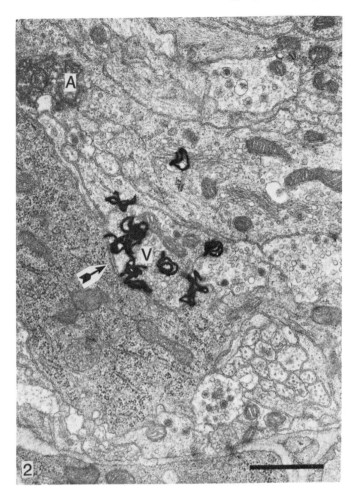

Fig. 2   Radioautographic localization of [^{3}H] 5-HT in the myenteric plexus of a mouse pretreated with 6-hydroxydopamine (100 mg/kg). A labelled varicosity (V), containing small lucent and large dense-cored vesicles, makes a synaptic contact (arrow) with a dendrite. A degenerating adrenergic axon (A) can be seen. Marker = 1.0 μm

aldehyde-fixed material but can readily be detected in tissue fixed with KMnO$_4$ (compare Figs. 6 and 7). Early lesions consist of membrane-enclosed regions of electron-dense cytoplasm containing synaptic vesicles within varicosities of affected axons. Progression of the lesions leads to complete axonal degeneration and phagocytosis by surrounding supporting cells (Figs. 4 and 5).

The sensitivity of the myenteric serotonin-accumulating mechanism to inhibition by drugs is qualitatively similar to that of central serotonergic neurones. Uptake of

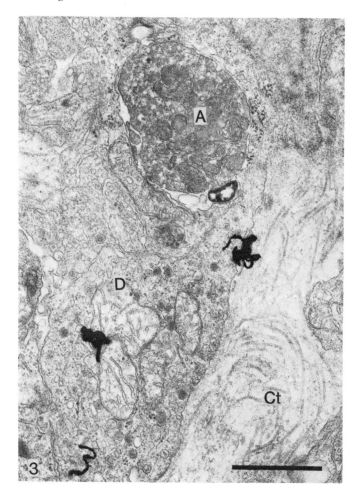

Fig. 3   Radioautographic localization of [³H] 5-HT in the myenteric plexus of the small intestine of a mouse pretreated with 6-hydroxydopamine (100 mg/kg). A dendritic process (D), containing ribosomes, has become labelled. The process also contains an irregular smooth membranous reticulum and numerous large dense-cored vesicles. A degenerating adrenergic axon (A) lies nearby. Ct = connective tissue. Marker = 1.0 μm

serotonin is inhibited by tricyclic antidepressants and amphetamines (Gershon *et al.*, 1976). The most potent tricyclic antidepressant drug is chlorimipramine while the least potent is desmethylimipramine. This is the reverse of the order of potency of these compounds in inhibiting the uptake of norepinephrine (NE). Amphetamines are more effective inhibitors of enteric serotonin uptake than are tricyclic antidepressants, although they differ from the tricyclics in not being competitive

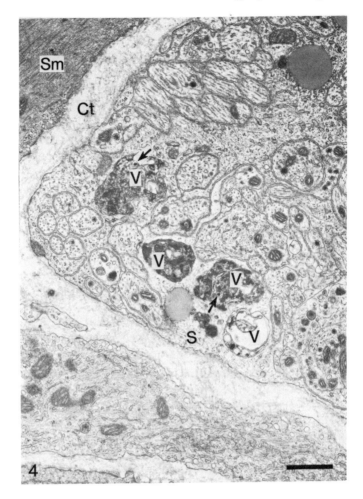

Fig. 4 The myenteric plexus, 24 h after the injection of 5,7-DHT (400 mg/kg) and desmethylimipramine (25 μm/kg), contains numerous degenerating axons (V). Large dense-cored vesicles (arrows) can still be identified within the electron-dense matrices of these terminals. The degenerating axons appear to have been phagocytized by supporting cells (S). Ct = connective tissue. Sm = smooth muscle. Marker = 1.0 μm

inhibitors of serotonin uptake. Neither of these classes of drugs are specific serotonin uptake inhibitors as they are more effective against the accumulation of NE than serotonin. These compounds, moreover, also release serotonin from enteric stores (Gershon and Jonakait, 1979). There is, however, at least one specific inhibitor of enteric serotonin uptake. The drug fluoxetine is a potent competitive antagonist. Fluoxetine, moreover, has little ability to inhibit uptake of NE, and, in

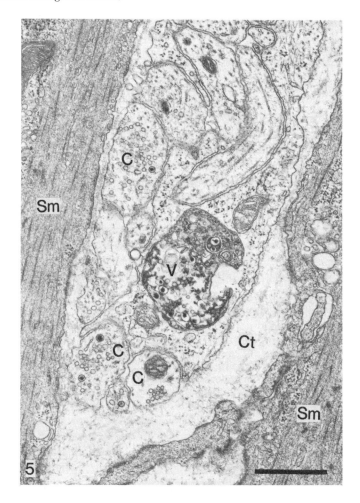

Fig. 5   A neural connective within the circular layer of smooth muscle (Sm) of the small intestine of a mouse. 5,7-DHT (400 mg/kg) and desmethylimipramine (25 μg/kg) were injected 24 h prior to removal of the tissue. A degenerating varicosity (V) is prominent. Note that large dense-cored and small round vesicles with electron-lucent cores can still be distinguished within it. Unaffected axonal terminal varicosities (C) uncovered by supporting cells appear to terminate on the connective tissue space (Ct). Marker = 1.0 μm

contrast to chlorimipramine, it does not release serotonin (Gershon and Jonakait, 1979).

It is important to consider the relationship of axons in the myenteric plexus that take up serotonin to NE. The myenteric plexus is known to contain nora-drenergic, sympathetic axons (Norberg, 1964; Jacobowitz, 1965). These axons,

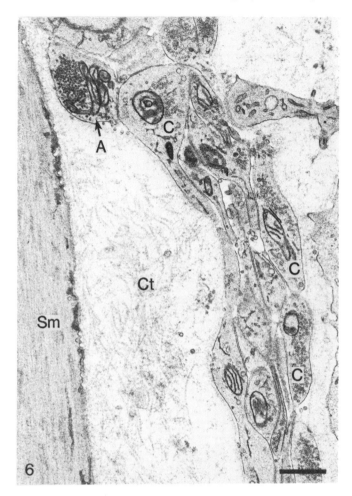

Fig. 6   KMnO$_4$ fixation. A neural connective passing from the myenteric plexus to the circular layer of smooth muscle (Sm) of the small intestine of a mouse. 5,7-DHT (600 mg/kg; 4 h) and desmethylimipramine (50 mg/kg; 4.5 h) were injected previously. A noradrenergic axonal varicosity (A) can be recognized by its content of small dense-cored synaptic vesicles. Neither it, nor other non-adrenergic (C) varicosities, appear to be have been affected by 5,7-DHT. Ct = connective tissue. Marker = 1.0 μm

with minor exceptions (Costa *et al.*, 1971), are all derived from perikarya that lie in ganglia outside the wall of the gut (Hamberger and Norber, 1965; Ahlman *et al.*, 1973; Furness and Costa, 1974, 1978; Furness *et al.*, 1979). Noradrenergic axons of the gut do take up serotonin (Drakontides and Gershon, 1972) and so the question arises as to what contribution sympathetic axons make to the net myenteric uptake of serotonin. The answer appears to be little or nothing. In the first place, the

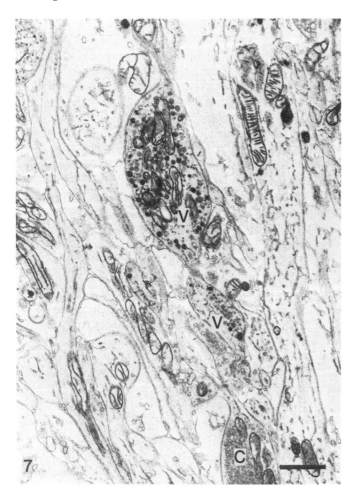

Fig. 7 $KMnO_4$ fixation. The myenteric plexus from the same animal as in Fig. 6. An axon in the centre of the field displays two vesicle-filled varicosities (V) and a thin, vesicle-free intervaricose segment. The axon contains pleomorphic large and small vesicles that appear to have filled with 5,7-DHT. Increased cytoplasmic density indicates early degenerative change. The varicosities of other types of axon (C) are unaffected by the drug. The vesicle population of the axons that fill with 5,7-DHT after desmethylimipramine pretreatment can readily be distinguished from that of noradrenergic axons. Compare with Fig. 6. Marker = 1.0 μm

affinity of noradrenergic axons for serotonin is very low; the Km for the uptake of serotonin into noradrenergic neurones is about 10^{-4} M (Thoa *et al.*, 1969), 100–1000-fold higher than the Km for the enteric uptake of serotonin. Moreover, NE does not inhibit the uptake of serotonin by the myenteric plexus, even when that amine is present in a concentration of 10–1000 times higher than that of

serotonin (Gershon and Altman, 1971; Rothman *et al.*, 1976). Uptake of serotonin, moreover, persists after enteric uptake of NE has been virtually abolished
by chemical sympathectomy with 6-hydroxydopamine (Gershon and Altman,
1971; Takiff and Gershon, 1970; see Figs. 1-5). Uptake of serotonin also precedes
uptake of NE by a substantial margin during ontogeny in rabbits (Rothman *et al.*,
1976), guinea pigs (Gintzler *et al.*, 1980), and mice (Rothman *et al.*, 1979). The
uptake of serotonin, furthermore, persists when the gut is grown for long periods
of time in organotypic tissue culture so that the extrinsic noradrenergic sympathetic innervation degenerates (Dreyfus *et al.*, 1977b). Similar persistence of
serotonin uptake follows extrinsic denervation of the bowel *in vivo* (Furness *et al.*,
1980). It seems clear, therefore, that the enteric neurones that are responsible for
uptake of serotonin are not sympathetic or noradrenergic. They represent a distinct subset of neurones. The tissue culture experiments (Dreyfus *et al.*, 1977b),
furthermore, indicate that this subset is intrinsic to the gut itself. In partial confirmation of this conclusion, no axons that specifically take up serotonin have been
found in the vagus nerves (Dreyfus *et al.*, 1977b).

Uptake of a substance, even an uptake process with as restrictive a set of molecular structural requirements as the enteric uptake of serotonin, cannot establish that
substance as the actual transmitter of the neurones that are defined by their posses-
sion of the uptake mechanism. The uptake of serotonin thus served to define a
unique population of peripheral neurones, thus far not found anywhere but in the
bowel and perhaps the pancreas (Koevary *et al.*, 1980; Nunez *et al.*, 1980), and as
an indication that enteric serotonergic neurones might exist. The uptake mechanisms
also provided a marker property that enabled the detection of this neuronal class in
studies of the ontogeny (Rothman *et al.*, 1976; Gintzler *et al.*, 1980; Gershon *et
al.*, 1981; Epstein *et al.*, 1980; Gershon *et al.*, 1980a) and phylogeny of the innervation of the gut (Goodrich *et al.*, 1980).

PRESENCE OF SEROTONIN IN ENTERIC NEURONES

Far more important than an uptake mechanism for the identification of a neurotransmitter is the demonstration that the substance in question is actually present
in, and released by, the neurones being investigated. Since the amount of 5-HT in
mucosal stores is so large, and the non-mucosal store by comparison is small (Feld-
burg and Toh, 1953), the demonstration of 5-HT in the ENS has been difficult.
Because, in addition, the ENS takes up 5-HT, care must be exercised in dissecting
the layers of the intestine, lest 5-HT released from the mucosa reach and be taken
up by intestinal nerves. Such an uptake, if it occurred, would falsely inflate a subsequently measured level of 5-HT in the neural tissue. This accidental uptake can be
avoided if the gut is dissected in the cold to take advantage of the very great temperature dependence of the neural 5-HT uptake mechanism (Gershon and Altman,
1971).

The concentration of 5-HT has been measured in preparations of longitudinal muscle with adherent myenteric plexus (LM-MP) dissected from the guinea-pig small intestine. These preparations contain ENS, uncontaminated by mucosal 5-HT. The 5-HT concentration in LM-MP strips has been reported to be between 80 and 110 ng/g (Juorio and Gabella, 1974; Robinson and Gershon, 1971; Gershon and Tamir, 1979, 1981). The myenteric plexus, however, constitutes a relatively small part of the mass of the LM-MP preparation. The bulk of the tissue is smooth muscle; therefore, if the 5-HT in the LM-MP preparation is exclusively limited to the myenteric plexus, its real concentration in the neural tissue would be much higher than that measured in the whole LM-MP. Since neither the myenteric nor the submucosal plexuses have even been isolated, histochemical or immunhisto-chemical studies are necessary to confirm the localization of 5-HT to the ENS.

Histochemical procedures for the demonstration of 5-HT exist, but they are inferior in quality to procedures for the demonstration of catecholamines (Fuxe and Jonsson, 1967). When the mammalian myenteric plexus is examined by for-maldehyde or glyoxylic acid-induced histofluorescence under control conditions, therefore, no evidence of the presence of 5-HT can be detected (Robinson and Gershon, 1971; Baumgarten *et al.*, 1973; Ahlman and Enerback, 1974; Dubois and Jacobowitz, 1974; Costa and Furness, 1979a). Curiously, 5-HT can readily be detected by histofluorescence in the ENS of lower vertebrates, such as cyclostomes (Baumgarten *et al.*, 1973; Goodrich *et al.*, 1980) and teleosts (Watson, 1979) and in the ENS of cephalochordates (Salimova, 1978). 5-HT has also been found recently in enteric neurones immunocytochemically, using an antiserum to 5-HT (H. Steinbusch, reported at the International Meeting, 'Le Neurone Serotoninergique', Marseilles, France, July, 1980; M. Costa, Tenth Annual Meeting of the Society for Neuroscience, Cincinnati, Ohio, November, 1980). The poor histofluorescence of 5-HT has been overcome with respect to central serotonergic neurones by increasing 5-HT levels pharmacologically. This has been done by administering the amino acid precursor of 5-HT, L-tryptophan, in combination with an inhibitor of monoamine oxidase (MAO) (Aghajanian and Asher, 1971). Administration of L-tryptophan appears to increase the 5-HT concentration only in serotonergic neurones and not, as occurs with 5-hydroxytryptophan, in other cells. This specificity depends on the restriction of tryptophan hydroxylase to serotonergic neurones (Kuhar *et al.*, 1971, 1972). In the CNS as well, tryptophan hydroxylase in cell bodies, but not axons and terminals, seems to be resistant to inhibition by parachlorophenylalanine (PCPA); therefore, administration of PCPA, together with L-tryptophan and an inhibitor of MAO, leads to enhanced 5-HT histofluorescence in the cell bodies, and a loss of 5-HT histofluorescence in the terminals, of serotonergic neurones (Agha-janian and Asher, 1971; Aghajanian *et al.*, 1973). Similar pharmacological treat-ments selectively enhance the detection of 5-HT in enteric neurones. 5-HT has been reported to be visible by histofluorescence in the myenteric plexus following inhi-bition of MAO (Robinson and Gershon, 1971; Feher, 1974, 1975). Histofluorescence is further enhanced by injection of animals with L-tryptophan and examining LM-MP whole mounts (Dreyfus *et al.*, 1977a). Unfortunately, the conditions necessary to enhance 5-HT histofluorescence in the myenteric plexus lead to a rather diffuse fluorescent pattern. This pattern may arise because individual serotonergic axons are too small to resolve individually with the light microscope; the pattern might also result from diffusion of 5-HT. In any case, 5-HT histofluorescence is limited to the myenteric plexus, and cell bodies can be visualized if PCPA is added to L-tryptophan and inhibition of MAO in the pretreatment regimen (Dreyfus *et al.*, 1977a). When PCPA is added, the diffuse pattern of fluorescence is lost but neuronal cell body fluorescence remains. Analogy to the results obtained with the CNS suggests that the diffuse pattern of 5-HT histofluorescence elicited by L-tryptophan and MAO inhibition reflects axonal storage of the amine. The histofluorescence, diffuse or in cell bodies, induced by L-tryptophan administration has identical activation and emission spectra to those of the authentic 5-HT-formaldehyde fluorophore (Dreyfus *et al.*, 1977a).

Another technique that has been used to demonstrate the histofluorescence of 5-HT in enteric neurones has been to grow the gut in organotypic tissue culture (Dreyfus *et al.*, 1977a,b). When hemisections of late foetal mouse gut are grown in culture the mucosa degenerates. These cultures, or companion cultures of just the muscularis externa, thus contain no enterochromaffin cells; however, they do contain neurones in which the histofluorescence of 5-HT is especially strong and visible without resort to pharmacological augmentation. A similar phenomenon occurs when the LM–MP preparation or the submucosa from late fetal guinea pigs are grown in culture; again, strong 5-HT histofluorescence is seen in cultured neurones (Gershon *et al.*, 1980b). In the guinea pig cultures neuronal cell bodies also become radioautographically labelled when [^{3}H] 5-HT is added to the culture medium. Such cell bodies are rarely labelled in the gut *in situ*. It is possible that the uptake mechanism for 5-HT is restricted to axons of serotonergic neurones and that cell bodies label in cultures because of retrograde transport of [^{3}H] 5-HT that was initially taken up by terminals. This is consistent with the relatively long exposure to [^{3}H] 5-HT used in the culture experiments and the relatively short axons of the cultured neurones. These considerations of neuronal geometry may also help account for the ease of demonstration of 5-HT histofluorescence in neurones in culture. Foreshortening of axons of cultured neurones may lead to 5-HTs achieving a higher concentration in cell bodies. Another plausible explanation is that cultured neurones may be inactive, leading to intraneuronal accumulation of transmitter. 5-HT can be measured in cultures (about 1.5 pmol/culture) of guinea-pig intestinal LM–MP strips (H. Tamir, M. Gershon, S. Erde and C. Dreyfus unpublished observations). These cultures contain an average of about 25 cells per culture that show 5-HT histofluorescence. Based upon 25 serotonergic cells per culture, each of these cells would contain approximately 60 fmol of 5-HT. Assuming an average cell-diameter of 25 μM, the average volume of a serotonergic neurone would be about 8 pl. The estimated concentration of 5-HT in each cultured serotonergic neurone would, therefore, be high, about 7 mM, accounting for the ease of histochemical demonstration of 5-HT in cultured neurones.

The high concentration of 5-HT in individual neurones raises a problem, not only for enteric serotonergic neurones but for other serotonergic neurones as well. This high concentration of 5-HT, if the 5-HT were free in solution, ought to have an osmotic effect on the neurones in which it is stored, drawing water into the 5-HT storage compartment. If 5-HT is even more concentrated within a subcellular organelle than it is in whole cells as, for example, within synaptic vesicles, the potential osmotic problem would be accentuated. A means must exist, therefore, to prevent osmotic swelling of the storage compartment. More than one possible mechanism to serve this function can be envisioned; however, a mechanism for which there is experimental evidence is the binding of 5-HT to a macromolecule.

SEROTONIN BINDING PROTEIN

Serotonin binding protein (SBP), first detected in synaptosomes obtained from the CNS (Tamir and Huang, 1974), is a putative osmo-protective storage macromolecule for 5-HT. The protein binds serotonin specifically and with a very high affinity. It is an intraneuronal protein that appears to be associated with serotonergic neurones; therefore, SBP activity is depleted from the forebrain when lesions are made in the median raphe to destroy the cell bodies of central serotonergic neurones (Tamir and Kuhar, 1975). In a similar fashion, SBP activity is lost distal to a lesion of the spinal cord (Tamir and Gershon, 1979); there are no serotonergic cell bodies in the spinal cord, and these lesions consequently interrupt the descending projections of caudal serotonergic raphe neurones. The spinal cord lesions, moreover, also lead to an accumulation of serotonin in the segment of the spinal cord immediately proximal to the lesions. The rate of this build-up of SBP activity in the segment of spinal cord above the lesions is rapid, suggesting that SBP is transported proximo-distally in serotonergic neurones (at the rate of about 100 mm/day) by the mechanism of fast (anterograde) axonal transport. Movement by fast axonal transport, in turn, suggests that SBP is associated with vesicular material destined for delivery to the terminal synaptic apparatus (Grafstein and Forman, 1980). These suggestions are supported by the observation that SBP, released under mild conditions from synaptosomes, is four-fold more concentrated in partially purified synaptic vesicles than in the synaptosomal supernatant (Tamir and Gershon, 1979). Further support for the localization of SBP, at least in part, in synaptic vesicles is derived from studies of the gut.

The gut, like the brain, contains SBP activity (Jonakait *et al.*, 1977). In its properties, the SBP derived from the gut resembles that of the brain. For example, both enteric and central SBP require Fe^{2+} (Tamir *et al.*, 1976; Jonakait *et al.*, 1977) for maximal 5-HT binding; binding of 5-HT to both is inhibited by reserpine and is enhanced by gangliosides; the molecular weight of the 5-HT-binding portion of the molecules is 50,000–60,000 Daltons and both have two dissociation constants of about 10^{-10} M and 10^{-7} M (Jonakait *et al.*, 1977; Bernd *et al.*, 1979; Tamir and Gershon, 1981). In fact, the properties of serotonin binding proteins extracted from all neuroectodermal derivatives that store 5-HT (brain, myenteric plexus, thyroid parafollicular cells) have been found to be similar to one another and to differ from the serotonin-binding proteins extracted from 5-HT-storing cells of mesodermal (platelets, mast cells) or entodermal (intestinal mucosa) origin (Tamir *et al.,* 1980a; 1980b; Tamir and Gershon, 1981; Bernd *et al.,* 1981).

The neuroectodermal-type of SBP of the gut is contained within the myenteric plexus and, during ontogeny, this SBP develops coincidentally with the appearance of the enteric nerves that specifically take up 5-HT (Jonakait *et al.,* 1977). It seems likely, therefore, that the SBP of the myenteric plexus is present within enteric serotonergic neurones. The loss of SBP activity from the myenteric plexus in animals given the indolic neurotoxin, 5,7-DHT, is consistent with this view. If SBP were to be released along with 5-HT when enteric nerves are stimulated, confirmation could be obtained of the storage of SBP along with 5-HT in synaptic vesicles (see below; Jonakait *et al.,* 1979b).

RELEASE OF SEROTONIN AND SBP FROM ENTERIC NEURONES

Release of 5-HT from the gut has been extensively studied; however, the demonstration that enteric nerves release the amine is made difficult by the presence of 5-HT in the enteroendocrine cells of the mucosa. For example. Paton and Vane (1963) showed that vagal stimulation released 5-HT from the stomach. Nevertheless, they could not identify the source of the 5-HT that was released, nor could they show Ca^{2+} dependence of the 5-HT release mechanism. It is likely that vagal stimulation does give rise to release of 5-HT from enterochromaffin cells (Tansy *et al.,* 1971; Ahlman *et al.,* 1976). This release is probably due to stimulation of descending vagal adrenergic axons that, in turn, innervate enterochromaffin cells. Bulbring and Gershon (1967) attempted to eliminate the mucosa as a possible source of 5-HT in their studies of the release of the amine by selectively removing the mucosa by asphyxiation prior to electrically stimulating the stomach. Following stimulation, after mucosal asphyxiation, Bulbring and Gershon (1967) still detected the release of 5-HT. They found, moreover, that the electrically induced release of the amine was abolished by tetrodotoxin and subsequent histological examination of their preparations confirmed that enterochromaffin cells had, in fact, been eliminated. They concluded that 5-HT had been released from nerves in the stomach. It was subsequently pointed out, however, that the nerves found by Bulbring and Gershon (1967) to have released 5-HT could conceivably have been loaded with the amine released from dying enterochromaffin cells during the period of mucosal asphyxiation that preceded nerve stimulation in their experiments (Costa and Furness, 1979a). On the other hand, circumstantial evidence has also been obtained for the release of 5-HT by nerves of the myenteric plexus of other regions of the gut, including the colon (Furness and Costa, 1973) and the small intestine (Dingledine *et al.,* 1974). In these instances, no opportunity was provided for artifactual loading of enteric nerves with mucosal 5-HT. Unfortunately, in these instances as well, no chemical identification of the amine was made; rather, electrical stimulation was found to release a substance that had 5-HT-like effects on the gut.

A different type of study of 5-HT release has been to incubate the whole gut or LM–MP strips with [^{3}H] 5-HT prior to electrical stimulation (Schulz and Cartwright, 1974a, Jonakait *et al.*, 1979b). This procedure has the drawback that endogenous 5-HT is not measured and the exogenous radioactive amine might not be taken up by neurones that utilize 5-HT as their transmitter; nevertheless, the prior labelling of 5-HT stores does permit the sensitive detection of the released amine and the effective study of the transmitter-release properties of at least that unique subset of enteric neurones defined by the specific 5-HT uptake mechanism (Gershon *et al.* 1976). [^{3}H] 5-HT can be released from LM–MP strips preloaded with the radioactive amine by electrical stimulation (Schulz and Cartwright, 1974a), elevating the ambient concentration of K$^+$, or by exposing the LM–MP strips to the Ca^{2+} ionophore, X-537A (Jonakait *et al.*, 1979b). In the latter two examples, however, release of [^{3}H] 5-HT is only slightly diminished in the absence of Ca^{2+}. Internal tissue Ca^{2+} might be mobilized by high K$^+$ or X-537A; these agents may be toxic, or they may act through a Ca^{2+}-independent mechanism. The LM–MP strip itself is not an ideal preparation to use to study 5-HT release because the ENS within it is necessarily disrupted by the dissection of the LM–MP from the rest of the wall of the bowel. The myenteric and submucosal plexuses are normally interconnected and these connections are severed when the longitudinal muscle is stripped from the wall of the gut; moreover, even the myenteric plexus that remains adherent to the longitudinal muscle is subjected to a certain amount of trauma. Another preparation, used to avoid these problems, has been the perfused, everted (turned inside out) guinea-pig small intestine (Jonakait *et al.*, 1979b). This preparation is mounted with the mucosal surface on the outside facing a large volume of bathing solution. The serosal surface is now internalized, lining a new lumen of small volume. The everted gut is perfused through the serosal lumen. When [^{3}H] 5-HT is added to the fluid perfusing the serosal surface, [^{3}H] 5-HT is taken up by enteric nerves and can be shown by radioautography to have labelled both the myenteric and the submucosal plexuses (Jonakit *et al.*, 1979b). Enterochromaffin cells, however, are hardly labelled at all. These results indicate that there are no barriers to the passage of [^{3}H] 5-HT between the serosa and the two ganglionated plexuses of the ENS. 5-HT released from these plexuses, therefore, will enter the serosal perfusate. On the other hand, the failure of [^{3}H] 5-HT to label the enterochromaffin cells could be due to the inability of these 5-HT-storing cells to take up [^{3}H] 5-HT (Rubin *et al.*, 1971) or to the failure of [^{3}H] 5-HT, applied to the serosa, to reach them. Application of [^{3}H] 5-HT (1.0 μM) for 15–30 min directly to the mucosa does lead to the radioautographic labelling of enterochromaffin cells in guinea-pigs and rabbits (personal observation). The radioautographic data, therefore, suggest that there is an intramural barrier to the passage of 5-HT.

The presence of an intramural barrier to 5-HT was previously inferred from the difference in action of 5-HT on peristalsis, depending upon which surface of the gut the amine is applied to (see above; Bulbring and Lin, 1958; Bulbring and Crema, 1959). The existence of this barrier has recently been demonstrated directly using the perfused, everted intestinal preparation. When [^{3}H] 5-HT is added to the serosal perfusate, very little [^{3}H] 5-HT appears in the fluid bathing the mucosal surface of the everted gut. More significantly, when [^{3}H] 5-HT is added to the mucosal medium, almost no [^{3}H] 5-HT appears in the serosal perfusate even after 15 min (Gershon and Tamir, 1979, 1981). When this data is considered together with the radioautographic tracing of the penetration of [^{3}H] 5-HT into the gut from the serosal surface, one can conclude that the barrier to the passage of 5-HT must lie either within the mucosa or between the mucosa and the submucosa. Enterochromaffin cells are known to continually release large amounts of 5-HT so that the concentration of 5-HT is higher in portal venous blood than it is is in the systemic circulation (Toh, 1954). 5-HT has profound effects on the neurones of the ENS (Gaddum and Picarelli, 1957; Brownlee and Johnson, 1963; Bulbring and Lin, 1958; Bulbring and Crema, 1959; Bulbring and Gershon, 1967; Hirst and Silinsky, 1975; Costa and Furness, 1979b; Johnson *et al.,* 1980b). Clearly, the ENS could not function if it were constantly to be bathed in a high concentration of 5-HT. The intramural 5-HT barrier probably serves to protect the ENS from the effects of mucosal 5-HT. The precise anatomical localization of the barrier has not yet been accomplished; nor has the relationship of the intramural 5-HT barrier to the blood-myenteric barrier (Gershon and Bursztajn, 1978) been established.

When preparations of perfused, everted intestine have been preloaded with [^{3}H] 5-HT from the serosal surface and stimulated electrically, [^{3}H] 5-HT is released into the serosal perfusate (Jonakait *et al.,* 1979b). This stimulation-induced release of [^{3}H] 5-HT, but not the lower level of spontaneous release of the amine, is Ca^{2+} dependent, and antagonized by elevated concentrations of Mg^{2+} or by tetrodotoxin. These preparations can also be shown, in the absence of preloading with [^{3}H] 5-HT, to release endogenous 5-HT (Gershon and Tamir, 1979, 1980). As is true of the release of the radioactive amine, the stimulated but not the resting release of endogenous 5-HT is abolished in Ca^{2+}-free media. The radioautographic results discussed above (Jonakait *et al.,* 1979b) leave little doubt but that the source of the [^{3}H] 5-HT released in experiments involving the preloading of tissue with the radioactive amine is the ENS; no other structures contain [^{3}H] 5-HT. Endogenous 5-HT, however, must also be derived from the ENS, since the intramural barrier to the passage of 5-HT prevents mucosal 5-HT from entering the serosal perfusate that is collected for measurement. When [^{3}H] 5-HT is used, the specific activity of radioactive 5-HT released by electrical stimulation is higher than that of the spontaneously released amine and is also higher than that of the [^{3}H] 5-HT stored in the tissue. Newly taken up [^{3}H] 5-HT, therefore, as is true of other neurotransmitters, appears to be preferentially released by nerve stimulation.

Not only is 5-HT released from the serosal surface of perfused, everted intestinal segments, but SBP is released as well (Jonakait *et al.,* 1979b). SBP appears to be released spontaneously by the perfused gut, but the release of the protein is increased greatly by electrical stimulation. The stimulated, but not the spontaneous release of SBP, like the release of 5-HT, is Ca^{2+}-dependent. Unlike SBP, the cytosol marker protein, lactate dehydrogenase, is not released from either the resting or stimulated perfused, everted intestine. The Ca^{+2}-dependent, stimulated release of SBP along with 5-HT suggests that exocytosis is the mechanism of transmitter release. This, in turn, supports the hypothesis, discussed earlier, that SBP is stored together with 5-HT, as a storage protein in synaptic vesicles. Since an increase in the release of 5-HT from the serosal surface of the rabbit gut has been found to accompany peristalsis (Gwee and Yeoh, 1968), it seems likely that the activity of enteric serotonergic neurones is increased during the peristaltic reflex. This conclusion is not inconsistent with the results of pharmacological analyses of the peristaltic reflex (Costa and Furness, 1976).

SYNTHESIS OF SEROTONIN

Strips of LM–MP removed from the guinea-pig small intestine convert [^{3}H] L-tryptophan to [^{3}H] 5-HT (Dreyfus *et al.*, 1977a). This synthesis is antagonized by the prior injection of the tryptophan hydroxylase inhibitor, PCPA. Since PCPA injection prior to dissection of LM–MP strips does not interfere with the neuronal uptake of L-tryptophan, it is likely that the antagonism of the drug to biosynthesis of 5-HT is attributable to its inhibition of tryptophan hydroxylase. Incubation of LM–MP strips with [^{3}H] L-tryptophan also leads to the production of the tritiated metabolite of 5-HT, [^{3}H] 5-hydroxyindoleacetic acid (5-HIAA).

Organotypic tissue cultures of mouse gut that contain neither enterochromaffin cells nor mast cells nevertheless convert [^{3}H] L-tryptophan to [^{3}H] 5-HT and [^{3}H] 5-HIAA (Dreyfus *et al.*, 1977a). These cultures contain neurones that manifest 5-HT histofluorescence but they do not contain noradrenergic neurites. The entirely extrinsic noradrenergic sympathetic innervation (Furness *et al.*, 1979) degenerates in the cultured explants (Dreyfus *et al.*, 1977a,b). PCPA markedly inhibits this conversion of [^{3}H] L-tryptophan to [^{3}H] 5-HT by the cultures, but because PCPA *in vitro* also inhibits the uptake of [^{3}H] L-tryptophan, it is hard to know to which of its actions PCPA-inhibition of 5-HT biosynthesis by cultures is due. These data, however, do indicate that there are intrinsic enteric neurones that synthesize 5-HT from L-tryptophan. This conclusion is supported by the immuno-cytochemical observation that enteric neurones contain material that reacts with an antibody to tryptophan hydroxylase purified from raphe neurones of rats (Gershon *et al.*, 1977). The enzyme has been found in cell bodies of intestinal neurones in mice, rats and guinea pigs. Immunoreactive tryptophan hydroxylase has also been found in neurones surviving in organotypic tissue culture for up to 3 weeks. Biosynthesis of 5-HT, therefore, appears to be due to the presence within the ENS of intrinsic neurones that contain tryptophan hydroxylase that is antigenically similar to the tryptophan hydroxylase of central serotonergic neurones. As would be expected, aromatic L-amino acid decarboxylase has also been found within intrinsic enteric neurones (Costa *et al.*, 1976). The presence of these biosynthetic enzymes for 5-HT within intrinsic enteric neurones is in marked contrast to the enzymes involved in the biosynthesis of catecholamines; these are entirely present in extrinsic neural processes (Furness *et al.*, 1979).

UPTAKE OF L-TRYPTOPHAN

Enteric neurones take up L-tryptophan (Dreyfus *et al.*, 1977a). The uptake mechanism for this amino acid is energy dependent, saturable, and has a relatively low Km of about 50 μM. A substantial portion of the [^{3}H] L-tryptophan taken up by LMMP strips that have been incubated with the amino acid and subsequently homogenized is concentrated in a subcellular fraction enriched with terminal axonal varicosities (Jonakait *et al.*, 1979a)

The uptake of [^{3}H] L-tryptophan is subject to neuromodulation (Gershon and Dreyfus, 1980). Uptake of the amino acid (50 μM) is stimulated up to three-fold by very low concentrations of [^{3}H] 5-HT (0.5 μM). This action of 5-HT appears to involve activation of neurones as it is blocked by tetrodotoxin and mimicked (to a lesser extent) by 50 mM K$^+$. Since the ability of 5-HT to stimulate tryptophan uptake is also antagonized by elevated concentrations of Mg^{2+} or in the absence of Ca^{2+}, the action of the amine may involve the release of another neurotransmitter. More recently, the uptake mechanism for L-tryptophan has been shown to be stereospecific (Gershon, 1981b). L-Tryptophan uptake is competitively inhibited by other amino acids in the following order of potency: phenylalanine > isoleucine > leucine. Many of the putative enteric neurotransmitters (Furness and Costa, 1980; Gershon, 1981a) fail to emulate 5-HT and stimulate uptake of L-tryptophan (Gershon, 1981b). These neurotransmitters (or related substances) that have been found not to stimulate tryptophan uptake include (in concentrations up to 10 μM) acetylcholine (ACh), ACh in the presence of eserine, carbamylcholine, ATP, norepinephrine, dopamine, leu-enkephalin, morphine, substance P, somatostatin, cholecystokinin and pentagastrin. A few putative enteric neurotransmitters, in addition to 5-HT, significantly stimulate tryptophan uptake. These include vasoactive intestinal polypeptide (VIP), bombesin, and neurotensin (all at 1. 0 μM); however, the action of these peptides, like that of 5-HT, is Ca^{2+}-dependent. All may thus release another compound(s) that finally act(s) on the neurones with the tryptophan uptake mechanism. In summary, some neurones of the myenteric plexus manifest a relatively high affinity uptake of L-tryptophan that is subject to modulation by putative enteric neurotransmitters. The relationship of the uptake of tryptophan to the biosynthesis of 5-HT remains to be determined; the neurones responsible for the uptake are still to be identified and the mechanism of neuromodulation has yet to be worked out.

ACTIONS OF 5-HT ON THE GUT

The action of 5-HT on the gut is complex (Costa and Furness, 1979b). In some species, and in some regions of the bowel, the smooth muscle is activated by 5-HT.

The fundus of the rat stomach, for example, is so sensitive to 5-HT that it is used as a bioassay preparation for the amine (Vane, 1957). The small intestinal longitudinal muscle coat in rabbits and guinea-pigs, on the other hand, is relatively insensitive to 5-HT (Gershon, 1967) and the circular muscle of the guinea-pig stomach (Bulbring and Gershon, 1967), ileum, and colon (Costa and Furness, 1979b) is virtually unresponsive. In contrast to smooth muscle, the neural tissue of the gut, in all regions and species of mammal thus far studied, has been found to be acted on by 5-HT.

Stimulation of cholinergic ganglion cells following exposure to 5-HT was found in early pharmacological experiments (Gaddum and Picarelli, 1957; Brownlee and Johnson, 1963). These cholinergic neurones may or may not be directly affected by 5-HT, but they are the final common excitatory neurones leading to the smooth muscle. Addition of 5-HT thus releases ACh (Vizi and Vizi, 1978) causing a cholinergic activation of both the longitudinal (Gershon, 1967) and circular (Harry, 1963; Costa and Furness, 1979b) muscle coats. Excitatory ganglion cells are not the only ones whose activity is increased following exposure to 5-HT. 5-HT also stimulates non-adrenergic non-cholinergic intrinsic inhibitory neurones (purinergic neurones—see Burnstock, 1979b; Bulbring and Gershon, 1967; Gershon, 1967; Drakontides and Gershon, 1968; Furness and Costa, 1973; Costa and Furness, 1979b); therefore, if a muscarinic antagonist is added to preparations of gut together with 5-HT, cholinergic excitation is blocked, and 5-HT will cause a relaxation of the enteric smooth muscle. There are, of course, several possible explanations for these observations. Within enteric ganglia, 5-HT may excite cholinergic and intrinsic inhibitory neurones by acting directly on 5-HT receptors on these cells; alternatively, the amine may activate interneurones that, in turn, stimulate the neurones of the final common excitatory or inhibitory pathways; finally, 5-HT may be inhibitory to other inhibitory interneurones, acting to release the final common neurones from a tonic inhibitory control. Obviously, intracellular recordings from the myenteric plexus are necessary to resolve the cellular mechanism of the enteric action of 5-HT; however, the consistency of the effect of the amine on ganglia, taken together with the observation that enteric serotonergic neurones are intrinsic to the gut (Dreyfus *et al.*, 1977a,b; Gershon *et al.*, 1977, 1980b), suggests that the serotonergic neurones are probably themselves interneurones within the ENS.

Electrophysiological studies have revealed that 5-HT does activate some enteric neurones (Sato *et al.*, 1974; Dingledine *et al.*, 1974; Dingledine and Goldstein, 1976). Using intracellular electrodes, two basic types of ganglion cell have been defined in the myenteric plexus. These have been called the type 1 or S cell and the type 2 or AH cell (Nishi and North, 1973; Hirst *et al.*, 1974; see review by Gershon, 1981a). The type 1 (S) cell responds to the prolonged injection of depolarizing current pulses with a continuous discharge of spikes. The type 2 (AH) cell, on the other hand, responds to a prolonged depolarizing current with only one or two action potentials, following which the cell manifests a prominent and prolonged hyperpolarizing after-potential. During the after-potential the type 2 (AH) cell is relatively inexcitable. These hyperpolarizing after-potentials are lacking in type 1 (S) cells. It seems likely that the rising phase of the action potential in type 2 (AH) cells is associated with the entry of Ca^{2+} into the cells (Hirst and Spence, 1973; North; 1973; North and Nishi, 1976). This inward movement of Ca^{2+} probably activates a Ca^{2+}-dependent K^+ conductance that, in turn, is responsible for the after-hyperpolarization that is the identifying characteristic of the type 2 (AH) cell (Grafe *et al.,* 1980). This mechanism is similar to that noted for cells of molluscan ganglia (Meech and Standen, 1975). Both of the physiologically identified classes of myenteric ganglion cell can be depolarized by 5-HT, although not every myenteric ganglion cell responds to the amine (Wood and Mayer, 1979b; Johnson *et al.,* 1980b). Evidence now exists, however, to implicate 5-HT as being the neurotransmitter of nerve fibres whose stimulation evokes a slow excitatory postsynaptic potential (EPSP) in type 2 (AH) cells (Wood and Mayer, 1979a).

Both the natural transmitter released by stimulation of a neural connective leading to a ganglion of the myenteric plexus and iontophoretically applied 5-HT produce slow EPSPs in type 2 (AH) cells (Wood and Mayer, 1979a,b). The slow EPSP is an extremely long-lasting membrane depolarization (over 1 min) that may persist 1000 times longer than the fast cholinergic EPSP, also manifested by type 2 (AH) neurones (Grafe *et al.,* 1979a; Johnson *et al.,* 1980a). The slow EPSP is associated with an increased membrane resistance and an abolition of the hyperpolarization that follows the action potential in type 2 (AH) cells (Wood and Mayer, 1979a,b). These effects lead to a pronounced augmentation of membrane excitability and make it possible for the type 2 (AH) neurone to fire repetitively when excited during the slow EPSP. The slow EPSP is probably due to antagonism, in response to the transmitter, of the Ca^{2+}-dependent K^+ conductance that mediates the after hyperpolarization of the type 2 (AH) cell (Grafe *et al.,* 1980; Johnson *et al.,* 1980b). All of the effects of the natural transmitter are precisely mimicked by the iontophoretic application of 5-HT to type 2 (AH) cells (Wood and Mayer, 1979b; Grafe *et al.,* 1979b; J. Wood, communicated at the Workshop on the Enteric Nervous System, Tenth Annual Meeting of the Society for Neuroscience, Cincinnati, November, 1980). Neither the naturally evoked slow EPSP nor the response to the iontophoretic application of 5-HT is affected by cholinergic agonists or antagonists (Wood and Mayer, 1979b); however, both responses are blocked by the 5-HT antagonist, methysergide, and by desensitization of 5-HT receptors by adding excess 5-HT to the ambient medium. Other substances, such as substance P, also produce a depolarization of type 2 (AH) cells associated with an increased membrane resistance (Katayama and North, 1978; Katayama *et al.,* 1979); nevertheless, substance P less precisely mimics the slow EPSP than 5-HT, and is not pharmacologically identical to the natural transmitter. Methysergide and desensitization block the slow EPSP and the effects of 5-HT, but they do not interfere with the action of substance P (Grafe *et al.,* 1979b). The slow EPSP, moreover, persists in preparations rendered unresponsive to substance P by desensitization (J. Wood, reported at the Workshop on the Enteric Nervous System, Tenth Annual Meeting of the Society for Neuroscience, Cincinnati, November, 1980). 5-HT, therefore, is the only putative enteric neurotransmitter that exactly duplicates the effects of the mediator of the slow EPSP. It has been reported that chymotrypsin antagonizes the slow EPSP (Morita *et al.,* 1980). This observation might suggest that a peptide, cleaved by that enzyme, is involved in generating the response; however, chymotrypsin is not a specific agent and its pharmacology is unknown. Its action on neurotransmission is thus subject to numerous interpretations. In order to confirm, finally, that 5-HT is the (or one of the) neurotransmitter(s) that mediate(s) the slow EPSP additional evidence will nevertheless be required. Identifying the transmitters at specific synapses within a local circuit such as exists within the myenteric plexus is difficult. There are serotonergic neurones in the system and 5-HT mimics the slow EPSP; 5-HT thus is certainly a good candidate to be the mediator. Missing, however, is the demonstration that enteric serotonergic terminals actually synapse on type 2 (AH) cells in which the slow EPSP can be evoked. An appropriate type of axo-somatic serotonergic synapse has been found in the myenteric plexus by radio-autography with [^{3}H] 5-HT (Fig. 2). If this technique could be combined with the physiological identification and anatomical marking of type 2 (AH) neurones, the appropriate evidence could be obtained.

NORADRENERGIC–SEROTONERGIC INTERACTIONS

The generation of the slow EPSP in type 2 (AH) neurones is inhibited by NE (Wood and Mayer, 1979c). Since NE does not correspondingly inhibit the action of ionto-phoretically applied 5-HT, it has been proposed that NE acts presynaptically to inhibit the release of 5-HT from the enteric serotonergic axons that provoke the slow EPSP. Exogenous NE, in support of this hypothesis, does inhibit the stimulated release of [^{3}H] 5-HT from the perfused, everted guinea-pig ileum (Gershon, 1980). This action of NE on 5-HT release is mediated through alpha adrenoreceptors and is blocked by phentolamine. Sympathetic nerve stimulation, however, does not act identically to exogenous NE. While presynaptic alpha adrenoreceptors inhibit 5-HT release, presynaptic adrenoreceptors appear to be facilitatory. Although exogenous NE has a predominantly alpha effect, inhibiting [^{3}H] 5-HT release, NE released from stimulated sympathetic nerve seems to have more of a beta action. This difference raises the possibility that the effects on the gut of the sympathetic nerves may be different from those of circulating catecholamines. Such a difference could be explained anatomically, if noradrenergic synapses, for example, covered beta adrenoreceptors, or if the route of penetration of exogenous NE into the myenteric plexus preferentially brought the amine into contact with alpha adrenoreceptors. Clearly, an anatomical investigation of the relationship between noradrenergic and serotonergic neural elements of the myenteric plexus is needed.

It is possible that an interaction between noradrenergic axons and serotonergic neurones can occur in the myenteric plexus through a more non-directive release of transmitter than occurs at most synapses, not requiring classical synaptic membrane specializations. Terminal axonal varicosities of sympathetic axons rarely, if ever, display the pre- and postsynaptic membrane specializations that characterize other more directive synapses (Burnstock, 1979a). A good example of a typical noradrenergic axonal terminal varicosity is shown in Fig. 6. This tissue was fixed with $KMnO_4$, a fixative that permits noradrenergic axons to be recognized by their content of small (about 50 nm diameter) dense-cored synaptic vesicles (Hokfelt, 1968; Kanerva *et al.*, 1980). Ordinarily, serotonergic axons and storage granules are not demonstrated with $KMnO_4$ (Kanerva *et al.*, 1980); however, when material from animals injected with the indolic neurotoxin, 5,7-DHT, is fixed with $KMnO_4$ within 4 h of injection, and examined electron microscopically, serotonergic axons appear to be filled with the neurotoxin and are identifiable (see Fig. 7) (Gershon and Sherman, 1981). In these experiments, uptake of 5,7-DHT by noradrenergic axons can be prevented, and the neurotoxin made useful as a serotonergic marker, by pretreating animals with desmethylimipramine (Gershon *et al.*, 1980c). The terminals that fill with 5,7-DHT contain a mixture of large and small synaptic vesicles. The vesicle population of these 5,7-DHT-filled varicosities differs from that of noradrenergic elements (compare Figs. 6 and 7). The mixture of large and small round synaptic vesicles found in axons that fill with 5,7-DHT corresponds to the vesicle population seen in terminals identified radioautographically in aldehyde-fixed gut as serotonergic by their uptake of [^{3}H] 5-HT (see Figs 1 and 2). A single morphological type of enteric axon thus seems to be able to take up either [^{3}H] 5-HT or 5,7,-DHT. This axon does not take up NE, differs morphologically from noradrenergic terminals, survives chemical sympathectomy with 6-hydroxydopamine, and its uptake mechanism is resistant to inhibition by desmethylimipramine. Since the axon identified with [^{3}H] 5-HT or 5,7,-DHT is probably serotonergic, the simultaneous recognition of serotonergic and noradrenergic elements of the myenteric plexus is possible (Gershon and Sherman, 1981). Axo-axonic synapses between the two have not yet been found. Such synapses, if they exist, therefore, must either be uncommon or lacking in classical synaptic structural specializations. Cell bodies and dendrites of enteric serotonergic neurones (Fig. 3) are more difficult than axons to find and identify in electron micrographs. It is thus still possible that axo-somatic or axo-dendritic noradrenergic-serotonergic synapses are more prevalent than axo-axonic ones.

NEURONAL ACTIONS OF 5-HT THAT HAVE NOT BEEN LINKED TO A SPECIFIC NEUROTRANSMITTER FUNCTION

Additional actions of 5-HT, besides its apparent mediation of slow EPSPs in type 2 (AH) neurones, have been described. These have been less well characterized than the slow EPSP and they have not, as yet, been linked to a specific neural property of the myenteric plexus. For example, 5-HT has been reported to be able to depress cholinergic fast EPSPs in both types 1 (S) and 2 (AH) cells (North *et al.*, 1980; Johnson *et al.*, 1980b). This effect has been attributed to a pre-synaptic inhibition by 5-HT of the release of ACh, since 5-HT does not alter responses evoked by the iontophoretic application of ACh. Rarely, 5-HT applied iontophoretically induces a hyperpolarization in occasional type 2 (AH) cells (Johnson *et al.,* 1980b). These hyperpolarizing responses to 5-HT are abolished in the absence of Ca^{2+}. They are probably, therefore, not due to a direct action of 5-HT on the type 2 (AH) cells, but may instead by a secondary effect of the local release by 5-HT of another transmitter. When the myenteric plexus is exposed to high concentrations of 5-HT, and if the NE uptake mechanism is not inhibited, noradrenergic nerves do take up 5-HT non-specifically; under these conditions, 5-HT can displace endogenous NE (Drakontides and Gershon, 1972). The rare hyperpolarizing responses to 5-HT thus may be due to the liberation of NE or another of the many enteric neurotransmitters (Furness and Costa, 1980; Gershon, 1981a).

ENTERIC 5-HT RECEPTORS

One of the problems that has stood in the way of determining the precise role of enteric serotonergic neurones in the physiological economy of the gut is the absence of a really effective antagonist, active at all of the neural 5-HT receptors. There are multiple receptors for 5-HT in the gut, although how many types of 5-HT receptor exist is not yet clear. Receptors for 5-HT found on the smooth muscle of those regions of the gut where the smooth muscle is sensitive to 5-HT are potently blocked by ergot derivatives, such as methysergide (Drakontides and Gershon, 1968; Costa and Furness, 1979a,b). Some neural actions of 5-HT are also blocked by these drugs (Costa and Furness, 1976; 1978a,b; Wood and Mayer, 1979b); however, often higher concentrations of methysergide must be used to block neural rather than muscular actions of 5-HT. Not all neural actions of 5-HT, moreover, are antagonized by methysergide (Drakontides and Gershon, 1968). In the mouse duodenum, for example, a neural receptor has been found that is quite different from 5-HT receptors on muscle and that is insensitive to methysergide. In this tissue, in fact, the smooth muscle is excited by 5-HT so that the neural effects of the amine must be studied in the constant presence of methysergide. The muscle receptors, but not the neural receptors for 5-HT, are also activated by tryptamine. The neurally mediated, tetrodotoxin-sensitive responses to 5-HT are ordinarily contractile; these contractions are potentiated by eserine and abolished by hyoscine and are thus cholinergically mediated. In the presence of hyoscine and methysergide, 5-HT induces a tetrodotoxin-sensitive relaxation of the mouse duodenum. All of these effects of 5-HT are first mimicked and then blocked by phenylbiguanide. The situation for 5-HT receptors in the mouse duodenum, therefore, is analogous to that for ACh. ACh too has neural (nicotinic) and muscle (muscarinic) receptors. Nicotine is to ACh as phenybiguanide is to 5-HT; muscarine is analogous to tryptamine and atropine is analogous to methysergide. These analogies should not be overdone. Phenylbiguanide does not have the same action on the guinea-pig gut as it does on the mouse duodenum (Costa and Furness, 1979b). Methysergide, on the other hand, does affect at least some neural responses to 5-HT in the guinea-pig gut (Wood and Mayer, 1979b). The multiplicity of 5-HT receptors is evident and probably explains these varying results. In fact, at least six classes of 5-HT receptor have been defined in molluscan ganglia (Gerschenfeld and Paupardin-Tritsch, 1974). 5-HT receptors, moreover, are quite distinct from those of ACh; nicotinic antagonists, such as hexamethonium or pentolinum, do not antagonize stimulation of 5-HT in neurones (Bulbring and Gershon, 1967; Drakontides and Gershon, 1968; Costa and Furness, 1979b). There is one nicotinic antagonist, however, that does block neural responses to the addition of 5-HT, D-tubocurarine (Hirst and Silinsky, 1975; Costa and Furness, 1979b). D-Tubocurarine is also a 5-HT antagonist at one (the 'A') of the receptors for 5-HT in ganglia of the mollusc, *Aplysia* (Gerschenfeld and Paupardin-Tritsch, 1974). In the absence of a clear definition of how many enteric 5-HT receptors there are, and in the absence of specific antagonists, desensitization to 5-HT itself has been the most useful means of blocking the action of 5-HT and testing the implication of the amine in various neurally mediated activities of the gut (Bulbring and Gershon, 1967; Furness and Costa, 1973; Costa and Furness, 1976, 1979a,b).

RESPONSES THAT MAY INVOLVE ENTERIC SEROTONERGIC NEURONES

The slow EPSP in type 2 (AH) cells is only the best studied of the neural effects in the mediation of which 5-HT has been pharmacologically implicated. Others include: vagal relaxation of the stomach in mice and guinea-pigs (Bulbring and Gershon, 1967), vagal relaxation of the lower oesophageal sphincter of the opossum (Rattan and Goyal, 1978), post-train synaptic excitation in the myenteric plexus (Dingledine and Goldstein, 1976), and ascending peristaltic excitation in the colon (Furness and Costa, 1973; Costa and Furness, 1976). Potential actions of enteric serotonergic neurones have also been suggested by the effect of the tryptophan hydroxylase inhibitor, PCPA. PCPA depresses intestinal motility (Welch and Welch, 1968; Weber, 1970; Breisch *et al.*, 1976; Saller and Stricker, 1978). The drug also renders the bowel supersensitive to 5-HT (Schulz and Cartwright, 1974b) as might be expected if PCPA were to chronically deplete 5-HT from the myenteric plexus. More recently, PCPA treatment was found to block the neural inhibiton of vagally evoked excitatory junction potentials that accompanies the descending inhibitory phase of the peristaltic reflex in the colon (Jule, 1980). This blockade, induced by PCPA, can be overcome by injection of a large dose of L-tryptophan. It is also produced by inhibiting the biosynthesis of 5-HT from 5-hydroxytryptophan. The antagonism of vagally evoked excitatory responses during descending inhibition is, furthermore, potentiated and prolonged in duration by inhibitors of the 5-HT re-uptake mechanism. The descending inhibitory phase of the peristaltic reflex, therefore, may have at least two neural components and 5-HT seems to be a mediator involved in one of them. One component is the relaxation of enteric smooth muscle by non-adrenergic, non-cholinergic, intrinsic, inhibitory neurones (Burnstock, 1979b; Costa and Furness, 1976; Jule, 1980). The other component is a parallel inhibition of potentially competing cholinergic excitation of the bowel (Jule,

1980). 5-HT apparently is one of the neurotransmitters of the enteric interneurones that are involved in the mediation of this portion of the reflex.

ONTOGENY AND PHYLOGENY OF ENTERIC SEROTONERGIC NEURONES

The EC cell system, with its large concentration of 5-HT (Erspamer, 1966), makes impossible, in species that have these cells, the assessment of the presence or absence of serotonergic neurones through the simple expedient of measuring the 5-HT content of the whole bowel. Nevertheless, enterochromaffin cells, are lacking in the bowel of many lower vertebrates, including both cyclostomes and teleosts (Erspamer, 1966; Watson, 1979). Despite the absence of enterochromaffin cells these lower vertebrates still have 5-HT in their intestines (Baumgarten *et al.*, 1973; Watson, 1979; Goodrich *et al.*, 1980). Dissection of the hagfish gut and measurement of 5-HT in its layers indicates that the amine is located within the myenteric plexus. 5-HT has also, as noted earlier, been found histochemically in hagfish, lamprey and teleost enteric neurones (Baumgarten *et al.*, 1973; Watson, 1979; Goodrich *et al.*, 1980). The similarity between the serotonergic neurones in the gastrointestinal tracts of primitive vertebrates and those of mammals is shown by the observation that the uptake mechanism for [^{3}H] 5-HT of hagfish enteric serotonergic neurones is identical to that of mammals (Goodrich *et al.*, 1980). In fact, even the synaptic vesicle population of axonal varicosities that label with [^{3}H] 5-HT in hagfish resembles that of mammals.

The distribution of enteric serotonergic neurones amongst vertebrates has been examined using the specific uptake of [^{3}H] 5-HT as a marker property. In addition to the guts of cyclostomes, these neurones have been found in the gastrointestinal tracts of teleosts and amphibia (Goodrich *et al.*, 1980), birds (Epstein *et al.*, 1980; Gershon *et al.*, 1980a), rabbits (Rothman *et al.*, 1976), guinea-pigs (Gershon and Altman, 1971; Gershon *et al.*, 1976), mice (Rothman *et al.*, 1979), rats (Gershon *et al.*, 1977), humans, and non-human primates (Goodrich and Gershon, 1977; Rogawski *et al.*, 1978). Tunicates and echinoderms, however, do not have enteric serotonergic neurones (Goodrich *et al.*, 1980). Since enteric serotonergic neurones have been reported in the gut and endostyle of amphioxus (Salimova, 1978), the vertebrate enteric serotonergic neurone may have evolved first in an ancestral cephalochordate.

It seems likely from these studies of phylogeny that enteric serotonergic neurones appeared early in vertebrate evolution and have persisted in birds and mammals. It may be fair to conclude that they are a constant member of the coterie of neurones that populate the vertebrate ENS. This early origin and evolutionary persistence suggests that these neurones are important to the functioning of the ENS.

The early origin of the enteric serotonergic neurone in evolution is parallelled by its similarly early appearance during mammalian and avian ontogeny. During the development of the gut in mammals, the enteric serotonergic neurones appear at the same time as, or just after, the appearance of enteric cholinergic neurones (Rothman *et al.,* 1976, 1979; Gintzler *et al.,* 1980; Gershon *et al.,* 1981). At the time these neurones can first be detected, the gut does not contain smooth muscle (Gershon *et al.,* 1981), no mechanical responses to nerve stimulation can be provoked (Gintzler *et a.,* 1980), and the noradrenergic innervation has yet to invade the bowel (Rothman *et al.,* 1976; Gintzler *et al.,* 1980; Gershon *et al.,* 1981). The gut does synthesize ACh from choline (Rothman *et al.,* 1979). If the mammalian gut is removed and explanted into tissue culture, or if the avian gut is explanted and grown as a graft on a host chorioallantoic membrane, before any neurones are morphologically apparent, a myenteric plexus will form in the explants (Rothman *et al.,* 1979, Gershon *et al.,* 1980a). The myenteric plexus in these explants has been found to contain intrinsic serotonergic neurones. It seems likely, therefore, that the neural crest progenitors that colonize the embryonic bowel and give rise to the ENS (LeDouarin, 1980) are not recognizable as they migrate to the gut. They thus acquire their definitive characteristics within the microenvironment of the gut itself. The role played by the embryonic enteric microenvironment in determining the choice of a serotonergic or other phenotype has yet to be established. That the microenvironment may be important is indicated by the observation that enteric serotonergic axons, unlike noradrenergic and cholinergic axons, fail to grow into congenitally aganglionic segments of the bowel in patients with Hirschsprung's disease (Rogawski *et al.,* 1978).

FUTURE TRENDS

In the past, much of the work done on enteric serotonergic neurones was designed to determine whether or not they existed. It seems reasonable to conclude now that they do exist. Research in the future will probably, therefore, address the issues of what these neurones do, how they relate to other enteric neurones, what specific properties they have, and what factors are important in determining their development.

Electrophysiological studies of the kind done to implicate 5-HT as the transmitter that mediates the slow EPSP in type 2 (AH) neurones (Wood and Mayer, 1979a,b) are just beginning. Anatomical techniques that permit the recognition of serotonergic neural elements have recently been developed. These include electron microscopic radioautography of material exposed to [3H]5-HT (Figs. 1-3), the loading of terminals with the indolic neurotoxin, 5,7-DHT (Gershon *et al.,* 1980c; Gershon and Sherman, 1981), and the use of an antibody to 5-HT to demonstrate the amine immunocytochemically. The combination of these anatomical techniques with electrophysiological exploration of the activity of enteric neurones promises to be a powerful tool to use to study the function of enteric serotonergic neurones. This approach will also be useful in investigating the relationships enteric serotonergic neurones have with other neurones of the ENS. Valuable too, in this regard, will be anatomical studies in which a combination of techniques is employed. For example, the immunocytochemical demonstration of enteric neuropeptides can be paired with the immunocytochemical, histochemical, or radioautographic demonstration of serotonergic elements. Additional work on the properties of enteric serotonergic neurones will certainly include work on the regulation of the biosynthesis and release of the amine. The neuromodulation that has been demonstrated of the uptake of L-tryptophan (Gershon and Dreyfus, 1980) and the noradrenergic control of 5-HT release (Wood and Mayer, 1979c; Gershon, 1980) are especially interesting in this regard. In addition, isolation and characterization of the types of neural 5-HT receptor present in the ENS would certainly be important. Further study of the subcellular storage mechanism for 5-HT is also to be expected. The synaptic vesicles of enteric serotonergic neurones have not yet been obtained, although isolated axonal varicosities (autonomic synaptosomes), including serotonergic varicosities, have successfully been prepared from the myenteric plexus (Jonakait *et al.*, 1979a). Considerable progress has already been made in developing techniques for the study of enteric serotonergic neurone development, especially tissue culture (Dreyfus *et al.*, 1977a,b; Rothman *et al.*, 1979; Gershon *et al.*, 1980b). In particular, the role of the enteric microenvironment and neuronal interactions in enteric neuronal development, seem ripe for study at this time.

SUMMARY

5-HT in the myenteric plexus has now satisfied all of the criteria necessary for the establishment of a substance as a neurotransmitter (Iversen, 1979; Gershon, 1979b). 5-HT is present in enteric neurones, and these neurones also synthesize 5-HT from the dietary amino acid, L-tryptophan. Enteric neurones also are modified to take up L-tryptophan by a high affinity mechanism and they contain a specific storage protein for 5-HT. This protein, SBP, is similar or identical to the SBP found in central serotonergic neurones but different from that of enterochromaffin cells, mast cells, or platelets. 5-HT, moreover, is released by enteric neurones upon stimulation. This release is Ca^{2+} dependent and is accompanied by the simultaneous release of SBP. It is likely, therefore, that the enteric neural release of 5-HT occurs by exocytosis. The specific uptake mechanism for 5-HT that is present in enteric neurones also resembles that of central serotonergic neurones and provides an adequate means of inactivating 5-HT acting as a neurotransmitter. 5-HT mimics several neural effects on the gut. The best studied of these is the slow EPSP present in type 2 (AH) neurones in the myenteric plexus. 5-HT not only duplicates the action of the naturally released transmitter, but pharmacological inhibition of the action of 5-HT also blocks the generation of the slow EPSP. Another action likely to involve enteric serotonergic neurones is the phase of descending inhibition of neurally evoked excitation that accompanies the peristaltic reflex in the colon. This descending inhibition is lost when 5-HT stores are depleted, returns when they are replenished, and is potentiated when 5-HT inactivation (re-uptake) is blocked. Enteric serotonergic neurones thus are well established. Their early appearance in vertebrate phylogeny and ontogeny indicate that they are basic to the function of the bowel. Much still needs to be learned about this function. More also needs to be learned about the function of the ENS. Future research should be directed at these functional problems as well as to the exploration of the cellular biology of the peculiarly accessible enteric serotonergic neurones.

REFERENCES

Aghajanian, G. K. and Asher, I. M. (1971). Histochemical fluorescence of raphe neurones: selective enhancement by tryptophan, *Science,* 172, 1159-1161.

Aghajanian, G. K., Kuhar, M. J., and Roth, R. H. (1973). Serotonin-containing neuronal perikarya and terminals: differential effects of p-chlorophenylalanine, *Brain Res.,* 54, 85-101.

Ahlman, H. and Enerback, L. (1974). A cytofluorometric study of the myenteric plexus in the guinea-pig, *Cell Tissue Res.* 153, 419-434.

Ahlman, H., Enerback, L., Kewenter, J., and Storm, B. (1973). Effects of extrinsic denervation on the fluorescence of monoamines in the small intestine of the cat, *Acta Physiol. Scand.* 89, 429-435.

Ahlman, H., Lundberg, J., Dahlstrom, A., and Kewenter, J. (1976). A possible vagal adrenergic release of serotonin from enterochromaffin cells in the cat, *Acta Physiol. Scand.* 98, 366-375.

Amin, A. H., Crawford, T. B. B., and Gaddum, J. H. (1954). The distribution of substance P and 5-hydroxytryptamine in the central nervous system of the dog, *J. Physiol.* (London), 126, 596-618.

Baumgarten, H. G., Bjorklund, A., Lachenmayer, L., Nobin, A., and Rosengren, E. (1973). Evidence for existence of serotonin-, dopamine- and noradrenaline-containing neurons in the gut of *Lampetra fluviatilis, Z. Zellforsch.* 141, 33-54.

Bayliss, W. M., and Starling, E. H. (1899). The movements and innervation of the small intestine, *J. Physiol.* (London), 24, 99-143.

Bernd, P., Gershon, M. D., Nunez, E. A., and Tamir, H. (1981). Separation of dissociated thyroid follicular and parafollicular cells: association of serotonin binding protein with parafollicular cells. *J. Cell Biol.* 88, 499-508.

Bernd, P., Nunez, E. A., Gershon, M. D., and Tamir, H. (1979). Serotonin binding protein: characterization and localization in parafollicular cells of the sheep thyroid, in *Molecular Endocrinology* (Eds I. MacIntyre and M. Szelke), pp. 123-132, Elsevier/North Holland, Amsterdam, New York, Oxford.

Boullin, D. J. (1964). Observations on the significance of 5-hydroxytryptamine in relation to the peristaltic reflex of the rat, *Br. J. Pharmacol.* 23, 14-33.

Breisch, S. T., Zemlan, F. P., and Hoebel, B. G. (1976). Hyperphagia and obesity following serotonin depletion by intravesicular p-chlorophenylalanine, *Science,* 192, 382-385.

Brownlee, G. and Johnson, E. S. (1963). The site of the 5-hydroxytryptamine receptor on the intramural nervous plexus of the guinea-pig isolated ileum, *Br. J. Pharmacol.* 21, 306-322.

Bulbring, E. and Crema, A. (1958). Observations concerning the action of 5-hydroxytryptamine on the peristaltic reflex, *Br. J. Pharmacol.* 13, 444-457.

Bulbring, E. and Crema, A. (1959). The release of 5-hydroxytryptamine in relation to pressure exerted on the intestinal mucosa, *J. Physiol.* (London), 146, 18-28.

Bulbring, E. and Gershon, M. D. (1967). 5-Hydroxytryptamine participation in the vagal inhibitory innervation of the stomach, *J. Physiol.* (London), 192, 823-846.

Bulbring, E. and Lin, R. C. Y. (1958). The effect of intraluminal application of 5-hydroxytryptamine and 5-hydroxytryptophan on peristalsis, the local production of 5-hydroxytryptamine and its release in relation to intraluminal pressure and propulsive activity, *J. Physiol.* (London), **140**, 381–407.

Burnstock, G. (1979a). Autonomic neuroeffector junctions, in *Neurosciences Research Program Bulletin,* Vol. 17, *Non-adrenergic, Non-cholinergic Autonomic Neurotransmission Mechanisms* (Eds G. Burnstock, M. D. Gershon, T. Hokfelt, L. L. Iversen, H. W. Kosterlitz, and J. H. Szurszewski), pp. 388–391, The MIT Press, Cambridge.

Burnstock, G. (1979b). Adenosine triphosphate, in *Neurosciences Research Program Bulletin,* Vol. 17, *Non-adrenergic, Non-cholinergic Autonomic Neurotransmission Mechanisms* (Eds G. Burnstock, M. D. Gershon, T. Hokfelt, L. L. Iversen, H. W. Kosterlitz, and J. H. Szurszewski), pp. 406–414, The MIT Press, Cambridge.

Costa, M. and Furness, J. B. (1976). The peristaltic reflex: an analysis of the nerve pathways and their pharmacology, *Naunyn-Schmiedeberg's Arch. Pharmacol.* **294**, 47–60.

Costa, M. and Furness, J. B. (1979a). Commentary: On the possibility that an indoleamine is a neurotransmitter in the gastrointestinal tract, *Biochem. Pharmacol.* **28**, 565–571.

Costa, M. and Furness, J. B. (1979b). The sites of action of 5-HT in nerve muscle preparations from guinea-pig small intestine and colon, *Br. J. Pharmacol.* **65**, 237–248.

Costa, M., Furness, J. B., and Gabella, G. (1971). Catecholamine containing nerve cells in the mammalian myenteric plexus, *Histochem.* **25**, 103–106.

Costa, M., Furness, J. B., and McLean, J. R. (1976). The presence of aromatic L-amino acid decarboxylase in certain intestinal nerve cells, *Histochem.* **48**, 129–143.

Dingledine, R. and Goldstein, A. (1976). Effect of synaptic transmission blockade on morphine action in the guinea-pig myenteric plexus, *J. Pharmacol. Exp. Ther.* **196**, 97–106.

Dingledine, R., Goldstein, A., and Kendig, J. (1974). Effects of narcotic opiates and serotonin on the electrical behavior of neurons in the guinea-pig myenteric plexus, *Life Sci.,* **14**, 2299–2309.

Drakontides, A. B. and Gershon, M. D. (1968). 5-HT receptors in the mouse duodenum, *Br. J. Pharmacol.* **33**, 480–492.

Drakontides, A. B. and Gershon, M. D. (1972). Studies of the interaction of 5-hydroxytryptamine (5-HT) and the perivascular innervation of the guinea-pig caecum, *Br. J. Pharmacol.* **45**, 417–434.

Dreyfus, C. F., Bornstein, M. B., and Gershon, M. D. (1977a). Synthesis of serotonin by neurons of the myenteric plexus *in situ* and in organotypic tissue culture, *Brain Res.* **128**, 125–139.

Dreyfus, C. F., Sherman, D., and Gershon, M. D. (1977b). Uptake of serotonin by intrinsic neurons of the myenteric plexus grown in organotypic tissue culture, *Brain Res.* **128**, 109–123.

Dubois, A. and Jacobowitz, D. M. (1974). Failure to demonstrate serotonergic neurons in the myenteric plexus of the rat, *Cell Tissue Res.* **150**, 493–496.

Epstein, M. L., Sherman, D. L., and Gershon, M. D. (1980). Development of serotonergic neurons in the chick duodenum, *Devel. Biol.* **77**, 20–40.

Erspamer, V. (1966). Occurrence of indolealkylamines in nature, in *Handbook of Experimental Pharmacology,* Vol. 19, *5-Hydroxytryptamine and Related Indolealkyamines.* (Ed. V. Erspamer), pp. 132–181, Springer-Verlag, New York.

Feher, E. (1974). Effect of monoamine oxidase inhibitor on the nerve elements of the isolated cat ileum, *Acta Morphologica Acad. Sci. Hung.* **22**, 249-263.

Feher, E. (1975). Effects of monoamine inhibitor on the nerve elements of the isolated cat's ileum, *Verh. Anat. Ges.* **69**, 477-482.

Feldberg, W. and Toh, C. C. (1953). Distribution of 5-hydroxytryptamine (serotonin, enteramine) in the wall of the digestive tract, *J. Physiol.* (London), **119**, 352-362.

Furness, J. B. and Costa, M. (1973). The nervous release and the action of substances which affect intestinal muscle through neither adrenoreceptors nor cholinoreceptors, *Phil. Trans. Roy. Soc.* Series B. **265**, 123-133.

Furness, J. B. and Costa, M. (1974). The adrenergic innervation of the gastrointestinal tract, *Ergebn. Physiol. Biol. Chem. Exp. Pharmakol.* **69**, 1-51.

Furness, J. B. and Costa, M. (1978). Distribution of intrinsic nerve cell bodies and axons which take up aromatic amines and their precursors in the small intestine of the guinea-pig, *Cell Tissue Res.* **188**, 527-543.

Furness, J. B., and Costa, M. (1980). Types of nerves in the enteric nervous system, *Neuroscience*, **5**, 1-20.

Furness, J. B., Costa, M., and Freeman, C. G. (1979). Absence of tyrosine hydroxylase activity and dopamine B-hydroxylase immunoreactivity in intrinsic nerves of guinea-pig ileum, *Neuroscience*, **4**, 305-310.

Furness, J. B., Costa, M., and Howe, R. C. (1980). Intrinsic amine-handling neurons in the intestine, in *Histochemistry and Cell Biology of Autonomic Neurons, SIF Cells and Paraneurons* (Eds O. Eranko, S. Soinila, and H. Paivarinta), pp. 367-372, Raven Press, New York.

Fuxe, K. and Jonsson, G. (1967). A modification of the histochemical fluorescence method for the improved localization of 5-hydroxytryptamine, *Histochem.* **11**, 161-166.

Gaddum, J. H. and Picarelli, Z. P. (1957). Two kinds of tryptamine receptor, *Br. J. Pharmacol.* **12**, 323-328.

Geffen, L. B. and Jarrott, B. (1977). Cellular aspects of catecholaminergic neurons, in *Handbook of Physiology. Section 1. The Nervous System*. Vol. 1, *Cellular Biology of Neurons* (Eds J. M. Brookhart, V. B. Mountcastle, E. R. Kandel, and S. R. Geiger), pp. 521-571, Am. Physiol. Soc., Bethesda, Maryland.

Gerschenfeld, H. M. and Paupardin-Tritsch, D. (1974). Ionic mechanisms and receptor properties underlying the responses of molluscan neurones to 5-hydroxytryptamine, *J. Physiol.* (London), **243**, 427-456.

Gershon, M. D. (1967). Effects of tetrodotoxin on innervated smooth muscle preparations, *Br. J. Pharmacol.* **29**, 259-279.

Gershon, M. D. (1977). Biochemistry and physiology of serotonergic transmission, in *Handbook of Physiology. Section 1. The Nervous System*. Vol. 1, *Cellular Biology of Neurons* (Eds J. M. Brookhart, V. B. Mountacastle, E. R. Kandel, and S. R. Geiger), pp. 573-623, Am. Physiol. Soc., Bethesda, Maryland.

Gershon, M. D. (1979a). The autonomic nervous system in *Neurosciences Research Program Bulletin*, Vol. 17, *Non-adrenergic, Non-cholinergic Autonomic Neurotransmission Mechanisms* (Eds G. Burnstock, M. D. Gershon, T. Hokfelt, L. L. Iversen, H. W. Kosteritz, and J. H. Szurszewski), pp. 384-388, The MIT Press, Cambridge.

Gershon, M. D. (1979b). Putative neurotransmitters: serotonin, in *Neurosciences Research Program Bulletin*, Vol. 17, *Non-adrenergic, Non-cholinergic Autonomic Neurotransmission Mechanisms* (Eds G. Burnstock, M. D. Gershon, T. Hokfelt, L. L. Iversen, H. W. Kosterlitz, and J. H. Szurszewski), pp. 414-424, The MIT Press, Cambridge.

Gershon, M. D. (1980). Storage and release of serotonin and serotonin binding protein by serotonergic neurons, in *Cellular Basis of Chemical Messengers in the Digestive Tract* (Eds M. I. Grossman, M. A. B. Brazier, and J. Lechago), pp. 285–298, Academic Press, New York.

Gershon, M. D. (1981a). The enteric nervous system, *Ann. Rev. Neurosci.* **4**, 227–272.

Gershon, M. D. (1981b). Neurotransmitter stimulatation of tryptophan uptake by the myenteric plexus of the guinea-pig ileum, *Proc. Eighth Int. Cong. Pharmacol., Tokyo, July.* (In press.)

Gershon, M. D. and Altman, R. F. (1971). An analysis of the uptake of 5-hydroxy-tryptamine by the myenteric plexus of the small intestine of the guinea-pig, *J. Pharmacol. Exp. Ther.* **179**, 29–41.

Gershon, M. D. and Bursztajn, S. (1978). Properties of the enteric nervous system: Limitation of access of intravascular macromolecules to the myenteric plexus and muscularis externa, *J. Comp. Neurol.* **180**, 467–488.

Gershon, M. D., Drakontides, A. B., and Ross, L. L. (1965). Serotonin: synthesis and release from the myenteric plexus of the mouse intestine, *Science*, **149**, 197–199.

Gershon, M. D. and Dreyfus, C. F. (1980). Stimulation of tryptophan uptake into enteric neurons by 5-hydroxytryptamine: a novel form of neuromodulation, *Brain Res.* **184**, 229–233.

Gershon, M. D., Dreyfus, C. F., Pickel, V. M., Joh, T. H., and Reis, D. J. (1977). Serotonergic neurons in the peripheral nervous system: identification in gut by immunohistochemical localization of tryptophan hydroxylase, *Proc. Natl. Acad. Sci., USA,* **74**, 3086–3089.

Gershon, M. D., Epstein, M. L., and Hegstrand, L. (1980a). Colonization of the chick gut by progenitors of enteric serotonergic neurons: distribution, differentiation, and maturation within the gut, *Devel. Biol.* **77**, 41–51.

Gershon, M. D., Erde, S. M., and Dreyfus, C. F. (1980b). Long term survival of serotonergic neurons in tissue cultures of myenteric plexus dissected from guinea pig ileum and duodenum, *Neurosci. Absts.* **6**, 274.

Gershon, M. D. and Jonakait, G. M. (1979). Uptake and release of 5-hydroxytryptamine by enteric serotonergic neurons: effects of fluoxetine (Lilly 110140) and chlorimipramine, *Br. J. Pharmacol.* **66**, 7–9.

Gershon, M. D., Robinson, R. G., and Ross, L. L. (1976). Serotonin accumulation in the guinea pig's myenteric plexus, ion dependence, structure activity relationship, and the effect of drugs, *J. Pharmacol. Exp. Ther.* **198**, 548–561.

Gershon, M. D. and Ross, L. L. (1966a). Radioisotopic studies of the binding, exchange and distribution of 5-hydroxytryptamine synthesized from its radioactive precursor, *J. Physiol.* (London), **186**, 451–476.

Gershon, M. D. and Ross, L. L. (1966b). Localization of sites of 5-hydroxytryptamine storage and metabolism by radioautography, *J. Physiol.* (London), **186**, 477–492.

Gershon, M. D. and Sherman, D. (1981). Interactions between noradrenergic and serotonergic elements in the myenteric plexus: electron microscopy of doubly labeled material and measurement of transmitter release, *Anat. Rec.* **199**, 92A–93A.

Gershon, M. D., Sherman, D., and Dreyfus, C. F. (1980c). Effects of indolic neurotoxins on enteric serotonergic neurons, *J. Comp. Neurol.* **190**, 581–596.

Gershon, M. D., Sherman, D., and Gintzler, A. (1981). An ultrastuctural analysis of the developing enteric nervous system of the guinea pig small intestine, *J. Neurocytol.* **10**, 271–296.

Gershon, M. D. and Tamir, H. (1979). Serotonin (5-HT) release from stimulated peripheral neurons, *Proc. Seventh Ann. Meeting Int. Soc. Neurochem.* p. 349.

Gershon, M. D. and Tamir H. (1981). Serotonin binding protein: role in transmitter storage in central and peripheral serotonergic neurons, in *Serotonin* (Eds B. Haber, S. Gabay, M. Issidorides, and S. Alivisatos), pp. 37-50, Plenum Publishing Co., New York.

Gintzler, A. R., Rothman, T. P., and Gershon, M. D. (1980). Ontogeny of opiate mechanisms in relation to the sequential development of neurons known to be components of the guinea pig's enteric nervous system, *Brain Res.* **189**, 31-48.

Goodrich, J. T., Bernd, P., Sherman, D. L., and Gershon, M. D. (1980). Phylogeny of enteric serotonergic neurons, *J. Comp. Neurol.* **190**, 15-28.

Goodrich, J. T. and Gershon, M. D. (1977). Serotonergic neurons in the enteric nervous system of human and sub-human primates, *Proc. Int. Union Physiol. Sci.* XIII, Paris, p2173.

Grafe, P., Mayer, C. J., and Wood, J. D. (1979b). Evidence that substance P does not mediate slow synaptic excitation within the myenteric plexus, *Nature* (London), **279**, 720-721.

Grafe, P., Mayer, C. J., and Wood, J. D. (1980). Synaptic modulation of calcium-dependent potassium conductance in myenteric neurons, *J. Physiol.* (London), **305**, 235-248.

Grafe, P., Wood, J. D., and Mayer, C. J. (1979a). Fast excitatory postsynaptic potentials in AH (type 2) neurons of guinea-pig myenteric plexus, *Brain Res.* **163**, 349-352.

Grafstein, B. and Forman, D. (1980). Intracellular transport in neurons, *Physiol. Rev.* **60**, 1167-1283.

Gwee, M. C. E. and Yeoh, T. S. (1968). The release of 5-hydroxytryptamine from rabbit small intestine *in-vitro, J. Physiol.* (London), **194**, 817-825.

Hamberger, B. and Norberg, K. A. (1965). Studies of some systems of adrenergic synaptic terminals in the abdominal ganglia of the cat, *Acta Physiol. Scand.* **65**, 235-242.

Harry. J. (1963). The action of drugs on the circular muscle strip from the guinea-pig isolated ileum, *Br. J. Pharmacol. Chemother.* **20**, 399-417.

Hirst, G. D. S., Holman, M. E., and Spence, I. (1974). Two types of neurones in the myenteric plexus of duodenum in the guinea-pig, *J. Physiol.* (London), **236**, 303-326.

Hirst, G. D. S. and Silinsky, E. M. (1975). Some effects of 5-hydroxytryptamine, dopamine and noradrenaline on neurones in the submucous plexus of guinea-pig small intestine, *J. Physiol.* (London), **251**, 817-832.

Hirst, G. D. S. and Spence, I. (1973). Calcium action potentials in mammalian peripheral neurones, *Nature* (London), **243**, 54-56.

Hokfelt, T. (1968). *In vitro* studies on central and peripheral monoamine neurons at the ultrastructural level, *Z. Zellforsch.* **91**, 1-74.

Iversen, L. L. (1979). Criteria for establishing a neurotransmitter, in *Neurosciences Research Program Bulletin,* Vol. 17, *Non-adrenergic, Non-cholinergic Autonomic Neurotransmission Mechanisms,* (Eds G. Burnstock, M. D. Gershon, T. Hokfelt, L. L. Iversen, H. W. Kosterlitz, and J. H. Szurszewski), p. 406, The MIT press, Cambridge.

Jacobowitz, D. (1965). Histochemical studies of the autonomic innervation of the gut, *J. Pharmacol. Exp. Ther.* **149**, 358-364.

Johnson, S. M., Katayama, Y., and North, R. A. (1980a). Slow synaptic potentials in neurones of the myenteric plexus. *J. Physiol.* (London), **301**, 505-516.

Johnson, S. M., Katayama, Y., and North, R. A. (1980b). Multiple actions of 5-hydroxytryptamine on myenteric neurones of the guinea-pig ileum, *J. Physiol.* (London), **304**, 459-479.

Jonakait, G. M., Gintzler, A. R., and Gershon, M. D. (1979a). Isolation of axonal varicosities (autonomic synaptosomes) from the enteric nervous system, *J. Neurochem.* **32**, 1387-1400.

Jonakait, G. M., Tamir, H., Gintzler, A. R., and Gershon, M. D. (1979b). Release of [^{3}H]serotonin and its binding protein from enteric neurones, *Brain Res.* **174**, 55-69.

Jonakait, G. M., Tamir, H., Rapport, M. M., and Gershon, M. D. (1977). Detection of a soluble serotonin binding protein in the mammalian myenteric plexus and other peripheral sites of serotonin storage, *J. Neurochem.* **28**, 277-284.

Jonsson, G., Fuxe, K., Hamberger, B., and Hokfelt, T. (1969). 6-Hydroxytryptamine—a new tool in monoamine fluorescence histochemistry, *Brain Res.* **13**, 190-195.

Jule, Y. (1980). Nerve-mediated descending inhibition in the proximal colon of the rabbit, *J. Physiol.* (London), **159**, 361-368.

Juorio, A. V. and Gabella, G. (1974). Noradrenaline in the guinea-pig alimentary canal: regional distribution and sensitivity to denervation and reserpine, *J. Neurochem.* **221**, 851-858.

Kanerva, L., Hervonen, A., and Fronblad, M. (1980). Observations on the ultrastructural localization of monoamines with permanganate fixation, in *Histochemistry and Cell Biology of Autonomic Neurons, SIF Cells, and Paraneurons* (Eds O. Eranko, S. Soinila, and H. Paivarinta), pp. 279-285, Raven Press, New York.

Katayama, Y. and North, R. A. (1978). Does substance P mediate slow synaptic excitation within the myeneteric plexus? *Nature* (London), **274**, 387-388.

Katayama, Y., North, R. A., and Williams, J. T. (1979). The action of substance P on neurons of the myenteric plexus of the guinea-pig small intestine, *Proc. R. Soc. Lond.* B, **206**, 191-208.

Koevary, S. B., McEvoy, R. C., and Azmitia, E. C. (1980). Specific uptake of tritiated serotonin in the adult rat pancreas, *Am. J. Anat.* **159**, 361-368.

Kuhar, M. J., Aghajanian, G. K., and Roth, R. H. (1972). Tryptophan hydroxylase activity and synaptosomal uptake of serotonin in discrete brain regions after midbrain raphe lesions: correlations with serotonin levels and histochemical fluorescence, *Brain Res.* **44**, 165-176.

Kuhar, M. J., Roth, R. H., and Aghajanian, G. K. (1971). Selective reduction of tryptophan hydroxylase activity in rat forebrain after midbrain raphe lesions, *Brain Res.* **35**, 167-176.

LeDouarin, N. M. (1980). The ontogeny of the neural crest in avian embryo chimeras, *Nature* (London), **286**, 663-669.

Meech, R. W. and Standen, N. B. (1975). Potassium activation in *Helix aspersa* neurons under voltage lamp: A component mediated by calcium influx. *J. Physiol.* (London), **249**, 211-239.

Morita, K., North, R. A., and Katayama, Y. (1980). Evidence that substance P is a neurotransmitter in the myenteric plexus, *Nature* (London), **287**, 151-152.

Nishi, S. and North, R. A. (1973). Intracellular recording from the myenteric plexus of the guinea-pig ileum, *J. Physiol.* (London), **231**, 471-491.

Norberg, K. A. (1964). Adrenergic innervation of the intestinal wall studied by fluorescence microscopy, *Int. J. Neuropharmacol.* **18**, 743-751.

North, R. A. (1973). The calcium-dependent slow after-hyperpolarization in myenteric plexus neurones with tetrodotoxin-resistant action potentials, *Br. J. Pharmacol.* **49**, 709-711.

North, R. A., Henderson, C., Katayama, Y., and Johnson, S. M. (1980). Electrophysiological evidence of presynaptic inhibition of acetylcholine release by 5-hydroxytryptamine in the enteric nervous system, *Neuroscience,* **5**, 581-586.

North, R. A. and Nishi, S. (1976). The soma spike in myenteric plexus neurons with a calcium-dependent after-hyperpolarization, in *Physiology of Smooth Muscle* (Eds E. Bulbring and M. F. Shuba), pp. 303–307, Raven Press, New York.

Nunez, E. A., Gershon, P., and Gershon, M. D. (1980). Serotonin and seasonal variation in the pancreatic structure of bats: possible presence of serotoninergic axons in the gland, *Am. J. Anat.* **159**, 347–360.

Paton, W. D. M. and Vane, J. R. (1963). An analysis of the response of the isolated stomach to electrical stimulation and to drugs, *J. Physiol.* (London), **165**, 10–46.

Rapport, M. M., Green, A. A., and Page, I. H. (1948). Serum vasoconstrictor (serotonin). IV. Isolation and characterization, *J. Biol. Chem.* **176**, 1243–1251.

Rattan, S. and Goyal, R. K. (1978). Evidence of 5-HT participation in vagal inhibitory pathway to opossum LES, *Am. J. Physiol.* **234**, E273–276.

Roberston, P. A. (1953). Antagonism of 5-hydroxytryptamine by atropine, *J. Physiol.* (London), **121**, 54P–55P

Robinson, R. and Gershon, M. D. (1971). Synthesis and uptake of 5-hydroxytryptamine by the myenteric plexus of the small intestine of the guinea pig, *J. Pharmacol. Exp. Ther.* **179**, 29–41.

Roche e Silva, M. and DoValle, R. (1953). Mechanism of action of serotonin upon guinea pig ileum, in *Abstracts of the 19th International Physiological Congress,* Montreal, p. 503.

Rogawski, M. A., Goodrich, J. T., Gershon, M. D., and Touloukian, R. J. (1978). Hirschsprung's disease: absence of serotonergic neurons in the aganglionic colon, *J. Pediat. Surg.* **13**, 608–615.

Rothman, T. P., Dreyfus, C. F., and Gershon, M. D. (1979). Differentiation of enteric neurons from unrecognizable precursors within the microenvironment of cultured fetal mouse gut, *Neurosci. Abst.* **5**, 176.

Rothman, T. P., Ross, L. L., and Gershon, M. D. (1976). Separately developing axonal uptake of 5-hydroxytryptamine and norepinephrine in the fetal ileum of the rabbit, *Brain Res.* **115**, 437–456.

Rubin, W., Gershon, M. D., and Ross, L. L. (1971). Electron microscopic radioautographic identification of serotonin-synthesizing cells in the mouse gastric mucosa, *J. Cell Biol.* **50**, 399–415.

Salimova, N. (1978). Localization of biogenic monoamines in Amphioxus *Brachiostoma lanceolatum, Dokl. Acad. Sci. USSR,* **242**, 939–941.

Saller, C. F. and Stricker, E. M. (1978). Decreased gastrointestinal motility in rats after parenteral injection of *p*-chlorophenylalanine, *J. Pharm. Pharmacol.* **30**, 646–647.

Sato, T., Takayanagi, I., and Takagi, K. (1974). Effects of acetycholine releasing drugs on electrical activities obtained from Auerbach's plexus in the guinea-pig ileum, *Jap. J. Pharmacol.* **24**, 447–451.

Schulz, R. and Cartwright, C. (1974a). Effect of morphine on the serotonin release from the myenteric plexus of the guinea pig, *J. Pharmacol. Exp. Ther.* **190**, 420–430.

Schulz, R. and Cartwright, C. (1974b). Supersensitivity with PCPA pretreatment on the sensitivity of normal and morphine tolerant muscle strips of the guinea pig ileum. *Fed. Proc.* **33**, 502a.

Takiff, J., and Gershon, M. D. (1970). Uptake of 5-hydroxytryptamine by noradrenergic elements of the myenteric plexus, *Anat. Rec.* **166**, 388.

Tamir, H. and Gershon, M. D. (1979). Storage of serotonin and serotonin binding protein in synaptic vesicles, *J. Neurochem.* **33**, 35–44.

Tamir, H. and Huang, I. L. (1974). Binding of serotonin to soluble protein from synaptosomes, *Life Sci.* **14**, 83–93.

Tamir, H. and Kuhar, M. J. (1975). Association of serotonin-binding protein with projections of the midbrain raphe nuclei, *Brain Res.* 83, 164–172.

Tamir, H., Klein, A., and Rapport, M. M. (1976). Serotonin binding protein: enhancement of binding by Fe^{2+} and inhibition by drugs, *J. Neurochem.* 26, 871–878.

Tamir, H., Bebirian, R. J., Casper, D., and Muller, F. (1980a). Differences between intracellular platelet and brain proteins that bind serotonin, *J. Neurochem.* 35, 1033–1044.

Tamir, H., Askenase, P. W., Theoharides, T. C., and Gershon, M. D. (1980b). Comparison of serotonin binding proteins of neuronal and mesenchymal (mast cell + basophil) origin, *Neuroscience Abst.* 6, 119.

Tamir, H. and Gershon, M. D. (1981). Intracellular proteins that bind serotonin in neurons, paraneurons, and platelets, *J. Physiol.* (Paris), 77. (In press.)

Tansy, M. F., Rothman, G., Bartlett, J., Farber, P., and Hohenleitner, F. J. (1971). Vagal adrenergic degranulation of enterochromaffin cell system in guinea-pig duodenum, *J. Pharm. Sci.* 60, 81–85.

Thoa, H. B., Eccleston, D., and Axelrod, J. (1969). The accumulation of C^{14}-serotonin in the guinea pig vas deferens, *J. Pharmacol. Exp. Ther.* 169, 69–73.

Toh, C. C. (1954). Release of 5-hydroxytryptamine (serotonin) from the dog's gastrointestinal tract, *J. Physiol.* (London), 126, 248–254.

Trendelenburg, P. (1917). Physiologische und pharmakologische Versuche uber die Dunndarm peristaltik. *Naunyn-Schmiedebergs Arch. Exp. Pathol. Pharmakol.,* 81, 55–129.

Vane, J. R. (1957). A sensitive method for the assay of 5-hydroxytryptamine, *Br. J. Pharmacol. Chemother.* 12, 344–349.

Vizi, V. A. and Vizi, E. S. (1978). Direct evidence for acetylcholine releasing effect of serotonin in the Auerbach plexus, *J. Neural Transmission,* 42, 127–138.

Watson, A. H. D. (1979). Fluorescent histochemistry of the teleost gut: evidence for the presence of serotonergic neurones, *Cell Tissue Res.* 197, 155–164.

Weber, L. J. (1970). *p*-Chlorophenylalanine depletion of gastrointestinal 5-hydroxytryptamine, *Biochem. Pharmacol.* 19, 2169–2172.

Welch, A. S. and Welch, B. L. (1968). Effect of stress and parachlorophenylalanine upon brain serotonin 5-hydroxyindole acetic acid and catecholamines in grouped and isolated mice, *Biochem. Pharmacol.* 17, 699–708.

Welsh, J. H. (1957). Serotonin as a possible neurohumoral agent: evidence obtained in lower animals, *Ann. N.Y. Acad. Sci.* 66, 618–630.

Wood, J. D. and Mayer, C. J. (1979a). Intracellular study of tonic-type enteric neurons in guinea pig small intestine, *J. Neurophysiol.* 42, 569–581.

Wood, J. D. and Mayer, C. J. (1979b). Serotonergic activation of tonic-type enteric neurons in guinea pig small bowel, *J. Neurophysiol.* 42, 582–593.

Wood, J. D. and Mayer, C. J. (1979c). Adrenergic inhibition of serotonin release from neurons in guinea pig Auerbach's plexus, *J. Neurophysiol.* 42, 594–603.

Biology of Serotonergic Transmission
Edited by N. N. Osborne
© 1982, John Wiley & Sons Ltd.

Chapter 16

Evidence for Serotonin being a Neurotransmitter in the Retina

N. N. OSBORNE

*Nuffield Laboratory of Ophthalmology, University of Oxford,
Walton Street, Oxford OX2 6AW, UK*

INTRODUCTION

The vertebrate retina is derived embryologically from the brain and offers considerable advantages as a model preparation for integrating the morphology, physiology and biochemistry of specific neuronal systems. Several compounds, such as acetylcholine, dopamine and GABA are considered to be neurotransmitters in the retina (Graham, 1974; Kramer, 1976; Neal, 1976; Bonting, 1976) as they are in the brain (see Iversen, 1978). A number of reports have also suggested that noradrenaline (Osborne, 1981a), glycine, taurine, and glutamate (see Graham, 1974; Voaden,

1976; Voaden *et al.*, 1980), enkephalin (Brecha *et al.*, 1979), substance P (Karten and Brecha, 1980), somatostatin (Yamada *et al.*, 1980) and cholecystokinin (Osborne *et al.*, 1981a) have specific functions in the retina. It is clear that substances which are thought to have neurotransmitter functions in the brain are also present in the retina. Even benzodiazapine receptors exist in the retina (Borbe *et al.*, 1980; Osborne 1980a; Regen *et al.*, 1980) which is puzzling in view of proposed involvement of benzodiazepine in emotional types of behaviour (Robertson, 1979).

During the past 4 years, a number of studies have focused on the issue of whether serotonin is a transmitter in the retina as it is in other nervous tissues (see Osborne and Neuhoff, 1980; Gershon, 1977). Until recently (see Osborne *et al.*, 1981b), it had been argued that serotonin is not a transmitter in the retina (Florén and Hansson, 1980; Ehinger and Florén, 1980) mainly because no serotonin-containing neurones could be demonstrated there, despite the fact that neurones in the retina can accumulate exogenous serotonin. Ehinger and co-workers therefore coined the term 'serotonin-accumulating neurones' in the retina and proposed that the true transmitter of these neurones might be a substance related to serotonin (see Ehinger, 1976, 1978; Ehinger and Florén, 1978a; Florén, 1979; Ehinger and Florén, 1980; Florén and Hansson, 1980). Recent evidence, however, clearly favours the opinion that serotonin has a function in the retina (Osborne *et al.*, 1981b) and the aim of this chapter is to discuss fully the data supporting this contention.

MORPHOLOGY OF THE RETINA

A qualitative rather than a quantitative description of the morphology of the vertebrate retina will suffice here. An extensive treatment of this subject is covered by the following authors: Dowling (1970), Stell (1972), Cajal (1972), Rodieck (1973), and Dubin (1974). Figure 1b is a fixed and stained section of the rat retina where seven main layers may be identified. Each layer corresponds to specific areas of different cells within the retina which is shown schematically in Fig. 1a. There are five principal neuronal classes in the retina, each subdivided into different types. They are photoreceptors (rods and cones), bipolar cells (midget type, flat type, rod bipolar), horizontal cells, amacrine cells and ganglion cells (midget type, diffuse type). In addition there are the large Müller or glial cells.

The various classes of neurones in the retina are interconnected by their processes, so that information from the photoreceptors can be relayed either by direct or lateral pathways to the optic nerve. This is structurally recognizable (see Fig. 1a). From morphological and electrophysiological findings it is thought that horizontal and bipolar cells integrate and coordinate the initial input signals generated from the photoreceptors. Furthermore, bipolar cells appear to conduct the integrated activity state of the rods and cones either directly towards the amacrine cells or towards the ganglion cells (Dowling and Boycott, 1966). The ganglion cells send the continuously changing information from the rods and cones, filtered and processed by the retinal network, via axonal fibres in the optic nerve to the lateral geniculate body. Information is then relayed to other parts of the optic system.

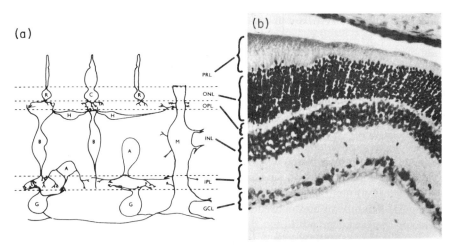

Fig. 1   Low-power magnification micrograph showing the organization of the rat retina (Fig. 1b). Six areas can be observed in the Nissel stained section. They are the photoreceptor layer (PRL), the outer nuclear layer (ONL), the outer plexiform layer (OPL), the inner nuclear layer (INL), the inner plexiform layer (IPL) and the ganglion cell layer (GCL). The distribution of the various cell types in the retina related to the different layers is shown in Fig. 1(b). R = rod, C = cone, H = horizontal cell, B = bipolar cell, A = amacrine cell, M = Müller cell, G = ganglion cell

ENDOGENOUS SEROTONIN IN THE RETINA

A variety of methods has been employed to determine the serotonin content in the vertebrate retina (see Table 1). The levels reported have generally been very low, resulting in the suggestion that the serotonin measured originates from exogenous sources e.g. blood platelets (Florén and Hansson, 1980). A way of disproving this idea is to measure the serotonin level in retinas which have been thoroughly washed free of blood or by showing that the amine is associated with specific areas of the retina. Generally, washed (perfused) retinas have been shown to contain less serotonin (see Table 1), though it could be argued that this decrease is not due to exogenous contamination, but to release of the amine from the tissue. It was therefore decided to look at the intraretinal distribution of serotonin in the retina. Sections of bovine retina, 20 µm thick, were cut from frozen tissue and dissected under microscopic vision into the following areas: inner nuclear and plexiform layers, ganglion cell layer, and outer nuclear and plexiform layers. The serotonin content in these different areas of the retina was then determined by a microdansylation procedure (see Osborne, 1974). A photograph of a microchromatogram showing a whole spectrum of substances in bovine retina which react with dansyl chloride is represented in Fig. 2. Of the substances involved in serotonin metabolism, only serotonin and tryptophan can be identified. By using labelled dansyl chloride, the radioactive dansyl products of serotonin were scraped from the chromatograms and the amount of serotonin subsequently determined. As shown

Table 1. Serotonin concentration in the retina of different species (ng/g wet weight). All values are ± SEM, $n < 4$ and > 20

	Suzuki et al. (1977b) Fluorometry	Thomas and Redburn (1979) Fluorometry	Florén and Hansson (1980)		Ehinger and Florén (1980) HPLC	Osborne (1980b) and unpublished data	
			HPLC	Radiochemical		Dansyl method	HPLC
Normal rabbits			24±5	36±6	36±6	30±6	30±6
Perfused rabbits					3.8±0.9	20±6	20±6
Frog						290±18	500±80
Pig					5.9±0.8		
Guinea-pig					6±2.6		
Chicken	176±12		34±4	67±17			
Chicken (new born)			71±5	90±8			80±9
Lizard							96±8
Bovine (whole retina)		100.3±10.1			26.2±7	31±7	39±4
Outer nuclear and plexiform layers						trace	
Inner nuclear and plexiform layers						48±10	
Ganglion cell layer						trace	

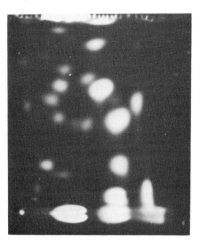 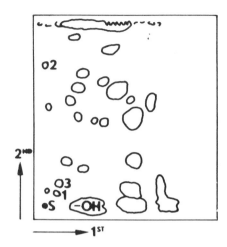

Fig. 2 Microchromatogramm (3 X 3 cm) of an extract of retina which had been reacted with dansyl chloride and developed in two chromatographic systems to separate most dansyl derivatives. A key is presented to assist in the identification of dansyl-tryptophan (spot 1), dansyl-*N*-5-HT (spot 3) and dansyl-bis-5-HT (spot 2). S = starting point. The direction of chromatography is indicated by the arrows. First direction, water/formic acid (100:3 v/v); second direction, benzene/acetic acid (9:1 v/v). From Osborne (1980b)

in Table 1, the small amount of serotonin found in the bovine retina is almost exclusively associated with the inner nuclear and plexiform layers, which shows that the serotonin in the retina is endogenous in origin. A case might also be made for its having special functions.

It has been proposed that the true indoleamine of the retina is not serotonin but a substance closely related to the amine (Ehinger and Florén, 1980; Florén and Hansson, 1980). Recent experiments have disproved this (see Osborne *et al.*, 1981b). Using perchloric acid extracts from retinas of either lizards, frogs, cows or rabbits, it has been shown by high performance liquid chromatography (HPLC) that serotonin is definitely present (see Fig. 3). If other indoleamines like 5,7-dihydroxytryptamine, 5,6-dihydroxytryptamine, 6-hydroxytryptamine or bufotenine exist in the retina, their concentrations are at least a thousand times lower than serotonin, because these substances were undetectable in the extracts. With the HPLC procedures used, serotonin was completely separated from the following substances: dopamine, noradrenaline, dihydroxyphenylacetic acid, 5,7-dihydroxytryptamine, 5,6-dihydroxytryptamine, homovanillic acid, 3-methoxytryptamine, tryptamine, 5-hydroxyindoleacetic acid, 5-methoxytryptamine, melatonin; bufotenine, and 6-hydroxytryptamine.

ENZYMES RELATED TO SEROTONIN SYNTHESIS

It has been known for some time that the enzymes 5-hydroxytryptophan decarboxylase and monoamine oxidase, which are necessary for the metabolism of serotonin,

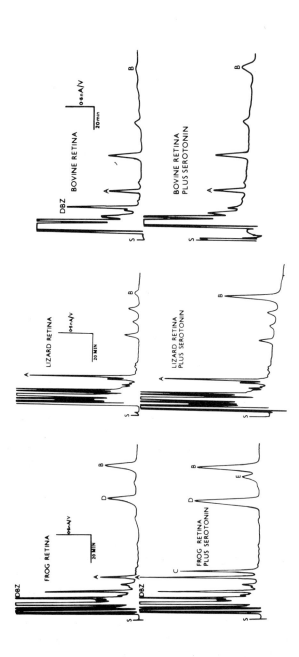

Fig. 3   High performance liquid chromotography (HPLC) to show the separation of various substances in perchloric and extracts of frog, lizard and bovine retina extracts. The apparatus was from DuPont, equipped with a C-18 reverse phase column and a LC4 amperometric detector. The electrochemical detector was set at 2 nA/V with a potential of 0.8 V. The mobile phase was 0.02 mM sodium citrate buffer pH 3.5 containing 0.2 mM octane-1-sulphonic acid and 6.5 per cent methanol. Dihydroxy-benzylamine (DBZ) was routinely, but not always, added to the perchloric acid extracts as an internal standard. In each instance the upper trace shows the separation of a retinal extracts to which was added DBZ. The lower traces are of extracts to which serotonin was added. In the case of the frog, the extract analysed in the lower trace also contained dopamine (A), 5-hydroxy-indoleacetic acid and 3-methoxytyramine (D), homovanillic acid (E). The result of the above analysis is to conclude that all the retinas contain dopamine (A) and serotonin (B). S = start. Partly from Osborne et al. (1981b)

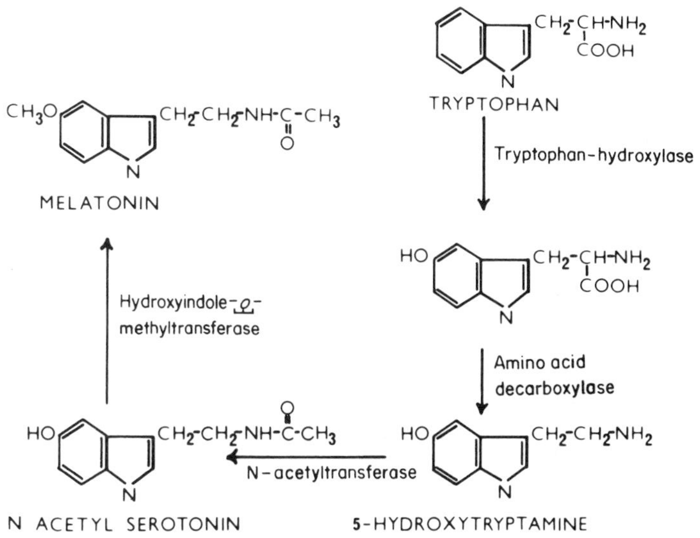

Fig. 4   The pathway for the biosynthesis of melatonin from tryptophan

exist in the retina (Baker and Quay, 1969; Smith, 1973; Suzuki *et al.*, 1977b; Graham, 1974). Both enzymes are, however, unspecific and do not prove that serotonin is necessarily found in the retina. The major, and rate-limiting enzyme involved in the synthesis of serotonin is tryptophan-hydroxylase (see Fig. 4) and attempts by some authors to demonstrate its existence in the retina have failed (Smith, 1973; Smith and Baker, 1974; Florén and Hansson, 1980). Osborne (1980b), using radioactive tryptophan, showed clearly that bovine retina *in vitro* can metabolize part of the amino acid to form 5-hydroxytryptophan, serotonin and 5-hydroxyindoleacetic acid (see Fig. 5). Furthermore, when *p*-chlorophenylalanine, an inhibitor of the enzyme tryptophan-hydroxylase (Koe and Weissman 1966), was included in the incubation medium, the amount of products produced from the radioactive tryptophan was drastically reduced. These data demonstrate unequivocally that a pathway for the synthesis of serotonin from tryptophan exists in the retina, as it is known to in the brain.

Other enzymes related to the synthesis of serotonin are also present in the retina. These are hydroxyindole-*O*-methyltransferase (Quay, 1965; Cardinali and Wurtman, 1972; Wainwright, 1979) and *N*-acetyltransferase (Binkley *et al.*, 1979; Miller *et al.*, 1980). This, together with the observation that melatonin exists in the retina (Pang *et al.*, 1977, Baker and Hoff, 1971; Bubenik *et al.*, 1976) shows that melatonin can be formed within the retina. In this respect the retina resembles the pineal organ (Wurtman *et al.*, 1968). The amount of melatonin in the retina is, however, very small in comparison with the serotonin content. In the frog it was calculated that the serotonin level is about 0.5 μg/g fresh weight (see Table 1), whereas with the same HPLC method, the melatonin level was not measurable.

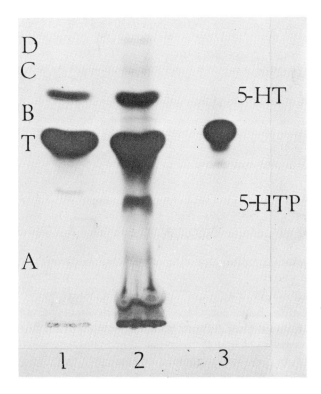

Fig. 5   Photograph of an autoradiogram showing the metabolism of [^{14}C] tryptophan by bovine retina. Tissue was incubated with 10^{-7} M [^{14}C] tryptophan for
30 min (no. 1) or 90 min (no. 2) at 37 °C. Thereafter amines and amino acids were
extracted and chromatographed. 3 = standard [^{14}C] tryptophan. It can be seen that
[^{14}C] tryptophan (T) is metabolized to form two major substances, 5-HTP and
5-HT. At a longer period of incubation with [^{14}C] tryptophan (90 min no. 2), the
amino acid is additionally incorporated into four minor substances (A)–(D), with
(D) being mostly produced. (D) was identified as 5-hydroxyindoleacetic acid.
From Osborne (1980b)

LOCALIZATION OF SEROTONIN

In only one study with the histochemical fluorescence method of Falck and Hillarp
(Falck and Owman, 1965) was a retinal neurone containing serotonin discovered
(Ehinger and Florén, 1978a; Ehinger and Florén, 1980). The one report is on a
special type of cell in the chicken retina which appears only during a limited phase
of development (Hauschild and Laties, 1973). It has therefore been proposed that
serotonin-containing neurones do not exist in the mature retina (see Ehinger and
Florén, 1980; Florén and Hansson, 1980), although some authors have been pre-

pared to argue in the light of other evidence that the serotonin level in retinal neurones is, for unknown reasons, very low and therefore cannot be detected by the histofluorescence method of Falck and Hillarp (see Osborne, 1980b).

The latter opinion has been affirmed by the use of a specific, highly sensitive, immunohistochemical procedure (see Steinbusch *et al.*, 1978) for the localization of serotonin neurones (see Figs. 6 and 7). Monoclonal antibodies, specific against serotonin, were prepared and used to localize serotonin as described by Cuello and Milstein (1981) and Consolazione *et al.* (1981). Immunoreactivity specific for serotonin was observed only in amacrine cell bodies situated in the inner nuclear layer and in processes throughout the inner plexiform layer in the frog retina (see Figs. 6 and 7). No immunoreactive somata were observed in the outer nuclear layer or ganglion cell layer, nor was immunoreactivity associated with the outer plexiform layer of Müller cells. Immunoreactive amacrine cells have only a single process which descends directly into the inner plexiform layer where it ramifies into a very broad band. Immunoreactivity for serotonin was observed in both central and peripheral retinal regions.

As shown in Table 2, the serotonin level in the frog retina is much higher than in the lizard, chick, and cow and so it was not surprising to find that there was far less serotonin-immunoreactivity in the chick retina than in the frog, although it is localized in the same regions (see Figs. 6 and 7). Any attempts to show the serotonin-immunoreactivity in rabbit, bovine and rat retinas have so far failed, presumably because of the low amounts of endogenous serotonin in these retinas.

UPTAKE OF SEROTONIN BY RETINA

A neurotransmitter substance released from presynaptic terminals may be inactivated in different ways: (a) by rapid diffusion away from the synaptic area, (b) by metabolic conversion, or (c) by cellular uptake from the region of the synpatic cleft, for example reuptake into the presynaptic terminals. The reuptake procedure is most probably the major way of inactivating serotonin in the vertebrate brain (Shaskan and Snyder, 1970). Recent experiments with retinas from chicken (Suzuki *et al.*, 1978), rabbit (Ehinger and Florén, 1978a) and cow (Osborne, 1980b; Thomas and Redburn, 1979, 1980), have demonstrated biochemically that an active uptake mechanism for serotonin occurs in this tissue and that this mechanism, which shows saturation kinetics, is sodium sensitive (see Table 3) and temperature dependent. Table 2 shows some data for the uptake of radioactive serotonin by isolated pieces of bovine retina. Analysis of these data by the iterative method of fitting them directly into rate equations with a digital computer revealed that the data fit a two-carrier model rather than a single model (see Osborne 1980b), i.e. a high-affinity system with a Km value of 1.3×10^{-7} M and a low-affinity process with a Km value of 1.2×10^{-5} M. The high affinity system is thought to represent a specific mechanism for the inactivation of a transmitter substance (Iversen, 1971). The high affinity uptake process is dependent on sodium ions and selectively inhibited by 3-chlorimipramine and Lilly 110140 (see Table 4). The specificity of the serotonin uptake by retinas has been demonstrated by comparing the entry of tryptamine and serotonin into the tissue (Osborne and Richardson 1980). Tryptamine, which is structurally similar to serotonin but lacks a hydroxyl group, was found to enter the bovine retina by a single mechanism with a Km value of 2×10^{-5} M. The total influx of tryptamine is, in principle, similar in character to the low-affinity uptake mechanism of serotonin, i.e. sodium insensitive (see Table 3) with a relatively low Km value. Further evidence for this is that tryptamine uptake, as well as the low affinity uptake mechanism for serotonin, is not influenced by a number of pharmacological agents and analogues of tryptamine (see Table 4). Since the low-affinity uptake mechanism is generally considered to be an unspecific process (see Iversen, 1971; Shaskan and Snyder, 1970) due to various factors such as passive accumulation, unspecific binding or incorporation into tissue, while a high affinity uptake mechanism is specifically associated with neuronal processes (see Iversen, 1971; Shaskan and Snyder, 1970), it would appear that the uptake of tryptamine by bovine retina is not a specific process. A mechanism therefore exists in the retina for the uptake (high affinity) of serotonin and this mechanism has a high specificity in being able to discriminate between molecules as similar as serotonin and tryptamine.

Conclusive support for the specific uptake of serotonin by retinal tissue has been provided by autoradiography and fluorescence histochemistry. The normal retina treated for fluorescence microscopy with the Falck-Hillarp technique (Falck and Owman, 1965) displays dopamine-containing cell bodies at the junction of the inner nuclear layer in all species investigated so far (for reviews see Ehinger and Florén, 1980) except in one instance—the chick retina at a special developmental stage (Hauschild and Laties, 1973). However, retinas pretreated with serotonin (either by injecting the amine directly into the eye or incubating the isolated retina

Fig. 6   (a) Shows the presence of serotonin immunofluorescence associated with a population of amacrine cell bodies (thick arrows) in the inner nuclear layer (INL) and perfuse dendritic ramifications in the inner plexiform layer (IPL, see (b)) of the frog retina. Immunofluorescence was not associated with the ganglia cell layer (double arrow), the outer plexiform layer (dotted arrow) or the outer nuclear layer (ONL). Specific immunofluorescence for serotonin was achieved by immersing retinas in 4 per cent paraformaldehyde prepared in 0.1 M phosphate buffer (pH 7.2) for 1–3 h. The retinas were then stored in 5 per cent sucrose in the same buffer and frozen sections of 10 micrometers prepared and mounted on gelatine coated glass slides. The sections were then incubated overnight at 4 °C with a rat monoclonal antibody (YC5/45 HL) (see Cuello and Milstein, 1981 and Consolazione *et al.*, 1981). The monoclonal antibody was developed by an antirat IgG FiTC-conjugated immunoglobin by incubation for 1 hr at 37 °C. The monoclonal antibodies and developing fluorescent antibodies were diluted 1:300 and 1:10 respectively, in phosphate buffered saline (PBS) containing 0.2 per cent Triton X-100. The glycerine/ PBS mounted sections were observed under microscope examination using epifluorescence optics. The positive immunoreaction was obliterated when the monoclonal antibody was absorbed with an excess of serotonin (1 mM) as shown in (b)

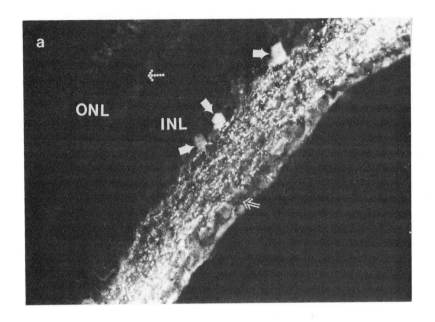

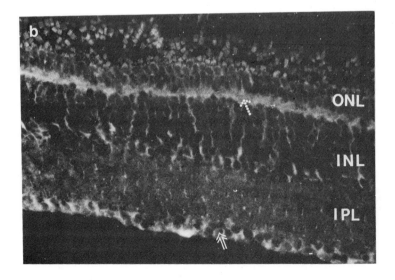

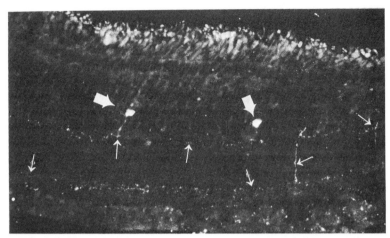

Fig. 7   Serotonin immunofluorescence in the chick retina. Amacrine cell bodies (thick arrows) and a sparse distribution of terminals (thin arrows) in the inner plexiform layer react positively for serotonin. For details of the procedure see Fig. 6. It can be seen that the chick retina contains relatively few processes and terminals in the inner plexiform layer when compared to the frog retina (see Fig. 6)

Table 2. [^{14}C] Serotonin (5-HT) uptake by bovine retina (from Osborne, 1980b). Retina pieces (approx. 4 mg) were preincubated at 0 °C and 37 °C for 10 min and then incubated for a further 10 min with radioactive 5-HT. The influx of 5-HT was then calculated (observed value) in subtracting the values at 0 °C from those at 37 °C. Each experimentally determined value is the mean ± SEM for six determinations. The predicted values are those for a two-carrier system with the following kinetic parameters: $Km_1 = 1.3 \times 10^{-7}$ M, $V_{max_1} = 0.65$ pmole/mg/10 min; $Km_2 = 1.2 \times 10^{-5}$ M, $V_{max_2} = 27$ pmole/mg/10 min; standard deviation 10 per cent

Exogenous [^{14}C] 5-HT (μM)	Observed	5-HT accumulated into retina (pmole/mg/10 min)		
		Predicted (V_{max_1})	Predicted (V_{max_2})	Predicted ($V_{max_1} + V_{max_2}$)
0.04	0.28±0.01	0.15	0.08	0.23
0.08	0.46±0.02	0.24	0.17	0.41
0.16	0.73±0.02	0.35	0.33	0.68
0.48	1.64±0.06	0.51	0.99	1.50
0.73	2.0±0.1	0.55	1.48	2.03
1.00	2.6±0.1	0.57	1.99	2.56
2.5	5.1±0.1	0.61	4.48	5.09
5.0	8.5±0.24	0.63	7.70	8.34
10.0	13.4±0.9	0.64	12.01	12.65
50.0	21.0±2.1	0.64	21.71	22.36
100.0	29.0±3.8	0.65	24.14	24.79

Table 3. The influence of various factors on the total uptake of [^{14}C] tryptamine and [^{14}C] serotonin by bovine retina. Data taken from Osborne (1980b) and Osborne and Richardson (1980)

Incubation condition	[^{14}C] Tryptamine influx (% of control)	[^{14}C] Serotonin influx (% of control)
Normal medium by at 5 °C	28±9	32±9
Sodium replaced by sucrose	98±6	50±5
Calcium replaced by sucrose	98±6	98±6
Magnesium replaced by sucrose	120±16	110±6
Potassium replaced by sucrose	99±4	71±6
Chloride replaced by sulphate	110±9	80±4

Values are ± SEM for five determinations. Retinal pieces were always preincubated for 10 min in the various conditions before the incubation phase with ^{14}C solution for a further 10 min. The normal incubation temperature was 37 °C.

Table 4. Comparison of the effect of various substances on the influx of [^{14}C]-tryptamine and [^{14}C] serotonin. Values are means ± SEM of five or more independent determinations

	[^{14}C] Tryptamine influx (% control)	[^{14}C] Serotonin influx (% control)
(a) Amines (10^{-4} M)		
5-Methoxytryptamine	81±5	76±8
Tryptamine	77±8	49±8
5,7-Dihydroxytryptamine	83±7	66±8
5,6-Dihydroxytryptamine	92±6	81±10
Dopamine	94±9	60±4
Noradrenaline	90±2	69±4
6-Hydroxytryptamine	88±3	59±6
Octopamine	99±7	70±3
5-HT	82±4	70±5
(b) Inhibitors (5×10^{-4} M)		
Dinitrophenol	100±6	92±6
Chlorpromazine	70±5	60±8
Desimipramine	70±4	38±7
Lilly 110140	69±3	29±5
KCN	71±3	94±7
Ouabain	98±6	80±10
Benztropine	47±4	41±5
Reserpine	97±5	50±7
Chlorimipramine	83±10	34±6

From Osborne (1980b) and Osborne and Richardson (1980)

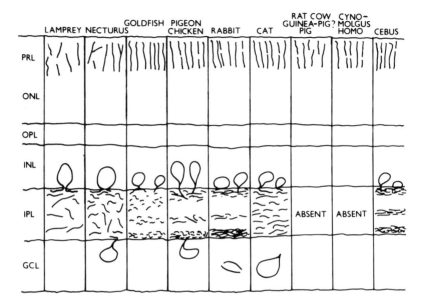

Fig. 8   Schematic drawing of the distribution of serotonin accumulating neurones in the retina of various retinas. PRL = photoreceptors, ONL = outer nuclear layer, OPL = outer plexiform layer, INL = inner nuclear layer, IPL = inner plexiform layer, GCL = ganglion cell layer. Redrawn from Ehinger and Florén (1980)

in a medium containing low amounts of serotonin) and processed by the fluorescence histochemical method revealed the presence of additional fluorescent perikarya and terminals (see Ehinger and Florén, 1980). This observation led Ehinger and collaborators to term these cells 'serotonin-accumulating neurones', i.e. neurones in the retina which apparently do not contain serotonin(or dopamine) but specifically take up exogenous serotonin (Ehinger and Florén, 1976). These serotonin-accumulating neurones were shown to occur in retinas of a variety of animals (see Fig. 8), and the uptake of amine into these cells could be blocked by various substances (Ehinger and Florén, 1978b).

When retinas are incubated with [³H]serotonin and then examined by autoradiography, specific processes and cell bodies have also been shown to take up the amine (see Fig. 9). Results involving the use of amine fluorescence histochemistry or autoradiography have generally been corroborative in showing that specific amacrine somata and processes in the inner plexiform layer take up exogenous serotonin (Ehinger and Florén, 1978a; Osborne and Richardson, 1980). An exception to this was the report by Ehinger and Florén (1980) (see Fig. 8) which states that serotonin was not taken up by specific elements in the bovine retina; the finding by Osborne and Richardson (see Fig. 9) clearly disproves this.

Biochemical studies have also demonstrated that the inner plexiform layer is the

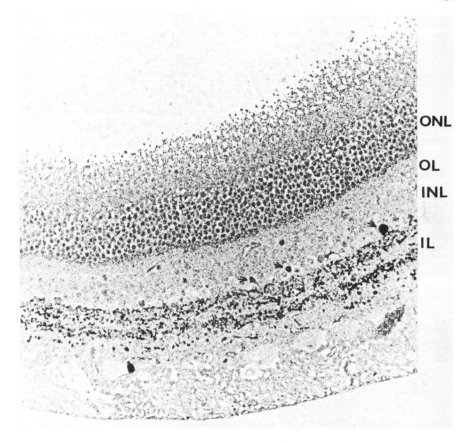

Fig. 9   Light microscopy autoradiography of [³H] serotonin uptake in bovine retina. Perikarya (arrows) in the inner nuclear layer (INL) and nerve endings in the inner plexiform layer (IL) are the sites which primarily take up radioactivity. OL = outer plexiform layer, ONL = outer nuclear layer

area which takes up exogenous serotonin. Thomas and Redburn (1980) and Osborne and Richardson (1980) produced subcellular fractions of the bovine retina, and the synaptosomal fraction derived primarily from the inner plexiform layer took up more exogenous serotonin than synaptosomes derived from the outer plexiform layer.

RELEASE OF SEROTONIN

It has been demonstrated that potassium induces the release of serotonin from retinas previously loaded with the amine and that this release is calcium dependent (see Thomas and Redburn, 1979, 1980; Osborne, 1980b; Osborne *et al.*, 1981b). Since all transmitter release mechanisms so far investigated are calcium dependent

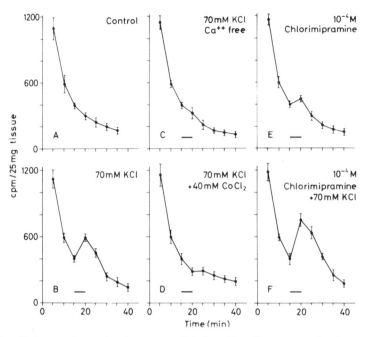

Fig. 10   Release of previously accumulated radioactive serotonin from bovine retina. Substances as indicated, were either added to or eliminated from the normal incubation media between the 15th and 20th minute. For details of experiment see Osborne (1980b)

(Simpson, 1969), the release data support the idea that serotonin is a transmitter agent in the retina. This suggestion would be more convincing if the release of serotonin could be demonstrated by a light stimulation rather than the depolarizing agent, potassium. Such experiments have, however, not yet been conducted, nor has it been demonstrated that potassium stimulation results in a release of endogenous serotonin.

Figure 10 demonstrates in a series of experiments that the potassium-induced release of serotonin from bovine retina is counteracted in the presence of cobalt ions. It is thought that cobalt ions interfere with the calcium entry into synaptic terminals (Weakly 1973). Furthermore, the release caused by potassium is apparently increased when chlorimipramine is present. This is due to chlorimipramine blocking any reuptake of serotonin into the presynaptic site so that the amount of serotonin released appears to be increased. The release of radioactive serotonin from retinas by potassium is dose-dependent (see Table 5) which proves that potassium depolarizes retinal neurones in a fairly specific manner. The finding by Thomas and Redburn (1980) that the potassium-induced release of radioactive serotonin from synaptosomes derived from the inner plexiform layer is greater than that from the synaptosomes from the outer plexiform layer supports the opinion that the release of loaded serotonin by potassium is associated with serotonergic neurones.

Table 5. Effect of high potassium (K^+) on the release of [^{3}H] serotonin from isolated frog retina

Stimulus	Evoked release of [^{3}H] serotonin (% of total tissue stored)
K^+ 10 mM	2.11±0.66
K^+ 20 mM	4.61±0.92
K^+ 35 mM	7.92±1.44
K^+ 35 mM, calcium-free medium	0.07±0.19
K^+ 35 mM + 35 mM $CoCl_2$	0.06±0.19

Retinas were incubated for 20 mins at 35 °C in Ringer (frog heart) solution containing 10^{-7} M [^{3}H] serotonin (Radiochemical Centre, Amersham, specific activity 29.1 Ci/mmole). After the incubation, individual retinas were transferred to 1 ml samples of Ringer solution at 35 °C for periods of 2.5 min and their radioactivity measured. During the 10 and 15 min periods, the Ringer buffer was modified in the manner shown above. After 30 min, the retinas were recovered, solubilized using Soluene-350 (Packard) and the total content of radioactivity in the tissues measured. In chromatographic experiments it was established that 90 per cent of radioactivity recovered in the tissue was unmetabolised [^{3}H] serotonin. Values are means ± S.E.M. (n = 8) calculated as the total evoked release of [^{3}H] serotonin in the 10 min period during and after the application of a 5 min stimulus. The evoked release was calculated by subtraction of the average value for spontaneous [^{3}H] serotonin efflux during the 7.5 min before the stimulus. From Osborne *et al.* (1981b)

SEROTONIN RECEPTORS IN THE RETINA

There are electrophysiological, biochemical and pharmacological data providing evidence that serotonin receptors exist in the vertebrate retina. Straschill (1968) was the first to report that serotonin injected into the carotid artery of cats produces a strong inhibition of ganglion cells. In rabbit retina serotonin was found to produce an enhancement of ganglion cell activity (Ames and Pollen, 1969). Micro-iontophoretic application of serotonin directly onto retinal neurones have generally been inconclusive, though the data suggested that serotonin receptors do exist in the retina (Straschill and Perwein, 1969, 1975). Recently Osborne (1981b) has demonstrated the presence of specific serotonin binding sites in bovine retina which are similar in character to those described for brain tissue (Bennett and Snyder, 1976; Peroutka and Snyder, 1979; Fillion *et al.*, 1978).

The specific binding of different amounts of [^{3}H] serotonin to bovine retina membranes reveals a hyperbolic curve which approaches saturation at [^{3}H] serotonin concentrations well over 20 nM (see Fig. 11). Scatchard analysis of the binding data showed a single binding site with a K_D value of 7.6 ± 0.5 nM and a B_{max} of 3.8 ± 0.3 pmole/g (see Fig. 12). Analysis of the saturation data by the Hill plot (Fig. 13) showed that there was no cooperativity in the binding.

Examination of the subcellular distribution of specific [^{3}H] serotonin binding showed that binding, in terms of specific activity, was greater in the fraction enriched with synaptosomes from conventional nerve endings (i.e. inner plexiform layer)

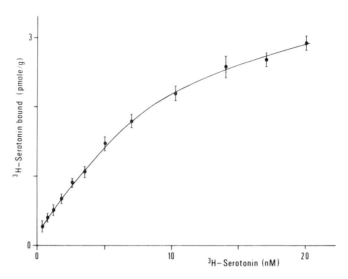

Fig. 11   Specific binding of various concentrations of [³H]serotonin to 30 mg equivalent of retinal membrane preparations. Binding studies were carried out at room temperature and the incubation time was 15 min. Each point is the mean of at least four experiments, each carried out in triplicate. Vertical lines show S.D. From Osborne (1981b)

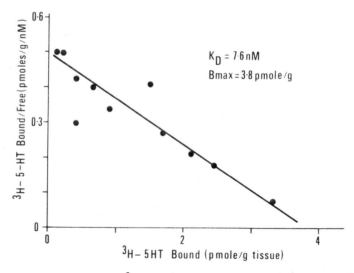

Fig. 12   Scatchard plot of [³H]serotonin bound to 30 mg equivalent retinal membranes. Membranes were incubated with various concentrations of [³H]serotonin. The data in the curve shown were from three separate experiments carried out in triplicate. From Osborne (1981b)

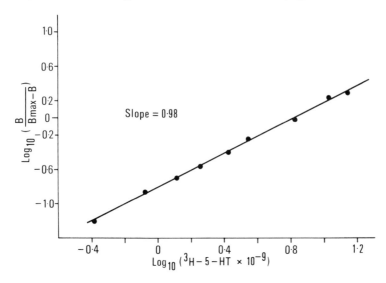

Fig. 13   Hill plot using the data from Fig. 12. The line of best fit was estimated by the method of least squares ($r = 0.98$). From Osborne (1981b)

(see Table 6). The fraction which contained nerve terminals derived from the photoreceptors (i.e. outer plexiform layer) possesses substantially fewer binding sites for [³H] serotonin.

The substrate specificity of [³H] serotonin binding to retinal tissue membranes provides strong evidence that the amine binds to physiological receptors. Only those substances which are known to interact neurophysiologically with serotonin receptors, e.g. 6-hydroxytryptamine, bufotenine and 5-methoxytryptamine have an appreciable affinity with the [³H] serotonin binding sites (see Table 7). The putative transmitter substance, e.g. dopamine, noradrenaline, histamine and acetylcholine show little affinity for the binding sites, strongly suggesting that [³H] serotonin binds to serotonin and no other neurotransmitter receptors (see Table 7). This opinion is enhanced by the finding that methysergide and methergoline, both known serotonin blockers, are very effective in diminishing the binding of [³H] serotonin to retinal membrane (see Table 7). The fact that chlorimipramine, a serotonin uptake inhibitor, competes for [³H] serotonin binding sites to a much lesser extent can be taken as proof that the ligand is attached primarily to postsynaptic receptors and not to transport sites of serotonin.

The effect of guanine and adenine nucleotides on specific binding of [³H] serotonin to bovine retinal membranes is shown in Fig. 14. At high concentrations, all the guanine nucleotides reduced specific binding, while the adenine nucleotides did not. This finding is of interest in view of the studies by Peroutka *et al.*, (1979), which have shown that a class of serotonin receptors in brain tissue is regulated by guanine nucleotides.

Table 6. Binding of [^{3}H] serotonin (7 nM) to subcellular fractions of retinal membranes

Fraction	f mole specifically bound per mg protein	Enrichment
Homogenate	8±1	1.0
P$_1$ fraction	6±1	0.75
P$_2$ fraction	13±2	1.6

Homogenate of retina was centrifuged at 150×g for 10 min to remove cell debris and nuclei. The resultant supernatant was centrifuged at 800×g for 10 min to obtain a P$_1$ pellet which is highly enriched with photoreceptor synaptosomes. Centrifugation of the supernatant at 25,000×g for 12 min produced a P$_2$ pellet which contains conventional synaptosomes derived primarily from the inner plexiform layer. The pellet, which was originally prepared in 0.32 M sucrose, was suspended in 0.05 M Tris-HCl before carrying out the binding assays. Results are given as the mean ± S.D. for six separate observations. From Osborne (1981b)

DISCUSSION

It has been suggested by Werman (1966) that neurotransmitter substances should fulfill the following five criteria.

1. The compound should be present in nerve endings.
2. The corresponding synthesizing enzymes should be present.
3. There should be a mechanism for the termination of the action of the compound on the postsynaptic membrane.
4. Stimulation of the neurone should lead to its release into the extra-cellular fluid.
5. The action of ths suspected transmitter should be identical in every way to the natural transmitter.

Inherent in the fifth criterion is that receptors to the suspected transmitters exist in the tissue. The data reported and reviewed can be interpreted as fulfilling all five criteria and demonstrating that serotonin is a transmitter substance in the vertebrate retina.

Criterion 1

Ehinger and collaborators (see Ehinger and Florén, 1980; Florén, 1979; Florén and Hansson, 1980) have repeatedly argued that the major cricitism against serotonin being a transmitter in the retina is first, the amine level is lower than would be expected for a transmitter substance and secondly, all fluorescence-histochemical attempts to localize serotonin-containing elements in the retina have failed, which would mean that the serotonin is non-neuronal in origin. The same authors proposed that the low amounts of serotonin found in retinas originate from exogenous blood and suggested that the true transmitter utilized by 'serotonin-accumulating

Table 7. Displacement of specifically bound [^{3}H] serotonin (from Osborne, 1981b). Retinal membranes were incubated at room temperature for 15 min with 5 nM [^{3}H] serotonin. The non-radioactive substance was added simultaneously with the radioactive ligand; the rest of the procedure was performed as described by Osborne (1981b).

IC_{50} values were determined from inhibition curves which were constructed from four to six values, each determined in triplicate, and are the means of three to five assays for each substance

Substance	IC_{50} value (M)
Tryptamines	
Tryptamine	3×10^{-6}
5-Hydroxytryptamine (Serotonin)	10^{-8}
6-Hydroxytryptamine	9×10^{-8}
5,7-Dihydroxytryptamine	$> 10^{-4}$
5,6-Dihydroxytryptamine	10^{-6}
5-Methoxytryptamine	3×10^{-7}
5-Methoxyindole	8×10^{-5}
Melatonin	10^{-5}
Bufotenine	9×10^{-8}
Serotonin drugs	
Methergoline	2×10^{-8}
Cyproheptadine	$> 10^{-4}$
Miaserine	$> 10^{-4}$
Methysergide	8×10^{-8}
Pizotifen	$> 10^{-4}$
Neurotransmitter candidates	
Dopamine	$> 10^{-4}$
Noradrenaline	8×10^{-5}
Histamine	$> 10^{-4}$
Acetylcholine	$> 10^{-4}$
Other Substances	
5-Hydroxytryptophan	$> 10^{-4}$
Desipramine	6×10^{-5}
Amitriptyline	9×10^{-6}
Benztropine	2×10^{-5}
Chlorpromazine	3×10^{-5}
Chlorimipramine	6×10^{-6}

cells' is a substance closely related to serotonin. Various reports have appeared in the literature that not only serotonin, but also other indoleamines exist in the brain and supposedly could occur in the retina. Björklund *et al.* (1970, 1971) concluded that 5-methoxytryptamine, 6-hydroxytryptamine and *N*-methylserotonin exist in the rat brain. The presence of 5-methoxytryptamine in brain tissue was confirmed

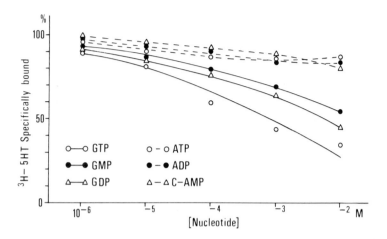

Fig. 14   Effect of increasing concentrations of nucleotides on the specific binding of 5 nM [³H]serotonin to retinal membrane preparations. Specific binding in the absence and presence of nucleotides was measured as described in the methods. Each point is the average of quadruplicate experiments. Values in three separate experiments varied between 10 and 18 per cent. (From Osborne (1981b)

by Green *et al.* (1973) and Koslow (1974). Two other indoleamines, tryptamine (Boulton and Majer, 1972; Snodgrass and Horn, 1973) and melatonin (Koslow, 1974) are known to exist in brain tissues.

It is now apparent from the results presented in this report that while its levels in the retina are generally low, serotonin is endogenous in origin. This conclusion is reached because, (a) the serotonin level varies in retinas of different species, (b) retinas thoroughly washed or perfused still contain substantial quantities of serotonin, and (c) the intraretinal distribution of serotonin in the retina shows the amine to be almost exclusively localized in the inner nuclear and plexiform layers.

Results involving the use of a HPLC procedure clearly show that the major indoleamine in the retina is serotonin (see Fig. 2 and 3). The procedure employed separated serotonin from various substances which include 5-methoxytryptamine, tryptamine, 5,6-dihydroxytryptamine, 5,7-dihydroxytryptamine, 5-hydroxytryptamine, bufotenine and melatonin. Should any of these substances exist in the retina, their concentrations would be far smaller than the serotonin level, as they could not be detected.

Immunohistochemical studies provide proof of a neuronal localization of serotonin in the retina. They show serotonin immunoreactivity to be associated with amacrine cell bodies and neuronal processes situated in the inner plexiform layer. Yet even with the immunohistochemical procedure used (Cuello and Milstein, 1981; Consolazione *et al.*, 1981) which is much more sensitive than the para-formaldehyde fluorescence technique at localizing serotonin (Falck and Owman, 1965), a clear observation of serotonin neurones was only possible in those retinas (i.e. frog) which contained large amounts of serotonin. One speculation is that as nervous systems evolved, neurones utilizing a specific transmitter became more efficient, with the result that these neurones required a lower transmitter level. This idea would explain the findings that amphibian retinas have the largest amounts of serotonin and as animals progress higher up the evolutionary scale, so the level of amine is decreased. This evolutionary tendency exists for brain serotonin (Brodie *et al.*, 1964, Welsh, 1968) and dopamine (see Holzbauer and Sharman, 1979) levels. Should this hypothesis be correct, then it could be argued that the low serotonin levels in the retina of the rat do not mean that only a few neurones in this tissue utilize the amine, but rather that those neurones which do use serotonin require and contain substantially less of the amine than the analogous cell in the frog retina. This would then explain why, even with the very sensitive immuno-histochemical procedure, serotonin cannot be located in retinas of the rat.

Criterion 2

Enzymes necessary for the synthesis of serotonin clearly exist in the retina. It has been shown that tryptophan is metabolized to form 5-hydroxytryptophan and serotonin (Osborne, 1980b) and that if *p*-chlorophenylalanine is present, it blocks the metabolism of tryptophan. Since *p*-chlorophenylalanine is a potent inhibitor of the enzyme tryptophan-hydroxylase (Koe and Weissman, 1966) in brain tissue, these data show that it is the same enzyme which metabolizes tryptophan to 5-hydroxytryptophan in brain and retina. Some attempts to show the existence of tryptophan hydroxylase in the retina have been unsuccessful (Florén and Hansson, 1980); this is difficult to understand. Monoamine oxidase, the enzyme necessary for the catabolism of serotonin to 5-hydroxyindoleacetic acid also exists in the retina (Smith 1973; Graham, 1974; Osborne, 1980b). The retina, like the brain, therefore contains the enzymes tryptophan-hydroxylase, 5-hydroxytryptophan decarboxylase and monoamine oxidase, and tryptophan is metabolized along the same pathway in both tissues to form 5-hydroxytryptophan, serotonin and 5-hydroxindole acetic acid.

The retina also has the necessary enzymes for producing melatonin from serotonin (see above). In this respect the retina shows a similarity to the pineal organ. It is not yet known whether melatonin is restricted to specific neurones in the retina.

Criterion 3

The ability of specific neurones in the retina to take up exogeneous serotonin is now clearly documented. Biochemically it has also been shown that serotonin is taken up by two mechanisms, a high affinity system which is dependent on serotonin and blocked by 3-chlorimipramine and a low affinity system which is independent of sodium and 3-chlorimipramine. High affinity uptake mechanisms are now generally thought to be the major method of inactivating released transmitter substances (Bennett and Snyder, 1976; Iversen 1971). High affinity uptake systems need not necessarily be associated with neurones, as there have been reports of glial cells possessing high affinity uptake systems (Iversen *et al.*, 1975; Wilkin *et al.*, 1974). However, in the case of the retina it has been demonstrated both by fluorescence histochemistry and autoradiography that exogenous serotonin is taken up by neuronal elements (see above). Significantly, when 3-chlorimipramine is present, the serotonin is not taken up into specific neurones as revealed by autoradiography (Osborne and Richardson, 1980).

The specificity of serotonin uptake into neuronal elements of the retina, i.e. the high affinity uptake system, has been demonstrated by using tryptamine. Tryptamine is structurally similar to serotonin, but lacks only a hydroxyl group. However, tryptamine is not taken up by any specific components in the retina as revealed by autoradiography, and it has been shown biochemically that tryptamine enters the retina by a single low-affinity component and is unaffected by 3-chlorimipramine or sodium. The characteristics of the total uptake of tryptamine in fact resemble the low-affinity uptake process of serotonin. Low-affinity uptake mechanisms are generally considered to be unspecific processes (see Iversen, 1971; Shaskan and Snyder, 1970) caused by various factors such as passive accumulation, unspecific binding, or incorporation into tissue.

While the precise function of the specific high-affinity uptake system of the serotonergic neurones in the retina has not been established, such a mechanism could function for the inactivation of the amine at post-synaptic sites. This would be consistent with the theory that serotonin is a neurotransmitter in the vertebrate retina.

Criterion 4

The demonstration of a potassium-induced release of serotonin from retinas previously loaded with the amine, which is calcium-dependent, support the idea that serotonin is a transmitter in this situation. All transmitter release mechanisms so far investigated are calcium-dependent (Simpson, 1969). This hypothesis is strengthened by the fact that cobalt ions counteract the release of serotonin by potassium stimulation. It is thought that they interfere with the calcium entry into synaptic terminals (Weakley, 1973). The demonstration that pulses of different concentrations of potassium chloride release proportionally varying amounts of radioactive serotonin shows that potassium chloride stimulates the retina in a physiological manner. The more natural stimulus would be light, but such experiments still remain to be tried.

Criterion 5

Evidence for the presence of serotonin receptors in the retina is quite convincing. The first electrophysiological findings from Straschill (1968), Straschill and Perwein (1969, 1975), and Ames and Pollen (1969) that serotonin receptors might exist in the retina have now been corroborated by Osborne (1981b). The fact that [³H] serotonin binds in a saturable fashion to membrane preparations of bovine retina with a dissociation constant of 7.6 nM is similar to what has been established for the serotonergic system in the brain (see Bennett and Snyder, 1976; Peroutka and Snyder, 1979; Fillion *et al.*, 1978). There is considerable structural specificity in the affinity of various tryptamines for the [³H] serotonin binding sites, and the degree of [³H] binding associated with subcellular fractions of the retina enriched with conventional synaptosomes is higher than with photoreceptor synaptosomes (see Table 6). Most of the conventional synaptosomes are derived from the inner plexiform layer, while the photoreceptor synaptosomes come from the outer plexiform layer (see Redburn, 1977). The localization of [³H] serotonin binding sites in the inner plexiform layer is therefore in that area of the retina which contains serotonin nervous elements (see Figs. 6 and 7). This is most efficient at taking up exogenous serotonin which can be released by potassium stimulation (Thomas and Redburn, 1980).

The discovery that 3-chlorimipramine, which is a potent inhibitor of serotonin uptake, competes to a slight exent with [³H] serotonin binding, shows, as would be expected, that the ligand is binding predominantly to both physiological receptors and the transport sites. This confirms the observation that other putative transmitter substances (dopamine, noradrenaline, histamine and acetylcholine) have little affinity for the binding sites, while methysergide and methergoline, both known serotonergic blockers, are very effective competitors for [³H] serotonin binding sites.

The finding that guanine nucleotides influence the [³H] serotonin binding sites in the retina is similar to what has been described for brain tissue (Peroutka *et al.*, 1979). In a recent review by Snyder and Goodman (1980), it was proposed that two types of serotonin receptor exist in the brain: one regulated by guanine nucleotides and the other not. The latter can be selectively labelled by [³H] spiroperidol. It would therefore appear that both retina and brain contain serotonin receptors linked to guanine nucleotides.

GENERAL CONCLUSION

The experiments reviewed in this article can be taken to support the view that serotonin is a neurotransmitter in the vertebrate retina. Although catabolism of serotonin to form 5-hydroxindoleacetic acid occurs in the retina, the breakdown is slow. The data substantiate the idea that a reuptake mechanism for serotonin into the neurone is the method for terminating the action. Further proof of a transmitter role for serotonin is provided by the fact that enzymes capable of synthesizing serotonin from tryptophan occur in the retina and that the retina can be stimulated by high potassium to cause a release of serotonin which is calcium-dependent. The serotonin found in the retina is present in the inner plexiform and nuclear layers and immunohistochemical methods have shown the amine to be restricted to some amacrine neurones and processes within the inner plexiform layers. The degree of [^{3}H] serotonin binding to membranes from synaptosomes derived from the innerplexiform layer is higher than to membranes from photoreceptor synaptosomes. The characteristics of [^{3}H] serotonin binding to retinal membranes strongly suggest that the [^{3}H] ligand is binding to serotonin physiological receptors. This is further confirmation of the claim that serotonin is a transmitter in the retina.

REFERENCES

Ames, A. and Pollen, D. A. (1969). Neurotransmission in central nervous tissue: a study of isolated rabbit retina, *J. Neurophysiol.* **32**, 424–442.

Baker, P. C. and Quay, W. B. (1969). 5-Hydroxytryptamine metabolism in early embryogenesis, and the development of brain and retinal tissues, *Brain Res.* **12**, 273–295.

Baker, P. C. and Hoff, K. (1971). Melatonin localisation in the eyes of larval *Xenopus*, *Comp. Biochem. Physiol.* **39**, 879–881.

Bennett, J. P. Jr. and Snyder, S. H. (1976). Serotonin and lysergic acid diethylamide binding in rat brain membranes: relationship to postsynaptic serotonin receptors, *Molecular Pharmacol.* **12**, 373–389.

Binkley, S., Hryshchyshyn, M., and Reilly, K. (1979). *N*-Acetyltransferase activity responds to environmental lighting in the eye as well as in the pineal gland. *Nature (Lond.)*, **281**, 479–481.

Björklund, A., Falck, B., and Stenevi, U. (1970). On the possible existence of a new intraneuronal monoamine in the spinal cord of the rat, *J. Pharmacol. Exp. Ther.* **175**, 525–532.

Björklund, A., Falck, B., and Stenevi, U. (1971). Microspectrofluorometric characterization of monoamines in the central nervous system: evidence for a new neuronal monoamine-like compound. in *Progress in Brain Research* (Ed O. Fränkö) Vol. 34, pp. 63–73, Elsevier, Amsterdam.

Bonting, S. L. (1976). *Transmitters in the visual process*, Pergamon Press, Oxford.

Borbe, H. O., Müller, W. E., and Wollert, U. (1980). The identification of benzodiazepine receptors with brain like specificity in bovine retina, *Brain Res.* **182**, 466–469.

Boulton, A. A. and Majer, J. R. (1972). Determination and quantitative analysis of some noncatechol primary amines, in *Research Methods in Neurochemistry*, (Eds N. Marks and R. Rodnight) Vol. 1, pp. 341–335, Plenum Press, New York.

Brecha, N., Karten, H. J., and Laverack, C. (1979). Enkephalin-containing amacrine cells in the avian retina: Immunohistochemical localisation, *Proc. Nat. Acad. Sci.* **76**, 3010–3014.

Brodie, B. B., Bogdanski, D. F., and Bonomi, L. (1964). Formation, storage and metabolism of serotonin (5-hydroxytryptamine) and catecholamines in lower vertebrates, in *Comparative Neurochemistry* (Ed. D. Richter), pp. 367–378, Pergamon Press, Oxford.

Bubenik, G. A., Browb, G. M., and Grota, L. J. (1976). Differential localisation of *N*-acetylated indolealkylamines in CNS and the Harderian gland using immunohistology, *Brain Res.* **118**, 417–427.

Cajal, S. R. (1972). *The structure of the retina* (Translated by S. A. Thorpe, and M. Glickstein), Thomas, Springfield, Illinois.

Cardinali, D. P. and Wurtman, R. J. (1972). Hydroxyindole-*O*-methyl transferases in rat pineal, retina and Harderian gland, *Endocrinology*, **91**, 247–252.

Consolazione, A., Milstein, C., Wright, B., and Cuello, A. C. (1981). Immuno-cytochemical detection of serotonin with monoclonal antibodies, *J. Histochem. Cytochem.* (In press.)

Cuello, A. C. and Milstein, C. (1981). Monoclonal antibodies against neurotransmitter substances, in *Monoclonal antibodies against neural antigens,* Cold Spring Harbor Workshop, New York. (In press.)

Dowling, J. E. (1970). Organisation of vertebrate retinas, *Invest. Ophth.* 9, 655-680.

Dowling, J. E. and Boycott, B. B. (1966). Organisation of the primate retina: Electron Microscopy. *Proc. R. Soc. Lond. B.* 166, 80-111.

Dubin, M. W. (1974). Anatomy of the vertebrate retina, in *The Eye* (Eds. H. Davson and L. T. Graham, Jr.) Vol. 6, pp. 227-256, Academic Press, New York.

Ehinger, B. (1976). Biogenic monoamines as transmitters in the retina, in *Transmitters in the Visual Process* (Ed. S. L. Bonting), pp. 145-163, Pergamon Press, Oxford.

Ehinger, B. (1978). Biogenic monoamines and amino acids as retinal neurotransmitters, in *Frontiers in Visual Science*, (eds. S. J. Cool and E. L. Smith III), pp. 42-53, Springer Verlag, Berlin, New York.

Ehinger, B. and Florén, I. (1976). Indoleamine accumulating neurons in the retina of rabbit, cat and goldfish, *Cell. Tiss. Res.* 175, 37-48.

Ehinger, B. and Florén, I. (1978a). Quantitation of the uptake of indoleamines and dopamine in the rabbit retina, *Exp. Eye Res.* 26, 1-11.

Ehinger, B. and Florén, I. (1978b). Chemical removal of indoleamine accumulating terminals in rabbit and goldfish retina, *Exp. Eye Res.* 26, 321-328.

Ehinger, B. and Florén, J. (1980). Retinal indoleamine accumulating neurons, *Neurochem. Int.* 1, 209-229.

Falck, B. and Owman Ch. (1965). A detailed methodological description of the fluorescence method for the cellular demonstration of monoamines. *Acta. Univ. Lund.* 7, 1-23.

Fillion, G. M. B., Rouselle, J., Fillion, M., Beaudoin, D. M., Going, M. R., Denian, J., and Jacob, J. J. (1978). High-affinity binding of [³H] 5-hydroxytryptamine to brain synaptosomal membranes: comparison with [³H] lysergic acid diethylamide binding, *Mol. Pharmacol.* 14, 50-59.

Florén, I. (1979). Arguments against 5-hydroxytryptamine as neurotransmitter in the rabbit retina, *J. Neural. Trans.* 46, 1-15.

Florén, J. and Hansson, H. C. (1980). Investigations into whether 5-hydroxytryptamine is a neurotransmitter in the retina of rabbit and chicken, *Invest. Ophthal.* 19, 117-125.

Gershon, M. D. (1977). Biochemistry and physiology of serotonergic transmission, in *Handbook of Physiology* Section 1, *The Nervous System* (Ed. E. R. Kandel), Bethesda. pp. 573-623, American Physiological Society.

Graham, L. T. Jr. (1974). Comparative aspects of neurotransmitters in the retina, in *The Eye* (Eds. H. Davson and L. T. Grahan, Jr.). Vol. 6, pp. 283-342, Academic Press, New York.

Green, A. R., Koslow, S. H., and Costa, E. (1973). Identification and quantitation of new indole alkylamine in rat hypothalamus, *Brain Res.* 51, 371-374.

Hauschild, D. C. and Laties, A. M. (1973). An indolamine containing cell in chick retina, *Invest. Ophthalmol.* 12, 537-401.

Holzbauer, M. and Sharman, D. F. (1979). The distribution of dopamine in vertebrates, in *The neurobiology of dopamine* Ed. A. S. Horn, J. Korf, and B. H. C. Westerink), pp. 357-379, Academic Press, London.

Iversen, L. L. (1971). Role of transmitter uptake mechanisms in synaptic neurotransmission, *Brit. J. Pharmacol.* 41, 571-591.

Iversen, L. L. (1978). Chemical messengers in the brain, *Trends in Neurosci.* 1, 15–16.

Iversen, L. L., Dick, F., Kelly, J. S., and Shon, F. (1975). Uptake and localization of transmitter amino acids in the nervous system, in *Metabolic Compartmentation and Neurotransmission* (Eds. S. Berl, D. D. Clarke, and D. Schneider), pp. 65–78, Plenum Press, New York and London.

Karten, H. J. and Brecha, N. (1980). Localisation of substance P immunoreactivity in amacrine cells of the retina. *Nature (Lond.)*, 283, 87–88.

Koe, B. K. and Weissman, A. (1966). *p*-Chlorophenylalanine, a specific depletor of brain serotonin. *J. Pharmacol. exp. Therap.* 154, 499–516.

Koslow, S. H. (1974). 5-Methoxytryptamine: a possible central nervous system transmitter, in *Advances in Biochemical Psychopharmacology* (Eds. E. Costa, G. L. Gessa, and M. Sandler), Vol. 11, pp. 95–100, Raven Press, New York.

Kramer, S. (1976). Dopamine in retinal neurotransmission, in *Transmitters in the Visual Process* (Ed. S. L. Bonting), pp. 165–198, Pergamon Press, Oxford.

Miller, L., Stier, M., and Lovenberg, W. (1980). Evidence for the presence of *N*-acetyltransferase in rat retina, *Comp. Biochem. Physiol.* 66C, 213–216.

Neal, M. J. (1976). Acetylcholine as a retinal transmitter substance, in *Transmitters in the Visual Process* (Ed. S. L. Bonting). pp. 127–143, Pergamon Press, Oxford.

Osborne, N. N. (1974). *Microchemical Analysis of Nervous Tissue*, Pergamon Press, Oxford.

Osborne, N. N. (1980a). Benzodiazepine binding to bovine retina, *Neuroscience Letters*, 16, 167–170.

Osborne, N. N. (1980b). *In vitro* experiments on the metabolism uptake and release of 5-hydroxytryptamine in bovine retina, *Brain Res.* 184, 283–297.

Osborne, N. N. (1981a). Noradrenaline—a transmitter candidate in the bovine retina, *J. Neurochem.* 36, 17–27.

Osborne, N. N. (1981b). Binding of ³H-serotonin to membranes of the retina, *Exp. Eye Res.* (In press.)

Osborne, N. N. and Neuhoff, V. (1980). Identified serotonin neurons, in *International Review of Cytology* Eds. G. H. Bourne and J. F. Danielli) pp. 259–290, Academic Press, New York.

Osborne, N. N. and Richardson, G. (1980). Specificity of serotonin uptake by bovine retina: comparison with tryptamine, *Exp. Eye Res.* 31, 31–39.

Osborne, N. N., Nicholas, D. A., Cuello, A. C., and Dockray, G. (1981a). Localisation of cholecystokinin in the retina, *Neuroscience Letters* 26, 31–35.

Osborne, N. N., Nesslehut, T., Nicholas, D. A., and Cuello, A. C. (1981b). Serotonin: A transmitter candidate in the vertebrate retina. *Neurochem. Internat.* 3, 171–176.

Pang, S. F., Brown, G. M., Grota, L. J., Chambers, J. W., and Rodman, R. L. (1977). Determination pf *N*-acetylserotonin and melatonin activities in the pineal gland, retina, Hardenian gland, brain and serum of rats and chickens. *Neuroendocrinology*, 23, 1–13.

Peroutka, S. J. and Snyder, S. H. (1979). Multiple serotonin receptors: differential binding of [³H]5-hydroxytryptamine, [³H]lysergic acid diethylamide and [³H]spiroperidol, *Mol. Pharmacol.* 16, 587–699.

Peroutka, S. H., Lebowitz, R. M., and Snyder, S. H. (1979). Serotonin receptor binding sites affected differentially by guanine nucleotides, *Mol. Pharmacol.* 16, 700–708.

Quay, W. B. (1965). Indole derivatives of pineal and related neural and retinal tissues, *Pharmacol. Rev.* 17, 321–345.

Redburn, D. A. (1977). Uptake and release of GABA from rabbit retina synaptosomes, *Exp. Eye Res.* 25, 255–275.

Regen, J. W., Roeska, W. R., and Yamamura, I. (1980). ^{3}H-Flunitrazepam binding to bovine retina and the effect of GABA thereon, *Neuropharmac.* **19**, 413–414.

Robertson, H. A. (1979). Benzodiazepine receptors in 'emotional' and 'non-emotional' mice: comparison of four strains, *Europ. J. Pharmacol.* **56**, 163–166.

Rodieck, R. W. (1973). *The Vertebrate Retina*, Freeman, San Francisco.

Shaskan, E. G. and Snyder, S. H. (1970). Kinetics of serotonin accumulation into slices from rat brain: relationship to catecholamine uptake, *J. Pharmacol. Exp. Therap.* **175**, 404–418.

Simpson, L. C. (1969). The role of calcium in neurohumoral and neurohormonal extrusion processes, *J. Pharm. Pharmacol.* **20**, 889–910.

Smith, M. D. (1973). 5-Hydroxytryptophan decarboxylase (5-HTPD) and mono-amine oxidase (MAO) in the maturing mouse eye, *Comp. Gen. Pharmacol.* **4**, 175–178.

Smith, D. B. and Baker, P. C. (1974). The maturation of indoleamine metabolism in the lateral eye of the mouse, *Comp. Biochem. Physiol.* **49A**, 281–286.

Snodgrass, S. R. and Horn, A. S. (1973). An assay procedure for tryptamine in brain and spinal cord using its [^{3}H] dansyl derivative, *J. Neurochem.* **21**, 687–696.

Snyder, S. H. and Goodman, R. R. (1980). Multiple neurotransmitter receptors, *J. Neurochem.* **35**, 5–8.

Steinbusch, H. W. M., Verhofstad, A. A. J., and Joosten, H. W. J. (1978). Local-isation of serotonin in the central nervous system by immunohistochemistry: description of a specific technique and some applications, *Neuroscience*, **3**, 811–819.

Stell, W. K. (1972). The morphological organization of the vertebrate retina, in *Physiology of Photoreceptor Organs* (Ed. M. G. F. Fuortes) Vol. VII/2, pp. 112–213, Springer Verlag, Berlin-Heidelberg-New York.

Straschill, M. (1968). Actions of drugs on single neurons in the cat's retina, *Vision Res.* **8**, 35–47.

Straschill, M. and Perwein, J. (1969). The inhibition of retinal ganglion cells by catecholamines and γ-aminobutyric acid, *Pflügers Arch. Eur. J. Physiol.* **312**, 45–54.

Straschill, M. and Perwein, J. (1975). Effects of biogenic amines and amino acids on the cat's retinal ganglion cells, in *Proceedings of the Golgi Centennial Symposium* (Ed. M. Santini) pp. 583–591, Raven Press, New York.

Suzuki, O., Noguchi, E., and Yagi, K. (1977a). Monoamine oxidase in developing chick retina, *Brain Res.* **135**, 305–313.

Suzuki, O., Noguchi, E., Miyake, S., and Yagi, K. (1977b). Occurrence of 5-hydroxy-tryptamine in chick retina, *Experientia*, **33**, 927–928.

Suzuki, O., Noguchi, E., and Yagi, K. (1978). Uptake of 5-hydroxytryptamine by chick retina, *J. Neurochem.* **30**, 295–296.

Thomas, T. N. and Redburn, D. A. (1979). 5-Hydroxytryptamine—a neurotrans-mitter of bovine retina, *Exp. Eye Res.* **28**, 55–61.

Thomas, T. N. and Redburn, D. A. (1980). Serotonin uptake and release of sub-cellular fractions of bovine retina, *Vision Research*, **20**, 1–8.

Voaden, M. J. (1976). Gamma aminobutyric acid and glycine as retinal neurotrans-mitters, in *Transmitters in the visual process* (Ed. S. L. Bonting). pp. 107–125, Pergamon Press, Oxford.

Voaden, M. J., Morjaria, B., and Oraedu, A. C. I. (1980). The localisation and metabolism of glutamate, aspartate and GABA in the rat retina, *Neurochem. Int.* **1**, 167–182.

Wainwright, S. D. (1979). Development of 5-hydroxyindole-*O*-methyltransferase activity in the retina of the chick embryo and young chick, *J. Neurochem.* **32**, 1099–1101.

Weakley, J. N. (1973). The action of cobalt ions on neuromuscular transmission in the frog, *J. Physiol.* **234**, 597–612.

Welsh, J. H. (1968). Distribution of serotonin in the nervous systems of various animal species, *Advanc. Pharmacol.* **A**, 171–188.

Werman, R. (1966). Criteria for identification of a central nervous transmitter system, *Comp. Biochem. Physiol.* **18**, 745–766.

Wilkin, G., Wilson, J. E., Balasz, R., Shon, F., and Kelly, J. S. (1974). How selective is high affinity uptake of GABA into inhibitory nerve terminals? *Nature* (Lond.), **252**, 397–399.

Wurtman, R. J., Axelrod, J., and Kelly, D. E. (1968). *The Pineal*, Academic Press, New York.

Yamada, T., Marshak, D., Basinger, S., Walsh, J., Morley, J., and Stell, W. (1980). Somatostatin-like immunoreactivity in the retina, *Proc. Nat. Acad. Science*, **77**, 1691–1695.

Biology of Serotonergic Transmission
Edited by N. N. Osborne
© 1982, John Wiley & Sons Ltd.

Chapter 17

Serotonin-containing Neurones within the Segmental Nervous System of the Leech

C. M. LENT
Center for Neuroscience, Brown University, Providence, Rhode Island 02912, USA

INTRODUCTION

In 1891, Gustav Retzius published his observations upon the morphology of the methylene blue-stained neurones within the segmental nervous system of two species of leeches. His description of the paired colossal ganglion cells (KZ) or Retzius cells (RZ) established them as perhaps the first neurones which could be repeatedly identified from ganglion to ganglion and animal to animal. Twelve years later, Poll and Sommer (1903) showed that the large Retzius cell somata, as well as those of some four smaller neurones, gave a positive reaction to chromium salts. This positive chromaffin reaction provided what is likely the first histochemical

Biology of Serotonergic Transmission

evidence for a class of identifiable and biochemically related neurones which differed distinctly from the remaining majority of other ganglionic neurones. It is now well established that serotonin (5-HT) which is found within the Retzius cells is responsible for their chromaffinity (Rude *et al.*, 1969). The extensive literature upon the biochemistry of Retzius cells was reviewed recently (McAdoo, 1977), and a more comprehensive review of the morphology, fine structure, histochemistry, physiology and pharmacology appeared in the same year (Lent, 1977). It is the purpose of this chapter to review the recent researches upon the population of ganglionic neurones in the leech nervous system which contain serotonin: this includes not only the large Retzius cells (80 μM) but also five or seven other small (10-20 μM) neurones (Fig. 1). This discussion will be developed with a perspective toward the comprehensive field of leech neurobiology.

SEGMENTAL NERVOUS SYSTEM

A characteristic of leeches (Annelida: Hirudinea) which offers distinctive research advantages is the segmental body plan of these worms, for they have one discrete ganglion per metamere. These ganglia together with their intercommunicating, longitudinal connectives comprise the ventral nerve cord. Each of the 21 segmental ganglia exchange information with the periphery *via* two pairs of lateral roots (Fig. 1). Most of the segmental ganglia are very similar except for the 5th and 6th which innervate the complex genitalia of these hermaphroditic worms. It is, therefore, possible to use any of the 19 different ganglia from the same animal as both the experimental and control preparations. Much larger ganglionic masses innervate the two suckers found at each end of the leech. The caudal ganglionic mass is composed of seven subganglia, each of which bears some superficial resemblance to the segmental ganglia, while the anterior sucker is innervated by both sub- and supraoesophageal ganglia. The suboesophageal ganglion is composed of four subganglia (each of which resemble segmental ganglia) and it has an embryological origin which is common to the remainder of the metameric nervous system. This system is therefore comprised of a total of 32 segments (4, 21 and 7; Fernandez, 1978). The supraoesophageal ganglion, which communicates with the suboesophageal by circumoesophageal connectives, arises from different embryological tissues and bears little resemblance to the segmental nervous system. During development, all the segmental ganglia are fused and those destined to innervate the middle segments of the body become separated by the intercalated growth of the connectives. The longest connectives are in the central regions of the leech and are shortened at each end. There is, of course, no growth of connectives within either the caudal or suboesophageal ganglia.

In *Hirudo medicinalis*, (a member of the family Gnathobdellidae: jawed leeches—which contain the most commonly utilized research species) each segmental ganglion is comprised of about 400 monopolar neurons while the sex ganglia have an additional 300 cells (Macagno, 1980). These spherical somata are arranged in a

Anterior

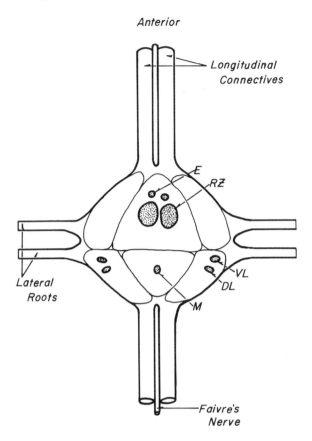

Fig. 1   A diagrammatic view of the ventral surface of a leech segmental ganglion. The topological relationship between the serotonin-containing neurones and their investing glial pockets is shown in this anterior ganglion. Every segmental ganglion has two colossal Retzius cells (RZ), a pair of ventro-lateral cells (VL), dorso-lateral cells (DL) and an unpaired medial cell (M). The first three segmental ganglia have an extra pair of cells (E) near the Retzius cells. The typical ganglion is 300–400 μm in diameter

cortical rind which is enveloped by six large glial packet cells and the entire structure is covered with an epineural sheath (Coggeshall and Fawcett, 1964). The four lateral and two ventromedial packets provide tell-tale landmarks for recognizing the positions of the individual neurosomata which are arranged in clusters of about 50–75 per packet (Macagno, 1980). Two additional glia in the segmental ganglia invest the medullary neuropile where most of the inter-neuronal synaptic events transpire. The axonal tracts of each of the paired connectives are enmeshed by yet other glial cells (Coggeshall and Fawcett, 1964). Each monopolar neurone sends its initial segment centrally toward the neuropile, wherein many fine processes are branched off and there interdigitate with the processes of other neurones making synaptic and/or electrotonic contacts. The axonal processes of sensory and effector neurones leave the ganglia by the lateral roots while those of interneurones egress only through the connectives.

IDENTIFIED NEURONES

Approximately one-third of the neurones in the leech ganglion have been identified with respect to their general position, electrophysiological characteristics, synaptic connectivity and functional roles. Three classes of mechanosensory neurones were found to respond with increased impulse activity to dermal manipulation: touch, pressure or noxious stimulation (Nicholls and Baylor, 1968). The six T cells, four P cells and four N cells, have their somata within the central ganglion at reasonably constant loci, and each of these sensory cells project axons ipsilaterally to its own discrete field of innervation. Action potentials are conducted toward the central somata following dermal stimulation and these cells seem to function as primary sensory neurones. Each class of sensory cells has its own characteristic complement of electrophysiological properties, and it was the description of these sensory cells and their highly-stereotyped properties which provided much of the impetus for what has become a rapid growth of interest in leech neurobiology.

Several effector cells have also been described, and include some 34 motor neurones, consisting of both excitors and inhibitors of the various muscle layers and bands which comprise most of the mass of the body wall of the leech (Stuart, 1970). These motor cells project contralaterally, conduct their impulses efferently and also have prescribed fields of innervation. Additional motor neurones have been found which are active during swimming behaviour (Ort, Kristan, and Stent, 1974), and others whose impulse activity excites the lateral tubular hearts into peristalic waves of contraction (Thompson and Stent, 1976).

The leech segmental nervous system has recognizable interneurones whose axons are always restricted to the ganglia and connectives. The first described was the S cell—a medial, unpaired neurosoma whose initial segment bifurcates into two large-calibre (10–15 μM) axons which project both anteriorly and posteriorly into the small medial longitudinal connective: Faivre's nerve (Gardner-Medwin, *et al.*, 1973; see Fig. 1). S-cell axons make electrical connections with S-cell axons from adjacent ganglia within the interganglionic region of Faivre's nerve and functionally constitute a single large axonal path having segmentally-interposed gap junctions (Muller and Carbonetto. 1979). This 'giant' axon conducts impulses along the length of the leech in either headward or tailward direction, and is probably associated with the shortening response exhibited by the leech to a variety of stimuli (Muller, 1979). Each ganglion has a quartet of interneurones whose inhibitory synaptic interactions appear to generate the swimming rhythm which is responsible for this sinuous behaviour which jawed leeches often display (Stent *et al.*, 1978). These 'oscillatory' interneurones restrict their connections to the central ganglia and their axons to the connectives. A population of interneurones (HN) is found within the seven most-anterior ganglia and their rhythmic impulse activity periodically inhibits the heart motor neurones (H.E.) (Calabrese, 1979). It is these interruptions of the tonic efferent activity of the motor neurones which provide the periodic control of the cardiac rhythm. These HN interneurones also restrict their axonal projections to the central nervous system (Shafer and Calabrese, 1980).

SYNAPTIC CONNECTIONS

For a comprehensive review of the anatomy, fine structure and the functional properties of the synapses in the leech nervous system, you should consult Müller (1979). The purposes of this chapter are better served by the following, brief treatment. Many of the identified neurones make functional interconnections which are mirrored in the behaviour of the leech. Thus, the sensory cells make mono-synaptic contacts upon some motor neurones, underlying their 'simple' reflexive behaviours which are seen both in the intact leech and mirrored by isolated single ganglion to body-wall preparations (Nicholls and Purves, 1972). Many pre- and post-synaptic neurones are accessible for simultaneous recording and/or control by multiple microelectrode experiments which can be carried out upon such isolated preparations. This is an attribute which offers significant research advantages for several fundamental neurobiological problems.

The N cell makes an excitatory chemical synapse upon the longitudinal motor neurone (L), and generates an EPSP which is sufficiently large to cause post-synaptic impulses and evoke shortening of the body wall of the leech after it is pinched. The touch cell evokes similar behaviour, but by means of an electrotonic connection which it makes onto the L motor neurone. The pressure-sensitive cell has a mixed synapse upon L, and is comprised of both chemical and electrotonic components. Inhibition is exemplified by the chemical synapses which the heart interneurones make upon motor neurones and by inhibitory synapses which function among the swimming pattern interneurones.

The strong intraganglionic electrical coupling which exists between the paired Retzius cells as well as between L motor neurones appears to insure a bilaterally symmetrical conduction of impulse traffic. This is in contrast to the interganglionic electrotonic coupling between S cells and between Leydig cells (Keyser, Frazer, and Lent, 1981) which seemingly functions in both instances to permit the bidirectional conduction of impulse traffic along the ventral nerve cord. Because the leech ganglion has an array of diverse types of synapses which occur between identified neurones, it is reasonable to consider them as identifiable synapses. The morphological properties of but a few of the many synapses known in the leech have been investigated. The major approach has utilized the intracellular injection of the pre- and post-synaptic neurones with fluorescent, opaque and/or electron-dense dyes followed by examination with the light and/or the electron microscope (Muller and McMahan, 1976).

REGENERATION AND DEVELOPMENT

When the axonal tracts between ganglia are interrupted by crushing, leech neurones have the capacity to re-establish interganglionic electrotonic and chemical connections with a surprisingly high degree of precision (Jansen and Nicholls, 1972; Müller and Carbonetto, 1979). Sensory neurones are also capable of re-establishing functional contacts with their original fields of innervation if their axonal tracts within the roots are crushed (Van Essen and Jansen, 1976). Much of this regeneration occurs by means of axonal growth along interrupted axon pathways and if the connections are disrupted either by transection or by removal of the axonal tracts, any re-establishment of function is both rare and highly aberrant. Nevertheless, these experiments suggest an interesting degree of recognition between neurones and their targets which normally have functional connectivity.

Leech nerve cell bodies survive and grow neurites under culture conditions. Retzius cell somata grown in culture send neurites toward other Retzius cells and form contacts which usually mediate electrical coupling (Ready and Nicholls, 1979). The efficacy of the coupling is highly variable, and on some occasions, the cells establish inhibitory chemical synapses, a condition which has not been seen within the nervous system itself. Retzius cells cultured *in vitro* failed to establish functional contacts with P cells—mirroring their *in vivo* lack of interconnections.

Raising leeches from egg through embryo to adult has become a laboratory routine which provides the opportunity to describe the development of the leech nervous system and, more importantly, to experimentally manipulate the developing nervous system in order to fathom the mechanisms which are involved in processes such as the establishment of synapses and the biochemical differentiation of identified neurones (Weisblat *et al.*, 1980).

CHEMICAL ORGANIZATION OF LEECH SEGMENTAL GANGLIA

An especially compelling property of the leech segmental ganglion is the differentiation of its neurones into discrete subsets which are functionally related by containing either identical neurochemicals or neurochemicals belonging to a particular class. The property of biochemical differentiation was first recognized by Poll and Sommer (1903) and the first attempt to codify their observations into a functional schema was made by Gaskell (1914), when he (erroneously) proposed that the chromaffin cells were involved in the cardio-regulation of the heart rate in the leech. Modern immunological experiments are bringing the biochemical specialization within leech ganglia into a sharper focus. For example, monoclonal antibodies have been raised against the entire nerve cord of the leech (Zipser *et al.*, 1980), and when neurones which bind these antibodies are stained, several individual classes of neurones can be seen. The number of neurones per class ranges from less than 5 to more than 25. Thus, the monoclonal antibody technique has the potential of identifying discrete classes of neurones which are distinguished by having either membranous or protoplasmic antigens in common.

Neurotransmitters

The massive longitudinal muscle of the leech body wall has long been known to be exquisitely sensitive to the vertebrate neuromuscular transmitter, acetylcholine (ACh) (Minz, 1932). The sensitivity of the leech body wall can be enhanced by anticholinesterases, making it useful as a bio-assay of ACh at concentrations as low as 1×10^{-9} g/ml (Feldberg and Gaddum, 1934). Kuffler (1978) showed that longitudinal muscle fibres have their highest sensitivity to ACh at foci which are identical to those where synaptic potentials are generated by the axon terminals of the L motor neurone. On this basis; observations of the blockage of transmission with curare; and the histology of identified nerve terminals; he concluded that the L cell very probably utilizes acetycholine as its transmitter chemical. Further, the L cell can synthesize ACh with choline acetyltransferase as can the annulus erector motor neurone (Sargent, 1977). These two excitatory motor neurones have levels of acetylcholine which are measureably higher than other classes of cells. Sargent also found that leech ganglia had the necessary enzymes to synthesize gamma-animobutyric acid, 5-HT, glutamic acid, dopamine and octopamine. However, the identity of the transmitter in the sensory neurones remains unknown, as it does in the vertebrate nervous system. Nor were any candidates proposed for the transmitter from inhibitory motor neurones.

Peptides and neurosecretory cells

The peptide leu-enkephalin has recently been detected in leeches with an immunocytochemical technique (Zipser, 1980), and this peptide is found within a single medially-positioned neurone within each of the ganglia in the posterior two-thirds of the leech. These enkephalin-reactive cells appear to have large varicosities within their ganglionic neuropiles.

The supraoesophageal ganglion (non-segmental) contains about 24 neurones which are bluish-white and contain 100–300 nM electron dense granules (Webb and Orchard, 1979). These cells are arranged in four distinctive groups based upon their morphology after the injection of the marker enzyme, horseradish peroxidase (Orchard and Webb, 1979). These cells are very probably neurosecretory and their endings project into specialized neurohaemal structures for which Webb (1980) proposes the term: intralamellar complexes.

Webb and Orchard (1979) noted that each ganglion in the segmental nervous system contains a pair of posterior-lateral somata which display a Tyndall blue-white coloration and which show some affinity for the dye orange G. These cells contain electron-dense granules which appear to be neurosecretory. These neurones are strongly coupled across the ganglion and are identical to those neurones referred to as Leydig cells (Nicholls and Baylor, 1968). The Leydig cells are spontaneously active and those in adjacent ganglia have a common axonal pathway along their ipsilateral connective (Keyser and Lent, 1980). The Leydig cells project toward the periphery via the roots of both their adjacent ganglia and not by the roots of their parent ganglia (Keyser, *et al.*, 1981). This morphology contrasts with all neurones in the leech CNS which have been described to date. These cells show intraganglionic coupling potentials, interganglionic impulse failures and reverberatory impulse patterns. Additionally, Leydig cell homologues have been found within the sub-oesophageal ganglion, but not within the caudal ganglion (Keyser and Lent, 1981). No functional role has been adduced for any peptide-containing or neurosecretory cells in the leech. However, Webb (1980) proposes that the neurosecretory cells in the supraoesophageal ganglion may play a role in spermatogenesis. For he finds that the removal of the anterior end of the leeches arrests sperm formation and that the injection of homogenates of the supraoesophageal ganglion reactivates spermatogenesis.

Monoamines (MA)

The researches upon the biochemistry of the monoamines within segmental ganglia of leeches is probably the best-focused and most-advanced of any of the work on their neurochemistry. The research upon these biogenic amines has the longest history and, of course, constitutes the major emphasis of this chapter. There is reasonable evidence for the existence of the ethanolamine-octopamine, the catecholamine-dopamine, and the indoleamine-serotonin within the CNS of the leech.

Both dopamine and octopamine can be synthesized by leech ganglia from either [^{14}C] tyrosine (Stuart, *et al.*, 1974) or tritiated tyrosine (Sargent, 1977). The former study also detected the presence of tyramine, but neither study detected norepinephrine (a catecholamine neurotransmitter often found in vertebrates). There is no evidence as to which leech neurones might contain octopamine; however, Webb and Orchard (1980) assayed octopamine in homogenates of ganglia, connectives, and roots using a microradioenyzymatic method. They found significantly more octopamine in those components of the nervous system which have cell bodies than those comprised exclusively of axon tracts. In further work, Webb and Orchard (unpublished) have found that leech ganglia have a high-affinity, sodium-dependent uptake system for octopamine.

The catecholamine, dopamine is very probably found in a pair of cell bodies which lie in the anterior roots; just proximal to its first major branch point (Fig. 2). These AR (anterior root) cells have a brilliant blue-green fluorophore with an emission maximum at 495 nM following the Falck–Hillarp histochemical treatment (Lent *et al.*, 1979). For this procedure, freeze-dried tissues are exposed to hot, vaporous formaldehyde which condenses with catechole and indoleamines to form dihydroisoquinolines and 3,4-dihydro-*B*-carbolines respectively. These two products act as fluorophores which emit at 480–490 nm and 520–530 nm. The AR cell body has obvious catecholamine-like fluorescence, and it appears to send its single axon into the ganglion wherein it branches repeatedly, and fills the ipsilateral neuropile with a blue-green fluorescence. Axon-like fluorescent processes leave the ganglia via both the anterior and posterior ipsilateral connectives. The intracellular

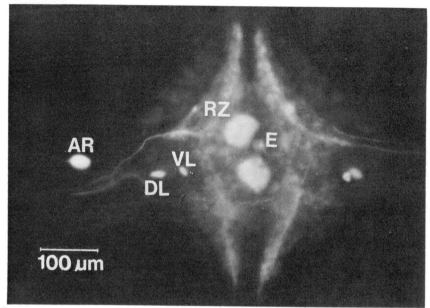

Fig. 2   A fluorescence micrograph of an anterior ganglion which was dissected from the leech *Macrobdella decora* and treated with the Falck–Hillarp method for monoamines. The neurosomata within the ganglion fluoresce yellow and include the Retzius cells, two pairs of lateral cells, and one pair of E cells. The M cell which usually has the lowest level of fluorescence is not visible in this micrograph. As blue-green fluorescent neurosoma is also visible in the left anterior root which is reflected back upon itself. Modified from Lent, *et al.* (1981)

injection of HRP into these AR somata, and the examination of their blackened cell processes with the light microscope, fully verifies this projection (Lent, 1981). In an attempt to measure dopamine in AR cells, individual somata, which were stained with neutral red, were dissected free by hand and assayed with a microradioenyzymatic technique which is sensitive to amounts of dopamine as low as 50 fentomoles (5×10^{-14} moles; C. Lent, J. Ono, and J. Eisen, unpublished observations). Dopamine was measured in 7 of 24 AR somata, but in none of the control cells. Each soma had about 0.2 pmole of dopamine, each axon segment about 0.3 pmole and the neuropile had about 1 pmole of dopamine. A fundamental problem with this study was the fact that staining with neutral red dye appeared to lower the amount of measureable dopamine. However, this effect might be explained by Nemes *et al.* (1979) who propose that the dye displaces amines from their receptors and binding sites.

SEROTONIN NEURONES IN THE LEECH CNS

As was pointed out previously, many facets of the research upon Retzius cells have been recently reviewed (Lent, 1977; McAdoo, 1977). Much of the earlier research will be omitted from this treatment. This portion of the chapter will deal primarily with the more-recent research upon the Retzius cells and upon the five or seven, small serotonin-containing neurones. Rude *et al.* (1969) provided the first conclusive evidence that the chromaffin positive Retzius cells in fact contain serotonin. This they did by means of four interdigitating studies; (a) microspectrofluorometry of sections as well as intact ganglia which had undergone the Falck–Hillarp reaction, (b) thin-layer chromatography of extracts of large numbers of individually dissected Retizus cells, (c) spectrofluorometry of similar pooled cell extracts, and (d) a chromaffin reaction with an electron-dense reaction product which could be localized in electron micrographs of Retzius cells. Their researches established beyond doubt that Retzius cells contained serotonin but yielded no fluorometric or chromatographic evidence for the presence of catecholamine. This research has since been verified with newly-developed micro-methods which are capable of measuring the amine levels found within single neurosomata. McCaman *et al.* (1973) developed a fluorometric method using neuronal extracts which had been eluted through a liquid chromatography column. With this technique, which could detect as little as 2 pmoles (2×10^{-12} moles) of 5-HT, they found an average value of 2.5 pmoles of 5-HT/Retzius cell—a value within the range of 0.8–5.8 pmoles/Retzius cell reported by Rude *et al.* (1969). More recently, gas chromatography was combined with mass spectrometry to measure the neurochemicals in individual Retzius cells because this method has an unusually high specificity and sensitivity (McAdoo and Coggeshall, 1976). Retzius cells had an average of 2.0 pmoles of 5-HT per cell with a range of 1.3–4.1 pmoles. This study also showed that individual Retzius cells had none of the following transmitters at levels above background: GABA, octopamine, dopamine and norepinephrine. They also calculated the Retzius neurosomata concentration of 5-HT to be at least 100 × more that within control neurone cell bodies.

Vital staining

The Retzius cells are easily identified in living preparations by virtue of their position within the anterior-ventral packet and by the fact that they are the largest neurosomata within the ganglion (up to 80 μM). This morphological property has made them the subject of many researches simply because of their size, position and experimental accessibility.

The remaining five or seven neurones, which share histofluorescence properties with these colossal cells are small (8-15 μM), widely scattered within the ganglion, and not recognizable as amine-containing cells in live ganglia. These cells were therefore inaccessible to researchers until Stuart *et al.* (1974) demonstrated that all of the amine neurones could be vitally and selectively stained with very dilute solutions of neutral red dye (Fig. 3). Neutral Red (and certain other basic dyes which contain a three-ring heterocyclic nucleus) appeared to selectively stain those cells which fluoresce following the Falck–Hillarp procedure. In order to prove that this was

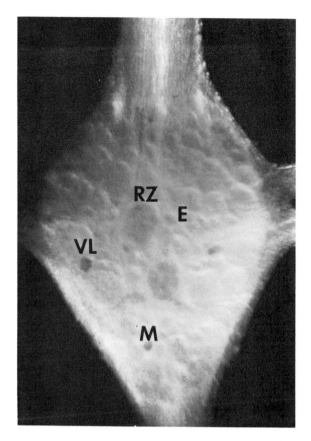

Fig. 3   A darkfield micrograph showing the ventral aspect of an anterior leech ganglion which was exposed to neutral red dye. The two Retzius cells, the ventro-lateral cell pair, the extra cell pair, and the unpaired medial cell have been selectively-stained by this procedure

indeed an isomorphic population, ganglia were stained with toluidine blue (a member of the family of basic dyes) prior to freeze-drying and formaldehyde condensation. When these ganglia were alternately viewed with white light and ultra-violet illumination, only the blue-stained neurones emitted the characteristic fluorescence. It is worth noting that this experiment cannot be done with Neutral Red because that dye is itself fluorescent, and even if selective staining of the amine cells has taken place, the entire ganglion becomes intensely fluorescent. This high background fluorescence masks any specific fluorescence which the formaldehyde condensation might produce (Lent, 1981). Stuart *et al.* (1974) further showed that red-stained Retzius cells exhibited action potentials, coupling potentials and synaptic potentials which were indistinguishable from those seen in unstained control Retzius cells. Thus, this study described a marker chemical which appeared to be capable of selectively and vitally staining all of the amine neurones in leech ganglion *in vivo* making the small, scattered cells experimentally accessible.

A most intriguing problem concerning vital staining is the mechanism of staining itself, for it is not know what class of chemicals or organelles bind or react with the Neutral Red. Ganglia which are treated with metabolic poisons or with fixatives continue to exhibit selective staining (Stuart *et al.*, 1974). If we examine the cytoplasmic organization of red-stained Retzius cells with Nomarski optics, the colouration is concentrated in two concentric perinuclear rings (Fig. 4A). This unusual arrangement is reminiscent of the two concentric rings of fluorescence which Retzius cells show following Falck–Hillarp treatment (Rude *et al.*, 1969). The Retzius cell cytoplasm is segregated into four layers (Gray and Guillery, 1963). Figure 4B shows layer I immediately adjacent to the reticulated nuclear membrane, as well as layers II and III. Layers I and III contain extensive Golgi apparatus and vesicles with electron dense cores which very probably contain serotonin (Rude *et al.*, 1969). Layers II and IV are similar in containing large amounts of endoplasmic reticulum. When the small amine cells are examined after Neutral Red treatment, they have but one dark red perinuclear sphere, and when they are examined after Falck–Hillarp treatment, they have but one fluorescent ring. These observations lead to the suggestion that the vital staining occurs in the region of the cytoplasm wherein the serotonin is located.

Biochemistry of amine cells

While it is abundantly clear that Retzius cells contain serotonin, the neurochemical within the small MA cells has been presumed to be serotonin primarily because of the similarity of its formaldehyde fluorescence properties to those of the RZ (Ehinger *et al.*, 1968). Microspectrofluorometry and a radioenzymatic assay for serotonin was employed to measure serotonin within single MA cells dissected from leech ganglia (Lent *et al.*, 1979). Ganglia from the American leech *Macrobdella decora* were treated with Neutral Red, and red-stained somata (VL, 10 μm) were dissected free in cold ethylene glycol. The enzymatic assay for serotonin is a modification of the method of Saavedra *et al.* (1973) whose sensitivity can be

Fig. 4   (a) A light micrograph (using Nomarski differential interference-contrast optics) of a pair of red-stained Retzius cells (about 75 μm in diameter) after fixation, dehydration, and clearing. Note the intensely-stained rings surrounding the clear nucleus. (b) An electron micrograph of a Retzius cell which shows the highly irregular and strikingly-hexagonal nature of the membrane around the nucleus (N). The dark region of cytoplasm immediately adjacent to the nucleus is layer I which contains granules, mitochondria, extensive Golgi, and vesicles. Layer II is next most peripheral and is composed, almost exclusively, of rough endoplasmic reticulum. Layer II is next to and is cytologically identical to layer I. Layer IV is not shown but is very similar to II and is bounded by the limiting plasmalemma. Layers I and III contain the serotonin of the RZ and appear to be the layers which are stained by neutral red

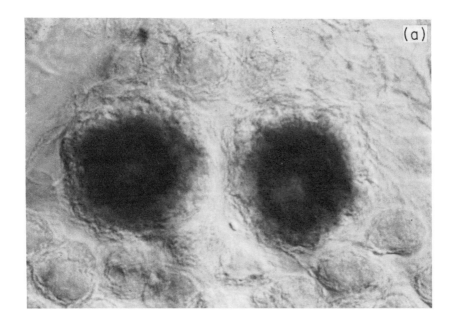

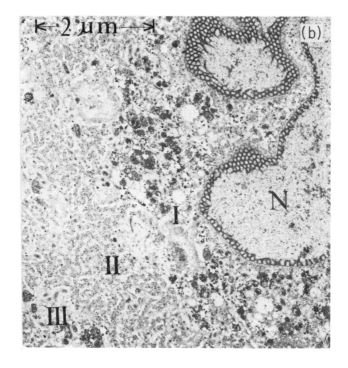

extended to 50 femtomoles (5×10^{-14} moles) of 5-HT. To find whether the staining procedure might alter the levels of 5-HT, measurements were also made of Retzius cells and whole ganglia in both unstained and stained conditions. The results of the assay are presented in Table 1 where it can be seen the Neutral Red staining does not significantly affect the levels of serotonin in either intact ganglia or Retzius cells.

The VL, as well as the other small MA, cells are only identifiable *in vivo* following Neutral Red staining. For the 18 VL cells measured, only one had no detectable serotonin and the average amount of serotonin for all 18 VL cells was 0.2 pmoles/cell. The VL cell somata which were removed for measurement had an average diameter of 10 μM. However, the ethylene glycol treatment is known to produce slight cell shrinkage. Therefore, we utilized an average cell diameter value of 15 μM and assumed that they were perfectly spherical. Under these assumptions, we calculated that VL cells have a minimal serotonin concentration of 100 mM. If no shrinkage occurred, then the serotonin concentration within VL cells approaches 400 mM.

None of the five control somata, which were of similar size as VL, had detectable serotonin. The controls had to be taken from stained ganglia to ensure that amine cells were not unintentionally selected. We also measured for serotonin in two DL cells: one had serotonin above sensitivity and the other did not.

For microspectrofluorometry, whole ganglia were treated according to the Falck–Hillarp procedure and examined on a Leitz MPV-2 microspectrofluorometer. Complete emission spectra for the ganglionic MA cells are shown in Fig. 5. Each of the four, vertically-displaced spectra are normalized such that its maximum represents 100 per cent of its emission value. A tissue spectrum (T) was taken over a non-fluorescing portion of the ganglion for each cell class, but is shown for the VL cell only. These tissue emissions are in the blue end of the spectrum at 495 nm and represent about 25 per cent of the peak values of the lateral cells, 40 per cent of the RZ, and 60 per cent of the M cell. The level of serotonin in *Macrobdella* Retzius cells is five times lower than the 2 pmoles measured in *Hirudo* RZ even though the cell diameters are similar. It would be premature to assume that these differences represent true species differences for both were not examined at the same time of year. Serotonin content within the nerve cord of *Hirudo* varies four-fold through-

Table 1 Serotonin in leech neural tissue

Tissue	Control	pmole Serotonin Neutral Red stained
Ganglia	9.5 ± 0.7 (11)	10.3 ± 0.5 (5)
Retzius cell	0.33 ± 0.11 (5)	0.43 ± 0.10 (8)
VL cell		0.20 ± 0.02 (18)
*Control cell		0.05 (5)

± Standard error of the mean (sample number) from Lent *et al.*, 1979.

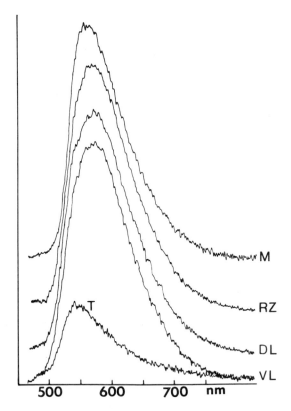

Fig. 5 The fluorescence emission spectra from four of the ganglionic neuro-somata following formaldehyde condensation. The curves have been normalized for equal amplitudes. (After Lent, *et al.*, 1979)

out the year (Stenzel and Neuhoff, 1976). Further, age, size, nutritional state or extrinsic factors could well be responsible for any observed differences.

The fact that serotonin can be measured within VL neurosomata strongly suggests that this neurochemical is responsible for its histochemical fluorescence. As the fluorescence peak of DL is indistinguishable from that of VL and similar to that of RZ, both of which contain serotonin, we infer that DL cells probably contain serotonin as well. This inference is strengthened by the single enzymatic measurement of 5-HT within a DL cell body. We were unable to measure 5-HT in M cells as they stain very lightly and were usually so faded by the ethylene glycol treatment as to be indistinguishable from non-amine cells.

Since it is not known what is stained by Neutral Red, it was interesting to note that the red staining seen in this study usually parallelled the fluorescence. That is, the lateral cells stained the darkest red and also had the most intense yellow fluorescence, while the M cell usually had the palest red stain and weakest fluorescence.

Retzius cells were usually intermediate in the degree of both their staining and fluorescence.

The emission peak of the RZ is shifted 7 nm toward the blue end of the spectrum while that of M is shifted by 14 nm. These spectra were taken from intact ganglia and it can be seen in Fig. 1 that both these cells are situated directly above the blue-green neuropile. The degree of blue shift is proportional to the relative intensity of the tissue emission. Thus, the M cell has the weakest yellow fluorescence, the highest relative amount of blue emission from underlying tissues and therefore the largest shift away from the serotonin fluorophore peak. Ehinger *et al.* (1968) found identical peaks from RZ and small cells when they examined histological sections of ganglia, which would, of course, have no underlying tissues with a blue fluorescence. We suspect, for these reasons, that all the MA cells within leech ganglia which stain with neutral red and fluorescence yellow contain serotonin.

Unstained (control) neurosomata had no measurable serotonin leading us to conclude that the 100 mM serotonin is located within the VL cell and is not an axonal contaminate from the neuropile. Were it such a contaminant, it also should appear in controls. The VL cell bodies have their serotonin concentrated at 100 mM, a level so high that this neurochemical is at its limit of solubility in aqueous media. This calculation of serotonin assumes that the chemical is equally distributed within all of the measured cell volume. Any regions which are unavailable to serotonin will raise the estimated concentration value. In Retzius cells, 5-HT has restricted membrane-bound neuroplasmic sites (Rude *et al.*, 1969). It is possible that VL cell serotonin is even more highly concentrated than the minimal calculation of 100 mM and therefore not dissolved in the aqueous phase of the cytoplasm.

Each leech ganglion has about 10 pmoles of serotonin. The two RZ somata contain about 1 pmole and if we assume that each of the seven small cell bodies has 0.2 pmole, we can account for 2.5 pmoles of serotonin. Thus, 25 per cent is located within the cell bodies, and 75 per cent of the measurable serotonin is extrasomatal. We presume that this extrasomatal serotonin is located within the extensively ramifying processes of these MA neurones within the neuropile.

Anatomy of serotonin cells

The essential features of the colossal cells have been well known since their original description by Retzius (1891). During the last decade, the anatomy of Retzius cells has been studied by the intracellular injection of a variety of marker chemicals including Cobalt (Smith *et al.*, 1975); Procion Yellow (Lent, 1973a) and horseradish peroxidase (Mason and Leake, 1978). The morphological conformation of a Retzius cell which was filled with the fluorescent dye Procion Yellow M4RS is illustrated in Fig. 6. The large soma sends its initial segment into the neuropile wherein several short processes emerge. This segment branches into two large-calibre axons which travel to the periphery *via* the paired ipsilateral roots. Smaller axons branch off the larger processes and egress by both ipsilateral connectives.

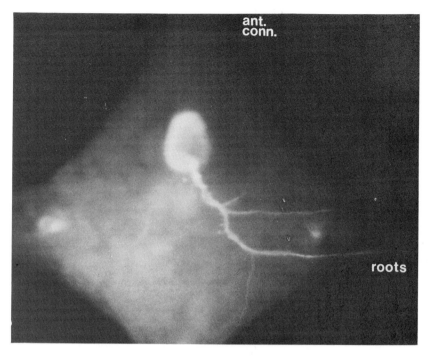

Fig. 6   A fluorescence micrograph of an 80 μm Retzius cell which has been ion-tophoretically filled with the dye Procion Yellow M4RS. The initial segment has several small processes within the neuropile. Large calibre axons egress via the ispilateral roots while smaller axons egress via both ipsilateral longitudinal connectives. (Lent, 1979, unpublished)

Fluorescence studies of Falck–Hillarp treated ganglia detected only one yellow axon in each root, that from RZ (Rude, 1969). Therefore, the small MA cells were suspected to be interneurones; however, their shapes and projection patterns remained unknown until recently (Lent, 1981). Neutral red-stained cells were pressure-injected with the enzyme horseradish peroxidase (Muller and McMahan, 1976) and stained with benzidine–a procedure which fills the neurone and its processes with a dark blue precipitate.

Darkly-stained cells were drawn with a *camera lucida* and representative cells are depicted by Fig. 7. The two pairs of lateral cells (ventral, VL; dorsal, DL), the unpaired medial cell (M) and the paired extra cells (E) from the three anterior-most ganglia all have their axons restricted to longitudinal connectives. Each cell type has a characteristic extensive arborization within its own ganglion and sends its axons into the connectives. Therefore, the small serotonin neurones within leech ganglia are interneurones with extensive neuropilar arborizations. They contrast sharply with Retzius cells which have limited arborization and major axonal tracts in the lateral roots.

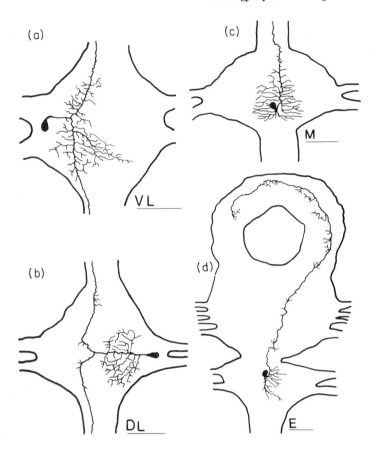

Fig. 7   *Camera lucida* drawings of the projections of the small serotonin-containing interneurones within the ganglia of *Haemopis marmorata*. (a) The VL cell has an extensive ipsilateral arborization and axons within both ipsilateral connectives. (b) The DL cell has an extensive ipsilateral arborization and axons within both contralateral connectives. (c) The M cell has a tapering, bilaterally-symmetrical arborization of its axon as it projects toward and into the anterior Faivre's nerve. (d) The E cell has a laterally-oriented arborization and it sends its axon through the suboesophageal ganglion and into the supraoesophageal ganglion wherein it terminates. The reference mark below each ganglionic cell abbreviation is a 200 μm calibration bar for that particular ganglion only. (Modified from Lent, 1981a)

The Falck–Hillarp method for demonstrating serotonin and catecholamines usually takes 24–48 hours, has many pitfalls and has highly variable results. A technique which involves ring condensations with glyoxylic acid (de la Torre and Surgeon, 1976) appears to be both rapid and straightforward. Ganglia from the leech can be treated with glyoxylic acid solution, heated, and viewed with fluorescence optics in less than 0.5 h (Lent *et al.*, 1981). The serotonin (and catecholamine) cell bodies are fluorescent as they are following Falck–Hillarp treatment, however, and more importantly, the axons and processes are more intensely fluorescent. An especially interesting aspect of this increased visibility is that the serotonin-containing axons within the longitudinal connectives appear to have regions of intense fluorescence which look like amine varicosities.

Functional properties of amine cells

The Retzius cells were first investigated because they were coupled by means of an efficacious electronic junction (Hagiwara and Morita, 1962). This junction passes current equally well in both directions, keeping both RZ at similar potential levels. Further, the currents from an impulse in either RZ is usually sufficient to excite the other, and therefore the impulses from the pair are usually conducted peripherally in a bilaterally-symmetrical fashion.

The first experiments which demonstrated any function for the Retzius cells were those which examined their role in the control of dermal mucus secretion (Lent, 1973b). The abundant mucus which is secreted from leech skin appears to be under neuronal control, for root transection abolishes secretion, and electrical stimulation of the longitudinal connectives increases the rate of secretion.

When a Retzius cell is stimulated by means of an intracellular micropipette, the amounts of mucus secreted from the skin is roughly proportional to the impulse number per unit time. This relationship persists when the ganglion is bathed in 20 mM Mg^{2+} a treatment which abolishes chemical synaptic transmission (Stuart, 1970). However, bathing the body wall in high Mg^{2+} saline blocks the increase in mucus secretion. Thus, the impulse activity of Retzius cells appears to control mucus secretion by releasing their serotonin onto the glands within the body wall. To test this possibility, deganglionated sections of body wall were exposed to differing concentrations of 5-HT. The amounts of mucus produced was a function of 5-HT concentration.

Serotonin is known to reduce the tension of longitudinal muscles (Schain, 1961), and RZ have the only serotonin-containing axons in the periphery (Ehinger *et al.*, 1968). In light of these facts, a role for Retzius cells in lowering the tension of the longitudinal muscles of the body wall was recently established (Mason *et al.*, 1979). An induced burst of action potentials by the Retzius cell causes a decrease in body wall tension after a 5–25 s delay, and the effect lasts for 10–15 min. Further, Retzius cell bursts appear to increase the rate of relaxation when a body wall is stimulated to contract with electrical current. These effects can be mimicked by bath application of 1–10 μM serotonin. Thus, Retzius-cell serotonin appears to be involved in a slow acting, long-duration relaxation of the longitudinal muscles of the leech body wall.

A third functional role for serotonin in activating the complex swimming motor programme has been recently described (Willard, 1981). The addition of micromolar serotonin to isolated nerve cords induces bursts of motor neurone impulses which are responsible for swimming episodes after several minutes of exposure. Those induced bouts continue for 2-3 h after washing away the serotonin. If a ganglion has its bath volume reduced to 50 μl and its Retzius cells are stimulated to fire a very large number of impulses, the cord can again be induced to produce prolonged episodes of swimming bursts. Finally, the injection of 100 mg of 5-HT into intact leeches caused them to increase the incidence of their spontaneous swimming from about once to about 14 times in a 20 min interval. This increase in spontaneous swimming lasted from 5-8 h after the injection of 5-HT.

Thus, the Retzius cell appears to play three rather diverse roles: a mucus secretogogue, a muscle relaxer and a modulator of the onset of a complex motor pattern. This neurone must therefore be considered to function as a multimodal effector cell. It is interesting that these large neurones have suprisingly diverse effects and each of which appears to be relatively slow in its onset and long-lasting.

It has been possible to study some of the properties of the entire population of serotonin neurones by utilizing Neutral Red vital-staining (Lent and Frazer, 1977). Stained neurones can be impaled singly and in pairs for recording and stimulating in order to ascertain any possible patterns of connectivity among these biochemically-related cells.

Stained Retzius cells have resting potentials (40-60 mV) and non-invading action potentials (30-55 mV) which are indistinguishable from those in unstained preparations. Further the stained pair of RZ remain coupled by efficacious electrotonic junction. The small MA cells have small action potentials (usually 5-15 mV) which do not actively invade their somata. We have recorded simultaneously from a Retzius cell and each of the small ganglionic MA neurones. The MA neurones usually have tonic impulse activities, post-synaptic potentials and frequently produce bursts of impulses. These bursts usually occur upon slowly-raising depolarizations of 5-15 mV. As can be seen by the examples in Fig. 8A, the bursts in the RZ and the small cell usually occur in temporal register: a pattern suggesting that both receive a common excitatory input. This pattern of nearly synchronous depolarizations and impulse bursts is seen between Retzius cells and both DL and M as well as VL and E. In order to map more completely the response of the MA cell population, we impaled and recorded simultaneously from these pairs of small cells: VL-VL, DL-DL, and E-E. In each case, the cells showed nearly synchronous depolarizations and generated impulse bursts. Pairs of non-homologous small cells showed similar synchronous pattern which also suggest common excitatory input.

Serotonin (10^{-5} M) neither enhances nor induces bursts by MA neurones; in fact, 5-HT hyperpolarizes RZ abolishing any spontaneous impulse activity. The serotonin receptor on the Retzius cell membrane appears to be identical with its dopamine receptor. Both neurochemicals work by binding to this site and inhibit the Retzius cell by increasing its membrane conductance to chloride ions (Sunderland *et al.*, 1980). Acetylcholine (ACh) may underlie these bursts since it is known to depolarize the RZ exciting them into impulse activity (Walker, 1967). Atropine is a competitive inhibitor of ACh at muscarinic sites and it effectively blocks the

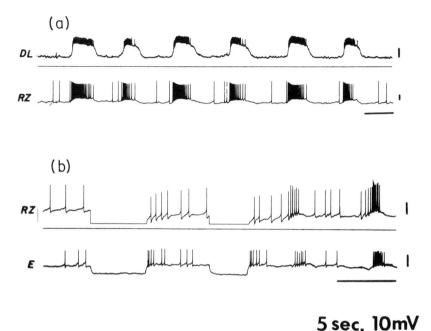

(a)

DL

RZ

(b)

RZ

E

5 sec, 10mV

Fig. 8 (a) Intracellular recording of the spontaneous activity of red-stained serotonin neurones from the leech *Haemopis marmorata*. These two traces show the slow depolarizations and action potential bursts within DL and RZ. (b) Electro-tonic interactions of RZ and E cells. Hyperpolarizing currents injected into RZ produce a DC shift in the membrane potential of E. (Modified after Lent and Frazer, 1977)

slow depolarizations and impulse bursts by RZ. The depolarizing bursts return after washing. Further, the anticholinesterase eserine (5×10^{-4} M) increases both the amplitude and duration of the spontaneous depolarizations seen in the RZ. These eserine-enhanced depolarizations can also be blocked by the addition of atropine. Therefore, the central synaptic input which depolarizes the MA neurones within the leech ganglion is very probably cholinergic.

We injected current into RZ while monitoring the effects upon small cells in order to determine the nature of any interconnections among the amine popu-lation. Inward current directly hyperpolarizes RZ and also hyperpolarizes the small cells to a lesser degree (Fig. 8B). These DC potential shifts persist in high Mg^{2+} and increase monotonically with increasing currents. Thus, the Retzius cell is coupled to each of the small cells by an ohmic junction. In addition, hyperpolarization of each of these small cells will also produce a DC shift in the RZ. The Retzius cells are coupled to one another by a low resistance pathway such that an impulse in one cell produces an electrotonic potential in the other which usually exceeds its impulse threshold, and the RZ pair maintains impulse synchrony. The small MA cells are coupled to one another and to the RZ by high resistance junctions. Each Retzius cell impulse can be detected within a small cell soma by an electronic potential of about 0.5 mV. These potentials are resistant to Mg^{2+} and have fixed latency of less than 1 ms. In 20 mm Mg^{2+}, a current-induced burst of RZ impulses will depolarize a small MA cell sometimes to impulse threshold. Electrotonic coupling can also be demonstrated between E-E, DL-DL, VL-VL, VL-DL, VL-M, and VL-E.

Thus, all nine monoamine-containing neurones within a ganglion appear to be electrotonically-intercoupled. The injection of currents into the MA cells can generate bursts; however, anodal break stimulation is more effective in generating a self-sustaining burst than is depolarization. These anodal break bursts can be induced in preparations showing no spontaneous bursts and need not coincide with the onset of the excitatory input in those showing spontaneous synchronous bursting. Occasionally, an anodal break burst can be generated by the MA neurones in high Mg^{2+} saline. Thus, the electrotonic coupling among the MA cells is responsible at least in part for burst formation.

The intraganglionic MA cells share a common cholinergic input, and are intercoupled. Either of these patterns of connectivity can produce bursts by the MA cells. Presumably the presence of both patterns assures concerted impulse activity.

CONCLUSIONS

The functional interactions of the serotonin neurones in the leech segmental nervous system can be studied because vital-staining techniques have rendered the entire population accessible. Every segmental ganglion has a pair of large Retzius cells which have extensive peripheral projections, and two pairs of lateral and one medial cell which have extensive intraganglionic arborizations and projections within the central nervous system. The three anterior ganglia have extra cells which project into the supraoesaphageal ganglion. Retzius cells are clearly implicated to the control of mucus secretion and muscle relaxation. Serotonin and Retzius cell impulse activity elicits expression of the swimming motor programme. As all the ganglionic neurones which contain serotonin are electrically interconnected, one cannot know which of them is primarily responsible for releasing 5-HT within the CNS. The extensive ganglionic processes of the small cells within the neuropile make them more likely candidates for this central role.

The studies of the serotonin neurones in leech ganglia provide evidence for a functional and morphological specialization within the population. Retzius cells are the only serotonin cells with projections and functions in the periphery. Each lateral and medial serotonin cell has a characteristic aborization and projection within the segmental nervous system. These small cells probably mediate the central requirements for 5-HT. The extra cells project into the supraoesophageal ganglion which has no serotonin-containing cell bodies of its own (Rude, 1969).

Further, the common synaptic input and electrotonic coupling suggest that the entire population of serotonin cells in leech are excited and inhibited as a functional unit. These properties make the leech nervous system a fruitful system for a maintained research effort which is designed to reach an understanding of the physiological and behavioural functions of serotonin.

ACKNOWLEDGEMENTS

I thank Dr Kent T. Keyser for a critical review and discussions of the manuscript. Supported in part by NIH grant NS-14482-02.

REFERENCES

Calabrese, R. L. (1979). The roles of endogenous membrane properties and synaptic interaction in generating the heartbeat rhythm of the leech *Hirudo medicinalis*, *J. Exp. Biol.* **82**, 163–176.

Coggeshall, R. E. and Fawcett, D. W. (1964). The fine structure of the central nervous system of the leech, *Hirudo medicinalis*, *J. Neurophysiol.* **27**, 229–289.

de la Torre, J. C. and Surgeon, J. W. (1976). A methodological approach to rapid and sensitive monoamine histofluoresence using a modified glyoxylic acid technique: the S.P.G. method, *Histochemistry*, **49**, 82–93.

Ehinger, B., Falck, B., and Myhrberg, H. E. (1968). Biogenic monoamines in *Hirudo medicinalis*, *Histochemie*, **15**, 140–149.

Feldberg, W. and Gaddum, J. H. (1934). The chemical transmitter at synapses in a sympathetic ganglion. *J. Physiol. (Lond.)*, **81**, 305–319.

Fernandez, J. (1978). Structure of the leech nerve cord: distribution of neurones and organization of fiber pathways, *J. Comp. Neurol.* **180**, 165–189.

Gardner-Medwin, A. R., Jansen, J. K. S., and Taxt, T. (1973). The 'giant' axon of the leech, *Acta. Physiol. Scand.* **87**, 30A–31A.

Gaskell, J. F. (1914). The chromaffine system of annelids and the relationship of this system to the contractile vascular system in the leech *Hirudo medicinalis*. A contribution to the comparative physiology of the contractile vascular system and its regulations, the adrenaline secreting system and the sympathetic nervous system, *Phil. Trans. Roy. Soc.* **205**, 153–211.

Gray, E. G. and Guillery, R. W. (1963). An electron microscopical study of the ventral nerve cord of the leech, *Z. Zellforsch. mikrosk. Anat.* **60**, 826–849.

Hagiwara, S. and Morita, H. (1962). Electrotonic transmission between two nerve cells in leech ganglion, *J. Neurophysiol.* **25**, 721–731.

Jansen, J. K. S. and Nicholls, J. G. (1972). Regeneration and changes in synaptic connections between individual nerve cells in the central nervous system of the leech, *Proc. Nat. Acad. Sci. (USA)*, **69**, 636–639.

Keyser, K. T. and Lent, C. M. (1980). Leydig cells: an electrical network within the segmental C.N.S. of leech, *Soc. Neurosci. Abstr.* **6**, 214.

Keyser, K. T. and Lent, C. M. (1981). Leydig cells in the cephalic and caudal ganglia of the leech (unpublished).

Keyser, K. T., Frazer, B. M., and Lent, C. M. (1981). Physiology and anatomy of Leydig cells within the segmental nervous system of the leech, *J. Comp. Physiol.* (in press.)

Kuffler, D. P. (1978). Neuromuscular transmission in longitudinal muscle of the leech, *Hirudo medicinalis*, *J. Comp. Physiol.* **124**, 333–338.

Lent, C. M. (1973a). Retzius cells from four species of leeches; Comparative neuronal geometry, *Comp. Biochem. Physiol.* **44A**, 35–40.

Lent, C. M. (1973b). Retzius cells: Neuronal effectors controlling mucus release by the leech, *Science*, **179**, 693–696.

Lent, C. M. (1977). The Retzius cells within the central nervous system of leeches, *Progr. Neurobiol.* **8**, 81–117.

Lent, C. M. (1981a). Morphology of neurons containing monoamines within leech segmental ganglia, *J. Exp. Zool.* **216**, 311–316.

Lent, C. M. (1981b). Fluorescent properties of monoamine neurones following glyoxylic acid treatment of intact leech ganglia. *Histochemistry* (in press).

Lent, C. M. and Frazer, B. M. (1977). Connectivity of the monoamine-containing neurones in central nervous system of leech, *Nature (Lond.)*, **266**, 844–847.

Lent, C. M., Ono, J., Keyser, K. T., and Karten, H. J. (1979). Identification of serotonin within vital-stained neurons from leech ganglia, *J. Neurochem.* **32**, 1559–1563.

Mason, A. and Leake, L. D. (1978). Morphology of leech Retzius cells demonstrated by intracellular injection of horseradish peroxidase, *Comp. Biochem. Physiol.* **61A**, 213–216.

Mason, A., Sunderland, A. J., and Leake, L. D. (1979). Effects of leech Retzius cells on body wall muscles, *Comp. Biochem. Physiol.* **63C**, 359–361.

Macagno, E. R. (1980). Number and distribution of neurons in leech segmental ganglia, *J. Comp. Neurol.* **190**, 283–302.

McAdoo, D. J. (1977). The Retzius cell of the leech *Hirudo medicinalis*, in *Biochemistry of Characterized Neurons* (Ed. N. N. Osborne), pp. 19–45, Pergamon, Oxford.

McAdoo, D. J. and Coggeshall, R. E. (1976). Gas chromatographic–mass spectrometric analysis of biogenic amines in identified neurons and tissues of *Hirudo medicinalis*, *J. Neurochem.* **26**, 163–167.

McCaman, M. W., Weinreich, D., and McCaman, R. E. (1973). The determination of picomole levels of 5-hydroxytryptamine and dopamine in *Aplysia*, *Tritonia* and leech nervous tissues, *Brain Res.* **53**, 129–137.

Minz, B. (1932). Pharmakologische Untersuchungen am Blutegelpräparat. Zugleich eine Methode zum biologische Nachweis von Azetylcholin be. Anwesenheit anderer pharmakologisch wirksamer körpereigener Stoffe, *Arch. Exp. Path. Pharmak.* **168**, 292–304.

Müller, K. J. (1979). Synapses between neurones in the central nervous system of the leech, *Biol. Rev.* **54**, 99–134.

Müller, K. J. and Carbonetto, S. (1979). The morphological and physiological properties of a regenerating synapse in the C.N.S. of the leech, *J. Comp. Neurol.* **185**, 485–516.

Müller, K. J. and McMahan, U. J. (1976). The shapes of sensory and motor neurones and the distribution of their synapses in the ganglia of the leech: A study using intracellular injection of horseradish peroxidase, *Proc. Roy. Soc. Lond.* B, **194**, 481–499.

Nemes, Z., Dietz, R., Luth, J. B., Gomba, S., Hackenthal, E., and Gross, F. (1979). The pharmacological relevance of vital staining with neutral red, *Experientia*, **25**, 1475–1476.

Nicholls, J. G. and Baylor, D. A. (1968). Specific modalities and receptive fields of sensory neurons in CNS of the leech, *J. Neurophysiol.* **311**, 740–756.

Nicholls, J. G. and Purves, D. (1972). A comparison of chemical and electrical synaptic transmission between single sensory cells and motoneurone in the central nervous system of the leech, *J. Physiol. (Lond.)*, **225**, 637–656.

Orchard, L. and Webb, R. A. (1979). The projections of neurosecretory cells in the brain of the North American medicinal leech, *Macrobdella decora*, using intracellular injection of horseradish peroxidase, *J. Neurobiol.* **11**, 229–242.

Ort, C. A., Kristan, W. B., Jr., and Stent, G. S. (1974). Neuronal control of swimming in the medicinal leech. II. Identification and connections of motor neurons, *J. Comp. Neurol.* **136**, 349–372.

Poll, H., and Sommer, A. (1903). Ueber phaeochrome Zellen in Centralnervensystem des Blutegels, *Arch. Anat. Physiol.* **10**, 549–550.

Ready, D. F. and Nicholls, J. (1979). Identified neurones isolated from leech CNS make selective connections in culture, *Nature (Lond).*, **281**, 67–69.

Retzius, G. (1891). Zur Kenntnis des centralen Nervensystems der Wurmer. Das Nervensystem der Annulaten, *Biol. Untersuch.* (NF) **2**, 1–28.

Rude, S. (1969). Monoamine-containing neurons in the central nervous system and peripheral nerves of the leech, *Hirudo medicinalis, J. Comp. Neurol.* **136**, 349–372.

Rude, S., Coggeshall, R. E., and Van Orden, L. S. (1969). Chemical and ultrastructural identification of 5-hydroxytryptamine in an identified neuron, *J. Cell Biol.* **41**, 832–854.

Saavedra, J. M., Brownstein, M., and Axelrod, J. (1973). A specific and sensitive enzymatic-isotopic microassay for serotonin in tissues, *J. Pharmacol. Exptl. Therap.* **186**, 508–515.

Sargent, P. B. (1977). Synthesis of acetylcholine by excitatory motorneurons in central nervous system of the leech, *J. Neurophysiol.* **40**, 453–459.

Schain, R. J. (1961). Effect of 5-hydroxytryptamine on the dorsal muscle of the leech (*Hirudo medicinalis*), *Br. J. Pharmacol.* **16**, 257–261.

Shafer, M. R. and Calabrese, R. L. (1980). Similarities and differences in the structure of segmentally homologous neurons controlling heartbeat in the leech, *Soc. Neurosci. Abstr.* **6**, 26.

Smith, P. A., Sunderland, A. J., Leake, L. D., and Walker, R. J. (1975). Cobalt staining and electro-physiological studies of Retzius cells in the leech *H. medicinalis*, *Comp. Biochem. Physiol.* **51A**, 655–661.

Stent, G. S., Kristan, W. B., Firesen, O., Ort, C. A., Poon, M., and Calabrese, R. L. (1978). Neuronal generation of the leech swimming movement, *Science*, **200**, 1348–1357.

Stenzel, K. and Neuhoff, V. (1976). Tryptophan metabolism and the occurrence of amino acids and serotonin in the leech (*Hirudo medicinalis*) nervous system, *J. Neurosci. Res.* **2**, 1–9.

Stuart, A. E. (1970). Physiological and morphological properties of motorneurons in the central nervous system of the leech, *J. Physiol. (Lond.),* **209**, 627–646.

Stuart, A. E., Hudspeth, A. J., and Hall, Z. W. (1974). Vital staining of monoamine-containing cells in the leech nervous system, *Cell. Tiss. Res.* **153**, 55–61.

Sunderland, A. J., Leake, L. D., and Walker, R. J. (1980). Evidence for an amine receptor on the Retzius cells of the leeches *Hirudo medicinalis* and *Haemopis sanguisuga*, *Comp. Biochem. Physiol.* **67C**, 159–166.

Thomspon, W. J. and Stent, G. S. (1976). Neuronal heartbeat control in the medicinal leech. II. Intersegmental coordination of heart motor neuron activity by heart interneurons, *J. Comp. Physiol.* **111**, 281–307.

Van Essen, D. and Jansen, J. K. S. (1976). Repair of specific neuronal pathways in the leech, in *The synapse, Cold Spring Harbor Symposium on Quantitative Biology*, Vol. XL pp. 495–502, Cold Spring Harbor Laboratories.

Walker, R. J. (1967). Certain aspects of the pharmacology of *Helix* and *Hirudo* neurons, in *Neurobiology of Invertebrates* (Ed. J. Salanki), pp. 227–253, Plenum, New York.

Webb, R. A. (1980). Intralamellar neurohaemal complexes in the cerebral commissure of the leech *Macrobdella decora* (Say 1824). An electron microscope study, *J. Morphol.* **163**, 157–165.

Webb, R. A. and Orchard, I. (1979). The distribution of putative neurosecretory cells in the central nervous system of the North American medicinal leech *Macrobdella decora, Can. J. Zool.* **57**, 1905–1914.

Webb, R. A. and Orchard, I. (1980). Octopamine in leeches. I. Distribution of octopamine in *Macrobdella decora* and *Erpobdella octoculata, Comp. Biochem. Physiol.* **67C**, 135–140.

Weisblat, D. A., Harper, G., Stent, G. S., and Sawyer, R. T. (1980). Embryonic cell lineages in the nervous systm of the glossophonied leech *Helobdella triserialis, Dev. Biol.* **76**, 58–78.

Willard, A. L. (1981). Effects of serotonin on the generation of the motor program for swimming by the medicinal leech, *J. Neuroscience,* **1**, 936–944.

Zipser, B. (1980). Identification of specific leech neurones immunoreactive to enkephalin, *Nature (Lond.),* **283**, 857–858.

Zipser, B., McKay, R., and Farrar, J. (1980). Individual nerve cells are immunologically distinguishable in the leech, *Soc. Neurosci. Abstr.* **6**, 338.

Biology of Serotonergic Transmission
Edited by N. N. Osborne
© 1982, John Wiley & Sons Ltd.

Chapter 18

The Serotonergic Cerebral Cells in Gastropods

V. W. PENTREATH
Department of Biology, University of Salford,
Salford M5 4WT, UK

M. S. BERRY
Department of Zoology, University of Swansea,
Swansea SA2 8PP, UK

N. N. OSBORNE
Nuffield Laboratory of Ophthalmology, University of Oxford,
Walton Street, Oxford OX2 6AW, UK

INTRODUCTION

The cerebral serotonin neurones of gastropods have been the subjects of extensive morphological, biochemical and physiological studies. The cells were first noted in *Helix pomatia* by Kunze (1921), and some of their electrical properties were studied by Kandel and Tauc (1966a,b). More general interest in the neurones was developed by Cottrell and Osborne (1970) who observed that, in both *Helix pomatia* and *Limax maximus*, they contained significant amounts of serotonin. Because of the size and accessibility of the cells it became evident that they might be useful for studying the roles of neuronal serotonin, perhaps providing clues concerning its proposed transmitter function in the nervous system of higher animals.

A large amount of data is now available on these neurones. They have been most thoroughly studied in *Helix* and *Aplysia*, but work has been extended to a dozen different opisthobranch and pulmonate genera which possess homologous neurones. Some of the findings are of general significance; among these are the localization, storage and transport of serotonin, the release of serotonin from nerve terminals, the variety of receptors mediating serotonin responses of different polarity and durations, and the modulatory effects caused by the release of serotonin from extensive projections. In addition it is possible to relate some of the physiological differences of the neurones in the different genera to the behavioural differences of the animals, thus providing the first steps towards elucidating behavioural adaptation at the level of individual serotonergic cells.

In the following pages we summarize and evaluate the information on serotonergic mechanisms in these neurones and on the functional roles of the cells. Detailed comparisons with the serotonin neurones in other animals have not been included because the information is available elsewhere in this volume.

LOCATION, ORGANIZATION AND PROJECTIONS OF THE SEROTONIN NEURONES

The serotonin neurone perikarya generally are located on the ventral surfaces of the cerebral ganglia. Each ganglion contains one such neurone. The neurones are bilaterally symmetrical and readily identifiable because of their large size relative to the other neurones, and their characteristic position and appearance (Fig. 1). The neurones have been identified in different species of the genera *Aplysia, Archachatina, Ariolimax, Arion, Helisoma, Helix, Lymnaea, Limax, Milax, Planorbis, Pleurobranchaea,* and *Tritonia* (Cottrell and Osborne, 1970; Osborne and Cottrell, 1971;

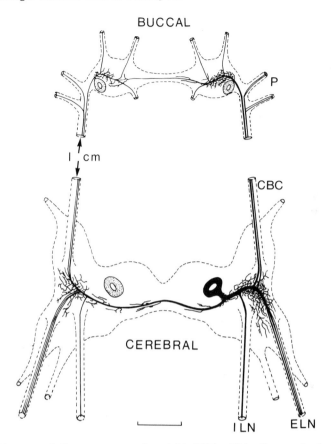

Fig. 1   Diagram of the processes of a right SCC within the cerebral and buccal ganglia of *Helix pomatia*. The presumed dendrites and terminal branches of the cell have been drawn schematically, the rest of the diagram is to scale. Two axon branches pass within each cerebro-buccal connective (CBC) to the buccal ganglia, where they divide into terminal processes. Other axon branches pass into the pharyngeal nerves (P), and other nerves leaving the buccal ganglia. The stippled neurone in each buccal ganglion is the 'middle buccal cell', which receives synaptic input from the SCC. Several axon branches run in each external lip nerve (ELN) and one axon branch runs in each internal lip nerve (ILN). The stippled neurone in the left cerebral ganglion is the left SCC. The scale is 0.5 mm. From Pentreath (1976); reproduced by permission of Chapman and Hall

Osborne and Cottrell, 1972b; Pentreath *et al.*, 1973a, b; Berry and Pentreath, 1976a; Sakharov and Zs-Nagy, 1968; Granzow and Kater, 1977; Sedden *et al.*, 1968; Weinreich *et al.*, 1973).

    The cells were originally called the metacerebral giant cells in *Helix* because of both their unusually large size (soma diameter 120–220 μm) and their obvious location in the metacerebral lobe of each ganglion (Kandel and Tauc, 1966a). Subsequently, when the studies were directed towards analysing the role of neuronal serotonin, the cells were termed giant serotonin neurones, or GSCs (Cottrell, 1970). Over the last 10 years several other names and abbreviations have been used by different groups for the cells in the different species. In *Aplysia* they have been termed metacerebral cells (MCCs; Weiss and Kupfermann, 1976) and giant cerebral neurones (GCNs; Goldman and Schwartz, 1974) and in *Lymnaea* cerebral giant cells (CGCs; McCrohan and Benjamin, 1980a,b). It is evident, as recently pointed out by Granzow and Fraser Rowell (1981), that a more general terminology would be useful. Furthermore in many of the genera the cells are not sufficiently large to be called giant, and the metacerebral lobes of the cerebral ganglia cannot be clearly distinguished. On these grounds it has been proposed that they be called the serotonergic cerebral cells (SCCs; Granzow and Fraser Rowell, 1981), and this is the terminology which we shall use for the cells in this review.

The SCCs are unipolar, as are most other central gastropod neurones. The perikarya are normally located at the edge of the ganglia and a significant proportion of their surface lies close to the vascular channels within the connective tissue (see Fig. 7b). The cells are separated from the haemolymph and connective tissue by a layer of glial cell processes. The glia completely envelope the SCC somata, axon hillocks and much of the primary axons (Pentreath *et al.,* 1973), but in *Helix* (and presumably the other genera) they do not restrict the movement of small molecules from the haemolymph to the surface of the neurone via the 10-20 nm extracellular spaces (Pentreath and Cottrell, 1970).

The cells have complex projections with main axons extending into the nerves which supply the buccal mass, lips and the buccal ganglia. Numerous dendritic branches arise within the cerebral ganglia, and terminal axonal processes are located in the buccal ganglia and feeding musculature (Figs. 1 and 2). Axons also continue through the buccal ganglia into nerves supplying the buccal mass and parts of the oesophagus and salivary glands (Fig. 2). There is some variation in the number and distribution of the main axon branches, and the degree to which they are bilaterally patterned, in the different species (Fig. 3). In all species at least one axon runs in the ipsilateral cerebrobuccal connective. In *Helix* (Kandel and Tauc, 1966a; Pentreath and Cottrell, 1974; Pentreath, 1976) and *Ariolimax* (Senseman and Gelperin, 1974) the cells also send axons along the contralateral cerebrobuccal connective, whilst in *Aplysia* (Weiss and Kupfermann, 1976), *Helisoma* (Granzow and Fraser Rowell, 1981), *Limax* (Senseman and Gelperin, 1974), *Lymnaea* (McCrohan and Benjamin, 1980a; J. T. Goldschmeding, unpublished observation) and *Pleurobranchaea* (Gillette and Davis, 1977) processes do not extend to the contralateral cerebral ganglion nor any of its nerves. In *Planorbis* a major axonal process projects to the contralateral cerebral ganglion via the buccal commissure (Berry and Pentreath, 1976a). In several species the two cells are, to varying extents, electrically coupled (Fig. 3; see also Granzow and Fraser Rowell, 1981, for a description of these variations). In *Tritonia diomedea* (Dorsett, 1967) a contralateral projection is absent, whereas in the closely related *Tritonia hombergi* (Bulloch,

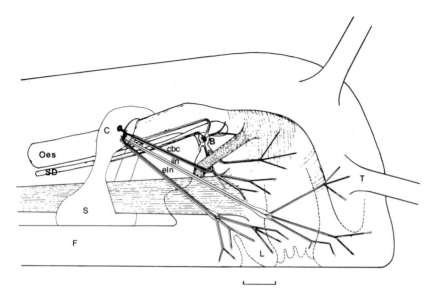

Fig. 2   Diagram showing the distribution of the axons of a right SCC to the anterior body of *Helix pomatia*. Only the ipsilateral branches are shown. The buccal mass is in the extended position. Branches of the SCC pass into the buccal ganglia (B) and into most of the nerves which supply the buccal mass and lips (L). The cell type(s) innervated by the SCC in the buccal mass and lips is not known. C, right cerebral ganglion; F, foot; Oes, oesophagus; S, suboesophageal ganglia; SD, salivary duct; T, inferior tentacle; cbc, cerebro-buccal connective; eln, external lip nerve; iln, internal lip nerve. The scale is 2 mm. From Pentreath (1976); reproduced by permission of Chapman and Hall

1977) it is generally, but not always present. In *Helix pomatia* several collateral branches run in both cerebrobuccal connectives and both external lip nerves (Fig. 1; Pentreath, 1976). In all species so far studied except for *Lymnaea* (McCrohan and Benjamin, 1980a,b) axons project to both ipsilateral and contralateral buccal ganglia and nerve trunks. Despite the individual anatomical variations the combined projections of both SCCs and their overall functions are generally similar in each species (see below p. 496). On the other hand the anatomical variations may be of great significance in relation to the evolutionary adaptation of the cells to the different feeding behaviours found in the various species (see below pp. 490 and 493 and Granzow and Fraser Rowell, 1981).

CHEMICAL ANALYSIS OF THE CELL SOMA

Isolation of the soma

The dissection and manipulation of single SCC perikarya has been carried out under a high resolution stereomicroscope using fine forceps, microscalpels and small-bore

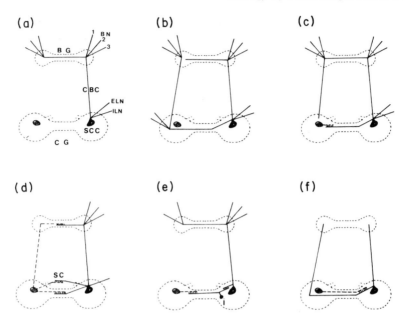

Fig. 3   Diagrammatic representation of the projections of the SCCs within the cerebral ganglia (cg), in the cerebro-buccal connectives (cbc), and to the buccal ganglia (bg) in different gastropod genera. (A) *Aplysia, Limax, Pleurobranchaea* and *Tritonia hombergi*; (B) *Helix* and *Tritonia diomedea*; (C) *Planorbis*; (D) *Lymnaea*; (E) *Helisoma*; (F) *Ariolimax*. ⊞⊞ indicates possible sites of electrical coupling between cells. SC, cerebral commissure; ELN, external lip nerve; ILN, internal lip nerve; i, presumed interneurone. For references see text and Granzow and Fraser Rowell (1981) from whom the diagram is reproduced, with modification, by permission of the Company of Biologists

pipettes attached by rubber tubing to a mouthpiece (see Osborne, 1974 for a detailed description). Optimal illumination is essential and is most satisfactorily obtained by the use of fibre glass or glass rod light guides. Some workers have employed loops of fine wire to handle the cell in its hillock region (McCaman and Dewhurst, 1970).

It is frequently mentioned in the literature that the SCCs (and other invertebrate neurones) isolated for chemical studies are not thought to be significantly contaminated by adhering satellite glial cells or other debris. Whilst this may be generally true it is important to realize that there are no adequate means for ensuring that all adhering remnants are actually removed. In a recent study Barnicoat and Pentreath (1981) found that it was virtually impossible to remove all the glial-cell processes from a variety of invertebrate neurones without serious damage to them. In *Helix pomatia* freshly isolated SCC perikarya are surrounded by large amounts of adhering glial cells and small neurones (Fig. 4). A proportion of these can be satisfactorily removed by microdissection (Fig. 4) but attempts to clean the neurones cause progressive damage. The somatoplasm becomes increasingly reduced because of nuclear swelling (Fig. 4B–D) and the structure of the cytoplasm changes markedly (Fig. 4E,F). Some glial cells resist all attempts to separate them from the SCC. The strength of the attachment appears to result from the infoldings of glial cells into the soma, rather than the presence of desmosomes or other attachment sites similar to those which occur between neurones and glia in the leech (Kai-Kai and Pentreath, 1981). It was estimated from planimetric measurements of sections of isolated and partially cleaned *Helix* SCCs (i.e. after removal of the outer glial layers but before excessive nuclear swelling) that the volume of glial tissue varied from 68 per cent to 102 per cent of the somatoplasm (Barnicoat and Pentreath, 1981). Whilst such contamination may not significantly interfere with analysis of serotonin chemistry, it is evident that great care must be taken over interpretation of the significance of trace amounts of substances, especially when there is evidence that they may be associated with glial cells in other situations (see below p. 467).

Proteins and phospholipids

Little is known about the protein composition of the SCCs. Osborne *et al.,* (1971), using microdisc electrophoresis, showed that the SCCs contained a large fast-migrating, water soluble, protein band which migrated to the anode. The protein band was only associated with nervous tissues of the snail, and this and other evidence led the authors to suggest that the band may represent the brain specific 14-3-2 protein (see Osborne *et al.,* 1971).

Using a gradient microelectrophoresis system to analyse the total protein composition of the SCCs, Osborne and Rüchel (1975) found that they contained a protein band which was absent from non-serotonergic cells (see Fig. 5). The microprocedure revealed a total of at least 18 definable components in the two types of cell analysed.

The phospholipid content of the SCCs has been analysed by a micro one-dimensional chromatographic method (Althaus *et al.,* 1973). A single SCC was found to contain $0.68\mu g$ phosphatidylcholine, $0.07\mu g$ phosphoinositides, $0.12\mu g$ phosphatidylserine and 0.57 μg phosphatidylethanoleamine.

Amino acids

The free amino acid composition of the SCCs of *Helix pomatia* has been determined by a microdansyl procedure (Briel *et al.,* 1971; Osborne, 1974; Osborne and Cottrell, 1972a). The following amino acids are present: tyrosine, tryptophan, ornithine, lysine, methionine, phenylalanine, leucine, isoleucine, valine, proline, alanine, glycine, glutamate, aspartate, glutamine, serine, threonine, arginine, histidine, cysteine and GABA. The analyses were only semi-quantitative, but it was demonstrated that the SCCs contain less ornithine and more glycine than certain cells located in the buccal ganglia with which they make synaptic contact. Although the significance of the difference is not yet clear it illustrates the great potential of the technique for studying the biochemical heterogeneities between the SCCs and other types of neurone in the gastropod brain.

Serotonin and other amines

The serotonin contents of the cell bodies of the SCCs of *Helix pomatia* and *Aplysia californica* have been determined by several sensitive fluorometric, radiochemical and microdansylation procedures (Osborne and Cottrell, 1972a; McCaman *et al.*, 1973; Weinreich *et al.*, 1973; Osborne, 1977). The concentrations of serotonin are similar in the two species: *Helix* 3.8 × 10^{-4} M, *Aplysia*, 4.2 × 10^{-4} M. Sensitive radiochemical methods have also been used to analyse the SCCs of *Helix pomatia* for other amines which could be potential transmitter substances (Osborne, 1977; see Table 1). Trace amounts of histamine and dopamine are present, although their significance is not clear. The SCCs of *Aplysia* have also been reported to contain low levels of histamine (Weinreich *et al.*, 1975; Weinreich, 1978).

Serotonin metabolism

Some properties of the uptake and metabolism of serotonin precursors by the SCCs have been analysed in *Helix pomatia*, *Aplysia californica* and *Tritonia diomedea*.

Fig. 4   Micrographs of isolated perikarya from *Helix pomatia*. The cells were prepared by hand dissection under a stereomicroscope and photographed using normal light microscopic optics. (a) Freshly isolated, 'uncleaned' soma (nucleus, n) with large amounts of adhering cell debris, glial cells and possibly other small neurones (e.g. asterisk). (b) Isolated and 'cleaned' soma, with most contamination removed, but with extensive nuclear (n) swelling and loss of cytoplasm. A part of the primary axon (arrow) is attached to the hillock, which appears dark because of the refractory properties of the large numbers of lysosomes and other granular structures located in this region (see Fig. 7). (c) and (d) Isolated and 'cleaned' somata which are typical of those used in microchemical studies. There is some nuclear swelling and glial contamination, but the SCC somatoplasm is still largely intact. (e) Electron micrograph of a part of the peripheral cytoplasm of a typical isolated SCC (boxed area in (d)) illustrating the close association of a satellite glial cell (g). The arrows mark areas where processes from this and other adjacent glial cells invaginate the neurone. Such sites almost certainly preclude complete separation of the two cell types without rupture of the SCC. The appearance of the SCC somatoplasm suggests that extensive damage has taken place (see (f); cf. Fig. 8). (f) electron micrograph of part of an isolated SCC. The arrows point to adjacent glial-cell processes. Although some structures (e.g. the lysosomes, l) appear normal, the majority of the cytoplasm contains extensive vacuoles and is drastically altered from the normal, intact cell (cf. Fig. 8; see also Pentreath *et al.*, 1973). Although there are no data concerning possible morphological correlates with the metabolic capabilities of the SCC (see text) it is likely that the changes in structure are associated with changes in the chemistry of the cell. The scales in (a)–(d) represent 50 μm, in (e) (f) represent 1.5 μm. Part (e) from Barnicoat and Pentreath (1981)

The results are similar in each species and demonstrate the specificity of serotonin synthesis and subsequent binding within serotonergic neurones.

Exogenous tryptophan is progressively accumulated by most of the neurones and glial cells in the ganglia (Pentreath and Cottrell, 1973), but the SCCs, as opposed to non-serotonergic neurones appear to have a specific ability to metabolize it to form 5-HTP and serotonin (Osborne, 1973b; Eisenstadt *et al.*, 1973; see Table 2). Pretreatment of *Helix* with p-chlorophenylalanine, which inhibits tryptophan hydroxylase (Koe and Weissmann, 1966) does not interfere with the uptake of tryptophan into the SCCs but prevents the formation of 5-HTP and serotonin (Osborne, 1973b). Tryptophan hydroxylase (the enzyme converting tryptophan to 5-HTP) is present in the SCCs and is responsible for determining the rate of formation of serotonin. 5,7-Dihydroxytryptamine, which has little influence on the uptake of tryptophan, also drastically reduces the amount of serotonin formed (Osborne and Pentreath, 1976). It was concluded that 5,7-dihydroxytryptamine,

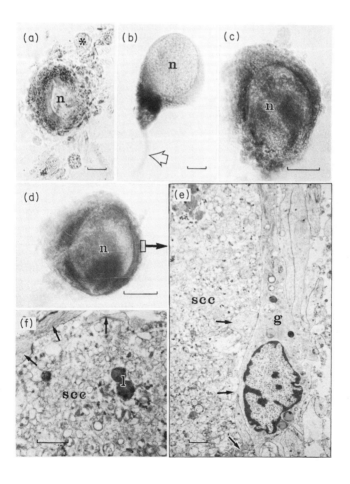

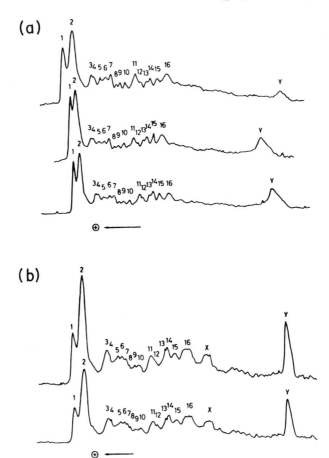

Fig. 5   Densitometric traces of micro-polyacrylamide gradient gels showing the separation of total proteins from single neurones of *Planorbis corneus* using the procedure described by Osborne and Rüchel (1975). Each set of traces is from different experiments with (a) non-serotonergic neurones and (b) the SCC. The patterns of protein fractionation achieved by this method are constant for each cell type. The nature of the proteins is not yet known, but the data clearly illustrate slight differences in the amounts of proteins 1–16 in serotonergic and non-serotonergic neurones, and also the presence of a protein band (X) which may be specific for the SCCs

like *p*-chlorophenylalanine exerts its influence upon the enzyme tryptophan hydroxylase. The fact that 5,7-dihydroxytryptamine has little effect on the 5-HTP formation has also been demonstrated in the vertebrate CNS (Baumgarten *et al.*, 1973).

Table 1. Serotonin and other transmitter candidates in the SCCs and a non-serotonergic cell (cell 21) of *Helix pomatia*

Substance	Concentration (M)	
	In SCC	In cell 21
Serotonin	3.8×10^{-4}	2×10^{-8}
Histamine	2×10^{-8}	10^{-8}
Octopamine	ND	ND
Dopamine	$8-10^{-9}$	3×10^{-8}
Noradrenaline	ND	ND
Ch. Ac.	0.2 pmole/cell/hour	61 pmole/cell/hour
Glutamate	6×10^{-4}	3×10^{-4}
Aspartate	7×10^{-5}	9×10^{-5}
Glycine	5×10^{-4}	5.9×10^{-4}
Taurine	6×10^{-6}	6×10^{-6}

The volume of each cell type was estimated by measuring the diameters of a number of cells (mean diameter of both cell types 120-140 μm) by light microscopy. It was found that the SCC and cell 21 had volumes of 1.2 nl. The molarity (results reported for five to seven determinations; in each experiment four neurones were pooled) of each substance in the two neurone types could then be calculated.
ND = not detected.
Data from Osborne, 1977.

The SCCs of *Helix* accumulate radiolabelled 5-HTP (Pentreath and Cottrell, 1973) and convert it to serotonin, both *in vitro* (Cottrell and Powell, 1971) and *in vivo* (Osborne, 1972). In *Aplysia* and *Tritonia* the cells have been shown to contain relatively high levels of aromatic amine decarboxylase (Weinreich *et al.*, 1973). On the other hand known cholinergic neurones in *Aplysia* are capable of converting injected 5-HTP to serotonin, although only the serotonergic neurones are capable of binding and transporting the serotonin (Goldman and Schwartz, 1974; see below p. 470).

Although whole ganglia of *Helix* can metabolize serotonin to 5-hydroxyindoleacetic acid (Osborne and Neuhoff, 1974), this does not appear to be a property of the SCC perikarya (Osborne, 1978). However, there is good evidence that serotonin is selectively taken up by an active process into the soma, axonal processes and nerve endings of the cell (Osborne, 1978; Pentreath and Cottrell, 1972, 1973; Osborne *et al.*, 1973). The evidence for the re-uptake of synaptically released serotonin in terminating its transmitter action is summarized below p. 483.

Do the SCC somata synthesize acetylcholine?

Some controversy has arisen regarding the possibility that in *Helix* the SCCs may form and release acetylcholine as well as serotonin, thus providing a possible excep-

Table 2. The formation of $[^3H]$ ACh and $[^3H]$ serotonin from $[^3H]$ choline or $[^3H]$ tryptophan in SCCs and non-serotonergic cells (cell 21) of *Helix pomatia*

	$[^3H]$ ACh (fmole)		$[^3H]$ 5-HT (fmole)	
	Without eserine	With eserine	Without pargyline	With pargyline
Per SCC	0.2±0.2[b] (12)	0.5±0.2[b] (12)	9±0.9 (5)	9±1 (5)
Per cell	18±3[b] (12)	56±6[b] (12)	0.2±0.2[a] (5)	0.2±0.2[a] (5)
Per cerebral ganglion	3000±105 (3)	4900±195 (3)	71±4 (3)	69±5 (3)

'Clean' neurones were incubated in 30 μmoles $[^3H]$ choline or $[^3H]$ tryptophan. All values are ± SEM with the number of determinations in parentheses (in each determination, four neurones were pooled). Values were calculated by taking into consideration the specific activity of the $[^3H]$ choline and $[^3H]$ tryptophan. $[^3H]$ Amine neurotransmitters were fractionated from other substances by two-dimensional chromatography on 5 × 5-cm pre-coated silica gel plates (Osborne, 1977), while $[^3H]$ ACh was separated from $[^3H]$ choline by one dimensional chromatography on 5 × 5-cm pre-coated cellulose plates using the solvent butanol-ethanol-acetic acid-water (8:2:1:3 v/v). $[^3H]$ ACh was visualized by iodine vapour and eluted from the chromatogram with methanol; the absolute amount of $[^3H]$ ACh was calculated by assuming the specific activity to be equal to $[^3H]$ -choline added to the ganglia at the start of the incubation. Recovery of $[^3H]$ choline and $[^3H]$ acetylcholine was 85 per cent.
[a]Not significantly different from blank values (Student's t-test, $p < 0.05$).
[b]Significantly different from experiments without eserine and with blank values (Student's t-test, $p < 0.05$).
Data from Osborne, 1977.

tion to the conventionally accepted and widely substantiated principle first postulated by Dale (1935). Emson and Fonnum (1972) described small amounts of choline acetyltransferase in the SCCs of *Helix aspersa,* and the 'cholinergic' nature of the cell in this species and also in *H. pomatia* was later reaffirmed by Hanley *et al.* (1974). The amounts of enzyme (approximately 20 pmole ACh synthesized/cell/h) were similar in the SCCs of both animals, and significantly greater than in other known serotonergic neurones, although not as high as in identified cholinergic neurones. These interesting findings were subsequently given more weight by Hanley and Cottrell (1974) and Cottrell (1977) who provided evidence that the SCCs of *H. pomatia* contained measurable levels of acetylcholine. Attempts were made to rule out the possibility of localization within adhering glial cells by rupturing the isolated SCC soma with a fine needle to exude the somatoplasm and nucleus, and subsequently assaying the two fractions with a sensitive bioassay procedure (Cottrell, 1977). The results showed that the cell contents contain about 63 pg acetylcholine, compared to 22 pg in the 'ghost'.

However, the subject has recently been re-investigated by Osborne (1977) using more careful dissection techniques (Table 2). The results of this study showed that only trace amounts of choline acetyltransferase were present in the SCC, which themselves were probably exogenous in origin (i.e. possibly due to glial contamination), because most of the other neurones also contained trace quantities of the enzyme. Presumed cholinergic neurones (cell 21; see Table 2) contained much larger amounts of the enzyme. It was concluded that the SCCs of *H. pomatia* could not synthesize acetylcholine (see Osborne, 1977, 1979, 1981). The SCCs of *Aplysia* contain only traces (Brownstein *et al.,* 1974) or no choline acetyltransferase (Weinreich *et al.,* 1973).

Thus the evidence seems to argue against the SCCs having a synthetic capability for acetylcholine. The contradictory nature of the results appears likely to be due to problems in the dissection of the individual cells. Glial investment, adhering small neurones and exogenous substances from adjacent damaged neurones are all possible sources of contamination (see p. 461). Although the role of the glial cells surrounding the SCCs and other invertebrate neurones is not clearly understood (see Radojcic and Pentreath, 1979), it is probable that rapid alterations in the chemical integrity of both neurones and glia take place during the dissection procedure. A cholinergic system of interaction may occur between the giant axons and their satellite glial cells in the squid (see Villegas, 1975) and if a similar system occurs in other molluscs such as the snail, it would evidently be highly susceptible to alteration, translocation or mixing between cells during their dissection. The experiments of Cottrell (1977) described above in fact show that satellite glial cells of SCCs contain acetylcholine. Whilst there is some additional electrophysiological evidence for a cholinergic component in the SCCs of *Helix* (p. 485), it appears at present that the most plausible interpretation of the results is that the somatoplasm of the cells does in fact only synthesize serotonin, and the cholinergic component resides with contamination. It is to be hoped that suitably refined techniques will become available to finally clarify these tantalizing findings of such potential importance.

The occurrence of cholecystokinin-like peptide in the SCCs

Osborne *et al.* (1981) have recently shown that the SCCs contain a cholecysto-kinin-like peptide but lack substance P, which occurs, nevertheless, in other specific neurones. This study is of major significance for it shows for the first time that transmitter-like peptides (cholecystokinin and substance P) do exist in specific neurones of the snail and, more important, that the SCC contains a cholecysto-kinin-like peptide. A series of consecutive frozen sections were taken through the SCCs and alternate sections were processed by immunocytochemical procedures to visualize serotonin, cholecystokinin or substance P (Osborne *et al.* 1981). As shown in Fig. 6, the SCCs stained positively for serotonin and cholecystokinin-like peptide but not for substance P. Since an immunoreactive material can be found in snail brain extract which has a similar elution profile on Sephadex G50 to synthetic cholecystokinin-8 (Osborne *et al.* 1981), it argues in favour of the idea that the cholecystokinin immunoreactivity found in the SCCs is similar, as has been demon-strated in the vertebrate CNS (Rehfeld, 1980). The fact that the SCCs contain a chole-cystokinin-like peptide makes it possible to test precisely the function of the substance for the first time. If cholecystokinin is a transmitter substance, as has been proposed (see Rehfeld, 1980), it will now be possible to examine the hypo-thesis that the SCCs utilize at least two transmitters, i.e. cholecystokinin and serotonin.

CYTOCHEMISTRY AND MICROSTRUCTURE OF SEROTONIN LOCALIZATION, TRANSPORT, AND TURNOVER

The SCCs are relatively large cells compared with others in their vicinity (Fig. 7a,b). In practice this is advantageous because identification of the cell in wax or plastic sections is unambiguous, and also because substances such as radiochemicals can be injected into the cells. A variety of cytochemical and autoradiographic experiments have been undertaken to elucidate the serotonergic storage and trans-port mechanisms. Much more detailed information has been obtained on the

Fig. 6   A series of consecutive sections taken through a single SCC (arrows) from *Helix aspersa*. Sections (a) (d) and (g) were processed for visualizing serotonin, sections (b) and (e) were processed for visualizing substance (P) and sections (c) and (f) were processed for visualizing cholecystokinin. For explanation on the methods used see Osborne *et al.* (1981). Monoclonal antibodies to serotonin and substance P and an antiserum to cholecystokinin were employed. It can be seen that the cytoplasm of the SCC reacts positively for cholecystokinin and serotonin but not for substance P

mechanisms than is yet possible in vertebrates. The studies described in this section have been made principally in *Helix* and *Aplysia*.

The soma of the cell emits a yellow fluorescence when processed by the form-

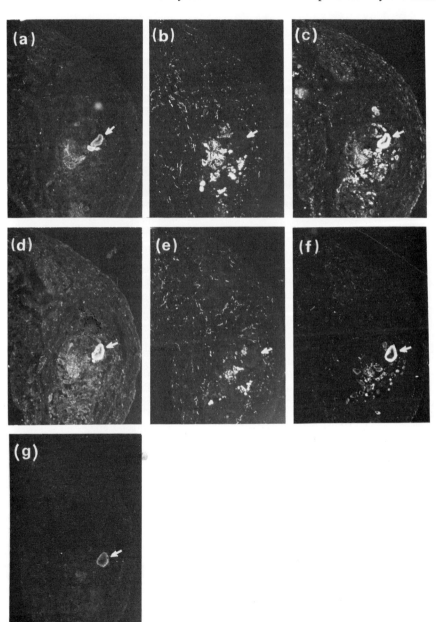

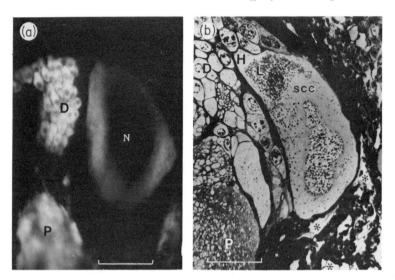

Fig. 7   (a) A section of the metacerebral lobe of a cerebral ganglion of *Limax maximus* which has been processed by the histochemical method for the demonstration of biogenic amines. The micrograph shows a section through the SCC, the cytoplasm of which fluoresces specifically (yellow) for serotonin. The nucleus (N) of the SCC does not fluoresce. A cluster of smaller-sized neurones (D), and a region of neuropile (P) fluoresce specifically for the catecholamines dopamine and/or noradrenaline (cf Fig. 7). The scale is 70 μm. From Pentreath, Osborne and Cottrell (1973); reproduced by permission of Springer-Verlag. (b) A section of the metacerebral lobe of a cerebral ganglion of *Limax maximus* which has been fixed in glutaraldehyde and osmium solutions, and embedded in Araldite. The section (2 μ thick) has been orientated for direct comparison with Fig. 7(a). A variety of granular and lysosome inclusions are concentrated in the axon hillock region (H) of the SCC. The smaller neurones (D) and neuropile (P) correspond to areas which fluoresce specifically for catecholamines. Layers of darkly-stained connective tissue and associated blood spaces (∗) lie to the right of the SCC. The scale is 70 μm. From Pentreath *et al.* (1973); reproduced by permission of Springer-Verlag

aldehyde histofluorescence technique (Fig. 7a; Falck and Owman, 1965; Cottrell and Osborne, 1970; Berry and Pentreath, 1976a). Extra support that the yellow fluorescence is specific for serotonin was obtained in *Helix pomatia* by pretreatment of the animals with drugs known to modify levels of the amine, and subsequent histochemical analysis. Only drugs which influence serotonin metabolism affected the yellow fluorescence (Table 3; Osborne, 1973a). The data provide extra evidence that the results of the microchemical experiments on serotonin metabolism described above apply to the somatoplasm.

    The cytoplasm of the perikarya of the SCCs contain a range of organelles and inclusions that are commonly found in neurones throughout the animal kingdom. These include Golgi structures, elaborate granular and smooth endoplasmic reticu-

Table 3. Summary of the effects of various drugs on the yellow fluorescence of the SCCs of *Helix pomatia*. Each drug (2 mg) was administered over a period of 30 h before observation

Name of drug	Effects	Effects on yellow fluorescence of giant cell
Reserpine	Depletes amines from molluscan nervous tissue	All fluorescence eliminated
p-Chlorophenyl-alanine	Reduces 5-HT content by inhibiting the enzyme tyrosine hydroxylase in vertebrates	Colour of fluorescence still yellow although intensity reduced
α-methyl-p-tyrosine	Reduces CA content by inhibiting the enzyme tyrosine hydroxylase in vertebrates	No change in colour and intensity of fluorescence
5-HTP	Precursor of 5-HT in molluscs	Intensity of yellow fluorescence increased
DOPA	Precursor of CAs in molluscs	No change in colour and intensity of fluorescence
Nialamide	Monoamine oxidase inhibitor in vertebrates	Slight increase in intensity of yellow fluorescence
NSD 1024	DOPA decarboxylase inhibitor in molluscs	Yellow fluorescence very slightly reduced

From Osborne, 1973a.

lum, small membrane-bound vesicles (diameter range 50–120 nm), and larger lysosomes and cytosomes (up to 6 μm diameter). Some of the structures are illustrated in Fig. 8. However, attention has been focussed on the vesicles and lysosomes because these are the structures which have been positively associated with the binding, transport and possible turnover of serotonin.

In the initial studies on *Helix* and *Limax* (Osborne, 1970; Cottrell and Osborne, 1970) tissues were processed by the methods of Wood (1965, 1966) and Cannata *et al.* (1968), in which serotonin may be visualized as an electron dense precipitate. In SCCs of snails examined during the summer months serotonin was found to be localized in small aggregates (Fig. 8, inset) whose distribution and size coincided with the dense-cored vesicles. Injection of the animals with reserpine greatly reduced the reaction product within the vesicles (Cottrell and Osborne, 1970). On the other hand tissues examined in the spring provided evidence that serotonin could also be associated with the lysosomes.

These studies were later refined by introducing radioactive serotonin or its precursor via micropipettes into the SCC perikarya by iontophoresis (Pentreath and Cottrell, 1974; Pentreath, 1976) or by pressure injection (Eisenstadt *et al.*, 1973; Goldman and Schwartz, 1974), and subsequently analysing the tissues by chemical and physical separation procedures or autoradiography. Such procedures also make

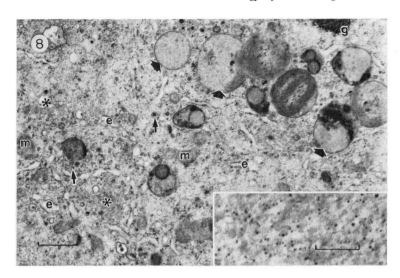

Fig. 8   Electron micrograph of part of the perikaryon of a SCC of *Helix pomatia*
which has been fixed in glutaraldehyde and osmium solutions. The micrograph
illustrates the characteristic inclusions of the somatoplasm; g, aggregate of glycogen
particles; asterisks, Golgi structures; large arrows, some of the lysosomes; small
arrows, some of the dense-core vesicles; m, mitochondria; e, endoplasmic reticulum.
The scale is 1 μ. The inset (bottom right) shows a part of the SCC cytoplasm of a
preparation from *Limax maximus* processed by the method of Wood (1965) for
the demonstration of amines. Note the similarity in the distribution and size of
the reacting sites with the distribution of the dense-core vesicles. From Pentreath,
Osborne and Cottrell (1973); reproduced by permission of Springer-Verlag

it possible to study the nature of the transport mechanisms for serotonin, and to
locate and identify with the electron microscope the terminal processes of the SCCs
which may lie at a considerable distance (1-3 cm) from the cell bodies (Fig. 9; Pentreath, 1976, 1977).

Fig. 9   Anatomical features of the SCC revealed by injection of radiolabelled
serotonin and autoradiography (see Pentreath, 1976). The diagram at top shows the
arrangement of the projections in *Helix pomatia*. In this species two large axons
run in the ipsilateral cerebro-buccal connective; these are illustrated in the middle
photomicrographs. The left middle micrograph is of an autoradiograph of the
sectioned nerve; to the right is an adjacent 2 μ Araldite section stained with Toluidene
Blue. The arrows in both mark the axons of the SCC. At bottom left is an electron
microscope autoradiograph of a part of a SCC axon shown above. The silver grains
are positively associated with lysosomes (e.g. large arrow). Other obvious axoplasmic
organelles are the dense-cored vesicles (e.g. small arrow). At bottom right is a light
microscope autoradiogram of a section through the neuropile of the ipsilateral
cerebral ganglion neuropile. This illustrates the extensive branching of the SCC
dendrites. See also Figs. 11-18. Scales; middle 40 μ, bottom left 1 μ, bottom right
10 μ

Injected serotonin is incorporated primarily into a particulate form within the cell bodies of the SCCs, but not in non-serotonergic neurones (Goldman and Schwartz, 1974; Osborne, 1978; Table 4). The evidence strongly suggests that the sites of binding are dense-cored vesicles, which are then transported by a fast

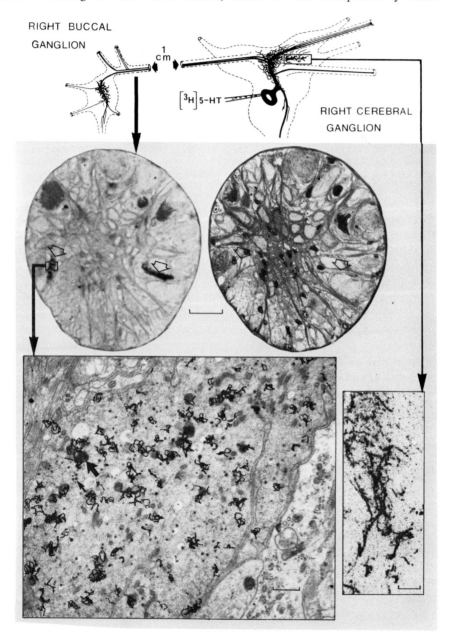

Table 4. Subcellular localization of injected [³H] serotonin in SCCs and buccal (control) neurones of *Helix pomatia*

Neurone-type	Amount of [³H]-serotonin injected into cells	Total [³H] serotonin in various fractions (%)		
		9000 g	110,000g	Supernatant
SCC	0.1 pmole	20	60	20
Buccal neurone	0.1 pmole	3	3	94

Neurones were injected with [³H] serotonin and the buccal ganglia and cerebral ganglia separately homogenized and subjected to differential centrifugation as described by Goldman and Schwartz (1974). For a single experiment at least four injected neurones (together with their ganglia) were combined. The individual fractions 9000 g, 110,000 g and supernatant were solubilized with Triton X-100 and the radioactivity measured. The values given are the mean of four different experiments.
From Osborne, 1978.

mechanism (up to 6 cm/day) into and along the axons (Schwartz *et al.*, 1975). Other experiments with *Aplysia* have shown that cutting one of the major paired axonal branches of the SCC doubles the amount and transport rate of serotonin in the other, unsevered axonal branch (transport rate 12 cm/day; Goldberg *et al.*, 1976). This intriguing finding was interpreted with a new model for fast axonal transport (Goldberg *et al.*, 1976; Goldman *et al.*, 1976). In this the presumed serotonergic vesicles are proposed as being transported at a fixed rate which is faster than that ever normally observed, but only when attached to essentially immobile tracks. The tracks may be microtubules, since the transport of serotonin is completely blocked by colchicine (Goldman *et al.*, 1976). However the vesicles are not always on tracks; their attachment is reversible. Additionally, and in order to account for the experimental evidence the model requires that the attachment to the tracks is concentration dependent. Thus net translocation may be analogous to enzyme reactions in which formation of the enzyme-substrate complex (the vesicle to track attachment) limits the appearance of product (the transported vesicle).

The fast transporting mechanism does not, in *Aplysia* SCCs, appear to be specific for serotonin. Other injected biogenic amines (dopamine, octopamine and histamine) are also transported in the same fashion, and in high concentrations these substances actually compete with serotonin for storage and transport, presumably because they are preferentially incorporated into the serotonergic vesicles (Goldberg and Schwartz, 1980; see below). It appears that two of the chemical criteria for fast transport in the SCC axons are the presence of an aromatic ring and a net positive charge, because other substances without either of these (e.g. γ-aminobutyric acid and a metabolite of choline, betaine) are not transported by this mechanism (Goldman and Schwartz, 1974; Goldberg and Schwartz, 1980). Metabolites of [³H]-choline injected into *Helix* SCCs are also not transported by fast mechanisms which are characteristic of transmitter substances (S. D. Logan, personal communication).

In *Aplysia* and *Helix* detailed electron microscope autoradiographic analysis of the axonal processes has provided valuable information about the vesicular inclusions which may be involved in the binding and transport of serotonin, although it must be appreciated that the technique cannot resolve the precise loci of radioactivity. The findings are generally similar in the two species and strongly implicate a population of dense-cored vesicles in serotonin storage. In *Aplysia* the vesicles may be elipsoidal, the majority ranging in diameter from 60–90 nm (Goldman *et al.,* 1976; Schwartz *et al.,* 1975). The studies have been refined by injecting tritiated *N*-acetylgalactosamine, which instead of labelling the vesicle contents of the neurone, labels their membranes. The sugar is incorporated into membrane glycoproteins (Ambron *et al.,* 1980), and this offers the potential advantage that not only serotonergic vesicles, but also other organelles transported along the axons become labelled, thus increasing the autoradiographic grain density over small-diameter axonal processes and hence the degree of certainty with which they can be identified under the electron microscope. Using this approach Shkolnik and Schwartz (1980) found that some of the terminal varicosities of the SCC of *Aplysia* were widely distributed around the neurones in the buccal ganglia, and that different types of vesicles were located in different regions of the cell. The cell body and axons contained compound vesicles 70–80 nm outside diameter (these vesicles were previously termed dense-cored by Schwartz *et al.,* 1975 and Goldman *et al.,* 1976), each of which were composed of a smaller lucent vesicle (the 'core') inside the other. However, the terminal varicosities in the buccal ganglia contained no compound vesicles, but dense-cored vesicles (96 nm) and large (86 nm) and small (53 nm) electron-lucent vesicles. The hypothesis was advanced by Shkolnik and Schwartz (1980) that the compound vesicles were formed in the soma by budding of the smooth endoplasmic reticulum, and were the precursors of the dense-core vesicles that reach maturity in the terminal varicosities, where they form a storage depot for serotonin. It was suggested that the dense-core vesicles that reach maturity in the terminal varicosities become transformed into the large and small electron-lucent vesicles after discharge of the core (the serotonin) by exocytosis; the inner membrane around the dense core becomes incorporated into the presynaptic membrane and is subsequently recycled into small-lucent vesicles, and the large lucent vesicles (the outer membrane of the vesicle) remain within the nerve terminals (Fig. 10). The small lucent vesicles were postulated as being responsible for the repeated release of serotonin. The terminal varicosities generally lacked areas which could be identified as active zones, as is the case for serotonergic terminals in other animals (see Pentreath and Berry, 1978). Hopefully the experimental evidence to verify these interesting suggestions will soon be forthcoming.

The work of Schwartz *et al.* (1979) has additionally shown that injected tritium-labelled serotonin becomes associated with lysosomes in the somatoplasm of the *Aplysia* SCC, as has previously been suggested by Cottrell and Osborne (1970) for *Helix pomatia.* The lysosomal association appeared to be specific; other injected

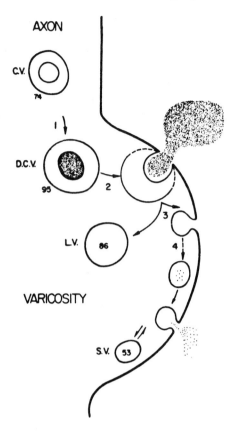

Fig. 10   Possible sequence of vesicle changes in terminal varicosities of the SCC in *Aplysia*. The compound vesicle (CV) is formed in the soma and transported to the terminal where it is converted into the dense-cored vesicle (DCV; first arrow) which represents a storage site for serotonin. After release of transmitter (2) the outer membrane of the DCV results in formation of the large lucent vesicle (LV,3) and the inner membrane (core) becomes the smaller lucent vesicle (SV,3,4) which is available for repeated release of transmitter. The mean diameter in nanometers of each vesicle type is indicated. See text for further details. From Shkolnik and Schwartz (1980), reproduced by permission of the American Physiological Society

transmitter substances were not localized in the lysosomes. A possible explanation is that the lysosomes contain membrane components that once belonged to serotonergic vesicles, and which still retain their ability to bind transmitter. It is also possible that the vesicles may be recycled from the terminal varicosities of the neurones to be deposited in the lysosomes (Schwartz *et al.*, 1979).

In *Helix pomatia* the distribution of the terminal processes of the SCCs in the buccal ganglia, as revealed by [3H] serotonin injection, together with the distribution of vesicles within the cell, are somewhat different from *Aplysia*. Terminal varicosities are largely confined to the neuropile of the ganglion (see Fig. 16; Pentreath, 1976), and do not form a basket-like net around the cell bodies. Terminal processes with the same general appearance and in the same location selectively accumulate exogenous [^{3}H] serotonin (Pentreath and Cottrell, 1972, 1973) in contrast to the SCC soma which readily takes up 5-HTP but not serotonin (p. 464). Large numbers of dense-core vesicles are present at all levels of the neurone, particularly in the main axons within the cerebro-buccal connectives (Figs. 9, 11, 12). The electron density of the core may vary slightly, but structures resembling compound vesicles are generally absent (Fig. 11). The terminal varicosities contain large numbers of electron-lucent vesicles (mean diameter 65 nm; Figs. 15, 17, 18) but these do not comprise two populations (small and large). Labelled lysosomes occur at all levels in the axons and preterminal processes, but not in the terminal varicosities (Figs. 9, 11-14; Pentreath, 1976). Some of the lysosomes contain dense-core vesicles, although these are not significantly labelled (Figs. 12, 13). It is possible that the lucent vesicles may be involved in the repeated release of serotonin, as suggested in *Aplysia* (Shkolnik and Schwartz, 1980), because they are frequently aggregated in groups against the edges of the varicosities. In these situations structures resembling active zones may, very occasionally, be found (see Fig. 18).

The injection experiments in conjunction with autoradiography have provided valuable information about the anatomical arrangements of the peripheral processes and the subcellular compartments which may bind and translocate serotonin in the SCCs of *Aplysia* and *Helix*. However, the experimental procedures introduce many factors which currently restrict clear interpretation of the results. Amongst these are:

1. The limited resolution of electron microscope autoradiography.
2. The possibility that impalement and injection of the soma cause alteration in vesicle and lysosome content and distribution throughout the SCCs.
3. Some of the injected serotonin may be washed out of its normal compartments during the histological processing.
4. The injected serotonin may become artifactually bound to the vesicles and/or lysosomes.
5. The autoradiographic silver grains may result from serotonin metabolites.

It will be necessary to preclude these in order to fully substantiate the interesting hypotheses concerning the genesis, turnover and fate of the serotonergic vesicles.

SYNAPTIC CONNECTIONS OF THE SEROTONIN NEURONES

Synapses with neurones in the buccal ganglia

Synaptic connections with buccal neurones have been found in *Helix, Aplysia, Helisoma, Lymnaea, Planorbis, Pleurobranchaea,* and *Tritonia* (Cottrell, 1977; Cottrell and Macon, 1974; Senseman and Gelperin, 1974; Granzow and Kater,

1977; Gerschenfeld and Paupardin-Tritsch, 1974b; McCrohan and Benjamin, 1980a; Berry and Pentreath, 1976a; Bulloch and Dorsett, 1979; Gillette and Davis, 1977). Details of synaptic transmission have been most thoroughly studied in *Aplysia,* where Gerschenfeld and Paupardin-Tritsch (1974a, b) found the each SCC makes excitatory or inhibitory connections with at least 13 neurones in the ipsilateral buccal ganglion. The connections were shown to be monosynaptic by the use of presynaptic injection of tetraethylammonium (Fig. 19) and by the persistence of the postsynaptic potentials when extracellular Ca^{2+} concentration was increased (see Berry and Pentreath, 1976b, for a review of these methods). Strong evidence has been presented that serotonin is the transmitter at each synapse; for example the substance is released from the SCCs when they are stimulated by intracellular currents (Gerschenfeld *et al.,* 1978), and the properties of the extrasynaptic serotonin receptors activated by iontophoresed serotonin appear identical to the receptors to the transmitter (Gerschenfeld and Paupardin-Tritsch, 1974a, b).

Receptor properties and ionic mechanisms underlying the postsynaptic potentials

Excitatory potentials. In *Aplysia,* EPSPs resulting from stimulation of the SCC were found to be caused by a selective increase in Na^+ permeability (Gerschenfeld and Paupardin-Tritsch, 1974b). Iontophoresis of serotonin demonstrated that there were two serotonin receptors: the A receptor producing a 'fast' depolarization and the A' receptor a longer-lasting 'slow' depolarization. The receptors could be separated pharmacologically; for example the A receptor was blocked specifically by

Figs. 11 and 12   Electron microscope autoradiographs of axonal processes of a SCC in the buccal ganglia of *Helix pomatia.* The cell soma had been injected with [^{3}H]serotonin 10 h before processing. Fig. 11 The open arrow (top) indicates a labelled lysosome. The smaller arrows mark some of the numerous dense-cored vesicles and illustrates their range in diameters (arrow at top left, 60 nm; at middle, 130 nm). Scale is 1 μ. Fig. 12 A group of labelled lysosomes, one of which (open arrow) contains vesicles. The vesicles appear to be (or represent previous) dense-cored vesicles, and may therefore be in a process of local degradation, perhaps destined for return to the SCC soma by retrograde transport. Whilst silver grains cannot be positively associated with this lysosome, there is high correlation of radioactivity with the two adjacent lysosomes (solid arrows), but which do not obviously contain dense-core vesicles or their remnants. The radioactivity in these structures may therefore be due to (i) radioactive serotonin or a metabolite(s) retained after vesicle ingestion and breakdown, (ii) radioactive serotonin or a metabolite(s) taken up by lysosomes in the cell soma which remove excess (injected) serotonin, and which were subsequently transported to the buccal ganglia, (iii) local uptake (and inactivation) of excess serotonin into lysosomes in axonal processes of the SCC after non-vesicular or non-lysosomal transport, (iv) artefactual binding of radioactive serotonin or a metabolite(s) to lysosomes induced by the preparative procedures. See text and Figs. 13–14. The scale is 0.5 μ

curare, 7-methyltryptamine and LSD 25 which had no effect on the A' receptor. These drugs partially antagonized the synaptic depolarizing potentials whereas bufotenine, which blocked both the A and A' receptors, completely abolished the synaptic potential (Table 5). Thus the postsynaptic membranes of the buccal neurones appear to contain both A and A' receptors to serotonin which are jointly responsible for the production of SCC-mediated EPSPs. (Gerschenfeld and Paupardin-Tritsch, 1974b).

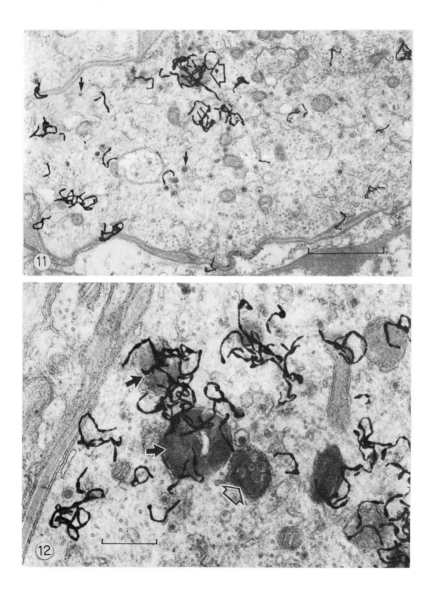

Voltage clamp studies of *Aplysia* neurones by Pellmar and Wilson (1977) suggest that the A′ response behaves in an unconventional way at lowered membrane potential; the Na^+ current is enhanced and accompanied by an apparent decrease in membrane conductance which seems to result from activation by serotonin of a voltage-sensitive Na^+ conductance in addition to the conventional voltage-independent conductance change. Work by Cottrell (1981) on the anterior buccal cell of *Helix* shows that EPSPs from the SCC may also be mediated by a mechanism which is voltage-sensitive.

Inhibitory potentials. Some SCC-mediated IPSPs in buccal ganglion neurones of *Aplysia* are hyperpolarizations produced by selective increase in K^+ conductance. The serotonin receptor (B receptor) responsible for these potentials is specifically blocked by 5-methoxygramine which has no effect on A and A′ receptors (Gerschenfeld and Paupardin-Tritsch, 1974a). The B receptor is also blocked by bufotenine (which antagonizes both A and A′ responses) and by LSD 25 and tryptamine (which block A responses); curare, an antagonist of the A receptor, is not effective.

In addition to the B responses there is an 'atypical' inhibitory potential which increases in amplitude with hyperpolarization and results from a *decrease* in both Na^+ and K^+ conductances (Fig. 19). This potential has been termed β response by Gerschenfeld and Paupardin-Tritsch (1974b). No specific blocking agents were found; the β response is not affected by bufotenine which blocks the B inhibitory response (and A and A′ depolarizing potentials).

Table 5 and Fig. 20 summarize the results of Gerschenfeld and Paupardin-Tritsch (1974a,b) on receptor properties and ionic mechanisms of SCC-mediated postsynaptic potentials in *Aplysia*.

Figs. 13–18   Autoradiographs of axonal and terminal processes of *Helix* SCC in the buccal ganglia. The cell had been injected with [^{3}H] serotonin 10 h before processing for autoradiography. Figs. 13 and 14 Preterminal processes containing groups of labelled lysosomes (1), and aggregates of dense-core vesicles (open arrows) which may represent their initial stages of breakdown. The possible nature of the radioactivity is discussed in the subscript to Fig. 12. The scales represent 0.5 μ. Fig. 15 Terminal SCC process containing small lucent vesicles (sv) and larger dense-cored vesicles. The scale is 0.5 μ. Fig. 16 Light microscope autoradiogram of a section of a buccal ganglion showing a neurone (nucleus, n) which receives synaptic input from the SCC. The micrograph shows the positional relationships of some of the axons and terminal processes (e.g. arrows) to this neurone. The radiolabelled processes are chiefly located in an area of neuropile close to the neurone, and presumably make synaptic contact with its dendrites in this region. The scale represents 25 μ. Figs. 17 and 18 Terminal processes of the SCC in the region of the buccal ganglion shown in Fig. 16. Note the large numbers of dense-cored vesicles of varying diameters (range 50–120 nm), and also the aggregates of small lucent vesicles (sv) especially in areas of possible synaptic thickenings (Fig. 8, arrow). The scales represent 0.5 μ. Figs. 15–17 from Pentreath (1976); reproduced by permission of Chapman and Hall

Serotonin inactivation

Imipramine, which blocks uptake of labelled serotonin by presumed serotonin-containing axon processes (Pentreath and Cottrell, 1972, 1973) potentiates transmission from the SCCs to buccal neurones in *Helix* (Cottrell, 1971a,b). Uptake blockers also potentiate and prolong SCC-mediated synaptic potentials in *Aplysia* (Gerschenfeld *et al.*, 1976, 1978). These results, and the fact that the neurone

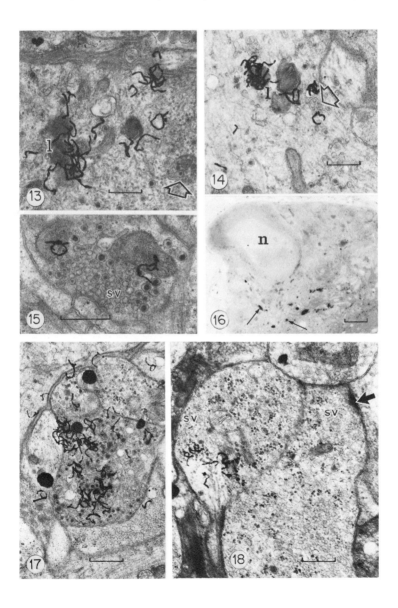

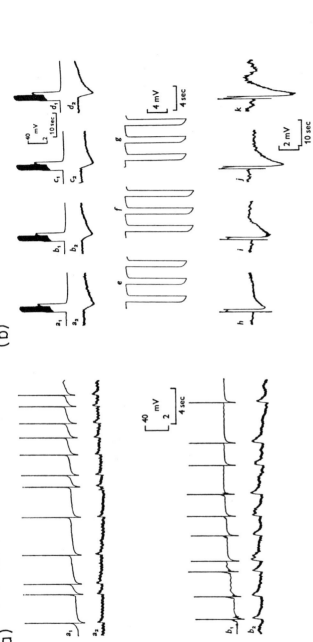

Fig. 19 (a) Example of the effects of an intracellular injection of TEA into a SCC in *Aplysia*. In a_1, a_2, the SCC is discharged at low frequency by passing a steady outward current across the membrane. Each spike evokes a discrete excitatory synaptic potential in a buccal neurone artificially hyperpolarized to 55 mV. In b_1, b_2, after injecting TEA into the presynaptic cell for 40 min the discrete synaptic potentials evoked by each presynaptic spike are much increased in amplitude, indicating the presence of a monosynaptic connection between the two neurones. (b) Atypical inhibitory synaptic potentials produced by the SCCs of *Aplysia*. a_1–a_2: presynaptic stimulation evokes an inhibitory potential in a buccal neurone hyperpolarized by inward current to 50 mV. b_1–b_2: when the buccal cell membrane potential is successively driven to 60 (b_2), 70 (c_2) and 80 mV (d_2), the inhibitory potential, instead of decreasing, increases in amplitude. e–g: square pulses of current are passed across the membrane of the buccal neurone to measure the input resistance. In e, control electronic potentials. In f, bath application of 10^{-5} M 5-HT causes an increase in the amplitude of the electrotonic potentials. In g, the effect of 5-HT disappears after removal of the amine, h–k: iontophoretic application of 5-HT on a buccal neurone responding to presynaptic stimulation by an 'atypical' potential. The hyperpolarizing response grows in amplitude when the cell membrane potential is successively driven from 50 mV (h) to 60 (i), 70 (j) and 80 mV (k). From Gerschenfeld and Paupardin-Tritsch (1974b); reproduced by permission of the Physiological Society

Table 5. Serotonin receptors and their antagonists in *Aplysia* and *Helix* neurones

Response		Ionic mechanism	Antagonists
Depolarizing	A (fast)	Increased gNa	Curare, LSD-25, tryptamine, 7-methyltryptamine, bufotenine
	A' (slow)	Increased gNa	Bufotenine
	α	Decreased gK	None found
Hyperpolarizing	B (slow)	Increased gK	LSD-25, tryptamine, bufotenine, 5-methoxygramine
	C (fast)	Increased gCl	Neostigmine
	β	Decreased gNa and gK	None

Four of these receptors (A, A', B and β) have been shown to be involved in serotonergic transmission from the SCCs of *Aplysia*.
From Gerschenfeld and Paupardin-Tritsch (1974a).

perikarya cannot metabolize serotonin (Osborne, 1978) suggest that reuptake is an important mechanism for terminating the action of released transmitter at SCC synapses.

Is acetylcholine a co-transmitter with serotonin in Helix SCCs?

In *Helix pomatia* and *H. aspersa* there is good evidence that serotonin is the transmitter at excitatory and inhibitory synapses between SCCs and buccal ganglion neurones (Cottrell, 1977; Cottrell and Macon, 1974; Cottrell *et al.*, 1974). However, experiments by Cottrell (1976, 1977) with the acetylcholine antagonist hexamethonium suggested that acetylcholine may be jointly responsible with serotonin for production of biphasic EPSPs in the so-called middle buccal cells (M cells) (Fig. 21); the first phase of the response was selectively antagonised by hexamethonium which had no effect on iontophoretically applied serotonin but abolished the response to acetylcholine.

The possible presence of acetylcholine in the SCC somata (Hanley and Cottrell, 1974; Hanley *et al.*, 1974) and the occurrence at SCC terminals of clear vesicles similar to those in other cholinergic neurones (see Pentreath, 1976) were used to add support to the possibility that both acetylcholine and serotonin may be released from the same synapses (Cottrell, 1977). However, as discussed above there is some likelihood that the acetylcholine is an artefact of contamination and the clear vesicles have subsequently been suggested to be involved in the repeated release of serotonin (p. 477). Unfortunately it was not possible to test certain other anti-cholinergic drugs such as curare and dihydro-βerythroidine in *Helix* because of their blocking action against serotonin (Cottrell, 1977). However, hexamethonium does seem to be selective in its antagonism of acetylcholine, and its observed effect on transmission between SCCs and middle buccal cells is of potential importance and requires further study.

Peripheral connections of the SCCs and the involvement of cyclic AMP

In addition to a central action, the SCCs send axons into nerves supplying the buccal musculature where they have modulatory effects on muscle contraction (Pentreath, 1973a; Kupfermann *et al.*, 1979). Peripheral function has been studied in detail in *Aplysia* by Weiss, Kupfermann and co-workers (Weiss and Kupfermann, 1976, 1977; Weiss *et al.*, 1975, 1978, 1979). They found that stimulation of the SCCs did not elicit any visible movements of the buccal musculature and did not produce synaptic potentials in the accessory radula closer muscle (ARC) which they were using to examine monosynaptic input from four identified buccal ganglion motor neurones. However, the contractions of the ARC in response to activation of any of the motorneurones were greatly increased in amplitude by stimulation of the SCC (Fig. 22). Typically the effect reached a maximum 30–40 s after a brief high-frequency train of SCC spikes, and it persisted for many minutes. Low frequency spike activity of the SCC at a rate that was comparable to that observed in chronic recording of SCC activity during feeding (2-4 spikes every 10 s) also produced potentiation of muscle contraction (Weiss *et al.*, 1978). This potentiation was shown to be a peripheral rather than central action of the SCCs by its persistence in a solution containing reduced Ca^{2+} and increased Mg^{2+} which was used to bathe the

Fig. 20   (a) Schematic drawing summarizing the central synaptic connections made by the SCCs of *Aplysia*. The SCC, which synthesizes and contains serotonin, releases this amine through different endings which establish synaptic connections with different buccal neurones. Serotonin thus released from some of the SCC terminals will activate A and A′ receptors, exciting some buccal neurones by causing an increase in their Na^+ conductance. At the same time, serotonin liberated from other endings will activate B receptors in other neurones, inhibiting them by causing an increase in K^+ conductance. It is also possible that the liberation of serotonin through other SCC terminals will activate B receptors in other neurones, inhibiting them by causing a decrease in both Na^+ and K^+ conductances. From Gerschenfeld and Paupardin-Tritsch (1974b); reproduced by permission of the Physiological Society. (b) Simplified membrane circuits representing the conductance changes induced by the activation of different 5-HT receptors in molluscan neurones. (a) Serotonin responses involving an *increase* in membrane conductance. The closing of either the A or the A′ switch or of both at the same time will introduce a shunt (Δg_{Na}) in the Na^+-conductance (g_{Na}), which becomes increased. In the same way, the closing of the B switch and the C switch by the activation of the homonymous receptors will cause an increase in the K^+ and the Cl^- conductances (g_K and g_{Cl}) respectively. (b) Serotonin responses caused by a *decrease* in the membrane conductance. The resting conductances to both Na^+ and K^+ are represented by two conductances in parallel (g_{Na} and Δg_{Na}; g_K and Δg_K). Serotonin produces an α-response by opening the switch which will remove the shunt Δg_K, and thus decreases K^+ conductance. The β-response will result from the opening of the coupled switches controlled by β, thus decreasing both Na^+ and K^+ conductances. The SCCs of *Aplysia* have been shown to mediate A, A′, B, and β responses in follower neurones of the buccal ganglia. From Gerschenfeld and Paupardin-Tritsch (1974a); reproduced by permission of the Physiological Society

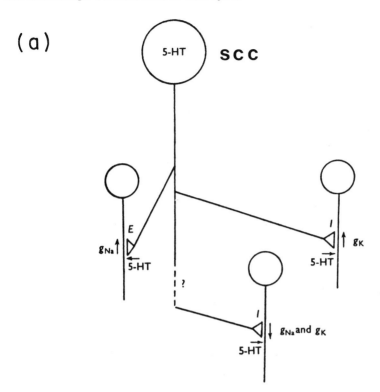

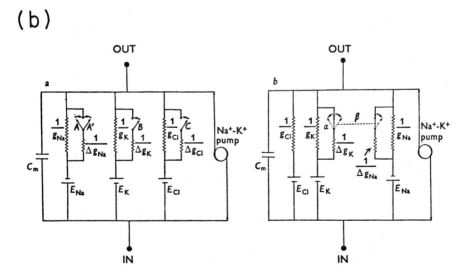

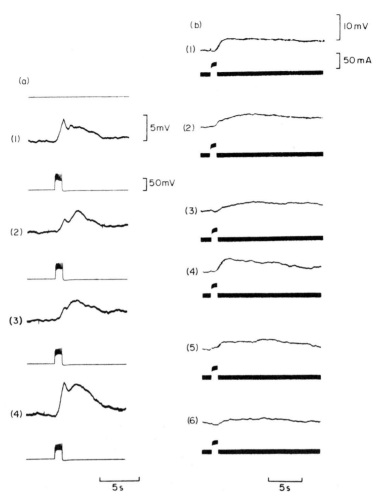

Fig. 21   (a) Effect of hexamethonium on the response of the 'M' neurones of *Helix aspersa* to repetitive firing of the SCC. The 'M' neurone, artificially hyperpolarized to 70 mV, was recorded on the upper trace, and the SCC on the lower trace of each pair. (1) Control response, (2) after 6 min, and (3) after 11 min exposure to hexamethonium bromide $(4 \times 10^{-4}$ M) and (4) after 6 min washing. Note that the response is biphasic, and the first phase is selectively reduced by hexamethonium. (b) Lack of effect of hexamethonium bromide $(4 \times 10^{-4}$ M) on responses of the M neurone of *Helix aspersa* to iontophoretic application of serotonin. Records (1–3) show control responses of the neurone (upper recording of each pair) to repeated applications of serotonin (iontophoretic current monitored in the lower traces). One second pulses of serotonin were applied to the neurone with intervals of 0.5 min between applications; some desensitization in the response is seen. Records (4–6) were made after 20 min exposure to hexamethonium bromide. The interval between applications was 0.5 min. The response was unaffected by hexamethonium and desensitization to repetitive application is still seen. From Cottrell (1977); reproduced by permission of Pergamon Press

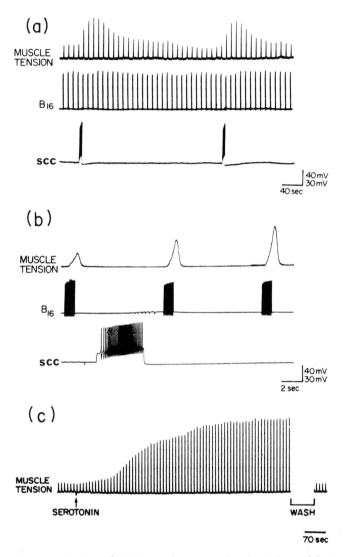

Fig. 22   SCC potentiation of ARC muscle contraction in *Aplysia*. (a) Contractions of the ARC muscle were produced every 10 s by firing a brief burst of spikes in the ARC motor neurone B16. In order to control the number and frequency of spikes, every spike was triggered by an individual depolarizing pulse. This stimulation procedure resulted in reliable and reproducible contractions of the ARC muscle. Immediately following a brief burst of SCC spikes there was an increase of the contraction elicited by motor neurone stimulation, and subsequent contractions continued to increase for 30–40 s. Contraction size then gradually returned to control over a period of 1–2 min. (b) Expanded version of the second SCC stimulation period shown in (a). (c) Effects of serotonin (10^{-8} M) on ARC muscle contraction resulting from stimulation of buccal motor neurone B16. Figures from Weiss *et al.* (1978); reproduced by permission of the American Physiological Society

buccal ganglia and block chemical synapses. Furthermore, 10^{-8} M serotonin added to the solution bathing the muscle mimicked the potentiating effects of SCC stimulation (Weiss *et al.*, 1978; Fig. 22).

Spikes in any of the four identified buccal ganglion motor neurones produce excitatory junction potentials (EJPs) in individual muscle fibres of the ARC. The EJPs show summation (and facilitation and post-tetanic potentiation) but do not trigger action potentials in the muscle; above threshold, the degree of contraction is directly dependent on the amount of depolarization produced by the summating EJPs. Weiss *et al.* (1978) found that increases in EJP amplitude may account for part of the enhancement of muscle contraction by the SCCs but that there was also a direct effect of the SCC on excitation-contraction coupling which accounted for the greater proportion of enhancement. This direct effect appeared to be mediated by cAMP (Weiss *et al.*, 1979). For example there was a dose-related increase in the synthesis and total levels of cAMP in the ARC in the presence of serotonin which did not appear to result from reduction of phosphodiesterase activity. Firing of a single SCC produced a statistically significant increase in the synthesis of cAMP (8 per cent increase per SCC spike) and the effect of SCC stimulation was enhanced by a phosphodiesterase inhibitor (Fig. 23). Phosphodiesterase-resistant analogues of cAMP enhanced ARC muscle contractions in spite of the fact that they reduced the size of EJPs in the muscle.

The SCCs in *Aplysia* thus exert an influence over the buccal musculature at two levels: centrally at the motorneurones (and possibly the pattern-generating pre-motor neurones) and peripherally, via cAMP, on excitation-contraction coupling.

BEHAVIOURAL ROLES OF THE SEROTONIN NEURONES

The extensive connections of the SCCs with buccal ganglion neurones and buccal musculature suggests an important role for the cells in feeding behaviour. Although there is considerable variation, even between closely related species, the SCCs generally seem to have a *modulatory* influence on feeding: they can potentiate buccal muscle contraction, and increase the rate and intensity of ongoing rhythmic motor output, but they do not form part of the pre-motor rhythm-generating network (Kupfermann *et al.*, 1979; Berry and Pentreath, 1976a; see Fig. 24).

Two features of modulatory synaptic actions are, (a) they are relatively prolonged, and (b) their effects depend on the occurrence of some other event; they produce no functional effect on their own but alter the excitatory or inhibitory effects of other inputs to the cell (Weiss *et al.*, 1979). The peripheral actions of the SCCs in *Aplysia* do seem to be genuinely modulatory (no SCC-mediated change is evident in the muscle in the absence of motor neurone input) but the central effects are more difficult to classify, having certain features of a command function; in some species and under certain conditions the SCCs can *initiate* behaviour and not merely modify a system that is already active (Granzow and Kater, 1977; Gillette and Davis, 1977). However, Weiss *et al.* (1978) suggest that because of the high-

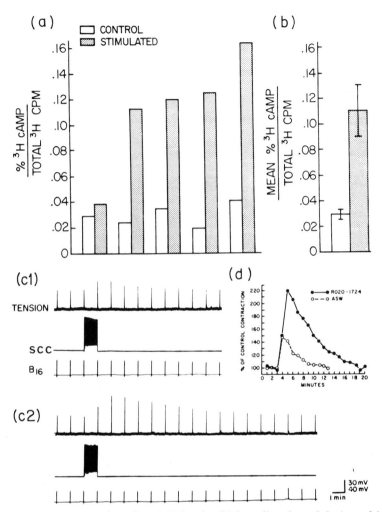

Fig. 23   Involvement of cyclic AMP in the SCC-mediated modulation of buccal muscle contractility in *Aplysia*. (a) Effect of stimulating individual SCCs on the accumulation of cyclic AMP in individual ARC muscles. The histograms show the results of five separate experiments. In each case the accumulation of [^{3}H]cAMP (from [^{3}H]adenosine in which the muscles were bathed) was higher in the muscle innervated by the stimulated SCC (dotted bar) than in the control contralateral muscle (open bar). (b) Means and standard errors of the five experiments shown in (a). (c) and (d) Effects of the phosphodiesterase inhibitor RO 20-1724 on the potentiation following stimulation of the SCC. (c) Continuous record of motor neurone, SCC, and muscle tension before (c1) and after (c2) the application of RO 20-1724. (d) Graphic representation of the results shown in part (c). The mean size of the three contractions preceeding the stimulation of the SCC was defined as 100 per cent of base line. Figures from Weiss *et al.* (1979); reproduced by permission of the American Physiological Society

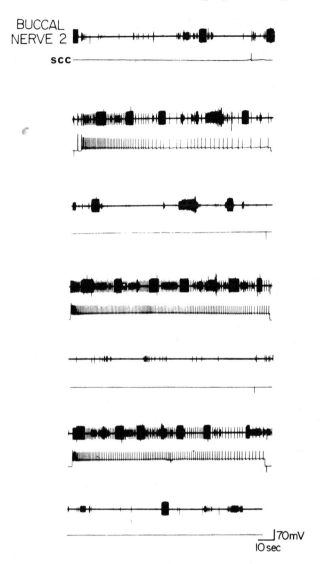

BUCCAL
NERVE 2

scc

70mV

10 sec

Fig. 24   Effect of SCC activity on spontaneous cyclic burst activity of the buccal ganglion of *Aplysia*. Fourteen min of continuous recording are shown. Each pair of traces represents 2 min of simultaneous recording from the SCC and buccal nerve 2. During alternate 2-min periods, by means of hyperpolarizing or depolarizing current, the SCC was either silenced or was made to fire at 2–5 spikes per s. During periods when the SCC was fired, burst activity clearly increased. One of the large spikes seen in nerve 2 is the extracellular record of an axon of the SCC and corresponds one for one with the intracellular spikes of the SCC. From Weiss *et al.* (1978); reproduced by permission of the American Physiological Society

frequency firing of the SCC required to initiate activity in a quiescent buccal ganglion in *Aplysia*, the effect may not be physiologically relevant. Furthermore, such an apparent command function does not necessarily involve a direct activation of the central pattern generator; it could occur by maintained depolarization of the motor neurones to allow phasic feeding inputs which are small and ineffective to exceed threshold (McCrohan and Benjamin, 1980b). In *Helix pomatia* there is evidence that the SCCs may normally be driven to fire at relatively slow rates (1/sec), but in synchronous bursts with a number of other neurones in the cerebral ganglion which also project to the buccal musculature (Pentreath, 1973b; Fig. 25). The peripheral modulatory effects (Pentreath, 1973a) in this species therefore appear to augment complex patterned output generated within the cerebral ganglia.

A hypothetical model of arousal of feeding in molluscs and the possible involvement of the SCCs has been proposed by Weiss and Kupfermann (1977; Fig. 26). The model is based on the observed synaptic connections of the SCCs and the results obtained from chronic recording of the SCCs in *Aplysia* which show that the cells become active when animals are exposed to food, and the level of activity correlates with the degree of arousal of the animal as judged by the latency to bite and strength of biting response (Kupfermann, 1979). Weiss and Kupfermann (1977) suggest that feeding may activate a central arousal system that in turn affects a large number of response systems. The SCCs are postulated to mediate the effects of this arousal system on biting (other similar cells may perform a similar function for other response systems). These authors point to the interesting facts that in the vertebrate nervous system, arousal also appears to involve biogenic amines and cAMP, and that there are central as well as peripheral actions. The peripheral effects of the SCCs appear similar to the functioning of the sympathetic nervous system in vertebrates, with serotonin taking the role that noradrenaline and other biogenic amines serve in vertebrates (Weiss *et al.*, 1978). Further work is needed to determine whether these similarities are coincidental or whether there are in fact unifying principles that apply to vertebrates and invertebrates, perhaps involving certain common molecular mechanisms of the biogenic amines (Weiss *et al.*, 1978).

DISCUSSION AND FUTURE TRENDS

Serotonin is one of the most extensively studied neurotransmitter substances in the invertebrates (see Leake and Walker, 1980). It is contained in a small proportion of neurones in most of the major invertebrate phyla, and evidently occurred at an early stage in the evolution of the nervous system. The large identifiable serotonergic cerebral cells in gastropods have been particularly suitable for elucidating some of the roles of the substance. There is a substantial literature on these cells, which itself could comprise a separate volume, and the data increase rapidly. It is probable that more is known about the biology of serotonergic transmission in the SCCs than in any other particular situation in any animal.

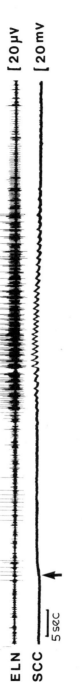

ELN [20μV

SCC [20mV

5 sec

Fig. 25   The generation of SCC activity and output to the external lip nerve (ELN) in the isolated cerebral ganglion of *Helix pomatia*. The top trace is an extracellular recording from the ELN 1 cm from the cerebral ganglion, the bottom trace is an intracellular recording of the ipsilateral SCC. The cerebral ganglia and ELN were isolated in saline for 6 h; the firing threshold of the SCC has become increased and action potentials are not elicited in this record. The records show a bursting series which occurs synchronously in the SCC and many other neurones which project into the ELN. Approximately 20 sec before the onset of the bursts the resting potential of the SCC is reduced by 7 mV (arrow), raising its potential nearer threshold. A series of EPSPs occurs in the SCC; these are synchronized with bursts of spikes in the ELN. The bursts last for approximately 30 s, following which the normal SCC resting potential is restored. The series of bursts, which occurred every 8 min in this preparation, may be a component of an arousal system analogous to that which has been proposed to occur in *Aplysia* by Weiss and Kupfermann (1977). (see also text and Fig. 26). From Pentreath (1973b)

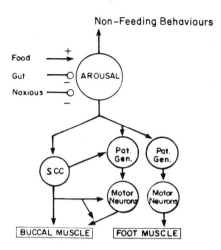

Fig. 26   Hypothetical model of the organization of the food-arousal system in *Aplysia*. A central arousal system is postulated to provide synaptic input to neural systems controlling feeding behavior or behaviors not directly related to feeding. The arousal system is excited by food stimuli and is inhibited by noxious stimuli to the skin and stretch of the gut. Many of the effects of the arousal system on the biting response are executed by means of the activity of the SCC. In addition to affecting buccal muscles, EJPs, and motor neurones, the SCC may affect a postulated pattern-generating system that drives the motor neurones of the buccal mass. From Weiss and Kupfermann (1977); reproduced by permission of the Society for Neuroscience

The chief reasons for the success in analysis of the SCCs are their large size and accessibility, their ability to survive for relatively long periods (at least 24 h) in the isolated ganglia in the appropriate physiological solutions, and their largely invariant morphological, physiological and pharmacological properties. In practice this allows correlations of biochemical, ultrastructural, electrophysiological and pharmacological information, and thus a progressively accurate picture is built up of the operations of the transmitter mechanisms within the cell. It is for example possible to make rigorous experimental tests of the criteria for serotonin being a transmitter, such as release of the substance on SCC stimulation, and its subsequent reuptake into nerve endings.

In the foregoing pages we have described and evaluated information on the SCCs in the different opisthobranch and pulmonate genera. Some of the data is summarized in Table 6. Whilst many areas have been discussed in the text some deserve further comment because their future study appears potentially valuable.

Regarding the chemical studies of the SCC soma, summarized in Fig. 27, it is evident that problems may arise because of contamination caused by adhering glial cells and other small neurones or cell debris from adjacent structures. Although these may not interfere significantly with analysis of serotonin metabolism because

Table 6. Comparison of SCC properties in gastropods

				Contains serotonin	Large soma size
Class: Gastropoda	Subclass: Pulmonata	Order: Basommatophora	Helisoma	+ (12)	+ (12
			Lymnaea	+ (24)	+ (17)
			Planorbis	+ (1, 16)	+ (1)
		Order: Stylommatophora	Helix	+ (7, 22, 25)	+ (14, 22)
			Limax	+ (7, 22, 25)	+ (22)
			Ariolimax		+ (26)
	Subclass: Opisthobranchia	Order: Anaspidea	Aplysia	+ (21, 28)	+ (28)
		Order: Nudibranchia	Tritonia	+ (28)	+ (8)
		Order: Notospidea	Pleurobranchaea	+ (11)	+ (11)

Monopolar cell	Descending CBC axons		Axons in other cerebral ganglia nerve trunks		Axons in buccal ganglia nerve trunks	
	Ipsilateral	Contralateral	Ipsilateral	Contralateral	Ipsilateral	Contralateral
+ (12)	+ (12)	− (12)	Frontal lip N. (12)	− (12)	+ (13)	+ (13)
+ (17)	+ (17)	− (17)	Labial artery N. (17)	− (17)	+ (17)	− (17)
+ (1)	+ (1)	− (1)			+ (1)	+ (1)
+ (22)	+ (21)	+ (21)	External and internal lip N. (21)	External and internal lip N. (21)	+ (21)	+ (21)
+ (22)	+ (26) + (26)	− (26) + (26)	External lip N. (26) External lip N. (26)			
+ (30)	+ (30)	− (30)	Posterior lip N. (30)	− (30)	+ (30)	+ (30)
+ (2)	+ (2, 8)	** (2, 8)	− (8)	− (8)	+ (8)	+ (8)
+ (11)	+ (11)	− (11)	Mouth N. (11)	− (11)	+ (11)	+ (11)

Table 6 (*Cont.*)

				Communication (electrical coupling) between right and left cells	Anomal rectifica
Class: Gastropoda	Subclass: Pulmonata	Order: Basommatophora	*Helisoma*	+ (13)	– (13)
			Lymnaea	+ (16)	
			Planorbis	+ (1)	+ (1)
		Order: Stylommatophora	*Helix*	– (14)	+ (15)
			Limax	– (26)	
	Subclass: Opisthobranchia		*Ariolimax*	– (26)	
		Order: Anaspidea	*Aplysia*	– (30)	+ (30)
		Order: Nudibranchia	*Tritonia*	– (2, 8)	
		Order: Notospidea	*Pleurobranchaea*	– (11)	+ (11)

Ionic basis of action potential	Spontaneous firing levels in dissected preparation	Spontaneous synaptic activity	Recurring synaptic input correlated with cyclical motor activity in BG
+ Na	1.5–2.5 i/s (13)	Tonic IPSPs, unitary PSPs common to both cells Phasic IPSPs, unitary PSPs common to both cells	Inhibitory during retraction (13)
	0.5–2.5 i/s (17)	Two types phasic compound I-ESPSs common to both cells Tonic EPSPs, common to both cells Long duration compound EPSPs, common to both cells	Excitatory during protraction, inhibitory during retraction (17)
	~1–2 i/s (1)	Phasic EPSPs and IPSPs, some common to both cells Some independent in two cells (1)	
	Quiescent (14)	Tonic EPSPs, unitary potentials common to both cells Phasic EPSPs, unitary potentials common to both cells (14)	
			Excitatory during retraction and inhibitory during protraction (9)
	Quiescent (30)	Tonic E-IPSPs, unitary potentials *not* common to both cells Phasic EPSPs, unitary potentials common to both cells (30)	
	1.1–1.7 i/s (8)		
	~0.5 i/s (11)		Excitatory during retraction and inhibitory during protraction (11)

Table 6 (*Cont.*)

				Synaptic responses to cerebral nerve stimulation
Class: Gastropoda	Subclass: Pulmonata	Order: Basommatophora	*Helisoma*	I; E-I; predominantly I (13)
			Lymnaea	
			Planorbis	E (1)
		Order: Stylommatophora	*Helix*	E (28); E-I, predominately E (6, 27)
			Limax	
			Ariolimax	
	Subclass: Opisthobranchia	Order: Anaspidea	*Aplysia*	E-I, predominately E (30)
		Order: Nudibranchia	*Tritonia*	I at resting potential (8)
		Order: Notospidea	*Pleurobranchaea*	

Connections with BG neurones	
Monosynaptic	Polysynaptic via BG interneurones
Excitatory with protractor motor neurone (12)	With protractor and retractor motor neurones (12), with cell 4 which innervates salivary gland and cell 5, the largest BG cell (13)
Excitatory with some protractor and retractor motor neurones and with cell 1 which innervates salivary gland (17)	With protractor and retractor motor neurones (17)
Excitatory with a few small BG cells (1)	With feeding motor neurones (1)
Excitatory with 2 of 3 largest BG cells (4, 5)	With large medial BG cell (4, 5, 26)
	With large medial BG cell (26)
	With salivary burster (23)
Excitatory with 2 of 3 largest BG cells (26)	With large medial BG cell (26)
Some excitatory and some inhibitory with many BG cells including motor neurones (10, 29)	
Some excitatory and some inhibitory with a few BG cells including motor neurones (2, 3)	With feeding motor neurones (2, 3)
Excitatory with a few BG cells including motor neurones (11)	With feeding motor neurones (11)

Table 6 (*Cont.*)

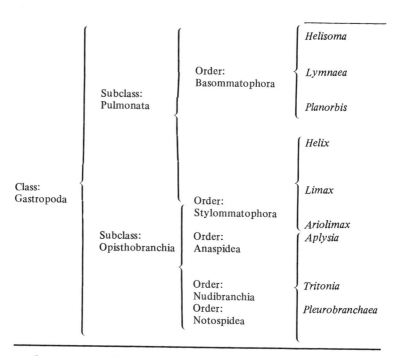

+ = that a particular cell possesses the property listed at the head of the column.
− = that a particular cell is known *not* to exhibit such a property; numbers in parentheses indicate the source of information as listed below.
(**Bulloch (1977) has reported contralateral CBC axons in some specimens of *Tritonia hombergi,* but they are absent in most; Dorsett (1967) has reported evidence for contralateral CBC axons in *Tritonia diomedea*.)

1. Berry and Pentreath 1976a
2. Bulloch 1977
3. Bulloch and Dorsett, 1979
4. Cottrell 1977
5. Cottrell and Macon, 1974
6. Cottrell *et al.,* 1972
7. Cottrell and Osborne 1970
8. Dorsett 1967
9. Gelperin and Forsythe, 1976
10. Gerschenfeld and Paupardin-Tritsch, 1974b
11. Gillette and Davis 1977
12. Granzow and Kater, 1977
13. Granzow and Fraser Rowell, 1981

Effects on feeding motor output of BG	Modulates at level of muscle
Can initiate and maintain rhythmical BG motor output, can speed up ongoing rhythm, can increase intensity of protractor motor neurone bursts (12)	
Can increase intensity of retractor motor neurone bursts; does not initiate rhythmical BG motor output nor influence rate of ongoing rhythm (17)	
Can initiate and maintain rhythmical buccal mass movements, although not complete feeding movements (1)	
	+ Lip muscle (20)
Can sometimes activate a few cycles of organized motor output from BG, but does not maintain rhythmical output (19)	
Can increase intensity of motor neurone bursts; in excitable preparations can speed up ongoing rhythmical BG motor output, can activate a single 'feeding cycle' but cannot maintain rhythmical BG motor output in otherwise quiescent preparations (29)	+ Buccal mass (29)
Can activate multicomponent post-synaptic responses in identified BG motor neurones typical of feeding motor output (2, 3)	
Can speed up ongoing rhythmical BG motor output; can activate a single 'feeding cycle' but cannot maintain rhythmical BG motor output in otherwise quiescent preparations (11)	

14. Kandel and Tauc, 1966a
15. Kandel and Tauc, 1966b
16. Marsden and Kerkut, 1970
17. McCrohan and Benjamin, 1980a
18. McCrohan and Benjamin, 1980b
19. Osborne and Cottrell, 1971
20. Pentreath 1973a
21. Pentreath and Cottrell, 1974
22. Pentreath *et al.*, 1973
23. Prior and Gelperin, 1977
24. Sakharov and Zs-Nagy, 1968
25. Sedden *et al.*, 1968
26. Senseman and Gelperin, 1974
27. Szczepaniak and Cottrell, 1973
28. Weinreich *et al.*, 1973
29. Weiss *et al.*, 1978
30. Weiss and Kupfermann, 1976
From Granzow and Fraser Rowell, 1981; reproduced by permission of the Company of Biologists.

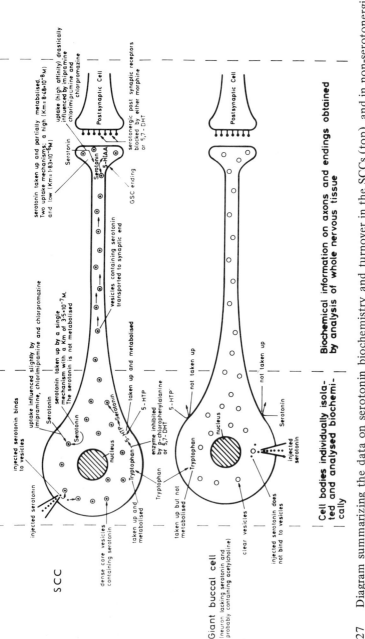

Fig. 27   Diagram summarizing the data on serotonin biochemistry and turnover in the SCCs (top), and in non-serotonergic buccal neurones (bottom). Both neurones accumulate tryptophan, but only the SCCs metabolize the substance to 5-HTP and serotonin. Electrical stimulation of the SCC increases the amount of serotonin formed from tryptophan whilst the 5-HTP content remains constant which suggests that tryptophan hydroxylase determines the turn-over rate of serotonin. Further support for this is provided by the effects of p-chlorophenylalanine and 5,7-DHT on tryptophan metabolism. A combination of biochemical and autoradiographic studies suggest that the uptake of serotonin by the SCC perikarya differs in kinetics and susceptibility to drug block from uptake in the cells' terminals, where a high affinity re-uptake is important for the substances' inactivation. Most properties of serotonin metabolism, transport, turnover and modification by drugs are specific for the SCCs and other serotonergic neurones. From Osborne (1978); reproduced by permission of Pergamon Press

the quantities of substances are sufficiently great for unambiguous measurement by current microtechniques (see Osborne, 1974), errors may arise when measuring trace quantities of other chemicals such as acetylcholine. A possible way of avoiding these is to rupture the cell, exude the somatoplasm and nucleus, and assay these and the remaining 'ghost' separately (Cottrell, 1977), but this procedure will almost certainly cause mixing of the glial cell and SCC cytoplasm at the site of rupture, and the exudate will also contain some of the extensive glial processes which extend into the cytoplasm of the cells (see Pentreath *et al.*, 1973). Alternatively, painstaking dissection to clean the SCC perikaryon without excessive damage may be the most satisfactory approach at present (Osborne, 1977). This procedure does not avoid the problem of glial infoldings into the SCC, but will maintain the metabolic integrity of the neurone. The application of immunohistochemical techniques which mark enzymes involved in transmitter metabolism (e.g. choline acetyltransferase) also appear potentially fruitful.

The studies on the ultrastructural correlates of serotonin binding, transport and turnover have provided valuable data on the mechanisms of the serotonin binding and transport, and have established the importance of the dense-cored vesicles and lysosomes in these processes. However, certain areas require further study, in particular the relationships of the different vesicle populations (electron-lucent and electron-dense) in the terminal regions, and their roles in serotonin release, re-uptake and recycling. The possible significance of the lysosomes in the retrograde transport and breakdown of the serotonergic vesicles also need clarifying; perhaps this may be achieved by ligature experiments of nerves containing SCC projections at different time periods after injection of the soma with [^{3}H] serotonin and other radioactive substances. It is also feasible with injection experiments and autoradiography to obtain data on the SCC terminals in the peripheral buccal musculature, the nature and distribution of active synaptic zones and terminal varicosities in relation to the post-synaptic buccal ganglion neurones, and on the variations of the terminal arrangements in different genera.

The electrophysiological experiments have shown beyond reasonable doubt that serotonin serves a neurotransmitter role at excitatory and inhibitory synapses between the SCCs and neurones located in the buccal ganglia in *Helix* and *Aplysia* (Cottrell and Macon, 1974; Gerschenfeld and Paupardin-Tritsch, 1974b). In *Aplysia* endogenous serotonin is released following SCC stimulation, and this is inactivated by re-uptake (Gerschenfeld *et al.*, 1976, 1978). Of further interest are the data concerning the numbers and types of serotonin receptors on neurones in the buccal ganglia of *Aplysia* (Gerschenfeld and Paupardin-Tritsch, 1974a; Cottrell, 1977). Of the six different types so far described four appear to mediate transmission from the SCCs. Such variety is impressive by comparison with serotonin receptors in the mammalian brain, though this may reflect an ignorance of mammalian serotonin receptors whose properties are more difficult to study (see Segal, 1980). It is difficult to predict the extent to which the data on gastropods may bear on the mammalian brain. On the other hand they may point towards multiple receptor discovery and classification, and thus argue for more extensive studies and re-examination of existing receptor data in the CNS. A recent report of multiple actions of serotonin on myenteric neurones of the guinea-pig (Johnson *et al.,* 1980) is further suggestive of the potential importance of such studies. Alternatively the multiplicity of serotonin receptors may be a parochial molluscan phenomenon by which the integrative potential of a relatively small number of large interneurones is optimized, i.e. the animals make maximum use of a few different transmitters by employing the same one (serotonin) to produce a variety of effects. Several other transmitters in molluscan ganglia also have multiple receptor and ionic mechanisms (i.e. dopamine, Berry and Cottrell, 1973, 1975; acetylcholine, Kehoe, 1972; histamine, Weinreich, 1978), but serotonin appears to be the most complex. Also it must be borne in mind that of the six receptor types described by Gerschenfeld and Paupardin-Tritsch, 1974a,b) two were somatic rather than synaptic. As discussed by Leake and Walker (1980) there is danger in extrapolating from the vertebrate CNS to the peripheral nervous system as well as from vertebrates to invertebrates, or vice versa.

Certain other features of the output transmitting mechanisms of the SCCs are also of great potential significance and require further study. Most of the serotonin-mediated EPSPs and IPSPs in the buccal neurones are of relatively small amplitude and have unusually slow rise and decay times. Even at high frequency firing of the SCC the evoked EPSPs do not appear to be capable of firing the post-synaptic neurone under normal conditions. In addition the SCCs have elaborate branching projections into the musculature of the buccal mass (see Fig. 2), where modulatory effects are exerted. The SCCs are therefore multi-action neurones which can influence a target organ (the buccal muscles) at more than one level: centrally via the neurones in the buccal ganglia (some of the neurones are motor to the buccal muscles) or peripherally at the muscles. The peripheral modulatory actions of the *Aplysia* SCCs are unconventional in the sense that they modify muscle contraction but in the absence of normal, electrophysiologically recordable signs. The studies by Kupfermann, Weiss, and co-workers (see Kupfermann *et al.,* 1979) have shown that SCC activity in *Aplysia* may produce enhancement of excitatory junctional potentials in the buccal muscles, but that the subsequent potentiation of muscle contraction is not associated with any detectable changes in membrane potential of the muscle cells. Their experiments provide good evidence that SCC activity has a direct potentiating effect on excitation-coupling, possibly mediated by cAMP, and further suggest that the cells may be involved in food arousal states where there is enhanced activity of the buccal musculature. Similar modulatory effects have been observed in *Planorbis* and *Helix* (Berry and Pentreath, 1976; Pentreath, 1973a,b). Unfortunately there are no comparable data on possible modulatory actions of serotonin in mammals. Whilst the serotonin-mediated effects on muscle may be specific to molluscs, it is nevertheless possible that the effects in the central ganglia may have a counterpart in the CNS.

Finally, the SCCs also offer potential for studies of a rather different approach.

These have not been discussed in detail in the text because they have not yet gathered sufficient momentum, but deserve attention. There is little doubt that the SCCs of the different gastropod species are homologous because of their unique combination of chemical, structural and physiological features (see Table 6; Weiss and Kupfermann, 1976; Granzow and Fraser Rowell, 1981). However there are also a large number of smaller variations superimposed on the general themes of these features; the number, arrangement and bilaterality of the projections, the degree of electrical coupling, the amount of chemical input, the precise number of neurones in the buccal ganglia receiving output, and the strength and polarity of the synaptic output to these neurones differ in each species. The differences all appear to reflect variations in the integrative roles of the cells, which because of their specific involvement in the control of feeding, may be directly associated with variations in the feeding behaviour of the animals. Since the gastropods differ widely in their habitats (e.g. marine, fresh water and terrestial groups) and feeding behaviour (e.g. *Tritonia* rapidly ingests relatively large pieces of animal matter, *Aplysia* grazes on macroalgae, *Helisoma* on microalgae) it may be possible to relate the differences in the cells' properties to the behavioural variations in the different species and genera, thus providing a correlation of evolution at the behavioural level with the evolution of the nervous system at the level of individual neurones. For example the SCCs of *Helisoma* and *Helix* fire continually but receive bursts of inhibitory input, whereas the cells in *Aplysia* and *Helix* are normally silent, but receive active bursts of initiation; this may be associated with the exposure of the former to a constant food source, but a discontinuous food source to the latter (Granzow and Fraser Rowell, 1981). Although the details of such relationships will take time to elucidate, they offer promise of a beginning in the understanding of evolutionary adaptation of individual neurones to different functional roles.

REFERENCES

Althaus, H. H., Osborne, N. N. and Neuhoff, V. (1973). Microchromatographische Extraktion und Fraktion von Lipiden einzelner Nervenzellen von *Helix pomatia, Naturwiss,* **60**, 553–554.

Ambron, R. T., Goldman, J. E., Shkolnik, L. J. and Schwartz, J. (1980). Synthesis and axonal transport of membrane glycoproteins in an identified serotonergic neuron of *Aplysia, J. Neurophysiol.* **43**, 929–944.

Barnicoat, B. F. and Pentreath, V. W. (1981). Glial-neuronal associations in 'isolated' neurons of selected invertebrates, *Cell and Tiss. Res.,* submitted.

Baumgarten, H. G., Bjorklund, A., Lachenmayer, L. L. and Nobin. A. (1973). Evaluation of the effects of 5,7-dihydroxytryptamine on serotonin and catecholamine neurons in the rat CNS, *Acta Physiol. Scand.* Supplement 391, 19 pp.

Berry, M. S. and Cottrell, G. A. (1973). Dopamine: excitatory and inhibitory transmission from a giant dopamine neurone. *Nature (New Biol.),* **242**, 250–253.

Berry, M. S. and Cottrell, G. A. (1975). Excitatory, inhibitory and biphasic synaptic potentials mediated by an identified dopamine-containing neurone, *J. Physiol. (Lond.),* **244**, 589–612.

Berry, M. S. and Pentreath, V. W. (1976a). Properties of a symmetric pair of serotonin-containing neurons in the central ganglia of *Planorbis, J. exp. Biol.* **65**, 361–380.

Berry, M. S. and Pentreath, V. W. (1976b). Criteria for distinguishing between monosynaptic and polysynaptic transmission, *Brain Res.* **105**, 1–20.

Briel, G. Neuhoff, V. and Osborne, N. N. (1971). Determination of amino acids in single identifiable nerve cells of *Helix pomatia, Int. J. Neurosci.* **2**, 129–136.

Brownstein, M. J., Saavedra, J. M., Axelrod, J., Zeman, G. H., and Carpenter, D. O. (1974). Coexistence of several putative neurotransmitters in single identified neurons of *Aplysia, Proc. Nat. Acad. Sci. USA*, **71**, 4662–4665.

Bulloch, A. G. M. (1977). A neurobiological study of feeding behaviour in the nudibranch mollusc *Tritonia hombergi*, Ph.D. Thesis, University of Wales.

Bulloch, A. G. M. and Dorsett, D. A. (1979). The integration of the patterned output of buccal motoneurons during feeding in *Tritonia hombergi, J. exp. Biol.* **79**, 23–40.

Cannata, M. A., Chiocchio, S. R. and Tramezzani, J. H. (1968). Specificity of the glutanaldehyde silver technique for catecholamines and related compounds, *Histochemie*, **12**, 253–265.

Cottrell, G. A. (1970). Direct postsynaptic responses to stimulation of serotonin-containing neurones. *Nature (Lond.)*, **225**, 1060–1062.

Cottrell, G. A. (1971a). Action of imipramine on 5-hydroxtryptaminergic transmission and on 5-hydroxytryptamine uptake in the snail (*Helix pomatia*) brain, *Br. J. Pharmacol.* **43**, 437P.

Cottrell, G. A. (1971b). Action of imipramine on a serotonergic synapse. *Comp. Gen. Pharmacol.* **2**, 125–128.

Cottrell, G. A. (1976). Does the giant cerebral neurone of *Helix* release two transmitters, acetylcholine and serotonin? *J. Physiol.* **259**, 44–45p.

Cottrell, G. A. (1977). Identified amine-containing neurons and their synapric connexions. *Neuroscience*, **2**, 1–18.

Cottrell, G. A. (1981). An unusual synaptic response mediated by a serotonergic neurone. *J. Physiol.* **310**, 26P.

Cottrell, G. A. and Macon J. B. (1974). Synaptic connexions of two symmetrically placed giant serotonin-containing neurones. *J. Physiol.* **236**, 435–464.

Cottrell, G. A. and Osborne, N. N. (1970). Subcellular localization of serotonin in an identified serotonin-containing neurone, *Nature* (Lond.), **225**, 470–472.

Cottrell, G. A. and Powell, B. (1971). Formation of serotonin by isolated serotonin-containing neurones and by isolated non-amine-containing neurons, *J. Neurochem.* **18**, 1695–1697.

Cottrell, G. A., Berry, M. S. and Macon, J. B. (1974). Synapses of a giant serotonin neurone and a giant dopamine neurone: Studies using antagonists, *Neuropharmacol.* **13**, 431–439.

Cottrell, G. A., Macon, J. and Szczepaniak, A. C. (1972). Glutamic acid mimicking of synaptic inhibition on the giant serotonin neurone of the snail, *Bri. J. Pharmacol.* **45**, 684–688.

Dale, H. H. (1935). Pharmacology and nerve endings, *Proc. Roy. Soc. Med.* **28**, 319–332.

Dorsett, D. A. (1967). Giant neurones and axon pathways in the brain of *Tritonia, J. Exp. Biol.* **46**, 137–151.

Eisenstadt, M., Goldman, J. E., Kandel, E. R., Koester, J., Koike, H. and Schwartz, J. H. (1973). Intrasomatic injection of radioactive precursors for studying transmitter synthesis in identified neurons of *Aplysia californica, Proc. natn. Acad. Sci. USA*, **70**, 3371–3375.

Emson, P. C. and Fonnum, F. (1972). Choline acetyltransferase, acetylcholinesterase and aromatic L-amino acid decarboxylase in single identified nerve cell bodies from snail *Helix aspersa, J. Neurochem.* 22, 1079–1088.

Falck, B., and Owman, C. (1965). A detailed methodological description of the fluorescence method for the cellular demonstration of biogenic monamines. *Acta. Univ. Lund.* Sect. II. No. 7, 1–23.

Gelperin, A. and Forsythe, D. (1976). Neuroethological studies of learning in molluscs, in *Simpler Networks and Behavior* (Ed. J. C. Fentress), pp. 239–246, Sinauer Associates Inc., Sunderland, Massachussetts.

Gerschenfeld, H. M. and Paupardin-Tritsch, D. (1974a). Ionic mechanisms and receptor properties underlying the responses of molluscan neurons to 5-hydroxytryptamine, *J. Physiol.* 234, 427–456.

Gerschenfeld, H. M. and Paupardin-Tritsch, D. (1974b). On the transmitter function of 5-hydroxytryptamine at excitatory and inhibitory monosynaptic junctions, *J. Physiol.* 243, 457–481.

Gerschenfeld, H. M., Hamon, M. and Paupardin-Tritsch, D. (1976). Release and uptake of 5-hydroxytryptamine (5-HT) by a single 5-HT containing neurone. *J. Physiol.* 260, 29–30p.

Gerschenfeld, H. M., Hamon, M., and Paupardin-Tritsch, D. (1978). Release of endogenous serotonin from two identified serotonin-containing neurons and the physiological role of serotonin re-uptake, *J. Physiol.* 274, 265–278.

Gillette, R. and Davis, W. J. (1977). The role of metacerebral giant neuron in the feeding behavior of *Pleurobranchaea, J. Comp. Physiol.* 116, 129–159.

Goldberg, D. J. and Schwartz, J. H. (1980). Fast axonal transport of foreign transmitters in an identified serotonergic neurone of *Aplysia californica, J. Physiol.* 307, 259–272.

Goldberg, D. J., Goldman, J. E. and Schwartz, J. H. (1976). Alterations in amounts and rates of serotonin transported in an axon of the giant cerebral neurone of *Aplysia californica, J. Physiol.* 259, 473–490.

Goldman, J. E. and Schwartz, J. H. (1974). Cellular specificity of serotonin storage and axonal transport in identified neurons in *Aplysia californica, J. Physiol.,* 242, 61–76.

Goldman, J. E., Kim, K. S., and Schwartz, J. H. (1976). Axonal transport of [^{3}H]-serotonin in an identified neuron of *Aplysia californica. J. Cell. Biol.* 70, 304–318.

Granzow, B. and Kater, S. B. (1977). Identified higher-order neurons controlling the feeding motor programme of *Helisoma, Neuroscience,* 2, 1049–1063.

Granzow, B. and Fraser Rowell, C. H. (1981). Further observations on the serotonergic cerebral neurons of *Helisoma* (Mollusca: Gastropoda): the case for homology with the metacerebral giant cells. *J. exp. Biol.* 90, 283–305.

Hanley, M. R. and Cottrell, G. A. (1974). Acetylcholine activity in an identified 5-hydroxytryptamine-containing neurone. *J. Pharm. Pharmac.* 26, 980.

Hanley, M. R., Cottrell, G. A., Emson, P. C. and Fonnum, F. (1974). Enzymatic synthesis of acetylcholine by a serotonin-containing neurone from *Helix, Nature (New Biol.),* 251, 631–633.

Johnson, S. M., Katayama, Y. and North, R. A. (1980). Multiple actions of 5-hydroxytryptamine on myenteric neurones of the guinea-pig ileum, *J. Physiol.* 304, 549–570.

Kai-Kai, M. A. and Pentreath, V. W. (1981). The structure, distribution and quantitative relationships of the glia in the abdominal ganglion of the horse leech, *Haemopis sanguisuga, J. Comp. Neurol.* (In press.)

Kandel, E. R. and Tauc, L. (1966a). Input organization of two symmetrical giant cells in the snail brain. *J. Physiol.* **183**, 269-286.

Kandel, E. R. and Tauc, L. (1966b) Anomalous rectification in the metacerebral giant cells and its consequences for synaptic transmission, *J. Physiol.* **183**, 287-304.

Kehoe, J. S. (1972). The physiological role of three acetylcholine receptors in synaptic transmission in *Aplysia, J. Physiol.* **225**, 147-172.

Koe, B. K. and Weissman, A. (1966). *p*-Chlorophenylalanine: A specific depletor of brain serotonin. *J. Pharmacol. exp. Therap.* **154**, 499-516.

Kunze, H. (1921). Zur Topographie und Histologie des Central Nervensystems von *Helix pomatia* L, *Z. wiss. Zool.* **68**, 539-566.

Kupfermann, I. (1979). Modulatory actions of neurotransmitters, *Ann. Rev. Neurosci.* **2**, 447-465.

Kupfermann, I., Cohen, J. L., Mandelbaum, D. E., Schonberg, M., Susswein, A. J. and Weiss, K. R. (1979). Functional roles of serotonergic neuromodulation in *Aplysia, Fed. Proc.* **38**, 2095-2102.

Leake, L. D. and Walker, R. J. (1980). *Invertebrate Neuropharmacology.* Blackie, Glasgow and London.

Marsden, C. A. and Kerkut, G. A. (1970). The occurrence of monoamines in *Planorbis corneus*: a fluorescence microscopic and microspectrometric study, *Comp. Gen. Pharmacol.* **1**, 101-116.

McCaman, R. E. and Dewhurst, S. A. (1970). Choline acetyltransferase in individual neurons of *Aplysia californica, J. Neurochem.* **17**, 1421-1426.

McCaman, M. W., Weinreich, D. and McCaman, R. E. (1973). The determination of picomole levels of 5-hydroxtryptamine and dopamine in *Aplysia, Tritonia,* and leech nervous tissues, *Brain Res.* **53**, 129-137.

McCrohan, C. R. and Benjamin, P. R. (1980a). Patterns of activity and axonal projections of the cerebral giant cells of the snail, *Lymnaea stagnalis, J. exp. Biol.* **85**, 149-168.

McCrohan, C. R. and Benjamin, P. R. (1980b). Synaptic relationship of the cerebral giant cells with motor neurons in the feeding system of *Lymnaea stagnalis, J. exp. Biol.* **85**, 169-186.

Osborne, N. N. (1970). Distribution, localization and functional significance of biologically active monoamines in gastropod molluscs. Ph.D. Thesis. St. Andrews University.

Osborne, N. N. (1972). The *in vivo* synthesis of serotonin in an identified serotonin-containing neuron of *Helix pomatia, Internat. J. Neuroscience.* **3**, 215-219.

Osborne, N. N. (1973a). Micro-biochemical and physiological studies on an identified serotonergic neuron in the snail *Helix pomatia, Malacologia,* **14**, 97-106.

Osborne, N. N. (1973b) Tryptophan metabolism in characterized neurons of *Helix, Brit. J. Pharmacol.* **48**, 546-549.

Osborne, N. N. (1974). *Microchemical Analysis of Nervous Tissue.* Pergamon Press, Oxford.

Osborne, N. N. (1977). Do snail neurons contain more than one transmitter? *Nature* (Lond.), **270**, 622-623.

Osborne, N. N. (1978). The neurobiology of a serotonergic neuron, in *Biochemistry of Characterised Neurons,* (Ed. N. N. Osborne), pp. 47-80, Pergamon Press, Oxford.

Osborne, N. N. (1979). Is Dale's principle valid? *Trends in Neurosci.* **2**, 73-75.

Osborne, N. N. (1981). Communication between neurons: A review. *Neurochem. Internat.* **3**, 3-16.

Osborne, N. N. and Cottrell, G. A. (1971). Distribution of biogenic amines in the slug. *Limax maximus, Z. Zellforsch. mikrosk. Anat.* **112**, 15–30.

Osborne, N. N. and Cottrell, G. A. (1972a). Amine and amino acid microanalysis of two identified snail neurons with known characteristics, *Experientia,* **28**, 656–658.

Osborne, N. N. and Cottrell, G. A. (1972b). The effect of optic tentacle removal on the transmitter content of the giant serotonin cell of *Helix aspersa, J. Neurochem.* **19**, 2363–2368.

Osborne, N. N. and Neuhoff, V. (1974). *In vitro* experiments on the metabolism, accumulation and release of 5-HT in the nervous system of the snail *Helix pomatia, J. Neurochem.* **22**, 363–371.

Osborne, N. N. and Pentreath, V. W. (1976). Effects of 5,7-dihydroxytryptamine on an identified 5-hydroxytryptamine-containing neurone in the central nervous system of the snail *Helix pomatia, Brit. J. Pharmacol.* **56**, 29–38.

Osborne, N. N. and Rüchel, R. (1975). Protein patterns of single neurons from *Planorbis corneus* as determined by microelectrophoresis on SDS-gradient-polyacrylamide gels, *J. Chromatography,* **105**, 197–200.

Osborne, N. N., Ansorg, R., and Neuhoff, V. (1971). Micro-disc electrophoretic separation of soluble proteins in neurons and other tissues of the gastropod *Helix pomatia. Int. J. Neurosci.* **1**, 259–264.

Osborne, N. N., Cuello, A. C. and Dockray, G. J. (1981). The specific localisation of substance P and cholecystokinin-like peptides in neurones of the snail *Helix* and the coexistence of cholecystokinin and serotonin in a defined giant neurone (GSC), *Science.* (In press.)

Osborne, N. N., Priggemeier, E. and Neuhoff, V. (1973). The effect of imipramine and nialamide on the accumulation of serotonin in snail (*Helix pomatia*) nervous tissue, *Experientia,* **29**, 1351–1353.

Pellmar, T. C. and Wilson, W. A. (1977). Unconventional serotonergic excitation in *Aplysia, Nature* (Lond.), **269**, 76–78.

Pentreath, V. W. (1973a). Effect of stimulating a central giant serotonin-containing neurone on peripheral muscles in the snail *Helix pomatia, Experientia,* **29**, 540–542.

Pentreath, V. W. (1973b). Studies on neuronal 5-hydroxytryptamine in a gastropod mollusc, *Helix pomatia* (L), Ph.D. Thesis, University of St. Andrews.

Pentreath, V. W. (1976). Ultrastructure of the terminals of an identified 5-hydroxytryptamine-containing neurone marked by intracellular injection of radioactive 5-hydroxytryptamine, *J. Neurocytol.* **5**, 43–61.

Pentreath, V. W. (1977). 5-Hydroxytryptamine in identified neurones, *Trans. Biochem. Soc.* **5**, 854–858.

Pentreath, V. W. and Berry, M. S. (1978). Radioautographic study of 5-hydroxytryptamine-containing nerve terminals in central ganglia of *Planorbis corneus*: Comparison with other species and characteristics of the serotonergic nerve terminal, *J. Neurocytol.* **7**, 443–459.

Pentreath, V. W. and Cottrell, G. A. (1970). The blood supply to the central nervous system of *Helix pomatia, Z. Zellforsch. mikrosk. Anat.* **111**, 160–178.

Pentreath, V. W. and Cottrell, G. A. (1972). Selective uptake of 5-hydroxytryptamine by axonal processes in *Helix pomatia, Nature,* (New Biol.), **239**, 213–214.

Pentreath, V. W. and Cottrell, G. A. (1973). Uptake of serotonin, 5-hydroxytryptophan and tryptophan by giant serotonin-containing neurons and other neurons in the central nervous system of the snail (*Helix pomatia*), *Z. Zellforsch. Mikrosk. Anat.* **143**, 21–35.

Pentreath, V. W. and Cottrell, G. A. (1974). Anatomy of an identified serotonin neurone studied by means of injection of tritiated 'transmitter', *Nature*, (Lond.), **250**, 665–658.

Pentreath, V. W., Osborne, N. N. and Cottrell, G. A. (1973). Anatomy of giant serotonin-containing neurones and other neurones in the cerebral ganglia of *Helix pomatia* and *Limax maximus*, *Z. Zellforsch.* **143**, 1–20.

Prior, D. J. and Gelperin, A. (1977). Autoactive molluscan neuron-reflex functions and synaptic modulation during feeding in terrestrial slug, *Limax maximus*, *J. Comp. Physiol.* A, **114**, 217–232.

Radojcic, T. and Pentreath, V. W. (1979). Invertebrate Glia. *Prog. in Neurobiol.* **12**, 115–179.

Rehfeld, J. F. (1980). Cholecystokinin, *Trends in Neuroscience*, **3**, 65–67.

Sakharov, D. A. and Zs-Nagy, I. (1968). Localization of biogenic monoamines in the cerebral ganglia of *Lymnaea stagnalis*, *Acta Biol. Hung.* **19**, 145–157.

Schwartz, J. H., Goldman, J. E., Ambron, R. T. and Goldberg, D. J. (1975). Axonal transport of vesicles carrying [^{3}H] serotonin in the metacerebral neuron of *Aplysia californica*. *Cold Spring Harbor Symposia, Quantitative Biology*, **40**, 83–92.

Schwartz, J. H., Shkolnik, L. J. and Goldberg, D. J. (1979). Specific association of neurotransmitter with somatic lysosomes in an identified serotonergic neuron of *Aplysia californica*, *Proc. Nat. Acad. Sci. USA*, **76**, 5967–5971.

Sedden, C. B., Walker, R. J. and Kerkut, G. A. (1968). The localizaation of dopamine and 5-hydroxytryptamine in neurons of *Helix aspersa*, *Symp. Zool. Soc. Lond.* **22**, 19–32.

Segal, M. (1980). The action of serotonin in the rat hippocampal slice preparation, *J. Physiol.* **303**, 423–439.

Senseman, D. and Gelperin, A. (1974). Comparative aspects of the morphology and physiology of a single identifiable neuron in *Helix aspersa*, *Limax maximus* and *Ariolimax californica*, *Malacolog. Rev.* **7**, 51–52.

Shkolnik, K. L. J. and Schwartz, J. H. (1980). Genesis and maturation of serotonergic vesicles in identified giant cerebral neuron of *Aplysia*, *J. Neurophysiol.* **43**, 945–967.

Szczepaniak, A. C. and Cottrell, G. A. (1973). Biphasic action of glutamic acid and synaptic inhibition in an identified serotonin-containing neurone. *Nature (New Biol.)*, **241**, 62–63.

Villegas, J. (1975). Characterization of acetylcholine receptors in the Schwann cell membrane of the squid nerve fibre, *J. Physiol.* **249**, 679–689.

Weinreich, D. (1978). Histamine-containing neurons in *Aplysia*, in *Biochemistry of Characterised Neurons*, (Ed. N. Osborne), pp. 153–175, Pergamon Press, Oxford.

Weinreich, D., Weiner, C. and McCaman, R. (1975). Endogenous levels of histamine in single neurons isolated from the CNS of *Aplysia californica*, *Brain Res.* **84**, 341–345.

Weinreich, D., McCaman, M. W., McCaman, R. E. and Vaughn, J. E. (1973). Chemical, enzymatic and ultrastructural characterization of 5-hydroxytryptamine-containing neurons from the ganglia of *Aplysia californica* and *Tritonia diomedia*, *J. Neurochem.* **20**, 969–976.

Weiss, K. R. and Kupfermann, I. (1976). Homology of the giant serotonergic neurons (metacerebral cells) in *Aplysia* and pulmonate molluscs, *Brain Res.*, **117**, 33–49.

Weiss, K. R. and Kupfermann, I. (1977). Serotonergic neuronal activity and arousal of feeding in *Aplysia californica*, *Soc. Neurosci. Symp.* **3**, 66–89.

Weiss, K. R., Cohen, J. and Kupfermann, I. (1975). Potentiation of muscle contraction: a possible modulatory function of an identified serontergic cell in *Aplysia, Brain. Res.* **99**, 381–386.

Weiss, K. R., Cohen, J. L. and Kupfermann, I. (1978). Modulatory control of buccal musculature by a serotonergic neuron (metacerebral cell) in *Aplysia, J. Neurophysiol.* **41**, 181–203.

Weiss, K. R., Mandelbaum, D. E., Schonberg, M. and Kupfermann, I. (1979). Modulation of buccal muscle contractility by serotonergic metacerebral cells in *Aplysia*: Evidence for a role of cyclic adenosine monophosphate. *J. Neurophysiol.* **42**, 791–803.

Weiss, K. R., Schonberg, M., Mandelbaum, D. E. and Kupfermann, I. (1978). Activity of an individual serotonergic neurone in *Aplysia* enhances synthesis of cyclic adenosine mononophosphate, *Nature* (Lond.) **272**, 727–728.

Wood, J. G. (1965). Electron microscopic localization of 5-hydroxytryptamine (5-HT). *Texas Rep. Biol. Med.* **23**, 828–837.

Wood, J. G. (1966). Electron microscopic localization of amines in central nervous tissue, *Nature* (Lond.), **209**, 1131–1133.

Index